Entrepreneurship and Technology Policy

The International Library of Entrepreneurship

Series Editor: David B. Audretsch
Max Planck Institute of Economics, Jena, Germany
and Ameritech Chair of Economic Development
Indiana University, USA

1. Corporate Entrepreneurship and Growth
   Shaker A. Zahra

2. Women and Entrepreneurship: Contemporary Classics
   Candida G. Brush, Nancy M. Carter, Elizabeth J. Gatewood, Patricia G. Greene and Myra M. Hart

3. The Economics of Entrepreneurship
   Simon C. Parker

4. Entrepreneurship and Technology Policy
   Albert N. Link

Future titles will include:

Entrepreneurship and Economic Growth
A. Roy Thurik and Martin Carree

Technological Entrepreneurship
Donald S. Siegel

New Firm Startups
Per Davidsson

Entrepreneurship Education
Patricia G. Greene and Mark P. Rice

Entrepreneurship and Global Capitalism
Geoffrey G. Jones and Rohit Daniel Wadhwani

International Entrepreneurship
Benjamin M. Oviatt and Patricia Phillips McDougall

Wherever possible, the articles in these volumes have been reproduced as originally published using facsimile reproduction, inclusive of footnotes and pagination to facilitate ease of reference.

For a list of all Edward Elgar published titles visit our site on the World Wide Web at
www.e-elgar.com

Entrepreneurship and Technology Policy

Edited by

Albert N. Link

Professor of Economics
University of North Carolina at Greensboro, USA
and
Research Professor
Max Planck Institute on Entrepreneurship, Growth and Public Policy, Germany

THE INTERNATIONAL LIBRARY OF ENTREPRENEURSHIP

An Elgar Reference Collection
Cheltenham, UK • Northampton, MA, USA

Published by
Edward Elgar Publishing Limited
Glensanda House
Montpellier Parade
Cheltenham
Glos GL50 1UA
UK

Edward Elgar Publishing, Inc.
136 West Street
Suite 202
Northampton
Massachusetts 01060
USA

A catalogue record for this book is available from the British Library.

ISBN-13: 978 1 84542 308 7
ISBN-10: 1 84542 308 9

Printed and bound in Great Britain by MPG Books Ltd, Bodmin, Cornwall.

Contents

Acknowledgements

The editor and publishers wish to thank the authors and the following publishers who have kindly given permission for the use of copyright material.

Association for Public Policy Analysis and Management for article: Ronald S. Jarmin (1999), 'Evaluating the Impact of Manufacturing Extension on Productivity Growth', *Journal of Policy Analysis and Management*, **18** (1), Winter, 99–119.

Elsevier for articles: Jérôme Doutriaux (1987), 'Growth Pattern of Academic Entrepreneurial Firms', *Journal of Business Venturing*, **2** (4), Fall, 285–97; Ian Moore and Elizabeth Garnsey (1993), 'Funding for Innovation in Small Firms: The Role of Government', *Research Policy*, **22** (5–6), November, 507–19; Erik Brouwer and Alfred Kleinknecht (1999), 'Innovative Output, and a Firm's Propensity to Patent. An Exploration of CIS Micro Data', *Research Policy*, **28** (6), August, 615–24; Albert N. Link and John T. Scott (2001), 'Public/Private Partnerships: Stimulating Competition in a Dynamic Market', *International Journal of Industrial Organization*, **19** (5), April, 763–94; Anthony Arundel (2001), 'The Relative Effectiveness of Patents and Secrecy for Appropriation', *Research Policy*, **30** (4), April, 611–24; Philip Shapira (2001), 'US Manufacturing Extension Partnerships: Technology Policy Reinvented?', *Research Policy*, **30** (6), June, 977–92; David B. Audretsch, Albert N. Link and John T. Scott (2002), 'Public/Private Technology Partnerships: Evaluating SBIR-Supported Research', *Research Policy*, **31** (1), January, 145–58; Benoît Leleux and Bernard Surlemont (2003), 'Public Versus Private Venture Capital: Seeding or Crowding Out? A Pan-European Analysis', *Journal of Business Venturing*, **18** (1), January, 81–104.

Institute for Operations Research and the Management Sciences (INFORMS) for article: Wesley M. Cohen, Richard R. Nelson and John P. Walsh (2002), 'Links and Impacts: The Influence of Public Research on Industrial R&D', *Management Science*, **48** (1), January, 1–23.

MIT Press Journals and the President and Fellows of Harvard College and the Massachusetts Institute of Technology for article: Bronwyn H. Hall, Albert N. Link and John T. Scott (2003), 'Universities as Research Partners', *Review of Economics and Statistics*, **85** (2), May, 485–91.

RAND Corporation for article: Scott J. Wallsten (2000), 'The Effects of Government-Industry R&D Programs on Private R&D: The Case of the Small Business Innovation Research Program', *RAND Journal of Economics*, **31** (1), Spring, 82–100.

Springer Science and Business Media for articles: Albert N. Link and John Rees (1990), 'Firm Size, University Based Research, and the Returns to R&D', *Small Business Economics*, **2** (1), 25–31; Renata Lèbre La Rovere (1998), 'Small and Medium-Sized Enterprises and IT Diffusion Policies in Europe', *Small Business Economics*, **11** (1), August, 1–9; Magnus Klofsten, Dylan Jones-Evans and Carina Schärberg (1999), 'Growing the Linköping Technopole – A Longitudinal Study of Triple Helix Development in Sweden', *Journal of Technology Transfer*, **24** (2–3), August, 125–38; Anne-Marie Maculan and Deborah Moraes Zouain (1999), 'Changes in Brazilian Public R & D Institutions Management: The National Institute of Technology Case-Study', *Journal of Technology Transfer*, **24** (2–3), August, 149–57; G. Bruce Hartmann and John Masten (2000), 'Profiles of State Technological Transfer Structure and Its Impact on Small Manufacturers', *Journal of Technology Transfer*, **25** (1), March, 83–8; David North, David Smallbone and Ian Vickers (2001), 'Public Sector Support for Innovating SMEs', *Small Business Economics*, **16** (4), June, 303–17; Spyros Arvanitis, Heinz Hollenstein and Stephan Lenz (2002), 'The Effectiveness of Government Promotion of Advanced Manufacturing Technologies (AMT): An Economic Analysis Based on Swiss Micro Data', *Small Business Economics*, **19** (4), December, 321–40; David B. Audretsch (2003), 'Standing on the Shoulders of Midgets: The U.S. Small Business Innovation Research Program (SBIR)', *Small Business Economics*, **20** (2), March, 129–35; Maryann P. Feldman and Maryellen R. Kelley (2003), 'Leveraging Research and Development: Assessing the Impact of the U.S. Advanced Technology Program', *Small Business Economics*, **20** (2), March, 153–65; Dennis Patrick Leyden and Albert N. Link (2004), 'Transmission of Risk-Averse Behavior in Small Firms', *Small Business Economics*, **23** (3), October, 255–9.

University of Chicago Press for article: Josh Lerner (1999), 'The Government as Venture Capitalist: The Long-Run Impact of the SBIR Program', *Journal of Business*, **72** (3), July, 285–318.

Every effort has been made to trace all the copyright holders but if any have been inadvertently overlooked the publishers will be pleased to make the necessary arrangement at the first opportunity.

In addition the publishers wish to thank the Library at the University of Warwick, UK and the Library of Indiana University at Bloomington, USA, for their assistance in obtaining these articles.

Introduction

Albert N. Link

The terms entrepreneur and entrepreneurship are common to the vocabularies of most people today, although definitions of these terms vary quite a bit depending on who is using the term and in what context. With regard to the nexus between entrepreneurship and technology policy (a term that will be defined in the following section), the terms entrepreneur and entrepreneurship take on a very specific and somewhat narrow meaning. In the context of this volume, entrepreneurship is narrowly defined, as is widely done, to refer to the innovative and risk-taking activities that are specific to small firms.

Indeed, an emphasis on small firms defines entrepreneurship very narrowly, even from an academic perspective. To juxtapose the concept of entrepreneurship and the concept of small-firm innovative activity, it is first important to overview briefly alternative meanings of entrepreneurship and to ground those meanings within an economics context; specifically within the context of the history of economic thought.

About Entrepreneurship

The history of economic thought holds diverse opinions about the nature and role of the entrepreneur.[1] At least twelve distinct themes have appeared within the economics literature about the entrepreneur. These themes are:

1. The entrepreneur is the person who assumes the risk associated with uncertainty.
2. The entrepreneur is the person who supplies financial capital.
3. The entrepreneur is an innovator.
4. The entrepreneur is a decision maker.
5. The entrepreneur is an industrial leader.
6. The entrepreneur is a manager or superintendent.
7. The entrepreneur is an organizer and coordinator of economic resources.
8. The entrepreneur is the owner of an enterprise.
9. The entrepreneur is an employer of factors of production.
10. The entrepreneur is a contractor.
11. The entrepreneur is an arbitrageur.
12. The entrepreneur is an allocator of resources among alternative uses.

From a history of economic thought perspective, there is very little if any foundation for entrepreneurship to be considered as synonymous with the innovative activities of small firms.

With regard to the twelve themes above, statements 2, 6, 8, and 9 describe static roles for the entrepreneur. In a static world the entrepreneur is a passive element because his actions merely constitute repetitions of past procedures and techniques already learned and implemented. The prevailing wisdom by those who study entrepreneurship from an evolutionary perspective eschew his static role. Only in a dynamic world does the entrepreneur become a robust figure, and a dynamic role for him is implied in the other eight thematic statements – but even then the innovative activities of small firms are not explicit.

To place the entrepreneur, and his dynamic entrepreneurial actions, in an economics concept, one must begin with the ideas proffered by Richard Cantillon (c. 1680–1734). From Cantillon three intellectual branches develop which are loosely referred to as: the German Tradition characterized by, among others, J.H. von Thünen (1785–1850) and Joseph A. Schumpeter (1883–1950); the Chicago School characterized by, among others, Frank Knight (1885–1972) and T.W. Schultz (1902–1998); and the Austrian Tradition characterized by, among others, Ludwig von Mises (1881–1973) and G.L.S. Shackle (1903–1992). Consider briefly the contributions of Cantillon, Schumpeter, Schultz and von Mises as background for the entrepreneurship-technology policy emphasis in this volume.

Richard Cantillon

In the eighteenth century, Cantillon outlined the framework of a nascent market economy founded on individual property rights and based on economic interdependency. He recognized three classes of agents: landlords, who are financially independent; entrepreneurs, who engage in market exchanges at their own risk in order to make a profit; and hirelings, who work for fixed wages.

Cantillon's entrepreneur is, within his framework, the central economic actor. He is someone who engages in exchanges, in the face of uncertainty, for profit. What Cantillon stressed is the function of the entrepreneur rather than his social status. Those who take chances in hopes of gain are entrepreneurs; even beggars and robbers who face uncertainty are entrepreneurial. It is the lack of perfect foresight that is the origin of the entrepreneurial spirit. If Cantillon's entrepreneur had constantly to exercise business judgment, and if he guessed wrong, he would be accountable for the result. This Cantillonesque notion of the entrepreneur would later be widened by both Knight and von Mises.

Joseph Schumpeter

The concept of the entrepreneur was unveiled by Schumpeter within the context of economic development. Quite simply, the entrepreneur was the *persona causa* of economic development; he was the mechanism of economic change.

To Schumpeter, competition involved mainly the dynamic innovations of the entrepreneur. Schumpeter used the theoretical concept of equilibrium, 'the circular flow of economic life', as his point of departure. Economic life proceeds routinely on the basis of past experience, he argued. In this equilibrium state, the entrepreneur is a non-entity because the relationship between inputs and production is invariant. The real question about economic growth and development is not how capitalism administers existing structures but how it creates and destroys them. This process – 'creative destruction' – is the essence of economic development. In other words, development is the disturbance of the circular flow, and the process of carrying out new combinations of inputs in production is carried out by, or is the role of, the entrepreneur.

T.W. Schultz

Schultz's theory of entrepreneurship is rooted in the economic theory of human capital. His points of departure rest on four criticisms of the earlier treatment of the entrepreneur: (1) the concept is generally restricted to businessmen, (2) it does not take into account the differences in allocative abilities among entrepreneurs, (3) the supply of entrepreneurship is not treated as a scarce resource, and (4) entrepreneurship is neglected in favor of general equilibrium considerations.

Schultz redefined the concept of entrepreneurship as the ability to deal with disequilibria, and he extended the notion of entrepreneurial activity to include non-market activities (e.g., household decisions, allocation of time) as well as market activities. He also produced evidence of the effects of education on individuals' ability to perceive and react to disequilibria. Schultz argued that Schumpeter's conceptualization of entrepreneurship did not go far enough, because it was confined to development disequilibria.

Ludwig von Mises

According to von Mises, economics is the study of human action, and human action that is distinctively economic takes place within a market framework. The nature of market activity is that it is an entrepreneurial process. The fundamental aspect of von Misesian human action is that it influences the future and is influenced by the future. Thus, participants in market activity make choices and cope with the subsequent uncertainties of the future. Within this von Misesian framework, every man is an entrepreneur because every man makes choices (or decisions) and is then subject to the uncertainty that they create.

Nowhere in the above listing of twelve themes of entrepreneurship is there a characteristic of an entrepreneur that is unique to small firms. Why, then, the generally accepted relationship between entrepreneurship and the innovative behavior of small firms?

I offer two hypotheses about the generally accepted relationship between entrepreneurship and the innovative actions of small firms. First, during the early 1970s, and then again in the late 1970s and early 1980s, there was a productivity slowdown in all major industrialized nations.[2] In the United States, as in other countries, a number of policy initiatives were introduced to stimulate the innovative activity of manufacturing firms, and special emphasis was placed on the role of small firms in technological growth. Productivity growth in the US industrial sector, for example, began to rebound in the mid-1980s, and in 1985 President Ronald Reagan attributed part of that rebound, led by small innovative firms, to the Age of Entrepreneurship. Second, there is a literature in economics and management that concludes, on the basis of empirical analyses, that under certain conditions small firms are more innovative than large firms. Thus, perhaps, innovativeness, meaning in a Schumpeterian sense the carrying-out of new combinations of inputs in production, is a, if not *the*, characteristic of entrepreneurial small firms.

My hypotheses aside, if one were to juxtapose entrepreneurship – conceptualized in terms of perception of and ability to react to disequilibria – and technology policy, one would quickly move from the behavioral parameters of economics and management into those associated with political theory, sociology, and psychology. Embracing those disciplines to understand how the needs of, and the forms of technology policy, are perceived is both interesting and important, but beyond the format of this volume since the extant literature on that topic represents a set of literature that is close to the null set.

The collection of papers herein treats technology policy as an independent phenomenon and focuses on various innovative-related behaviors of small firms that are affected by technology policy. This directional influence – from technology policy to the innovative behavior or reaction of small firms – is not void of dynamic characteristics, because small firms, with limited resources, must perceive how technology policy affects them, and when and how they must react to that disequilibrium within the context of dynamic competition.

About Technology Policy

Since the mid-1950s, when economists and other scholars began to quantify the dimensions of what was called technological change, researchers began to measure presumed-to-be-correlated, and hence causal, activity such as patenting activity and investments in industrial research and development (R&D). Not surprisingly, public policies intended to stimulate technological change – technology policies – were aimed in many industrialized nations at encouraging patenting activity and R&D expenditures.[3] Over time, stimulated in large part by the slowdown in total factor productivity growth during the early 1970s and then again in the late 1970s and early 1980s, attendant policies took on a broader dimension and included fiscal incentives, collaborative research policies to stimulate R&D, and public/private partnerships.

Figure 1 is a schematic of a technology-based model of entrepreneurial activity for a representative firm.[4] Within this model the strategic direction of the firm and the competitive pressures that it faces from market activity motivate an entrepreneurial response. Thinking of entrepreneurship as the perception of opportunity and the ability to act upon that perception,[5] the firm, on the basis of its perceptions of competitive market conditions – domestic and global – and its determined strategic direction, both perceives opportunities for growth and acts upon those perceptions. This action is referred to in Figure 1 as the firms 'entrepreneurial response'.

From left to right in the upper portion of the figure, the firm evaluates competitive market conditions in light of its own strategic directions, perceives opportunities, and responds to those opportunities through innovation. The results of the innovation process are new or modified products or processes, and they add value to the firm. This is, within the economics and management literatures, a traditional technology-based production process.

A number of investment activities lead to new or improved technologies that leverage or enhance the firm's entrepreneurial response. One activity is in-house research and development (R&D). There are other additional activities that yield knowledge and leverage the response. Leveraging the firm's R&D activity is its relationship with other research organizations and with the external environment. One such relationship is the firm's involvement in research partnerships, with other firms or perhaps with either a university or a federal laboratory.[6] Research partnerships provide a number of advantages to R&D-active firms including reduced research time and less redundant research leading to less R&D costs. Of course, countering these advantages is the possible lack of appropriability of the research results.

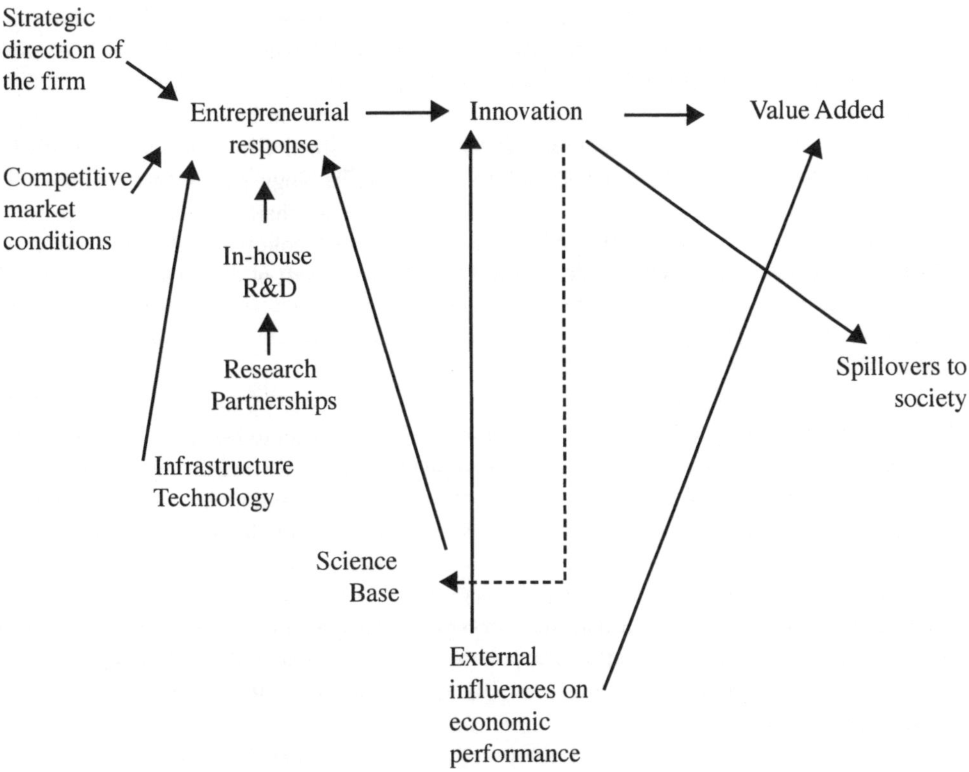

Figure 1. Technology-Based Model of Entrepreneurial Activity

Investment in infrastructure research is an important contribution to infrastructure technology that increases the efficiency of a firm's R&D. Infrastructure technology represents what economists call 'public good' technology, meaning that no one firm in an industry gains a competitive advantage from using it, but using it increases R&D efficiency. Infrastructure technology emanates from federal laboratories, such as, for example, the National Institute of Standards and Technology (NIST),[7] or from the environment created by, for example, being located in or near a research park.[8] Examples of infrastructure technology include test methods, measurement standards, and standard reference materials.

Finally, there is the science base, which consists of the stock of knowledge generated from investments in basic research at universities, for the most part, and also at federal laboratories. The science base resides in the public domain – and the public domain is international in scope – generally in the form of scientific journals but it is also in part embodied in a university's or institute's indigenous human capital – its researchers, scientists, and students. It is important to emphasize that in the model in Figure 1, the science base provides general knowledge to the firm to enhance its ability to respond to perceived opportunities. In the technology-based model of entrepreneurial activity specific to the biotechnology industry, discussed below, the science base has a different role.

The result of the entrepreneurial process is an innovation. An innovation will generate value added if it is accepted in the marketplace, and it will be diffused into society and generate spillover benefits to other firms both within the industry and in outside industries that ultimately use the innovation.[9]

In Figure 1 the dashed arrow coming back to the science base represents internal feedback. Once an innovation exists, knowledge has been created and it too will then reside in the public domain.

Also influencing the nexus between the firm's entrepreneurial response and innovation are other external influences such as federal and state science and technology policies. Examples of such include, but are certainly not limited to, federal or state policies to encourage innovation such as changes in patent regulations, or purposeful activities designed to grow a regional biotechnology industry.

The usefulness of this model is as a means to highlight the myriad sources of scientific and technical information that firms rely on to support their innovative activity, and thus to define relevant technology policies. Such attendant technology policies include the government's direct support of firm R&D through such activities as grants programs. These technology policies also include other support mechanisms that are indirect, but nevertheless also leverage firm R&D, such as programs that overcome barriers caused by both market and technical risk which prevent firms from investing in selected projects. Other technology polices relate to the government's provision of infrastructure technologies that have public good characteristics, and the availability of such infrastructure technologies is yet another leveraging mechanism to enhance firm R&D. In addition, all of these policies are formulated in an industrial nation under the umbrella of a pervasive patent policy.

Each of these broad policies falls under the rubric of public/private research partnerships.[10] In such partnerships the public sector invests financial resources, research resources, and infrastructural resources toward one or more of three economic objectives: to leverage public R&D, to leverage private R&D, and to enhance industrial competitiveness. That said, it is important to understand the economic foundation upon which the public sector bases its role in such partnerships, in particular, or in the overall innovation process, in general.

The government should have an important role to play in fostering innovation, especially private-sector innovation.[11] This intuitive conclusion logically follows from these facts:

- Innovation leads to technology.
- Technology is the prime driver of economic growth.
- In the absence of government intervention, firms will underinvest in technology, especially in basic research.
- Government has a responsibility to address this underinvestment or market failure by providing incentives to conduct additional R&D, which will enhance economic growth.

Such a sequential reasoning to justify the role of government in innovation has dominated the history of public-sector involvement in the innovation process, but a more complete justification, from an economic perspective, rests on a comparison of the efficiency of market resources with and without government intervention, or more broadly on what economists and policy makers call 'market failure'.

Market failure refers to conditions under which from society's perspective the market, including those performing R&D and those adopting the R&D outputs of others, underinvests in any particular technology. Such underinvestment occurs because of conditions that prevent firms from fully realizing or appropriating the benefits expected from their investments.

Stated alternatively, firms underinvest in R&D when they determine, on the basis of their expectations of post-innovation activity, that their private return is less than their private hurdle rate (minimum acceptable rate of return on their R&D investment). This is of public concern when the R&D investment not undertaken is socially desirable.

There are a number of factors that explain why a firm might perceive its private return to be less than its hurdle rate. These factors are what we call 'barriers to technology' and they relate to technical and market risk, where risk is defined as measuring the possibility that an actual outcome will deviate from an expected one.

Also, a firm may believe that it cannot appropriate a sufficient return on its R&D investment, even if it can overcome technical uncertainty. This may be due to a firm's perceived inability to maintain proprietary control of the technology, enabling rivals to imitate their invention and reducing any resulting profitability.

Individually or in combination, the following factors explain why a firm may perceive a private rate of return as being less than its hurdle rate:

- Because of high technical risk, the outcome of the R&D may not solve the technical problem adequately to meet perceived needs.
- Even if the R&D is technically successful, the market may not accept the technology because of competing alternatives or interoperability concerns.
- Even absent technical and market risk, it may be difficult to assign intellectual property rights to the technology, so that it might be quickly imitated and the innovator may not receive adequate return on the R&D investment.

These factors create barriers to investing in technology, which thus lead to market failure.

From an economic perspective, the role of government is to correct for these market failures in those instances where society will benefit from the technology. This latter situation occurs when the rate of return to society is greater than the social hurdle rate (minimum accepted rate of return for society from investments of resources with alternative social uses).

Concluding Statement

There is a burgeoning academic literature that looks at how such specific types of technology policies affect the innovative behavior of small firms. Scholars have not addressed all technology policies, owing to the relative newness of this area of investigation and to the obvious lack of systematic data. It has primarily been data limitations that have caused the late-blooming of this literature. Thus, most of the journal articles in this volume – which fall under four categories: government's direct support of R&D, government's leveraging of R&D, government's infrastructure policies, and knowledge flows from universities and laboratories – are relatively new, being published since the mid- to late 1990s. However, these articles represent what I believe will become the cornerstone for all future research related to entrepreneurship and technology policy.

Notes

1. This section draws from Hébert and Link (1988, 1989).
2. See Link (1987) and Link and Seigel (2003) for a review of the academic and policy literature related to these productivity slowdowns.
3. A special issue of *Research Policy* was published in 2000, guest edited by A.N. Link and D. Roessner, on the economics of technology policy.
4. This model of technology-based entrepreneurial activity was first set forth by Link and Tassey (1987), and later elaborated upon in Tassey (1997), in Feldman, Link, and Siegel (2002), and then again in Tassey (2005).
5. See Hébert and Link (1988).
6. For a review of the literature on research partnerships, see Hagedoorn, Link, and Vonortas (2000).
7. See Tassey (1992, 1997).
8. See Link (1995, 2002, 2003) and Link and Scott (2003a, 2003b, forthcoming a, forthcoming b).
9. Following Bozeman and Link (1983, p. 4): 'The concepts commonly used in connection with innovation are deceptively simple. *Invention* is the creation of something new. An invention becomes an *innovation* when it is put in use.' Innovations may be new products, new processes, or new organizational methods that are novel and add value to economic activity. Thus, at a general level, invention parallels the concept of science and innovation parallels the concept of technology.
10. See Link (1999).
11. What follows draws on Feldman, Link, and Siegel (2002) and the references cited therein.

References

Bozeman, B. and A.N. Link (1983), *Investments in Technology: Corporate Strategies and Public Policy Alternatives*, New York: Praeger.

Feldman, M.P., A.N. Link and D.S. Siegel (2000), *The Economics of Science and Technology*, Boston: Kluwer Academic Publishers.

Hagedoorn, J., A.N. Link and N.S. Vonortas (2000), 'Research Partnerships', *Research Policy*, **29**(4–5): 567–86.

Hébert, R.F. and A.N. Link (1988), *The Entrepreneur: Mainstream Views and Radical Critiques*, New York: Praeger.

Hébert, R.F. and A.N. Link (1989), 'In Search of the Meaning of Entrepreneurship', *Small Business Economics*, **1**(1): 39–49.

Link, A.N. (1987), *Technological Change and Productivity Growth*, London: Harwood Academic Publishers.

Link, A.N. (1995), *A Generosity of Spirit: The Early History of Research Triangle Park*, Research Triangle Park: Research Triangle Foundation of North Carolina.

Link, A.N. (1999), 'Public/Private Partnerships in the United States', *Industry and Innovation*, **6**(2): 191–217.

Link, A.N. (2002), *From Seed to Harvest: The Growth of the Research Triangle Park*, Research Triangle Park: Research Triangle Foundation of North Carolina.

Link, A.N. (2003), 'University-Related Research Parks', *ISSUES in Science and Technology*, Fall: 79–81.

Link, A.N. and J.T. Scott (2003a), 'U.S. Science Parks: The Diffusion of an Innovation and Its Effects on the Academic Mission of Universities', *International Journal of Industrial Organization*, **21**(9): 1323–56.

Link, A.N. and J.T. Scott (2003b), 'The Growth of Research Triangle Park', *Small Business Economics*, **20**(2): 167–75.

Link, A.N. and J.T. Scott (2005), 'Universities as Research Partners in U.S. Research Joint Ventures', *Research Policy*, **34**(3): 385–93.

Link, A.N. and J.T. Scott (2006), 'U.S. University Research Parks', *Journal of Productivity Analysis*, **25**(1): 43–55.

Link, A.N. and G. Tassey (1987), *Strategies for Technology-Based Competition: Meeting the New Global Challenge*, Lexington, MA: Lexington Books.

Tassey, G. (1992), *Technology Infrastructure and Competitive Position*, Norwell, MA: Kluwer Academic Publishers.

Tassey, G. (1997), *The Economics of R&D Policy*, Westport, CT: Quorum Books.

Tassey, G. (2005), 'Underinvestment in Public Good Technologies', *Journal of Technology Transfer*, **30**(1–2): 89–113.

Part I
Government's Direct Support of R&D

[1]

Funding for innovation in small firms: The role of government

Ian Moore and Elizabeth Garnsey

Judge Institute of Management Studies, University of Cambridge, Mill Lane, Cambridge CB2 1RX, UK

Final version received June 1992

This paper examines some of the issues facing innovative small firms. The shortage of enterprise funding for such small firms can be explained by the uncertainty associated with such ventures, uncertainty which is magnified in the case of high-technology firms. We characterise the uncertainty as an information gap (both technological and commercial) which is exacerbated by an asymmetry of interest between the founders of a firm and private sector sources of finance. We explore the sources of this uncertainty and examine the effect that one Government scheme, the Small Firms Merit Awards for Research and Technology (SMART), has had in offsetting such problems. Some suggestions for improvements in policy in the light of the experience of SMART are put forward.

1. Introduction

It is generally accepted that a pool of high-technology firms emerging from early growth into phases of consolidation and expansion is required to promote innovation and the capacity for industrial regeneration [31]. Such a pool is not a substitute for the competitive strength of large firms, but complements their existence through the generation of new technology either supplied to or in competition with existing firms [28,29]. In the UK in the 1980s, while the culture of enterprise in general and high-technology enterprise in particular was much encouraged by the Government, in practice support for small companies seeking funding for start-up and expansion has been found to be deficient [1]. However, unless start-ups are able to reach the stage of viability by obtaining necessary initial finance, the pool from which new successful innovative firms can grow will be depleted at the outset, reducing the range of firms among which successful expansion can be achieved.

The problem facing innovative small firms has previously been identified as one of matching providers of enterprise funding to seekers [18], but more recently it has been recognised that there is an absolute shortage of seed-corn capital and investment funding suitable for firms in the early stages of growth [2,13]. This shortage can be explained by the uncertainty associated with the early years of a business enterprise; such uncertainty is magnified in high-technology enterprises.

In this paper, we explore the sources of this uncertainty and examine in detail the effects of one Government scheme – the Small Firms Merit Awards for Research and Technology (SMART) – that provides support for innovation to small firms, in order to gain a better understanding of the needs and experience of such companies. We discuss the nature of the information gap which inhibits the assessment of risk and uncertainty in this area, discouraging the capital markets from promoting such ventures. The empirical analysis draws upon case studies made of 15 firms which have won SMART awards, with particular attention paid to the effect that the SMART award has had on the business. The analysis of this instance of Government support for innovation helps reveal the nature of the difficulties facing high-technology ventures in attracting funds. Not only is there an information gap; there is also an asymmetry of interests between private funders of enterprise and the principles in a new venture, which affects the supply of and demand for private investment funds.

Correspondence to: I. Moore, Judge Institute of Management Studies, University of Cambridge, Mill Lane, Cambridge CB2 1RX, UK.

Research Policy 22 (1993) 507–519
North-Holland

0048-7333/93/$06.00

We argue that given the effects of the information problem in funding small firms, an important positive impact of the SMART scheme is the accreditation of the new venture. There is also a danger in this though, namely that given the difficulty of making reliable assessments of the prospects of technology-based small firms, the accreditation through SMART may become necessary in order to attract further funds. This would imply that the awards do not represent a net expansion in the total pool of resources for high-technology ventures, but redirect resources in favour of firms winning a SMART award. Some suggestions to rectify this situation are put forward.

2. Uncertainty and new ventures

The uncertainties associated with the early years of small companies in general are linked to the general vulnerability of small firms. In small high-technology companies, these uncertainties are augmented by the unpredictability of leading edge research and development. The timescale of a research project is difficult to predict; embodying scientific or technological progress in products which are commercially relevant may be an even greater challenge. For potential outside investors, the problems of assessing timescales and feasibility will be greater than for those in the development team. Where such developments are taking place in a recently founded company without any track record in commercial work of this kind, assessment of the potential project must also encompass the commercial experience of the team. Hence the claim by venture capitalists that they place as much weight on the management capabilities of the firms as on the viability of the development work being undertaken.

Moreover, uncertainties on the supply side are complemented by uncertainties on the demand side. Where the market is very new, yet to be uncovered or faced with competing and incompatible technologies, the rate of take-up of new technology cannot be predicted with any degree of confidence.

The example of video technology is pertinent here. Technology developed by Sony – the Betamax standard – was widely held to be superior to that developed principally by JVC – the VHS standard. In addition, RCA entered the market with Videodisc. However, a series of "chance" factors, such as the fact that the VHS standard was originally capable of holding a longer length of tape, swung the market in favour of VHS, such that Sony has now licensed VHS technology from JVC, and RCA has completely withdrawn from the market [7,16]. The widely differing diffusion rates in different countries for video technology are also illustrative of the difficulties of accurately establishing future market prospects for new technology.

Inevitably, therefore, the costs and revenues of the start-up are hard to anticipate with precision, though on the whole the costs and difficulties of a manufacturing start-up are considerably greater than for a service- or consultancy-based firm [5]. However, projections for costs and revenues are precisely what are required in a business plan, the preparation of which is an essential part of gaining external investment in the firm. The difficulties outlined above suggest that an information problem is likely to occur, and we return to this below.

3. The need for further finance and options available

It is also in the nature of high-technology markets that products have a short life cycle. Thus, start-up problems are rapidly followed by those of consolidation and growth. The cashflow profile of the firm is unlikely to generate internal funds soon enough for further investment in the development of successor products or the application of the technology to new markets. New product development is at the same time crucial to the further expansion of the firm (and even its long-term survival, if absolute size does not increase) and the principal constraint on the profitability of the firm. At this point, as at start-up, the growth-oriented firm committed to technological innovation will need an injection of finance to cover operating costs as well as the commitment to further development.

The continual need for development finance in technology-related start-ups contrasts with the main sources of finance available. If the internal profitability of the firm is insufficient (as is likely until consolidation has taken place), it must turn to external sources. We can distinguish between equity finance and non-equity finance.

Equity finance is available from private investors, venture capital companies or corporate investors. The scale of private investment in the UK was encouraged by the Business Expansion Scheme which offered tax relief against investment, but recent changes to the scheme allowing property-related businesses to qualify have diverted almost all of the investment away from technology-related businesses. In addition, private investors are much less prevalent as a source of finance in the UK than in the US [25]. The venture capital market is the most developed of any European country [2], but studies have repeatedly shown that the pattern of investments is biased away from early stage investments (for example, see [13]). Further, research has clearly shown that there is an expectations gap between the owners of firms and venture capitalists, in terms of the scale of returns required and the size of the equity stake demanded [19]. Corporate investment might be expected to overcome some of these problems, but there has been relatively little research on the impact of either corporate venturing or on the importance of takeovers as a means of financing growing firms. [1] For technology-related start-ups, the proportion of firms in a position to contemplate a stock-market flotation by the time further funds are needed to finance future growth is so small as to be negligible.

Non-equity finance is typically obtained from either revenue-generating activities incidental to the firm's proprietary technology, or from bank loans. Examples of the former include contract R&D and, particularly in the scientific instruments industry, the sale of imported equipment under licensing and marketing deals. While these activities can be very successful in generating cashflow and can also have important technological or marketing benefits for the firm, they have the major drawback of diverting resources – particularly management time – away from the firm's own research and development activities, upon which future growth or survival is dependent.

[1] For example, Hughes has reviewed the literature on small-firm takeovers. He concludes that most of the work on takeovers has concerned large firms and that very little is known about small-firm takeovers, either as acquisitors or as targets [24].

Bank loans are not an automatic solution either, as they are typically secured against assets rather than lent against future prospects. In the absence of such assets, founders often have to pledge their own homes as securable assets, and an unwillingness to do so may deter many prospective borrowers from seeking the funds they require to expand their business.

We have drawn attention to the negative aspects of the available forms of finance for technology-related businesses, although there are undoubtedly many cases where such finance has been sought and obtained without conflict of objectives. However, it remains the case that there is often a gap between what the firm is able or willing to relinquish (be it equity or personal assets) and what potential investors require in order to safeguard their interest in an uncertain environment. For example, the Bank of England, commenting on the high fixed costs in evaluating and monitoring investment in start-up and expansion of a new business, stated:

> The provision of equity finance for new and developing businesses can be especially unattractive to potential investors for two reasons – it is extremely risky and the costs of evaluating and monitoring the investments can swamp expected returns. While the industry disagrees on the extent to which, if at all, there is a gap in the availability of second-stage growth capital in amounts of £100000 to £250000, there is considerable agreement that seed-corn capital of less than £100000 is hard to obtain. [2, p. 82]

These problems are inherent to the various types of finance under consideration. However, the difficulties faced by small firms because of the unsuitability of finance packages are compounded in two further ways: by an information gap between themselves and potential investors; and by the asymmetry of interests that exists between the two parties.

The information gap arises from the different positions of the firm's principals and outside investors. Where risk is quantifiable, it is insurable through portfolio investment. But where the issue is uncertainty (which is not insurable), investors are reluctant to venture. The information gap arises in the sense that venture capitalists are unable to evaluate the technical strengths and potential of development of the project as they will inevitably be less versed in the technological details than the firms' principals, while information on how to minimise risk and develop fall-back strategies through effective management procedures may well be unavailable to founders, particularly those from a technical background. ACOST [1] addressed the issue of an information gap, and stressed the importance of closing this gap. Thus the report suggests that better preparation in strategic planning and in drawing up business plans for technical managers would be beneficial. However, we have already suggested that the business plan itself forms part of the information problem. Grieve-Smith and Fleck, addressing this issue, concluded as follows:

> The financial institutions providing second-stage finance after the limits of bank-lending had been reached often seemed to demand business plans with an obviously unrealistic degree of quantification and with little regard for the costs of the scarce managerial time involved. [18, p. 67]

The second problem is the asymmetry of interest between the two parties, and this is not so easily reconcilable through the dissemination of best practice as suggested for the information gap. Founders often do not wish to lose control over their company to outside investors, and often feel threatened by the size of stake, typically 40%, demanded by venture capitalists. Further, the return to venture capital investment comes from capital growth not dividends, but this necessitates an exit route from the investment, which is often required over a 3–7 year timescale. A further change in ownership is thus an inevitable consequence of venture capital investment, and this too diminishes the control over the firm of the original founders.

Thus there are two main problems relating to the provision of venture capital. One is the information gap, the other is the asymmetry of interests between providers and seekers of such funding. Government schemes need to address the problem of information like any private investor, but there is not normally the same conflict of interest as found with market sources of finance.

Government finance as currently available does not pose the same threats to the independence and future rewards for founders of high-technology firms. Further, the timescale over which the firm can develop into a viable entity is not constrained by the need to find an exit route for venture capital investment. However, the scale and type of funds available from government sources is normally directed at support of individual projects, and is likely to be inadequate to finance start-up and further development activity completely; venture capital, for example, more typically produces overall finance for the firm rather than specific to one project. Despite this, small companies value highly government grants as a means of financing growth through innovation, as these do not require either an equity stake (unlike venture capital investments) nor security against tangible assets (as bank loans do).

4. UK Government support for enterprise and innovation

Given the deficiencies in the provision of finance to small firms and to technology-related firms in particular, there is a strong argument in favour of an active role for government. At the minimum, government should be strengthening market mechanisms to ensure the efficient functioning of markets, but there is also a role for government in intervening when markets fall. As Freeman has said,

> It is important [to provide] a corrective to some of the simplistic and short-term thinking about "markets" in the United Kingdom and United States, which sees "deregulation" and "privatisation" as the solution to all economic and social problems. This simplistic approach does violence to the technological sophistication of a century of neo-classical economic theory on market imperfections, market failure, market regulation, social costs and benefits and externalities. [14, p. v]

In the UK in the 1980s though, the political environment stressed the role of the market in seeking its own solutions and the need to avoid government intervention. Given this background, and in particular the revisions to the Department of Trade and Industry's role set out in the 1987 White Paper [21], existing programmes such as Support for Innovation (SFI) were curtailed. New schemes such as LINK stressed support for collaborative and pre-competitive research, as opposed to the single-company projects supported under SFI. Near-market research was no longer to be supported by the Government with the explicit exception of research in small firms. [2] The sums of money made available under LINK (£210 million over 5 years) and SMART (£29 million over 4 years) indicate the extent to which the bulk of government support for industrial R&D was to be channelled through LINK. [3] However, the acknowledged failure of LINK to spend the sums allocated to it means that DTI spending on support for innovation has probably been more effective through SMART than through LINK.

Thus, the launch of SMART in 1986 was not only an exception to the general pattern of support for industry by the government, but its formulation as a competition for government awards was the most acceptable form of intervention in the political context. The example in the US of the success of such a scheme where firms competed for government awards, namely the Small Business Innovation Research scheme [15], was instrumental in the development of the principles behind SMART. However, there is one major difference between SBIR and SMART which needs to be stressed at the outset, for it affects the analysis of the success of the scheme and also the suggestions that follow for changes to innovation support in the UK.

SBIR is funded by requiring that 1.25% of all government agencies' extramural research budgets are set aside for competitive tender by small firms. [4] Under SBIR, federal agencies identify their research needs, and invite competitive tenders from small firms on the basis of innovation. Under SMART, small firms submit their preferred R&D projects for the competitive allocation of funds. [5] This obviously allows small firms greater latitude in the choice of R&D projects, but it removes the clear identification of the innovative technology with an articulated need. Thus the distinction between the two schemes is that under SBIR firms are competing for research contracts, whereas under SMART they are competing for seed-corn finance. Put another way, SBIR mimics competition for industrial research contracts, whereas SMART mimics competition for financial backing.

5. The impact of the SMART scheme on winning firms

The objectives of SMART are described by the DTI as follows:

> to bring forward highly innovative but commercially viable projects, now dormant because existing sources of finance do not wish to support them;
> to encourage the formation of small firms which will develop and market new ideas in selected areas of new technology;
> to help these small firms to mature sufficiently for private sources of funds to take a practical interest. [10]

These objectives thus recognise the problem of financing early-stage technology-related ventures, which we have characterised above as mainly revolving round an information gap and asymmetry of interests. Specifically, these aims imply that in successful cases SMART awards are intended to advance firms through early-stage growth such that information surrounding their prospects can be gauged from a track record of achievement, and such that firms will have grown to the point where the types of finance available for funding future growth better match the circumstances they are in.

The details of the operation of the SMART scheme have not changed much since the scheme was first started in 1986. Firms, who must employ less than 50 (reduced from less than 200 from 1989 onwards), compete for the awards by submitting details of the innovative research and development for which they are seeking funding. The DTI operates a four-stage evaluation process covering the following aspects: the business plan, the technology, a Patent Office search, and the ability of the firm to carry out the project. Further, firms have to demonstrate additionality in their proposals, that is, the proposed R&D project would be additional to existing R&D activity

[2] A more detailed description of the changing objectives of the DTI's support for industry can be found in [23, pp. 64–79].

[3] If comparison is made with the annual expenditure on military R&D, then both sums pale into insignificance: MoD expenditure on R&D in 1988/1989 totalled £1,935m [22].

[4] These are defined as firms with less than 500 employees, and "other than the dominant firms in the field in which they are proposing to carry out SBIR projects" [15].

[5] Small firms are defined as those with less than 50 employees. This was a reduction from the limit of less than 200 up to and including the 1988 SMART competition. However, only 2% of SMART winners had more than 50 employees in 1988 [12].

Table 1
SMART awards 1986–1990

	Stage I		Stage II	
	Applicants	Awards	Applicants	Awards
1986	124	20		
1987			15	11
1988	930	140		
1989	826	150	92	81
1990	1415	180		

Source: [8,9,11,12].

and would not go ahead in that form without the award.

Winners receive a Stage I SMART award, which provides 75% funding of eligible project costs in the first year, currently up to a maximum of £45 000 (total project costs can exceed £60 000, but there is a ceiling on the funding provided by the DTI). This is a cash grant, of which the first instalment (£15 000) is paid in advance. At the end of the first year, all Stage I winners can apply for a Stage II award, which provides further funding of 50% of eligible project costs up to a maximum of £60 000 (again, total project costs can exceed £120 000). Approximately half the Stage I winners progress onto a Stage II award. The numbers of awards given by the DTI has been increasing over time (see Table 1), but the level of interest from small firms has increased dramatically over the last year. [6]

The steadily increasing numbers, and the perceived need of the DTI to increase the number of awards due to the number of proposals worthy of funding under SMART, are indicative of the size of the financing gap that SMART is partially filling. It is estimated that the number of proposals of high enough quality to be funded under SMART is around 200, [7] and this only reflects those firms which actually apply. The overall numbers of technology-based firms in a position to require some sort of financing must exceed this estimate considerably, for example not all firms will know of SMART while others will find that their development timetables do not match the once-a-year allocations of awards.

[6] Further details on the operation of the SMART scheme are given in the Appendix below.

[7] DTI official – personal communication.

The scale of impact of SMART in winning firms can be gauged in two ways. First, the level of additionality indicates the extent to which the firm was operating under constraints with respect to research and development activity, and the extent to which those constraints have been ameliorated. Second, the success or otherwise in negotiating stages of growth by the firm indicates the extent to which seed-corn or early-stage funding of this sort can overcome some of the problems associated with market sources of finance discussed above.

5.1. Additionality

As was stated above, additionality occurs when the proposed R&D project would be additional to existing R&D activity and would not go ahead in that form without the award. The simplest cases are of 100% additionality, that is either the project would not have gone ahead at all without the award, or, in the case of non-winners, that without the award the project has in fact not gone ahead. Obviously there are many cases where the R&D is partially additional, and in fact measuring additionality often relies on qualitative judgements as to what would have happened in different circumstances.

The DTI's official evaluation of the SMART scheme [12] estimated additionality for the 1988 winners. This was done using a methodology employed in previous DTI evaluations (for example of support to small firms under SFI), which enabled comparisons to be drawn. The results are reproduced in table 2. The evaluation also contacted 1988 non-winners, and found that 54% claimed that progress had been "considerably delayed" through not winning an award. The DTI concluded as follows:

> These findings together tend to confirm that winning a Stage I award provided finance which could otherwise not have been available to carry out a project. Even in established companies the award seems to have allowed resources, not available otherwise, to be dedicated to the project to expedite progress. [12, p. 11]

The evidence presented by the DTI in its report has been confirmed in our case studies, all of which were carried out before the DTI report was published. Details of the 15 firms studied are given in the Appendix. Most of the firms we interviewed were able to demonstrate additionality, for example through the employment of additional staff, either in an R&D function or in replacing an employee redeployed to work on the SMART project. In some cases, the fact that the additionality of the project was total was explained in terms of the constraints the firm was operating under. Thus, the SMART award allowed the firm's commitment to R&D to be extended beyond that to which it was already committed and was financing from other sources. Evidence from the DTI evaluation suggests that in non-winning firms, projects had been permanently put on ice or attempted in much reduced fashion, because of the inability either to generate sufficient internal finance for the project or to attract market sources of finance. However, from our case studies there is also evidence that SMART awards may sometimes have additionality greater than 100% (see also [26]). There are three ways this can happen.

The first is that the firm commits other resources to the project that would not otherwise have been deployed (the DTI report also refers to this: see the quote above). In these cases, we argue that the role of SMART goes beyond financing the one project through which the firm hopes to derive proprietary products, but also educates the firm into a permanently higher commitment to own-risk R&D than was the case before the SMART award.

The second route by which SMART awards can have additionality greater than 100% occurs because of the way the finance is administered. The first instalment of a SMART I award is paid in advance, and this is clearly highly beneficial to small firms with constrained cashflow. The receipt of funds in advance of expenditure allows materials to be ordered or new employees taken on without having to fund the activities in the short term until repayment is made. In terms of initiating R&D projects, this was found to be valued very highly by the firms. However, the receipt of finance from the Government was also identified in some case studies as having been responsible for the financial survival of the firm. That is, the firm would have folded without the SMART award because of cashflow problems. In this sense, the whole R&D activity of the firm is in some sense attributable to SMART funding in that none of the subsequent R&D could have happened in that firm if it had folded.

The third example of greater than 100% additionality arises through the impact of successful R&D on profitability in the firm, and subsequent growth which in turn finances future R&D (the issue of continual need for finance discussed in section 3 above). We discuss this in detail below as it is related to stages of growth in small firms.

Therefore, the additionality of SMART awards appears to be very high. The DTI evaluation, with a maximum level of additionality of 100%, refers to-overall additionality for the scheme of 80% [12, p. 28], but from the evidence for greater than 100% additionality we conclude that the Government funds spent through SMART have probably been totally additional in terms of increased R&D. Given the very high rating for SMART in terms of the innovativeness of proposals funded [12], the scheme appears to have met its objective "to bring forward highly innovative but commercially viable projects" [10] (see above).

Table 2
DTI assessment of additionality ratings

Category		% of Stage I winners	% of Stage II winners
Non-additional	Project would have gone ahead in similar way without award	0	1
Marginal	In light of company's overall resources, projects minimum acceleration or lack of risk	3	38
Reasonable	But not overriding grounds for accepting additionality	38	41
Strong	Grounds for accepting that project was wholly additional or that substantial acceleration/increase in scale took place	59	20

Source: [12].

5.2. Barriers to growth

One of the objectives of SMART, as listed above, was "to help these small firms to mature sufficiently for private sources of funds to take a practical interest". This is clearly related to the concept of stages of growth in a firm's life which is well established in the literature (for example, [4,17,32]), and to the ideas related to barriers to growth that need to be overcome if the firm is to survive its growth [3]. In other words, SMART is viewed as seed-corn capital which could be sufficient for a firm to develop its own products, thus enhancing both its track record and its profitability such that the information gap is considerably narrowed.

There is strong evidence from SMART winners that this is indeed happening. Several of the firms studied had subsequently received injections of finance to allow further growth and development of the firm through innovation. Others either had already found or expected in the future that their internal generation of funds would have improved sufficiently to fund further R&D projects without the need for recourse to market sources of finance. We consider these two situations in turn.

With respect to the increased external appeal of SMART winners, the operation of SMART awards appears to have created a kind of accreditation in terms of small firms' ability to secure market sources of finance. Many of the firms studied reported that they had received unsolicited enquiries from venture capital firms *after* receiving the award, and some firms did indeed take up these offers. In one case, the firm had sought venture capital investment for several years before winning SMART without success, but subsequent to winning a SMART award received six offers of venture capital finance from which it was able to choose. In another case, the success of the firm in winning SMART was the final indication that persuaded a multinational company to take over the firm, with a substantial injection of finance and managerial resources.

There are several explanations as to why SMART winners may find it easier to secure market sources of finance after their award than before. First, the fact that the firm is a SMART winner is one more piece of information available about the firm for investors to evaluate. Indeed, investors may take the fact that the firm has won the award as some kind of endorsement of the firm's prospects from the government. Of course, it is not a formal endorsement any more than the SMART award is a guarantee that the project will not run into technical difficulties, but it may be interpreted in this way by potential investors. Indeed the flooding of SMART winners with offers of finance from venture capital funds strongly endorses the idea that SMART is taken as a form of accreditation necessary for such investment.

Second, the SMART award may genuinely improve the firm's future prospects. If the development project is successful it should lead to proprietary products, which in turn should improve the firm's profitability. A quantified evaluation of this was not possible in our case studies, but this was undertaken in the DTI evaluation of SMART. Based on the demonstrated profitability of projects supported in the 1986 SMART awards, and the expectations for future profits from 1988 winners, the report concluded:

> We have calculated ... that the 1988 round of SMART which cost about £9 million may give rise eventually to profits of about £20 million in present value terms. If only four of the 140 Stage I winners were to achieve their predictions the scheme's contribution to profits might exceed its costs for that round. [12, pp. 11–12] [8]

This is clear evidence that technical success with a SMART project is likely to substantially increase a firm's profitability, thus making it a better investment for outside partners. As well as narrowing the information gap relating to the firm's prospects, it may also narrow the asymmetry of interests in that the investor's expectations of profits, and hence the size of the equity stake desired, may conform more closely to the expectations of the firm's owners. On the other hand,

[8] The DTI figures for the profitability of the 1988 SMART projects are for the whole set, and take account of the fact that not all projects will meet their technical objectives, let alone be successful in the market place. Indeed they assume that half the projects will fall by the wayside, and apply a further factor of one-quarter to take account of the relatively long timescales and probable overoptimism from firms about their future prospects. The assumptions are described in the analysis as heroic, but conservative.

the very success of the SMART project may lead to a widening of the asymmetry of interests, as small-firm owners see the scope for achieving growth and profitability without the need to involve external investors. It is far from clear which of these changes is likely to dominate.

However, from these figures it is clear that in those cases where the project does succeed technologically and commercially, the profitability of the firm is likely to be substantially enhanced. In these circumstances, the firm may thus be able to fund future growth requirements from the profitability of the SMART project and other activities. In these cases there is therefore no need to seek external funding from market sources as future expansion costs are covered.

This was found to be the case in the firms in our study where the commercial success had already been achieved. In general, the reaction of the firm was to plough further resources into other development projects, thus giving rise to additionality levels attributable to SMART in excess of 100% (as discussed above). In a few cases, though, the profitability of the SMART project enabled the firm's owners to retain control of the firm, even if this meant not taking advantage of all the market opportunities now open to them. We believe it inevitable that the returns to the SMART investment will be internalised by some firms in this way, but that the large majority of firms will externalise the return through expansion (and hence increased exports, or employment growth for example) and further commitments to R&D.

From this evidence, we conclude that SMART has succeeded in assisting winning firms in negotiating some of the early stages of growth facing technology-related small firms. SMART winners appear to be able to tap market sources of finance more readily as well as having a decreased need for such finance due to increased profitability. In addition, the example of one successful product can increase a company's reputation in its markets, increasing the market potential of future development projects. Thus the long-term financial viability of the firm does appear to be enhanced by the injection of finance for innovation through SMART where the R&D project is subsequently a technological and commercial success.

6. Conclusions

We have shown in this paper that the successful operation of SMART effectively demonstrates the case for Government support for innovation in small firms. In particular, the two main problems facing small firms seeking to raise external finance (the information gap between small firms and potential investors and the asymmetry of interests between the two parties) are reduced to a minimum in the SMART scheme. Specifically, the existence of a systematic vetting procedure ensures that SMART awards are allocated to innovative projects that are not likely to be hampered by other constraints internal or external to the firm. Thus through the application and vetting procedures the Government can gather what information it requires with efficiency. Furthermore, the fact that there is no conflict of interest between the firms and the "investor" considerably reduces the information gap.

This absence of conflict of interest at very early stages in a firm's growth is important. By providing some finance for small-scale innovation projects, the very gap in the provision of finance from market sources most clearly identified, SMART enables firms to develop their future prospects to the point where the information gap is much reduced (in addition, in some cases the success of SMART is sufficient to reduce or eliminate the need for access to external finance). In this process, a key component is the effective accreditation of the firm through the winning of a SMART award. However, in this context the DTI's procedures do not amount to a strategy of "picking winners" since the level of support to individual firms is relatively small and the support is very widely dispersed. [9] Moreover, the policy is one of nurturing a large number of firms in the knowledge that there will be some failures but that the numbers being supported are large enough to ensure that there should be some outstanding successes while the majority will be of minor benefit. Nonetheless, SMART has the appearance of an effective success indicator if

[9] Previous policies of picking winners have usually involved the large-scale backing of one or two cases, for example see Hendry on the NRDC's strategy in the 1950s and 1960s for the UK computer industry [20].

one allows for the inevitable failure rate of small firms.

This conclusion raises problems when there is a policy commitment to avoiding intervention in market processes as there has been in the UK. However, the rationale for the scheme is the failure of market processes to provide adequate funding for innovation in small firms, and thus it falls to the Government to rectify the matter through a scheme such as SMART. Given the effectiveness of SMART, and in particular the ability of winning firms to attract external investment, there is a strong case for expanding the scheme.

We suggest that to capitalise on the benefits already demonstrated, and in particular to maximise the size of the pool of growing firms from which successful large firms can emerge, a logical first step would be to expand the scheme so that all worthy applications are funded, and to operate the scheme twice yearly. It makes sense for the scale of the scheme to be expanded sufficient to fund all projects meeting the criteria established in the DTI vetting procedures, rather than limit funding to the number of awards that can be financed out of any one year's tranche of money. Evidence from our case studies as well as the DTI's own evaluation confirm that levels of additionality in winners are very high, and non-winners are usually unable to proceed with their projects successfully without SMART. Also, windows of opportunity for innovation tend to be fast moving, and very few firms will be in a position to delay starting a project for up to a year while waiting for the next SMART competition to come round, if they are unable to finance the project any other way.

The issue of the accreditation role of SMART is not merely incidental but is intrinsic to the nature of funding innovation in small firms. The lack of reputation of new firms is a feature of the information gap, and SMART clearly closes this gap substantially in the case of winning firms. However, non-winning firms are not so accredited, and such a firm may in fact face further conflicts of interest with potential investors if it is judged as a non-winner; that is, the terms under which finance is offered may be more stringent because potential investors perceive that the firm has been "judged" by the DTI as not up to the standard of SMART winners. If SMART were expanded so as to finance all proposed projects which meet its criteria, then the problem, which is essentially one that can be characterised as the relative scarcity value attached to SMART awards, would be alleviated. If the call on Government funds becomes too great, new criteria might be added, such as that applied in the US by the GAO in the operation of SBIR, namely that firms receiving awards must be other than the market leader in that technology.

There is clearly the possibility that an expanded SMART scheme would reach the stage whereby the additionality of the scheme is reduced. In particular, it is possible that the overall additionality of SMART is low because the increased R&D under SMART is being offset by reduced R&D activity elsewhere in the economy, for example within the large-firm sector. Under circumstances of restricted supply of R&D resources, for example qualified scientists and engineers, this might be the case, but as yet there is no evidence that this is happening. Indeed, the sums currently being spent on SMART are so small as to be insignificant at the macroeconomic level. However, the impact of these sums is much higher than that achieved through selective support of large firms [20,30], and so there is a strong argument for redirecting technology policy support into those areas with highest rates of return.

In conclusion, it is clear that in the short period in which the scheme has been operating, SMART has has been effective for many firms in filling the financing gap that faces small firms trying to raise funds for innovative research and development. We have argued that the reasons that the SMART has been so successful in helping small firms are to do with the nature of funding offered under the scheme, specifically that it does not have the same information gap and asymmetry of interest between small firms and external investors that afflicts market sources of finance. However, it is also clear that the SMART has only partially filled this gap, and may in fact have created new problems arising from the accreditation role it appears to perform. We therefore recommend an expansion of the scheme so that the relative scarcity value of SMART is eliminated, and so that all viable proposals are funded. The net benefit to the economy from the existing numbers of SMART awards has been seen to be high. Further expansion of the scheme should ensure that such benefits are not passed over in other cases where, for example, the timing of product development does not match the deadlines of the existing annual competition.

In this paper we have examined the reasons why small, innovative firms seeking funding for start-up and expansion encounter difficulties obtaining the resources they need. It is generally recognised that there is uncertainty surrounding the commercial prospects of new products and services, but there is also uncertainty surrounding the managerial competence of a new team operating in a start-up setting. The effects of these uncertainties are however exacerbated by an information gap between those seeking funding and those capable of providing it. Training high-technology managers in the skills of writing a business plan, or increasing the technological evaluation skills of venture capitalists, is insufficient to bridge this gap. An asymmetry of interests between entrepreneurs and providers of venture capital makes the matching of demand for and supply of investment capital hard to achieve. Again uncertainty is implicated in this mismatch, since to offset uncertainty venture capitalists impose stringent terms and conditions on recipients. For many founders of innovative small firms the motive for start-up is often independence, and they find the conditions on which outside funding is available unattractive, especially when it means they have to relinquish ownership and control of their company. Yet growth requires resources and the alternative may be stagnation or collapse. Under these circumstances the provision of government finance for innovation in small firms in the form examined in this paper is highly attractive to those seeking funding. We have shown that the scheme is also a relatively inexpensive way of allocating funding for innovation since the accreditation role of the scheme attracts further finance to recipient firms over and above the award itself. It appears that there is a need for technological and market assessments of the potential of new high-technology companies independent of the management itself, and the operation of the SMART scheme in effect fulfils this need. Given that a wide package of innovation support measures is required to raise the level of investment in innovation in the UK economy [23], we believe that an expansion of SMART and the introduction of other measures operating on the same principles would provide a highly effective starting point for UK technology policy.

References

[1] ACOST, *The Enterprise Challenge: Overcoming Barriers to Growth in Small Firms* (HMSO, London, 1990).

[2] Bank of England, Venture Capital in the United Kingdom, *Bank of England Quarterly Bulletin* 30(1) (1990) 78–83.

[3] J. Barber, J.S. Metcalfe and M. Porteous (Editors), *Barriers to Growth in Small Firms* (Routledge, London, 1989).

[4] N.C. Churchill and V.L. Lewis, The Five Stages of Small Business Growth, *Harvard Business Review* May–June (1983) 30–50.

[5] D. Connell, The Management of Growth in High Technology Companies (paper presented to Institute of Electronic and Radio Engineers Diamond Jubilee Conference, 1985).

[6] M. Daly, The 1980s – A Decade of Growth in Enterprise, *Employment Gazette* 98(11) (1990) 553–565.

[7] P. David and S. Greenstein, The Economics of Compatibility Standards: An Introduction to Recent Research, *Economics of Innovation and New Technology* 1(1) (1990) 3–41.

[8] Department of Trade and Industry, SMART Ideas win Awards (DTI Press Notice 23 October 1986).

[9] Department of Trade and Industry, *Small Firms Merit Awards for Research and Technology Directory of Winners, 1988 Stage 1* (DTI, London, 1988).

[10] Department of Trade and Industry, *Application and Guidance Notes for the SMART Competition* (DTI, London, 1989).

[11] Department of Trade and Industry, *Small Firms Merit Awards for Research and Technology, Directory of Winners, 1989 Stage 1* (DTI, London, 1989).

[12] Department of Trade and Industry, *The SMART Scheme: Evaluation Report*, DTI Assessment Paper No. 13, Research and Technology Policy Division (DTI, London, 1991).

[13] M. Dodgson and R. Rothwell, Financing Early-Stage Innovation in Small Firms (paper presented to European Conference, Enterprise Innovation and 1992: Innovation Support Services in Europe, Nice, 26–27 October 1989).

[14] C. Freeman, Preface, in: M. Dodgson (Editor), *Technology Strategy and the Firm: Management and Public Policy* (Longman, Harlow, 1989).

[15] General Accounting Office, *Federal Research: Assessment of Small Business Innovation Research Programs* (United States GAO, Washington DC, 1989).

[16] M.B.W. Graham, *The Business of Research: RCA and the Videodisk* (CUP, Cambridge, 1988).

[17] L. Greiner, Evolution and Revolution as Organisations grow, *Harvard Business Review* July–August (1972) 37–46.

[18] J. Grieve-Smith and V. Fleck, Business Strategies in Small High-Technology Companies, *Long Range Planning* 20(2) (1987) 66–68.

[19] J. Grieve-Smith and V. Fleck, Strategies of New Biotechnology Firms, *Long Range Planning* 20(3) (1988) 51–58.
[20] J. Hendry, *Innovating for Failure: Government Policy and the Early British Computer Industry* (MIT Press, Cambridge, MA, 1990).
[21] HMSO, *Civil Research and Development*, Cm185 (HMSO, London, 1987).
[22] HMSO, *Annual Review of Government Funded R&D 1990* (HMSO, London, 1990).
[23] House of Lords Select Committee on Science and Technology, *Innovation in Manufacturing Industry*, HL Paper 18-1 (HMSO, London, 1991).
[24] A. Hughes Small Firms' Merger Activity and Competition Policy, in: J. Barber, J.S. Metcalfe and M. Porteous (Editors), *Barriers to Growth in Small Firms* (Routledge, London, 1989).
[25] C.M. Mason and R.T. Harrison, Informal Risk Capital: A Review of US and UK Evidence (paper presented to 12th National Small Firms Policy and Research Conference, London, November 1989).
[26] I. Moore, Government Finance for Innovation in Small Firms: The Impact of SMART, *International Journal of Technology Management*, forthcoming (1993).
[27] K. Pavitt, M. Robson and J. Townsend, The Size Distribution of Innovating Firms in the UK: 1945–1983, *Journal of Industrial Economics* 35(3) (1987) 297–316.
[28] M. Robson, J. Townsend and K. Pavitt, Sectoral Patterns of Production and Use of Innovations in the UK, *Research Policy* 17(1) (1988) 1–14.
[29] R. Rothwell, Small Firms, Innovation and Industrial Change, *Small Business Economics* 1(1) (1989) 51–64.
[30] M. Sharp, Farewell to National Champions, in: M. Sharp and P. Holmes (Editors), *Strategies for New Technology: Case Studies from Britain and France* (Philip Allan, London, 1989).
[31] D. Storey, K. Keasey, R. Watson and P. Wynarcyk, *The Performance of Small Films* (Routledge, London, 1987).
[32] C. Tiler, J.S. Metcalfe and D. Connell, The Management of Growth: Negotiating Transitions (paper presented to Symposium, Growth and Development of Small High-Tech Businesses, Cranfieid Institute of Technology, 1–2 April 1990).

Appendix – Further details on the operation of the SMART scheme

Having been piloted in 1986, and repeated on a larger scale in 1988, SMART was relaunched for a further three years in 1989. Funding of £29 million was guaranteed, sufficient to fund 150 Stage I awards each year with approximately half expected to result in Stage II awards as well. As a result of the 70% increase in the number of Stage I applicants in 1990, the DTI obtained from the Treasury a one-off extra allocation of money that allowed an extra 30 awards to be made.

Table A1
Distribution of SMART awards by sector

	Number of Stage I awards	
	1988	1989
Instrumentation	27	33
Computing	22	17
Electronics/electrical engineering	14	16
Biotechnology	15	15
Materials	10	12
Medical	9	12
Mechanical engineering	12	17
Other	35	28
Total	140	150

Source: Authors' estimates based on [9,11].

In February 1991, two small changes were announced. First, the number of awards for 1991 was set at 180, in line with 1990, as further money had again been obtained from the Treasury. Secondly, the value of maximum eligible costs in the Stage I award was increased from £50000 to £60000 to restore its value in real terms. The Government contribution was increased from £37500 to £45000. Maximum eligible costs for Stage II awards were increased from £100000 to £120000. The 1991 Stage I and 1992 Stage II competitions are the last to be funded under the arrangements set up in 1988. Recently, the DTI has received clearance from the European Commission to fund a further programme over 1992–1994, with 180 Stage I awards each year and 100–110 Stage II awards, a higher proportion than in the past, following on. The 1992–1994 programme is costed at £42 million.

The sectoral distribution of the 290 SMART Stage I awards made in 1988 and 1989 is illustrated in Table A1. This shows a large number of awards going to projects in the instrumentation sector. Other sectors with large shares are computing, electronics and biotechnology. The concentration of SMART awards in these sectors is not unduly surprising. The limit in the size of R&D project and the initial timescale of 1 year limits the type of R&D that can be performed to small, product development projects. In addition, the concentration of awards the instruments and electronics sectors reflects the high proportion of small firms in these industries [6] and the concen-

Table A2
Sectoral and age distribution of case-study firms

Age of firm at receipt of first SMART award	Sector	
	Instruments	Biotechnology
Up to 2 years	4	1
2–4 years	2	4
4–6 years	1	–
Over 6 years	–	3

tration of innovation in the small-firm sector in these industries [27].

Table A2 gives the sectoral and age distribution of the 15 firms in our case studies. Ages are given as the age of the firm when it first received a SMART award (some firms have won repeated awards). Although the sectoral distribution is concentrated wholly in instruments and biotechnology, these are two sectors strongly represented in Table A1. The phenomenological evidence drawn from the case studies is unlikely to be affected by the sectoral distribution. The age distribution shows that 11 of the case studies were less than four years old, precisely the target group for SMART awards. Seven of these firms had secured venture capital funding before the receipt of a SMART award, but nonetheless were dependent upon SMART for the particular research programmes funded by the award. Only two of the 15 firms had developed sufficiently to be self-financing, and the level of additionality in these two firms was much lower than in other case study firms. Those with venture capital backing for example had not yet developed into profitable sustainable businesses.

[2]

Josh Lerner
Harvard University and National Bureau of Economic Research

The Government as Venture Capitalist: The Long-Run Impact of the SBIR Program*

Government programs to finance small firms have attracted little empirical attention. This article examines the largest U.S. initiative, the Small Business Innovation Research program. Using a unique database, I show that program awardees grew significantly faster than matched firms over a decade and were more likely to attract venture financing. The superior performance of awardees was confined to firms in regions with substantial venture capital activity and was pronounced in high-technology industries. Multiple awards did not increase performance. These results suggest that awards played an important role in certifying firm quality but also that distortions of the award process occur.

I. Introduction

Despite economists' interest in interactions between governments and firms, the public subsidization of small firms has attracted virtually no scrutiny. This article seeks to address this omission by examining the largest such program in the United States, the Small Business Innovation Research (SBIR) program, which has provided over $7 billion to small high-technology firms between 1983 and 1997.

Addressing this omission is important for three reasons:

* I thank Fran Featherston, Lowell Mininger, and Mark Nadel of the U.S. General Accounting Office, Jesse Reyes of Venture Economics, and David Gleba and Rolf Selvig of VentureOne for access to data; Horace Crouch and Robert Wrenn of the U.S. Department of Defense for assistance in arranging interviews with awardees; and Zoltan Acs, Michael Barclay, Lewis Branscomb, Douglas Diamond (the editor), Kenneth Flamm, Paul Gompers, Adam Jaffe, Steven Kaplan, Scott Wallsten, several current and former SBIR program managers, an anonymous referee, and participants in seminars at Georgia Tech, Harvard University, the University of California, Davis, the University of Mannheim, the University of Maryland, and Virginia Tech, as well as in the July 1995 and July 1996 NBER Science and Technology Policy program meetings, the November 1996 NBER Corporate Finance program meeting, and the February 1998 National Academy of Sciences "Symposium on the SBIR Program" for helpful comments. Chris Allen, Cedric Escalle, Mannie Liu, and Kathleen Ryan provided research assistance. I thank the Center for Science and International Affairs at the Kennedy School of Government and the Division of Research at Harvard Business School for financial support. All errors are my own.

(*Journal of Business*, 1999, vol. 72, no. 3)

0021-9398/99/7203-0001$02.50

First, it provides an opportunity to examine empirically theoretical models of interactions between the public and private sectors. This has been a topic of enduring interest in the economics literature for several decades.

Second, public programs to subsidize small high-technology firms have represented a significant but little-studied area of public expenditures. (Table 1 lists the programs in the United States over the past several decades. See Organization for Economic Cooperation and Development [1996] for tabulations of efforts in Europe and Asia.) For instance, the sum of the financing provided through, and guaranteed by, the programs listed in table 1 in 1995 was at least $2.4 billion, compared to the $3.9 billion disbursed by private venture capital funds in that year.

Finally, a number of these programs are reputed to have had a positive and significant effect on economic growth. Some of America's most dynamic companies received support through federal programs while privately held entities, including Apple Computer, Chiron, Compaq, Federal Express, and Intel. In addition to funding firms, publicly sponsored funds during the 1960s provided early experience for many of the individuals who later went on to lead independent venture organizations. Overseas, much of the recent growth in high-technology firms in such nations as Israel, Singapore, and Taiwan has been attributed to government venture capital initiatives (see Organization for Economic Cooperation and Development 1996).

Economic analyses suggest at least two sets of rationales for governments to offer subsidies to small high-technology firms. First, the social returns from the firms' R & D expenditures may exceed their private returns. Second, in making such awards, knowledgeable government officials may certify firms to private investors and address the informational asymmetries that might have otherwise precluded investments. An extensive literature on regulatory capture, however, suggests that government involvement may be distorted by the desire of interest groups—or of the politicians themselves—to maximize their own private benefits.

This article assesses the long-run success of firms participating in the SBIR program. I examine the employment and sales growth of 1435 firms over a 10-year period. Two-fifths of the sample received one or more awards in the first three cycles of the program; the others were matching firms chosen to closely resemble the awardees. The analysis relied heavily on a unique data set of awardees compiled by the U.S. General Accounting Office (GAO).

Over this period, the SBIR awardees enjoyed substantially greater employment and sales growth than the matching firms. The growth of the awardees did not appear to have been due to the receipt of procurement contracts by these firms or other alternative explanations. These

TABLE 1 U.S. Public Venture Capital Initiatives, 1958–97

Sponsoring Organization	Program Name	Brief Description	Span
Small Business Administration	Small Business Investment Company Program	Provides capital to federally sponsored funds that make debt and equity investments in growth firms.	1958–97
Department of Commerce	State Technical Services Program	Supported various government programs to help high-technology companies (especially new firms).	1965–69
Department of Housing and Urban Development, Model Cities Administration	Venture Capital Development Assistance	Demonstration projects in selected cities financed businesses begun by residents of targeted neighborhoods.	1967–71
At least 30 states	At least 43 state venture funds or SBIC programs	Make investments into funds supporting new enterprises, which often focus on high-technology firms.	1970–97
Department of State, Agency for International Development	At least 13 developing country venture funds	Provided loans to financial intermediaries that made equity and debt investments in new enterprises in over 30 countries.	1971–93
Small Business Administration	Specialized Small Business Investment Company Program	Provides capital to federally sponsored funds that make debt and equity investments in growth firms owned by disadvantaged individuals.	1972–97
Department of Commerce, National Bureau of Standards	Experimental Technology Incentives Program	Catalyzed new public programs (across agencies) to encourage industrial research and venture capial.	1972–79
National Science Foundation	Federal Laboratories Validation Assistance Experiment	Funded assessments by national laboratory personnel of prototype products and processes developed by entrepreneurs.	1972–75
National Science Foundation and Small Business Administration	Innovation Centers Experiment	Provided assistance to high-tech entrepreneurs through incubation centers, subsidies, and technical assistance.	1973–81
Department of Energy, Office of Energy-Related Inventions	Energy Related Inventions Program	Provides financing to individual inventors and small firms to commercialize energy-conserving discoveries.	1975–97

TABLE 1 ***(Continued)***

Sponsoring Organization	Program Name	Brief Description	Span
Small Business Administration	Small Business Development Centers Program	Funds university-based centers to assist small businesses and encourage technology transfer.	1976–97
Department of Commerce	Corporations for Innovation Development Initiative	Designed to fund state and regional corporations to provide equity financing to new firms; only one such corporation funded.	1979–81
Department of Commerce, Minority Business Development Agency	Technology Commercialization Program	Financed minority technology-oriented entrepreneurs, as well as centers to assist such entrepreneurs.	1979–82
At least 15 states	At least 107 business incubators	Provide office and manufacturing space, support services, and often financing to start-up businesses.	1980–96
11 federal agencies	Small Business Innovation Research Program	Provides awards to small technology-oriented businesses (also predecessor programs at 3 agencies, 1977–82).	1982–97
Department of Energy, Office of Energy Research	At least 6 contractor-organized venture funds	Make equity investments in spinouts from national laboratories (funds organized by prime or subcontractors at laboratories with Department's encouragement).	1985–97
At least 30 states	State Small Business Innovation Research Programs	Make SBIR-like grants, often in conjunction with federal SBIR awards.	1987–97
Department of Commerce, National Institute of Standards and Technology	Advanced Technology Program	Awards grants to develop targeted technologies to firms and consortia; some emphasis on small businesses.	1988–97
Department of Defense, Defense Advanced Research Projects Agency	Experimental venture capital investment program	Designed to make investments in private high-technology firms in exchange for equity or royalties; program only made one investment.	1989–91

continued overleaf

Department of State, Agency for International Development	Enterprise Fund Program	Oversees 12 federally funded venture funds investing in Eastern Europe, the former Soviet Union, and Africa.	1990–97
Overseas Private Investment Corporation	Venture capital fund guarantees	Guarantees full or partial return of capital to investors in at least 16 private venture funds in developing countries.	1990–97
Department of Housing and Urban Development, Community Relations and Involvement Office	Tenant Opportunity Program	Funds new businesses and other initiatives by public housing residents (other aspects of program had begun in 1987).	1993–97
Department of Energy, Office of the Under Secretary	Defense Programs Small Business Initiative	Provides funding, technological assistance, and national laboratory access to small high-technology businesses.	1993–97
11 federal agencies	Small Business Technology Transfer Program	Finances cooperative research projects between small high-technology firms and non-profit research institions.	1994–97
Department of Defense, Cooperative Threat Reduction Program	Defense Enterprise Fund	Finances an independent venture fund investing in defense conversion projects in the former Soviet Union.	1994–97
Department of the Treasury	Community Development Financial Institutions Fund	Invests in and provides assistance to community development venture capital and loan funds.	1995–97
Department of Defense	"Fast Track" Program	Provides matching funds for private financing raised by SBIR awardees.	1995–97
Department of Agriculture, Rural Business and Cooperative Development Service	Intermediary Relending Program (as amended)	Permits program managers to guarantee returns of investors in rural venture funds.	1997

SOURCE.—Eisinger (1988); Organization for Economic Cooperation and Development (1996); U.S. Small Business Administration (1996); and various news stories in the LEXIS/NEXIS database and government reports.

NOTE.—The table summarizes programs sponsored by state and federal organizations in which equity investments or equity-like grants were made into privately held companies or into funds that made such investments. If a program had multiple names, I report the name in May 1997. If a program was terminated prior to this date, I record its name at the time of the termination. If an organization sponsoring a program changed its name, or if responsibility for the program was transferred between organizations, I report the name of the sponsoring organization in May 1997. If the program was terminated prior to this date, I record the sponsoring organization at the time of the program's termination.

firms were also more likely to subsequently receive venture capital financing. The growth, however, was not uniform. The superior performance was confined to awardees in areas with substantial new firm creation, as measured by early-stage venture capital disbursements. The relationship between SBIR awards and growth was much stronger in high-technology industries. The SBIR awardees receiving large subsidies did not perform better than those receiving smaller subsidies. The results suggest that these awards played an important certification function. At the same time, distortions in the award process may have adversely affected the program's effectiveness, especially in regions with fewer high-technology firms and among firms receiving large subsidies.

The analysis had two important limitations. The most critical of these was that I was not able to assess the effect of these programs on social welfare. Only with much more detailed data could the relative costs and benefits of the initiative be fully assessed. Second, my focus on a single program limited my ability to discuss the effect of alternative program designs.

The plan of the article is as follows. In Section II, I discuss economic perspectives on the subsidization of small high-technology firms and their empirical implications. The third section reviews the structure of the SBIR program and the key insights gained through interviews. Section IV discusses the challenges associated with empirically evaluating such programs. The fifth section describes the data set. Section VI presents the analysis, and the final section concludes the article.

II. The Subsidization of Small High-Technology Firms

A. *Theoretical Motivations*

The economics literature suggests at least two sets of rationales for governments to offer subsidies to small high-technology firms. Other works, however, argue that government involvement may be distorted by the desire of interest groups or politicians to maximize their own benefits. These suggest a more skeptical view of such programs.

Public finance theory emphasizes that subsidies are an appropriate response to activities that generate positive externalities. Such investments as R & D expenditures and pollution control equipment purchases may have positive spillovers that benefit other firms or society as a whole. But because the firms making these investments are unlikely to capture the entire surplus, public subsidies may be an appropriate response.

An extensive literature (reviewed in Griliches [1992] and Jaffe [1996]) has documented the presence of R & D spillovers. These spillovers take several forms. The profits associated with new innovations may accrue to competitors who rapidly introduce imitations, to developers of complementary products, or to consumers of the new products. Whatever the mechanism of the spillovers, the consequence is the same: firms will tend to invest below the socially optimal level of R & D. After reviewing a wide variety of studies, Griliches argues that the social rate of return exceeds the private rate by a considerable amount. He estimates that the social rate of return is between 150% and 200% of the private rate of return. Few studies have examined how these gaps vary with firm characteristics. A number of case-based analyses (Jewkes, Sawers, and Stillerman 1958; Mansfield et al. 1977), however, suggest that spillover problems are particularly severe among small firms, which are often unable to defend effectively their intellectual property or to extract most of the rents in the product market.

A second rationale for public subsidies for small high-technology firms lies in the fact that they may convey information to other potential investors. Informational asymmetries may make raising external capital very expensive for entrepreneurs or even preclude it entirely. Myers and Majluf (1984) and Greenwald, Stiglitz, and Weiss (1984) demonstrate that equity offerings of firms may be associated with a "lemons" problem. If the manager is better informed about the investment opportunities of the firm and acts in the interest of current shareholders, then he only issues new shares when the company's stock is overvalued. Indeed, numerous studies have documented that stock prices decline on the announcement of equity issues, largely because of the negative signal that it sends to the market. Information problems have also been shown to exist in debt markets by Stiglitz and Weiss (1981).

These problems are a consequence of the information gaps between the entrepreneurs and investors. If the information asymmetries could be eliminated, financing constraints would disappear. Financial economists argue that specialized financial intermediaries, such as venture capital organizations, can address these problems by intensively scrutinizing firms before providing capital and then monitoring them afterward. Nonetheless, venture capitalists back only a tiny fraction of the technology-oriented businesses begun each year. In 1997, a record year for venture disbursements, 707 companies received venture financing for the first time.[1] By way of comparison, the Small Business Administration estimates that 885 thousand businesses were started in 1997. If federal awards could effectively certify firms to their potential investors, these problems could be addressed.[2]

The arguments outlined above implicitly assume that, once a market

1. Summary data on this and other years are available at http://www.ventureone.com.

2. Consistent with this view, several directories of high-technology firms denote companies that have won SBIR awards, and at least 167 press releases concerning SBIR awards were issued by firms in 1996.

failure is identified, the government can dispassionately address it. An extensive political economy and public finance literature, however, has emphasized the distortions that may result from government subsidies, as interest groups or politicians seek to direct subsidies in a manner that benefits themselves. As articulated by Olson (1965) and Stigler (1971), and formally modeled by Peltzman (1976) and Becker (1983), the theory of regulatory capture suggests that direct and indirect subsidies will be captured by groups that stand to gain substantial benefits and whose collective political activity is not too difficult to arrange (i.e., when "free-riding" by coalition members is not too large a problem). As Stigler (1971) points out, even very small firms (which have historically dominated industries such as trucking and the licensed professions) can organize to benefit from public largess. Many of the initial works in this literature treated politicians and government bureaucrats as passive implementers of policies desired by private coalitions. More recent models (summarized in Laffont and Tirole [1993]) have incorporated voters, politicians, and/or bureaucrats as distinct actors, often in a principal-agent relationship.

These distortions may manifest themselves in several ways. One possibility (discussed, e.g., in Eisinger [1988]) is that firms may seek transfer payments that directly increase their profits. Politicians may acquiesce to such transfers to politically connected companies. A more subtle problem is discussed by Cohen and Noll (1991) and Wallsten (1996): officials may seek to select firms based on their likely success, regardless of whether the government funds are needed. In this case, they can claim credit for the firms' ultimate success, even if the marginal contribution of the public funds was very low.

B. Empirical Implications

These contrasting views have different empirical implications for the relative performance of small high-technology firms receiving public subsidies. In this section, I review four sets of predictions about the effect of public subsidies on firm performance.

Overall effect of subsidies. The first rationale was that subsidies are required to encourage small firms to undertake research because of spillover problems. In this case, subsidy recipients should perform at least slightly better than their peers, even if much of the surplus created by their R & D accrues to other firms in their industry or to consumers. Similarly, if public awards can certify firms to investors, the awardees may be able to attract additional capital and grow more rapidly. If, however, these awards are captured by politically connected firms, the funds may be squandered and not translated into long-run growth. The final view delineated above—that federal agencies are selecting firms that are likely to grow rapidly in any case—also suggests that SBIR firms should perform better than a set of matched firms.

Variation in subsidy effect across regions. The externality and certification views suggest no clear implications for where the regional benefits of the program should be greatest. The more political economy-based views, however, raise the possibility of different geographic effects. In particular, it may be that, in regions without large numbers of high-technology firms and venture capital organizations, these awards may be captured by other firms than the intended population: politicians may encourage awards to politically connected firms with few growth prospects (e.g., consulting and contract research firms). Presumably, these types of distortions in the selection process are much less likely in regions with active communities of high-technology firms and venture investors, as these groups are likely to have a considerable degree of political influence. This suggests that the relative effect of the program may be greater in regions with concentrated populations of venture capitalists and high-technology firms.[3]

Variations of subsidy effect with industry characteristics. Industries vary tremendously in their degree of transparency to outside investors. In some cases, the financial statements traditionally analyzed by bankers and accountants provide much valuable information. In other instances, the firms' primary assets are intangible intellectual property, which does not show up on balance sheets and may yield returns only after many years. These differences can again be exploited in the empirical analysis:

- The certification hypothesis suggests that if public awards convey information to investors, these signals are likely to be particularly valuable in technology-intensive industries where traditional financial measures are of little use. Specialists at the National Institutes of Health or Department of Defense may have considerable insight into which biotechnology or advanced materials companies are the most promising, while the traditional financial statement analysis undertaken by bankers would be of little value. In these cases, the signal that a government award provides might be particularly significant.
- The hypothesis that federal officials are selecting firms likely to perform in a superior fashion suggests a different pattern. Even though the insights of federal officials may give them an advantage relative to private investors (and thus make an award a valuable signal), they may not find the selection of successful firms in high-technology industries to be easy. Predicting firm success in high-technology in-

3. A natural question is why any awards are made in regions with few high-technology firms. While this issue has not been discussed at great length in the public finance literature, a substantial body of work on political institutions documents that even in programs that disproportionately benefit particular regions (such as Air Force procurement contracts), some federal funds are spread across a wide number of states to insure congressional support. See Carsey, Rundquist, and Fox (1997) for a review of this literature.

dustries is very difficult. For instance, Audretsch (1995, table 4.11) analyzes the growth of 7,000 start-ups over a 10-year period. Both firm and industry characteristics have considerable explanatory power in his regressions analyzing the success of firms in low- and medium-technology industries. In a regression analysis of high-technology firms, however, the coefficients are individually and jointly statistically insignificant. Thus, if the crucial criterion for selecting awardees is the ultimate success of the firm, it is not clear that federal officials will be particularly successful in high-technology industries. I anticipated that federal officials should be equally able—or even more likely—to chose successful firms in low-technology industries. This hypothesis suggests that when comparing SBIR awardees to a sample of matching firms, the superior performance of awardees should not be confined to high-technology industries. In fact, the relative success of low-technology awardees—where identifying firms that will ultimately be successful is substantially easier—may be even greater.

Variations of subsidy effect with the amount of subsidies received. If the main role of public subsidies is to address the externality posed by small firms' inability to capture the bulk of the surplus generated by their innovations, there should be a positive relationship between the amount of subsidy received and the effect on the firm. As long as firms undertaking R & D capture at least some of the benefits from their research, they should grow more quickly as subsidies increase. The certification hypothesis suggests that there may not be a positive relationship between the amount of subsidies and firm growth. In particular, the marginal value of the signal that the tenth subsidy provides should be considerably lower than that of the first. Meanwhile, the pursuit of these awards may become increasingly costly to the firm as the size (or number) of subsidies increases and management attention is diverted from commercialization.

III. The Structure and Implementation of the SBIR Program

The Small Business Innovation Development Act, enacted by Congress in July 1982, established the SBIR program. The program mandated that all federal agencies spending more than $100 million annually on external research set aside 1.25% of these funds for awards to small businesses. When the program was reauthorized in 1992, Congress increased the size of the set-aside to 2.5%. In 1997, this represented annual funding of about $1.1 billion.

While the eleven federal agencies participating in the program are responsible for selecting awardees, they must conform to the guidelines stipulated by the act and the U.S. Small Business Administration.

TABLE 2 Volume of SBIR Awards

Year	SBIR Awards: $ Amount	SBIR Awards: No. of Phase II Awards	Size of External R & D Set-Aside (%)
1983	.07	0	.2
1984	.15	388	.6
1985	.27	407	1.0
1986	.40	564	1.25
1987	.45	768	1.25
1988	.49	711	1.25
1989	.52	749	1.25
1990	.53	837	1.25
1991	.53	788	1.25
1992	.55	916	1.25
1993	.71	950	1.5
1994	.71	961	1.5
1995	.86	1,263	2.0
1996	.89	1,191	2.0
1997	1.08	1,404	2.5

SOURCE.—U.S. Small Business Administration (1996); and personal communications with SBA officials.

NOTE.—The table indicates the amount of SBIR awards (in billions of 1995 dollars), the number of Phase II awards, and the share of external R & D spending that all agencies spending more than $100 million on external R & D were required to set aside for the program. The subsidy was phased in over a number of years. The Department of Defense was allowed an extra year to reach the 1.25% target and consequently had lower set-aside levels between 1983 and 1986.

Awardees must be independently owned, for-profit firms with less than 500 employees, at least 51% owned by U.S. citizens or permanent residents. Promising proposals are awarded Phase I awards (originally no more than $50,000; today, $100,000 or smaller), which are intended to allow firms to determine the feasibility of their ideas. (Typically about 10 Phase I applications are received for every award made.) Approximately one-half of the Phase I awardees are then selected for the more substantial Phase II grants. Phase II awards of at most $750,000 (originally, $500,000) are transferred to the small firm as a contract or grant. The government receives no equity in the firm and does not own the intellectual property that the firm develops with these funds. Table 2 displays the annual expenditures on the SBIR program, the number of Phase II awards, and the percentage set-aside.

Conversations with federal program managers, as well as assessments by the U.S. General Accounting Office (1985, 1987*b,* 1995), suggest that the program is to a certain extent insulated from the political pressures to make awards to particular firms encountered in other programs. In particular, three factors limit the pressures to select politically connected firms:

First, the awards are reasonably small in size, not exceeding $750,000. Consequently, an award is less likely to be the subject of an intense lobbying activity than that of a program making larger disbursements.

Second, the awards are not made by a single body in each agency but, rather, allocated according to each subunit's external R & D spending. Within the Department of Defense, for instance, hundreds of midlevel officials (whose primary responsibility is overseeing other research and development efforts) decide on SBIR awards. Thus, it is harder to influence the selection process than in programs where a single committee is responsible for making all the awards.

Finally, the scoring systems used to rank applicants largely focus on the technological merit of the proposals, rather than on more subjective assessments of commercial potential. This may leave evaluators with less discretion to choose well-connected applicants.

Despite the presence of these protections, it is not clear that the program always makes awards in a manner that addresses the externalities induced by spillover problems or certifies firms to outside investors. Firms that have participated in the SBIR program, former program managers, and other observers suggest several deviations:

First, the program managers have faced pressure from congressional officials to make geographically diverse awards. The share of funds going to firms in California and Massachusetts—which has consistently been about 40%—has attracted scrutiny from congressmen unhappy about the awards' geographic concentration. One indication of the political sensitivity of the allocation of SBIR awards may be the fact that in almost every recent fiscal year, all 50 states have received at least one SBIR award (U.S. Small Business Administration 1996). In private conversations, several former program managers indicated that interest from congressmen representing districts with few awards went beyond questions about the overall distribution of awards to include in some cases inquiries about the status of particular SBIR applications.

Second, particular companies have demonstrated an ability to capture a disproportionate number of awards. Evaluations of the SBIR program (e.g., U.S. General Accounting Office 1992) suggest that these "SBIR mills"—many of which have staffs in Washington that focus on identifying opportunities for applications—appear to commercialize projects at a significantly lower rate than other firms.

Third, many program officers focus on making awards to firms that are most likely to meet agency needs, with little regard to the effect of awards on the pace of technological innovation or even to the long-run sustainability of the awardees as businesses.

Finally, the contravening pressure exists to back firms that are likely to be successful. As Wallsten (1996) discusses, the Small Business Administration and program officers at many agencies compile stories of successful awardees, which are frequently presented in speeches and reports. This may lead to pressures to select firms that would succeed even without the award.

IV. The Challenge of Evaluation

While both government agencies and academic economists have sought to assess the efficacy of federal programs to aid technology-intensive industries, they have tended to focus on different questions. Most federal evaluations of these programs have consisted of case studies and surveys of the commercial activity *directly* attributable to the awards. Examples of such evaluations include Myers, Stern, and Rorke (1983), Price, Waterhouse (1985), and U.S. General Accounting Office (1987*a*, 1989, 1992).

Such surveys of direct commercial effects, however, are subject to several biases. In the case of the SBIR program, small business advocates conducted a protracted lobbying campaign to expand the program. Executives active in this effort may have overstated the benefits from the awards. At the same time, some firms may have downplayed the commercial benefits of the program, lest they attract unwelcome attention from reporters and politicians seeking to expose ''corporate welfare.'' Finally, because many firms pursued SBIR projects closely related to their core technologies, determining the direct effect of an SBIR award may have been difficult.

Studies of federal technology programs by academic economists, beginning with Levy and Terleckyj (1983), have tended to focus on the short-run effects of these efforts. In particular, they often ask whether federal funds substitute for or stimulate private R & D spending. For instance, Irwin and Klenow (1996) show that semiconductor manufacturers substantially reduced their own R & D spending while participating in the Sematech consortium. In the study most closely related to this one, Wallsten (1996) shows that the subset of SBIR awardees that were publicly traded reduced their own spending on R & D in the years immediately following the award.

However valuable a framework when examining the macroeconomic effect of public expenditures, it is less clear that this analytic approach is appropriate when assessing public programs for small high-technology firms. In many cases, small high-technology firms are organized around one key scientist or engineer and his or her research laboratory or product development team. It may not be possible to accelerate the project's progress by ''scaling up'' the project through the addition of more researchers or technicians. In this case, it is rational for the firm not to increase its rate of spending but rather to use the funds to prolong the time until it needs to seek additional capital. To interpret a short-run reduction in internal research spending by awardees as an indication of program ineffectiveness is very problematic. Rather, the long-run effect of the awards is the crucial question.

Once the need for an assessment of the long-term performance of the SBIR awardees was clear, identifying the appropriate metric was still problematic. Ideally, I would have measured both the social and private effects of the program. While case studies of a small number of participants in the SBIR program (along the lines, e.g., of Mansfield et al. 1977) might shed some light on the social benefits, such an effort would have been impossible for a larger sample.[4]

Even determining an appropriate measure of private benefits was difficult. Ideally, I would have examined the relationship between program participation and firms' valuations. Only 2% of the SBIR awardees in the first three program cycles, however, were publicly held at the time of their initial Phase I awards. I focused instead on two alternative measures: sales and employment growth. The valuations assigned to private firms by venture capitalists have been shown to be highly correlated with these two measures (Gompers and Lerner, in press). In a supplemental analysis discussed in Section VI*B* below, I show that the changes in valuations for the subset of publicly traded firms were consistent with the employment and sales results.

V. The Construction of the Data Set

In this analysis, I relied on a unique database prepared by the GAO. The GAO was mandated in the legislation establishing the SBIR program to evaluate the initiative periodically. In late 1985, the agency generated a sample of 933 firms that received SBIR awards in the first three program cycles from the 12 federal agencies participating in the program at the time. This sample included all firms that had received Phase II awards to that date, as well as a sampling of the firms that had only received the preliminary Phase I awards. The U.S. General Accounting Office surveyed these firms in early 1986 and mid-1988. Eight hundred and thirty-five firms responded to at least one survey. Of these, 541 received Phase II awards in the first three program cycles, while only 294 received Phase I awards.

In addition to the sample of 294 firms that received only Phase I awards, I constructed two matching samples of 300 firms each that did not receive SBIR awards in the first three program cycles. In each case, I selected firms that closely resembled the Phase II awardees using Corporate Technology Information Services' *Corporate Technology Directory* (1996), the most comprehensive directory of U.S. high-

4. These analyses were complemented, however, with eleven interviews of Massachusetts firms that had received at least one Phase II award from the Department of Defense, as well as numerous informal conversations with SBIR program managers.

technology firms. This directory employs a highly detailed industry and geographic classification scheme, which allows quite precise matches.

One set was matched on industry and firm size. I recorded the primary industry classification of each Phase II awardee in the 1986 edition of the *Corporate Technology Directory* or the first subsequent volume in which the firm was listed. (In all, 74% of the Phase II awardees were listed.) I then pulled a random selection of 300 of these awardees. For each selected firm, I identified in the 1986 *Corporate Technology Directory* the firm in the same classification with the closest sales level.[5] I excluded as possible matches firms that were units or subsidiaries of other concerns or firms that had received any SBIR awards in the first three program cycles (determined through U.S. Small Business Administration [1986]).

The second set was matched on geographic location and firm size. Following the same procedure as above, I identified in the 1986 *Corporate Technology Directory* the firm in the same city and state with the closest sales level to a random sample of 300 of the 541 Phase II awardees. If there were no matches within the same city, I chose a firm from the next city with the same three-digit ZIP code (in the same state) listed in the volume. I determined whether the awardees or matching firms had received venture financing using the Venture Economics and VentureOne databases (described in Gompers and Lerner [1999]).

The samples are summarized in tables 3 and 4. The firms were similar in most respects. The matching firms had slightly greater sales in 1985; the awardees, greater employment. The awardees were slightly more likely to be located in California or Massachusetts (35% vs. 30%).

A final step in constructing the sample was determining the employment and sales of the SBIR Phase II awardees and the matching firms at the end of 1995.[6] For publicly traded firms, I used Security and Exchange Commission filings. For private firms, I employed a variety of printed and electronic databases.[7] For firms where I could not ascertain employment and sales, I checked the news stories compiled in various

5. While the standard industrial classification (SIC) scheme places all software firms in class 7372, this directory distinguishes between firms that manufacture educational software used for teaching typing and for music instruction. If there were no matches within the class, I chose a firm in a related industry: e.g., if there was no appropriate match for a firm classified as a plastic composite laminate manufacturer, I matched the firm to another composite laminate manufacturer.

6. In a few cases, firms did not disclose 1985 sales. Furthermore, SBIR awardees were only requested to report ranges for 1985 sales. Consequently, in all cases where firms did not report sales or where reported sales were in one of the ranges above $1 million, I ascertained 1985 sales using the 1986 editions of the same sources used to determine 1995 sales.

7. Where possible, I used the *Corporate Technology Directory*, one of the few business directories with a rigorous procedure to identify (and correct) problematic responses. I repeated the analyses below, eliminating observations that relied on other commercial data sources. These omissions had little effect on the magnitude of the results.

TABLE 3 Construction of Sample of SBIR Phase II Awardees and Matching Firms

	No. of Firms
General Accounting Office Survey:	
Number of firms surveyed	933
Respondents to 1986 General Accounting Office survey	750
Respondents to 1988 General Accounting Office survey	729
Respondents to either General Accounting Office survey	835
Of respondents to either General Accounting Office survey:	
Received one or more Phase II awards in first three program cycles	541
Did not receive Phase II award in first three program cycles	294
Matching firms selected through 1986 *Corporate Technology Directory:*	
Industry- and sales-matched	300
Location- and sales-matched	300
Final sample:	
Phase II awardees from survey	541
Matching firms from survey and *Corporate Technology Directory*	894

TABLE 4 Comparison of SBIR Phase II Awardees and Matching Firms

	Mean	SD	Minimum	Maximum
A. SBIR Phase II awardees:				
Year founded	1977.5	6.9	1935	1985
1985 sales	4.1	8.4	.1	87.9
1985 employment	46.6	85.8	0	600
No. of Phase II awards in first three cycles	2.8	3.5	1	30
Located in California?	.21		0	1
Located in Massachusetts?	.14		0	1
No. of early-stage venture financings, 1983–85, in:				
State	230.0	293.3	0	770
ZIP code	3.2	8.1	0	46
B. Matching firms:				
Year founded	1976.6	9.6	1899	1985
1985 sales	4.2	8.9	.1	86.9
1985 employment	33.2	68.9	0	650
No. of Phase II awards in first three cycles	0	0	0	0
Located in California?	.20		0	1
Located in Massachusetts?	.10		0	1
No. of early-stage venture financings, 1983–85, in:				
State	216.2	286.4	0	770
ZIP code	2.2	6.6	0	46

NOTE.—Panel A provides summary statistics on the 541 firms that received SBIR Phase II awards in the first three program cycles; panel B, the 894 matching firms in the sample. All sales figures are in millions of 1995 dollars.

electronic databases. These searches revealed that many firms for which I was unable to obtain 1995 sales and employment data from the published and on-line directories had gone out of business. In cases where firms were liquidated prior to the end of 1995, I recorded the firm's 1995 employment and sales as zero.

Acquired firms posed another challenge. In many cases, their operations were folded into those of the acquirer, precluding a determination of the original firm's level of activity in 1995. Consequently, I did not include in the analysis the employment and sales of firms that had been acquired unless the companies continued to be operated as separate subsidiaries through the end of 1995. Another exception was cases where the purchase of the firm was an asset sale as part of a bankruptcy, in which case employment and sales were recorded as zero. In some cases, I was only able to determine a range in which sales fell. In these cases, I employed the midpoint of this range.

VI. Empirical Analysis

A. *Overall Effect of Subsidies*

I first compared the SBIR awardees with the three groups of matching firms. Panel A of table 5 shows that the mean sales increase from the end of 1985 to the end of 1995 was greater for the awardees ($4.0 million vs. $1.1 million, both in constant 1995 dollars). Panel B displays a similar employment increase (a boost of 26 employees vs. six). For the mean SBIR awardee, this represented a 98% boost in sales (in inflation-adjusted dollars) and a 56% increase in employment. In both cases, the differences in means were statistically significant.

Both distributions were highly skewed: among the SBIR awardees, for instance, only about one-fifth of the observations were above the mean level of employment growth. This skewness was consistent with evidence on the financial and operating performance of other small firms, such as portfolio companies of venture capitalists or those completing initial public offerings. At each reported percentile, the changes in employment and sales were more positive for the awardees. The differences between the awardees and the matching firms were relatively modest, however, until the top one-third of the distribution. I tested—and rejected—the null hypothesis of the equality of medians, thus demonstrating that a few outliers did not cause the differences.

In unreported analyses, I compared the sales and employment growth between the three subsets of matching firms. These patterns did not appear to be a consequence of peculiarities in one of the groups of matching firms. The growth of sales and employment differed little between the 294 firms that received SBIR Phase I awards and the 600 firms taken from the *Corporate Technology Directory*. For instance,

TABLE 5 Growth of SBIR Phase II Awardees and Matching Firms

	SBIR Phase II Awardees	Matching Firms	*p*-Value from Comparison
A. Change in sales, 1985–95:			
Mean	4.03	1.14	
90th percentile	8.29	3.68	
75th percentile	1.40	.33	
Median	−.07	−.40	
25th percentile	−.64	−1.00	
10th percentile	−4.01	−7.19	
SE	1.29	.81	
p-value, comparison of means			.046
p-value, comparison of medians			.000
No. of observations	493	836	
B. Change in employment, 1985–95:			
Mean	26.20	5.78	
90th percentile	66	29	
75th percentile	10	2	
Median	−1	−3	
25th percentile	−8	−10	
10th percentile	−45	−50	
SE	10.32	4.47	
p-value, comparison of means			.038
p-value, comparison of medians			.000
No. of observations	499	846	

NOTE.—The table summarizes the change in the sales and employment of the 541 firms that received SBIR Phase II awards in the first three program cycles and the 894 matching firms in the sample. All sales figures are in millions of 1995 dollars.

the mean employment for the Phase I awardees increased by 4.2 jobs, while for the directory firms, the average increase was 6.5. (The average employment increase was 6.2 among the directory firms matched by industry and 6.8 among those matched by geography.) Joint *F*-tests of the null hypothesis that the three groups did not differ, as well as pair-wise comparisons of means and medians, did not identify any significant differences. The differences between the SBIR Phase II awardees that were significant in tables 5 and 6 remained so, at least at the 10% confidence level, when the Phase II awardees were compared only to the Phase I awardees or to the directory firms.

An alternative explanation for the superior long-run growth of the SBIR awardees was that winners cultivated relationships with politicians or federal program managers and were consequently likely to receive subsequent procurement contracts. To examine this possibility, I restricted the above analysis to the two agencies that made significant numbers of SBIR awards but were unlikely to undertake procurement contracts with awardees: the National Science Foundation (NSF) and the Department of Health and Human Services (HHS). (The HHS's external research, and consequently their SBIR awards, overwhelm-

TABLE 6 Growth of SBIR Phase II Awardees and Matching Firms, by Location

Firms Located in a ZIP Code	SBIR Phase II Awardees	Matching Firms	*p*-Value
A. With an early-stage venture financing, 1983–85:			
Change in sales, 1985–95:			
Mean	7.71	1.06	
Median	−.06	−.42	
SE	3.09	1.23	
p-value, comparison of means			.026
p-value, comparison of medians			.001
No. of observations	189	271	
Change in employment, 1985–95:			
Mean	47.43	3.35	
Median	−1	−4	
SE	22.99	7.99	
p-value, comparison of means			.039
p-value, comparison of medians			.001
No. of observations	190	277	
B. Without an early-stage venture financing, 1983–85:			
Change in sales, 1985–95:			
Mean	1.74	1.18	
Median	−.07	−.40	
SE	.81	1.04	
p-value, comparison of means			.718
p-value, comparison of medians			.000
No. of observations	304	565	
Change in employment, 1985–95:			
Mean	13.14	6.96	
Median	−2	−3	
SE	8.80	5.39	
p-value, comparison of means			.527
p-value, comparison of medians			.057
No. of observations	309	569	

NOTE.—The table summarizes the change in the sales and employment of the 541 firms that received SBIR Phase II awards in the first three program cycles and the 894 matching firms in the sample. Firms are divided by whether their headquarters in 1985 were in a ZIP code with one or more seed or early-stage financings between 1983 and 1985. All sales figures are in millions of 1995 dollars.

ingly came from the basic research-oriented National Institutes of Health.) The gap between the firms that received Phase II awards from these agencies and those that only received Phase I awards was even greater than the dispersion in the sample as a whole. For instance, firms that received Phase II awards from NSF and HHS in the first three program cycles grew by 32.6 employees between 1985 and 1995; those that only received Phase I awards grew by 2.1 employees. These results suggested that the growth of the awardees was not a consequence of them subsequently obtaining procurement contracts.[8]

8. I also compared the awardees from HHS to matching firms specializing in biotechnology and medical devices that had received no SBIR awards, with similar results. An alternative explanation for the superior growth of the SBIR awardees was that winners of SBIR

B. Variation in Subsidy Effect across Regions

I next examined the regional disparities in these growth patterns. To measure new venture activity in the region where the firm was located, I tabulated seed and early-stage financings by private venture capital funds in the years 1983–85. I determined the dollar volume and number of early-stage financings in each five-digit ZIP code using the Venture Economics database.[9]

Panels A and B of table 6 present comparisons similar to those in table 5 but also divide firms by whether their headquarters at the end of 1985 was in a ZIP code with at least one early-stage venture financing in the years 1983–85. The differences between the SBIR awardees and the matching firms in the ZIP codes with venture capital activity were pronounced. For instance, employment increased by 47 for these awardees, as opposed to increasing by three for the others. For the average SBIR awardee in a ZIP code with venture activity—which in 1985 had $5.1 million in sales (again in 1995 dollars) and 57 employees—inflation-adjusted sales over the next decade grew by 151% and employment by 83%. The differences were much smaller in other areas (employment rose by 13 for the awardees, as opposed to by seven for the matching firms). The differences in means were statistically significant only among the firms in regions with venture activity. Medians displayed a similar pattern. The difference in median sales, however, was also highly significant when awardees and nonawardees were compared in areas without venture capital activity.

Table 7 examines the growth of employment and sales in a regression framework. The table examines the change in employment and sales, as well as the percentage change in these variables.[10] Among the independent variables were 1985 sales or employment, the volume (in millions of dollars) of early-stage venture capital activity in the firm's ZIP code between 1983 and 1985, a dummy variable indicating whether the firm received any Phase II awards in the first three SBIR funding

awards were more likely to win subsequent SBIR awards. While this may have been true in the initial years of the program (e.g., U.S. General Accounting Office 1987*b*), conversations with federal officials suggest that it has been much less of an issue in recent years. Not only were SBIR awards to firms that had received multiple previous grants more likely to be scrutinized by the GAO, resulting in a reluctance to make such grants, but the 1992 reauthorization of the SBIR program (Public Law 102-564) added additional criteria for the evaluation of SBIR applications of firms that had received awards in the past.

9. Five-digit ZIP codes are a fine gradation: the city of Cambridge, Massachusetts, for instance, is divided into six five-digit ZIP codes. I repeat the analyses in Section VI using three-digit ZIP codes (the code 021 encompasses the city of Boston and the nearby suburbs). I also repeated the analyses using the *number* of seed and early-stage financings in the ZIP code. These alternative measures had little effect on the results. In an analysis later in this section, I explore the use of broader geographic divisions.

10. Because some of the firms had minimal employment and sales in 1985, I calculated the percentage change variable as 100 * (1995 value − 1985 value)/(1985 value + 1).

TABLE 7 Regression Analyses of the Change in Employment and Sales between 1985 and 1995

A. OLS Regressions, with Change and Percentage Change in Sales as the Dependent Variables

	Change	% Change
1985 sales	−.18 [2.67]	−2.28 [2.46]
Value of early-stage financings in ZIP code	.07 [2.12]	1.90 [2.56]
Any Phase II SBIR awards in first three cycles?	1.88 [.80]	25.05 [.80]
Value of financings * any Phase II awards in first three cycles	.16 [4.75]	2.41 [3.43]
Constant	1.63 [1.26]	60.34 [2.07]
Root mean square error	25.37	564.93
Adjusted R^2	.02	.01
No. of observations	1,329	1,329

B. OLS Regressions, with Change and Percentage Change in Employment as the Dependent Variables

	Change	% Change
1985 employment	−.19 [3.32]	−.46 [5.13]
Value of early-stage financings in ZIP code	.29 [.85]	.39 [.31]
Any Phase II SBIR awards in first three cycles?	14.70 [1.00]	−12.54 [.41]
Value of financings * any Phase II awards in first three cycles	1.43 [11.84]	5.11 [5.45]
Constant	10.83 [3.80]	110.08 [4.89]
Root mean square error	173.04	930.47
Adjusted R^2	.02	.002
No. of observations	1,345	1,345

C. Median Regressions, with Change and Percentage Change in Sales as the Dependent Variables

	Change	% Change
1985 sales	−.62 [5.37]	−1.69 [5.02]
Value of early-stage financings in ZIP code	.003 [4.09]	.11 [.43]
Any Phase II SBIR awards in first three cycles?	.39 [18.42]	18.96 [3.08]
Value of financings * any Phase II awards in first three cycles	.01 [10.67]	.27 [.69]
Constant	−.03 [1.89]	−17.38 [4.37]
Sum of absolute weighted deviations	7,918.87	160,923.20
Pseudo R^2	.07	.01
No. of observations	1,329	1,329

D. OLS Regressions, with Change and Percentage Change in Sales as the Dependent Variables, Using Only Firms Located in ZIP Codes without Early-Stage Venture Financings

	Change	% Change
1985 sales	−.20 [3.91]	−1.75 [4.23]
Was state's venture capital financings above mean?	−1.07 [.65]	−36.75 [.94]
Any Phase II SBIR awards in first three cycles?	−.53 [.37]	−8.50 [.21]
State venture dummy * any Phase II awards in first three cycles	2.56 [1.98]	84.57 [6.10]
Constant	2.38 [1.26]	69.61 [1.89]
Root mean square error	21.62	571.10
Adjusted R^2	.01	.002
No. of observations	869	869

TABLE 7 ***(Continued)***

E. OLS Regressions for Firms in Industries with High and Low Intangible to Total Asset Ratios, with Change in Sales as the Dependent Variable

	Industry Intangible to Total Asset Ratio	
	Above Median	Below Median
1985 sales	−.13 [1.64]	−.29 [4.89]
Value of early-stage financings in ZIP code	.12 [1.62]	.02 [1.67]
Any Phase II SBIR awards in first three cycles?	3.26 [.96]	−.07 [.04]
Value of financings * any Phase II awards in first three cycles	.44 [5.73]	−.08 [1.01]
Constant	.90 [.98]	3.19 [1.61]
Root mean square error	24.97	25.67
Adjusted R^2	.05	.002
No. of observations	789	531
p-value, *F*-test of equality of SBIR coefficients in two regressions		.000

F. OLS Regressions for Firms in Industries with High and Low R & D to Sales Ratios, with Change in Employment as the Dependent Variable

	Industry R & D/Sales Ratio	
	Above Median	Below Median
1985 employment	−.15 [1.63]	−.19 [7.51]
Value of early-stage financings in ZIP code	.10 [.21]	.60 [1.67]
Any Phase II SBIR awards in first three cycles?	10.13 [.60]	19.50 [1.14]
Value of financings * any Phase II awards in first three cycles	3.60 [6.71]	−1.19 [2.31]
Constant	16.08 [1.69]	3.98 [.60]
Root mean square error	207.27	116.07
Adjusted R^2	.03	.02
No. of observations	735	600
p-value, *F*-test of equality of SBIR coefficients in two regressions		.000

G. OLS Regressions with Change and Percentage Change in Sales as the Dependent Variables, and Additional Independent Variables for Multiple SBIR Awardees

	Change	% Change
1985 sales	−.19 [2.64]	−2.28 [2.28]
Value of early-stage financings in ZIP code	.07 [2.13]	1.90 [2.56]
Any Phase II SBIR awards in first three cycles?	.11 [.08]	19.83 [.60]
Multiple Phase II SBIR awards in first three cycles?	4.02 [1.66]	15.61 [.99]
Value of financings * any Phase II awards in first three cycles	.48 [2.69]	8.71 [2.22]
Value of financings * multiple Phase II awards in first three cycles	−.58 [2.24]	−10.93 [1.93]
Constant	1.67 [1.27]	60.35 [2.05]
Root mean square error	25.23	562.73
Adjusted R^2	.02	.01
No. of observations	1,329	1,329
p-value, *F*-test of insignificance of multiple awardee variables	.018	.144

TABLE 7 ***(Continued)***

H. OLS Regressions, with Change and Percentage Change in Employment as the Dependent Variables, and Additional Independent Variables for Multiple SBIR Awardees

	Change	% Change
1985 employment	−.19 [3.05]	−.42 [5.86]
Value of early-stage financings in ZIP code	.29 [.85]	.38 [.30]
Any Phase II SBIR awards in first three cycles?	11.72 [.79]	.59 [.02]
Multiple Phase II SBIR awards in first three cycles?	8.05 [.78]	−23.99 [2.44]
Value of financings * any Phase II awards in first three cycles	3.96 [7.41]	13.46 [4.40]
Value of financings * multiple Phase II awards in first three cycles	−4.39 [4.19]	−14.25 [3.16]
Constant	10.75 [4.08]	108.89 [4.74]
Root mean square error	171.81	928.16
Adjusted R^2	.03	.01
No. of observations	1,345	1,345
p-value, F-test of insignificance of multiple awardee variables	.000	.007

NOTE.—The dependent variables in the regressions are the change and percentage change in sales (in millions of 1995 dollars) and employment between the end of 1985 and the end of 1995. The percentage change is computed as 100 * [(1995 value − 1985 value)/(1985 value + 1)]. The independent variables are sales (in millions of 1995 dollars) and employment at the end of 1985, the volume (in millions of dollars) of seed and early-stage venture capital transactions between 1983 and 1985 in the ZIP code in which the firm was located in 1985, a dummy variable measuring whether the firm received one or more Phase II SBIR awards in the first three program cycles, and an interaction between the venture capital and SBIR variables. Panel A presents ordinary least squares (OLS) regressions with change and percentage change in sales as the dependent variables; panel B presents similar OLS employment regressions. Panel C uses a median regression specification to analyze the change and percentage change in sales. Panel D uses only those firms based in ZIP codes without a seed and early-stage venture capital transaction between 1983 and 1985. This regression uses a dummy variable denoting whether the state had above the mean level of *per capita* venture financings instead of the dollar volume of disbursements in the ZIP code. Panels E and F analyze the change in sales and employment, estimating separate OLS regressions for firms whose industries are above or below the median in intangible to total assets and R & D to sales ratios at the end of 1985. F-tests examine the null hypothesis that the coefficients on the SBIR variables are the same across the sub-samples. Panels G and H add to the basic OLS specification a dummy variable denoting whether the firm had received more than one SBIR award in the first three program cycles, and an interaction between this and the venture capital activity in the firm's ZIP code between 1983 and 1985. F-tests examine the null hypothesis that the two new variables do not differ from zero. The sample (in all regressions except panel D) consists of 541 firms that received SBIR Phase II awards in the first three program cycles and 894 matching firms. Absolute t-statistics (computed using heteroskedastic-consistent standard errors in all but panel C) are in brackets.

cycles, and an interaction between the venture capital activity and SBIR awardee variables.

In panel A, I examined the change in sales and the percentage change in sales (both in constant 1995 dollars) in ordinary least squares (OLS) regressions. Because the firms were of very different sizes, a heteroskedasticity problem potentially existed. Consequently, I divided the firms into groups on the basis of 1985 sales and calculated heteroskedastic-consistent standard errors. (A similar procedure was used in all the OLS regressions reported in this table.) Panel B undertakes a similar analysis of the change in employment.

The regressions strongly supported the suggestion that the superior performance of SBIR awardees was not uniform. The presence of an SBIR award alone had little relationship with sales and employment growth. Rather, only the interactions between the SBIR indicator and venture activity in the ZIP code were consistently significant. The SBIR awards were associated with firm growth in areas with considerable venture financing activity, but no relationship emerged elsewhere. These patterns were consistent with suggestions that politically connected firms with few growth prospects may have captured awards in regions without large numbers of high-technology firms or venture capital organizations.

While the coefficient of the interaction variable was consistently significant, the overall goodness-of-fit in the regressions was disappointing. The poor fit reflected the fact that much of the success and failure of small firms is driven by idiosyncratic, firm-specific factors, difficult to control for in a large-sample study. This pattern raised the concern, however, that a few extreme outliers might drive the results. In order to address this issue, I employed a median regression specification, which reduced the influence of outliers. As panel C of table 7 reports, these modifications did have some effect on the results: the magnitude and statistical significance of the coefficient of the SBIR awardee dummy variable increased, while the interaction term became less important. This pattern was consistent with the median analysis of sales reported in table 6, which suggested a significant difference between awardees and nonawardees across all regions.

If the geographic disparities were a consequence of political pressures, then we might expect also to see differences among the firms located in ZIP codes without venture capital. Political pressures to allocate SBIR funds to firms located in states with high levels of venture capital activity (even if concentrated elsewhere) should have been considerably less than in states with very little venture activity at all. Put another way, state and congressional officials were much more likely to conduct a concerted lobbying campaign for awards to firms in Idaho than in northeastern California.

To address this suggestion, I undertook regression analyses confined to the SBIR awardees and matching firms that were located in ZIP codes without a seed or early-stage venture capital financing round between 1983 and 1985. I compared the awardees and matching firms located in those states with above and below the mean level of per capita venture capital disbursements during this 3-year period. Panel D of table 7 presents two regression analyses, which employ the change in sales as a dependent variable, as well as the percentage change. The independent variables were 1985 sales of the firm, a dummy variable indicating whether the firm received any Phase II awards in the first three SBIR funding cycles, a dummy variable denoting whether the state received above the mean amount of per capita venture capital disbursements in the years 1983–85, and an interaction between the dummy variables denoting SBIR awardees and above-average venture capital activity in the state.

The political pressure hypothesis suggests that the disparity between awardees and matching firms should have been greater in the states with high levels of venture capital activity. In states with low levels of venture activity, the distortions in the award process may have led to the selection of firms that could not benefit from the funds. Consistent with this suggestion, the interaction between the SBIR awardee dummy and that denoting above-average venture capital activity was significantly positive. Consider, for instance, the coefficient of 2.56 on the interaction term in the leftmost regression. This implied that the difference in the sales growth of the SBIR awardees and the matching firms (in ZIP codes without venture capital activity) was $2.6 million greater in states with above the mean level of venture activity.

I examined the robustness of the geographic regressions in several additional unreported analyses. First, I was concerned that the results could be an artifact of the two measures of firm success that I used, employment and sales. I consequently examined the changes in valuations, under a quite stark assumption: that the valuation of privately held or acquired firms was nominal when compared to the publicly traded firms. If this supposition held, changes in public market valuations could provide a reasonable indication of the overall change in these firms' valuations. I summed the value of those SBIR awardees and matching firms that were publicly traded in 1985 and 1995 (determining these valuations through the databases of the Center for Research in Security Prices and the several *Moody's* stock guides). I then examined the difference between the total valuation at the beginning and end of this period. The difference in the growth of the SBIR awardees and the matching firms was dramatic. The increase in the public market valuation of the 541 SBIR awardees between 1985 and 1995 was $9.5 billion. The valuation of the 894 matching firms, in contrast, only increased by a total of $3.0 billion during this period. Consistent with the results involving employment and sales, the increase in the valuation was the greatest for the awardees that were located in a ZIP code with an early-stage venture financing in the years 1983 to 1985. The 209 awardees located in such ZIP codes increased in value by $8.7 billion; the 332 awardees located in other ZIP codes increased in value by under $800 million.

Second, I was concerned about my matching approach. As discussed above, I constructed two populations of matching firms. An alternative approach would have been to undertake a single matched set, based on both region and industry: that is, rather than matching a Cambridge genome sequencing company to either another Cambridge firm or another genome firm, I could have selected a matching firm in the same industry and region. I chose not to employ this procedure owing to fears that it would introduce systematic biases. In particular, when I attempted to do such a matching on two criteria, I found that the quality of the matches differed dramatically by region. While quite precise matches could be obtained in California and Massachusetts, in some other regions the matching firms had very different characteristics (e.g., they tended to be much larger than the awardees). This was a consequence of the fact that there were few potential matches in many states: for example, in Hawaii, Idaho, and Kentucky, there were only a handful of entries in the 1986 *Corporate Technology Directory.* I was concerned that such a procedure would bias the results in ways that would be very difficult to address in the subsequent analyses. I undertook various tests in order to address concerns that one of the subpopulations of matching firms that I constructed might still be biased in some not readily apparent manner. I thus reestimated the equations only using the Phase II SBIR awardees and the *Corporate Technology Directory* matching firms, as well as only the Phase II and Phase I SBIR awardees. I similarly used only the *Corporate Technology Directory* firms matched on industry and those matched on location. The smaller sample sizes led to lower significance levels but had little effect on the magnitudes or signs of the coefficients. I also added controls for the age of the firm, whether it was venture-backed by the end of 1985, and for the two-digit SIC code of the firm, in case the results were shaped by some unobserved differences between the SBIR awardees and the matching firms. These had little effect on the coefficients or significance levels.

Third, I explored the use of alternative econometric specifications. Because the dependent variable in the regressions analyzing the percentage change in sales and employment was bounded at -100% (corresponding to the liquidation of a firm), I repeated these analyses employing a Tobit specification. This had little effect. The magnitude and significance of the coefficients changed little when I estimated systems of two equations, with the second equation examining the propensity of firms to win Phase II awards in the first three program cycles. I also estimated a two-stage Heckman regression, looking first at whether the firm survived and then at the extent of its growth. The interaction terms remained strongly significant. An alternative way to control for outliers was to divide the dependent variable in each regression into 10 deciles: that is, I created 10 equal-sized groupings based on the change in sales and employment. The results with this ordered logit specification very closely resembled those in the median regressions reported above.

C. *Industry Patterns*

The next analysis examined whether these patterns were especially pronounced in certain industries. As discussed above, the contrasting explanations for the growth of the SBIR awardees—the suggestions that SBIR awards certified firms to investors and that federal officials attempted to select successful firms for SBIR awards, whether the funds were needed or not—implied different empirical patterns. To address these concerns, I divided the firms into those in higher- and lower-technology industries and reestimated the regressions in panels A and B of table 7.

To divide the firms, I examined the characteristics of the firm's industry. Because I could not observe the financial condition of the private firms, I used the characteristics of the average public firm in the same industry. I computed the ratios of intangible assets to total assets, R & D spending to sales, and market to book equity values for all firms with a primary assignment in Compustat to the same four-digit SIC class as a company in the sample during the fiscal year ending in 1985 or at the end of the fiscal year ending in 1985. If there were fewer than four firms assigned to a four-digit SIC code, I used all firms with a primary assignment to the same three-digit SIC code. I then computed the average of these ratios.

I present the results of two of the analyses in panels E and F of table 7.[11] I report separate regressions for the observations from industries above and below the median on these proxies. I also present the p-values from F-tests of the null hypothesis that the two coefficients measuring the effect of SBIR awards (the dummy variable and the interaction term) were identical in the two regression analyses. In the regression employing the change in sales as the dependent variable (panel E), the SBIR dummy variable was positive, and the interaction term was positive and significant, for the firms in industries with more intangible assets. Among the other firms, the two variables were negative and insignificant. The null hypothesis of no difference was rejected. In panel F, where the change in employment was used as the dependent variable and firms were divided by industry R & D to sales ratios, a similar pattern held for firms with R & D spending above the median. The null hypothesis of no difference was once again rejected. This pattern was consistent with the view that SBIR awards provided certification to firms.

D. *Number of Awards*

The fourth empirical analysis examined how the effect of the awards changed with the number of awards. The distribution of awards in the first three program cycles was quite skewed. While the median SBIR Phase II awardee firm received one Phase II award in the first three

11. The results dividing firms on the basis of the third measure (the ratio of market to book equity values) was similar. The analyses were also robust to checks described in the final paragraphs of Section VI*B*.

program cycles, four firms received a total of 90 awards. (The mean firm received 2.8 awards in this period.) I added two independent variables that controlled for firms that received multiple SBIR awards. As discussed above, if the main role of public subsidies is to address the externality created by small firms' inability to capture the bulk of the rents from their innovations, more awards should translate into faster growth. As long as the firms capture some of surplus generated by the research, they should continue to benefit from additional subsidies. The certification hypothesis suggests a more ambiguous relationship, with the marginal value of subsequent awards declining sharply. Furthermore, the pursuit of these awards might introduce increasing distortions as the size (or number) of subsidies increases.

Panels G and H of table 7 present four regressions. The regressions were identical to those in panels A and B, with the addition of a dummy variable denoting firms that received more than one Phase II award in the first three program cycles and an interaction between this new dummy and the measure of venture activity in the firm's ZIP code. I also present the p-values from F-tests of the null hypothesis that these two new variables were jointly equal to zero. The dummy denoting frequent awardees was never significantly positive and in one regression was negative in sign. The interaction term between the multiple awardee dummy and the amount of venture activity in the ZIP code was significantly negative in all regressions. The joint test of the two variables rejected the null hypothesis of no difference from zero in three out of the four regressions.[12] The pronounced superior performance of the SBIR awardees in regions with venture activity was confined to firms receiving modest subsidies. Additional awards appear to have had minimal positive benefits, and the pursuit of these awards may even have had detrimental effects on the firms.

E. SBIR Awards and Venture Capital Financing

An important group to whom SBIR awards may provide information is the venture capital community. As discussed above, venture capitalists fund only a handful of the high-technology businesses begun each year. While the number of companies funded in this manner is small, the amount of capital provided is significant. Furthermore, these firms account for a disproportionately large share of the companies going public each year (Gompers and Lerner 1999) and of successful patent applications (Kortum and Lerner 1998). The importance of venture capitalists suggests an alternative test of the certification hypothesis. In this section, I examine not the overall growth of the SBIR awardees

12. The results were robust to using other cut-off points to denote multiple SBIR awardees, such as firms that received more than five Phase II awards. In unreported analyses, I also examined the robustness of the results to the controls described in Section VI*B*.

and the matching firms but, rather, the relationship between the receipt of SBIR awards and venture capital financing.

The interpretation of the results may be quite different, depending on the relative timing of the awards and the venture financing. If the SBIR awards played a certification function, then we might expect to see awardees disproportionately receiving venture capital financing in the years after the awards. The awards themselves might have served as a signal to venture capitalists or, alternatively, might have allowed the firm to attract high-quality managers or individual investors who in turn attracted venture investors. It would be harder to argue that the awards played a certification role if awardees were no more likely to receive venture financing than matching firms. The reverse relationship would also be problematic: that is, if venture capital financing *predicted* the receipt of an SBIR award. This might suggest—in the spirit of the models of political capture delineated in Section II above—that the venture capitalists were using the SBIR program as a subsidy for firms already in their portfolios. Rather than refinancing firms, they may have pushed them to obtain public funds.

Table 8 examines these issues in univariate and regression analyses. Panel A presents the probability that the awardees and the matching firms received venture financing in the years before and after the first three program cycles. First, I present the probability that the firms received venture financing through the end of 1981, the year immediately previous to the first SBIR awards. While the probability that SBIR awardees received venture financing was greater, the difference was not statistically significant. I then present the probability that the firms received venture financing for the first time after 1987, when the results of the first three award cycles were largely known. (Awardees were typically notified of whether they received a Phase II award within a year of completing their Phase I project.) In this calculation, I excluded firms that had already attracted venture financing or were publicly traded by the end of 1987. The difference between the awardees and the matching firms was larger in magnitude and statistically significant. These patterns were robust to the selection of other break points (e.g., 1980, 1982, and 1986). Thus, the univariate comparisons suggested that the awards might have certified firms to venture capitalists.

One concern with these univariate results is that they may reflect unobserved differences between the firms. In order to address these concerns, I examined these patterns in logit regression analyses. I first examined whether the presence of a venture capital investor explained the probability that a firm received a SBIR award. I used each of the 1,435 firms as an observation. The dependent variable was a dummy that took on the value of one if the firm received an SBIR Phase II award in the first three program cycles. Independent variables were a dummy denoting whether the firm received venture financing through

TABLE 8 Univariate and Regression Analyses of SBIR Awards and Venture Financings

A. Univariate Analysis of Probability of Venture Capital Financings for First Time

	SBIR Phase II Awardees (%)	Matching Firms (%)	*p*-Value
Funding prior to end of 1981	2.2	1.3	.210
Funding after 1987	2.5	.9	.020

B. Logit Regression Analyses

	Dependent Variable	
	Did Firm Receive SBIR Award in First Three Cycles?	Did Firm Receive Venture Funding after 1987?
Value of early-stage financings in ZIP code	.01 [2.04]	−.02 [.77]
Was firm based in California?	−.001 [.00]	.11 [.16]
Was firm based in Massachusetts?	.25 [1.27]	.62 [.88]
Did firm receive venture capital prior to end of 1981?	.34 [.73]	
Any Phase II SBIR awards in first three cycles?		1.39 [2.60]
1985 employment		.002 [.54]
χ^2-statistic	173.33	24.01
p-value	.000	.020
Log likelihood	−431.46	−81.35
No. of observations	1,435	1,193

NOTE.—Panel A summarizes the probability that the SBIR awardees and the matching firms received venture capital financing. The panel presents the probability that the firms received venture financing for the first time before the end of 1981 and after 1987. (Firms that received venture capital financing or went public in 1987 or before are excluded from the second analysis.) Panel B presents two logit regression analyses. The dependent variable in the first regression is a dummy coded as 1.0 if the firm received an SBIR Phase II award in the first three program cycles. The dependent variable in the second regression is a dummy coded as 1.0 if the firm received venture financing for the first time after 1987. The independent variables are the volume (in millions of dollars) of seed and early-stage venture capital transactions between 1983 and 1985 in the ZIP code in which the firm was located in 1985, dummy variables denoting companies based in California and Massachusetts in 1985, dummy variables denoting the firm's two-digit standard industrial classification code (not reported), a dummy variable denoting whether the firm received venture financing prior to the end of 1981 (in the first regression only), a dummy variable measuring whether the firm received one or more SBIR Phase II awards in the first three program cycles (in the second regression only), and the firm's employment at the end of 1985 (in the second regression only). The sample consists of 541 firms that received SBIR Phase II awards in the first three program cycles and 894 matching firms. (Firms that received venture capital financing or went public in 1987 or before are excluded from the second regression.) The first regression employs a fixed-effect logit specification; the second, a standard logit specification. Absolute *t*-statistics are in brackets.

the end of 1981 and controls for whether the firm was located in Massachusetts or California, the extent of venture capital activity in the ZIP code, and the two-digit SIC code. Because I constructed the data set by matching the universe of SBIR awardees to a sample of nonawardees, a simple logit analysis would have led to inconsistent and biased estimates (Palepu 1986). To address this problem, I estimated a fixed-effects logit regression (Chamberlain 1980). As the leftmost regression in panel B reports, once the firm and industry characteristics were controlled for, the presence of an earlier venture capital round did not predict whether the company received an SBIR award. The results were robust to modifications in the set of regressors.

I then examined the probability that the firm received venture capital financing for the first time in the years after the three SBIR award cycles (i.e., after 1987). I excluded from the analysis any firms that received venture financing or went public in earlier years.[13] I once again controlled for the characteristics of the firm (including now the employment level at the end of 1985) and added a dummy variable denoting whether the firm had received any Phase II SBIR awards in the first three program cycles. In the reported regression—as well as in others using modifications of the dependent and independent variables—the SBIR awardee variable was significantly positive. The SBIR awardees were more likely to receive venture financing. At the mean of the other independent variables, the predicted probability of a nonawardee obtaining venture financing was 0.8%, while that of an SBIR awardee was 3.1%.

VII. Conclusions

This article examined the long-run performance of high-technology firms receiving funds from a major public venture capital initiative, the SBIR program. I compared the growth of awardees to a set of matching firms. The SBIR awardees enjoyed substantially greater employment and sales growth. While the awardees and matching firms did not differ significantly in the likelihood of receiving venture capital in the years prior to the awards, in subsequent years the awardees were significantly more likely to receive such financing.

This pattern, however, was not uniform. The superior growth of SBIR awardees was confined to firms based in ZIP codes with substantial venture capital activity. These patterns were more pronounced in high-technology industries. No increase of performance was associated with larger subsidies. These patterns were consistent with the awards playing an important role in certifying firm quality but also with some

13. The sample also shrank because of missing employment data and because some independent variables completely determined the outcome.

distortions of the award process. These findings raised questions about the practice of awarding large numbers of SBIR awards to a few firms, as well as about congressional efforts to insure greater geographic dispersion of awards.

Two important limitations should be acknowledged, which in turn point to future research opportunities.[14] First, this analysis has not sought to assess the social benefits of the program.[15] Numerous studies have suggested that, because of knowledge spillovers, social rates of return to R & D are often much higher than the private returns that the firms performing the research enjoy. This analysis has focused exclusively on private returns, as roughly measured through sales and employment growth. The differentials between the private and social benefits of the SBIR awards might be particularly large because many projects involve very early-stage technologies (where spillovers to other firms may be more frequent) or those important to national defense or public health. The detailed case studies prepared by the National Science Foundation (Tibbetts 1996) suggest that awards in many cases led to substantial spillovers. At the same time, the gains in employment and sales of the typical awardee could have come at the expense of small losses for many competitors. Furthermore, many academic critics (e.g., Mervis 1996) argue that the SBIR set-aside has led to a reduction in funding for academic research, which may have even greater social benefits. While challenging, a detailed computation of the social costs and benefits must be a crucial element in assessing these programs.

14. Another issue is that political pressures and economic changes may have led to a deterioration of the SBIR program's effectiveness over time. For instance, as discussed above, the set-aside for SBIR awards has increased dramatically. This may have led to more efforts to influence the award process. The pool of venture capital also greatly increased during the 1980s and 1990s. The expansion of the private equity market may have obviated the need for such a program, and the long-run competitive advantage that SBIR awards confer today may be much lower. No performance differential can be seen in the relative success of firms that received SBIR awards in the first and third program cycles (venture disbursements had increased sharply between 1982 and 1984). But questions about the changes in the effectiveness of the program over longer time spans will only be answerable in the future.

15. While a rough calculation indicates that this program has had an attractive return for the federal government, these estimates must be approached with a degree of skepticism. For instance, given that the typical SBIR awardee 10 years later has $2.9 million in additional sales and 20 more employees than a matching firm, under plausible assumptions (30% corporate and individual tax rates, $50,000 annual salary per employee, and 15% pretax profit margins), the federal government would receive in additional taxes in 1 year an amount almost equal to the (nominal) amount of a Phase II award. If this level of enhanced sales and employment were sustained for a little more than 2 years, the government's rate of return would exceed the 10-year Treasury bond rate. Any more sustained increase in the firm's activity would make the government's return even more attractive. The returns would increase if there were significant spillovers to other firms and consumers or decrease if much of the growth was at the expense of other firms (resulting in a smaller marginal effect on aggregate economic activity or employment).

A second important unanswered question relates to the design of public venture capital programs such as the SBIR initiative. Because this study has focused on a single program that has been essentially unchanged since its inception, it is difficult to assess whether the program could be altered in significant ways. For instance, in light of evidence suggesting that a key benefit of SBIR awards is the certification that they provide, it might be possible that a program that offered much more modest subsidies could also be effective in certifying the quality—and spurring the growth—of small high-technology firms. Two crucial assumptions, however, are that such recognition would be sufficient to motivate firms to seek awards and that the integrity of the selection process could be insured. The recent wave of experimentation with public venture capital programs, both in the United States and abroad, suggests that a cross-program analysis of these and other questions could yield important insights.

References

Audretsch, D. B. 1995. *Innovation and Industry Evolution.* Cambridge, Mass.: MIT Press.

Becker, G. S. 1983. A theory of competition among pressure groups for political influence. *Quarterly Journal of Economics* 98 (August): 371–400.

Carsey, T.; Rundquist, B.; and Fox, S. 1997. Defense spending and economic development in the American states, 1970–89: Untangling a complex relationship. Working paper. Chicago: University of Illinois at Chicago.

Chamberlain, G. 1980. Analysis of covariance with qualitative data. *Review of Economic Studies* 47 (January): 225–38.

Cohen, L. R., and Noll, R. G., eds. 1991. *The Technology Pork Barrel.* Washington, D.C.: Brookings Institution.

Corporate Technology Information Services. 1996 (and earlier years). *Corporate Technology Directory.* Woburn, Mass.: Corporate Technology Information Services.

Eisinger, P. K. 1988. *The Rise of the Entrepreneurial State: State and Local Economic Development Policy in the United States.* Madison: University of Wisconsin Press.

Gompers, P. A., and Lerner, J. In press. Money chasing deals? The impact of fund inflows on private equity valuations. *Journal of Financial Economics.*

Gompers, P. A., and Lerner, J. 1999. *The Venture Capital Cycle.* Cambridge, Mass.: MIT Press.

Greenwald, B. C.; Stiglitz, J. E.; and Weiss, A. 1984. Information imperfections in the capital market and macroeconomic fluctuations. *American Economic Review Papers and Proceedings* 74 (May): 194–99.

Griliches, Z. 1992. The search for R&D spillovers. *Scandinavian Journal of Economics* 94, suppl.: S29–S47.

Irwin, D. A., and Klenow, P. J. 1996. High tech R&D subsidies: Estimating the effects of Sematech. *Journal of International Economics* 40 (May): 323–44.

Jaffe, A. B. 1996. *Economic Analysis of Research Spillovers: Implications for the Advanced Technology Program.* Washington, D.C.: U.S. Department of Commerce, National Institute of Standards and Technology, Advanced Technology Program.

Jewkes, J.; Sawers, D.; and Stillerman, R. 1958. *The Sources of Invention.* New York: St. Martin's.

Kortum, S., and Lerner, J. 1998. Does venture capital spur innovation? Working Paper no. 6846. Cambridge, Mass.: National Bureau of Economic Research.

Laffont, J.-J., and Tirole, J. 1993. *A Theory of Incentives in Procurement and Regulation.* Cambridge, Mass.: MIT Press.

Levy, D. M., and Terleckyj, N. E. 1983. Effects of government R&D on private R&D

investment and productivity: A macroeconomic analysis. *Bell Journal of Economics* 14 (Autumn): 551–61.

Mansfield, E.; Rapoport, J.; Romeo, A.; Wagner, S.; and Beardsley, G. 1977. Social and private rates of return from industrial innovations. *Quarterly Journal of Economics* 91 (May): 221–40.

Mervis, J. D. 1996. A $1 billion "tax" on R&D funds. *Science* 272 (May 17): 942–44.

Myers, S. C., and Majluf, N. 1984. Corporate financing and investment decisions when firms have information that investors do not have. *Journal of Financial Economics* 13 (June): 187–221.

Myers, S.; Stern, R. L.; and Rorke, M. L. 1983. *A Study of the Small Business Innovation Research Program.* Lake Forest, Ill.: Mohawk Research Corporation.

Olson, M. 1965. *The Logic of Collective Action.* Cambridge, Mass.: Harvard University Press.

Organization for Economic Cooperation and Development. 1996. *Government Programmes for Venture Capital.* Paris: Organization for Economic Cooperation and Development.

Palepu, K. G. 1986. Predicting takeover targets: A methodological and empirical analysis. *Journal of Accounting and Economics* 8 (May): 3–35.

Peltzman, S. 1976. Towards a more general theory of regulation. *Journal of Law and Economics* 19 (August): 211–40.

Price Waterhouse. 1985. *Survey of Small High-Tech Businesses Shows Federal SBIR Awards Spurring Job Growth, Commercial Sales.* Washington, D.C.: Small Business High Technology Institute.

Stigler, G. 1971. The economic theory of regulation. *Bell Journal of Economics* 2 (Spring): 3–21.

Stiglitz, J. E., and Weiss, A. 1981. Credit rationing in markets with incomplete information. *American Economic Review* 71 (June): 393–409.

Tibbetts, R. 1996. 50 examples of SBIR commercialization. Photocopied. Washington, D.C.: U.S. National Science Foundation.

U.S. General Accounting Office (GAO). 1985. *Implementing the Small Business Innovation Development Act—the First Two Years.* Washington, D.C.: GAO.

U.S. General Accounting Office. 1987*a*. *Federal Research: Small Business Innovation Research Participants Give Program High Marks.* Washington, D.C.: GAO.

U.S. General Accounting Office. 1987*b*. *Federal Research: Effectiveness of Small Business Innovation Research Program Procedures.* Washington, D.C.: GAO.

U.S. General Accounting Office. 1989. *Federal Research: Assessment of Small Business Innovation Research Programs.* Washington, D.C.: GAO.

U.S. General Accounting Office. 1992. *Federal Research: Small Business Innovation Research Shows Success But Can Be Strengthened.* Washington, D.C.: GAO.

U.S. General Accounting Office. 1995. *Federal Research: Interim Report on the Small Business Innovation Research Program.* Washington, D.C.: GAO.

U.S. Small Business Administration. 1986 (and earlier years). *Listing of SBIR Awardees for FY 1985.* Washington, D.C.: Government Printing Office.

U.S. Small Business Administration. 1996 (and earlier years). *Small Business Innovation Development Act: Twelfth-Year Results.* Washington, D.C.: Government Printing Office.

Wallsten, S. J. 1996. The Small Business Innovation Research program: Encouraging technological innovation and commercialization in small firms? Working paper. Stanford, Calif.: Stanford University.

[3]

RAND Journal of Economics
Vol. 31, No. 1, Spring 2000
pp. 82–100

The effects of government-industry R&D programs on private R&D: the case of the Small Business Innovation Research program

Scott J. Wallsten*

I ask whether government-industry commercial R&D grants increase private R&D. Regressing some measure of innovation on the subsidy can establish a correlation between grants and R&D, but it cannot determine whether grants increase firm R&D or whether firms that do more R&D receive more grants. Using a dataset of firms involved in the Small Business Innovation Research (SBIR) program, I estimate a multi-equation model to test these hypotheses. Firms with more employees and that appear to do more research win more SBIR grants, but the grants do not affect employment. Moreover, I find evidence that the grants crowd out firm-financed R&D spending dollar for dollar.

1. Introduction

■ Government programs that subsidize commercial R&D are justified on the grounds that profit-maximizing firms underinvest in R&D. Government funding of private R&D projects, however, can increase R&D effort only if the subsidies cause firms to undertake projects that would otherwise be unprofitable. Empirical studies to measure the effect of these government grants typically regress some measure of innovation or firm productivity (e.g., R&D spending or employment) on the subsidy. Many of these studies find a positive correlation between government R&D funding and private R&D effort and employment (e.g., Levy and Terleckyj, 1983; Robson, 1993; Nadiri, 1993; Irwin and Klenow, 1996; and Lerner, 1996). While this approach can establish a correlation between government grants and firm R&D, it cannot determine whether grants cause firms to do more R&D or whether firms that do more R&D receive more grants. Using a unique, extensive dataset, I estimate a multi-equation model that permits separate tests of these hypotheses. I find that firms with more employees and patents win more federal R&D grants from the Small Business Innovation Research (SBIR) program, but that the grants do not appear to affect employment. Moreover, using this methodology

* Stanford University and The World Bank; wallsten@stanford.edu.

I have benefited from, and am grateful for, the thoughtful comments of many people, including Michael Ash, Timothy Bresnahan, Linda Cohen, Bret Dickey, Joshua Holmes, Michael Mazzeo, Roger Noll, Rob Porter, Michael Pries, Paul Romer, Geoff Rothwell, Frank Wolak, other members of the Stanford IO seminar, and three anonymous referees. All mistakes are, of course, my own. I gratefully acknowledge financial support from the Markle Foundation and the Alfred P. Sloan Foundation.

I also find evidence that these R&D grants crowd out firm-financed R&D dollar for dollar.

Until recently, most federally funded industrial R&D was directed at government needs, such as large weapon systems. While federal support for private R&D dates back to the 19th century, government-industry R&D programs have become increasingly popular in recent years. Many of these programs aim to help firms commercialize innovations by subsidizing their R&D. Cohen and Noll (1995a) and Nelson and Romer (1996) document the increasing popularity of these programs and the reasons for their rise. Others examine individual government-industry R&D programs.[1] The economic justification for these programs is clear: because the social returns to private R&D are often higher than the private returns, some research projects would benefit society but would be privately unprofitable. By lowering the cost to the firm, a subsidy can make these projects privately profitable as well.

A theoretically optimal government response does not mean that government can, in practice, rectify the market failure. It is well established that government programs to mitigate market failures may generate incentives that lead the private sector to undo many of the programs' intended benefits (Stiglitz (1988) provides an overview of this literature). This literature, however, generally does not explore how incentives faced by government agencies that implement the programs may also affect the degree to which the programs can mitigate market failures. Such incentives may be especially prevalent in today's government-industry technology programs, where firms apply for subsidies for research aimed at commercializing products and the government chooses which research projects to subsidize. If government agencies face incentives to fund the most commercially promising proposals they receive, they will be inclined to support projects that would be privately profitable—and thus would be undertaken anyway—rather than projects that would benefit society but are privately unprofitable. If government funds research that would be funded privately, it will fail to increase R&D.

Lichtenberg (1987) demonstrates that econometric estimates of the effect of government subsidies on firm R&D may be biased upward if the estimation does not control for government purchases from firms whose R&D it sponsors. In today's government-industry technology programs, where firms apply for subsidies and government may not be the final consumer of R&D output, it is necessary to control for the government's decision about which firms to subsidize. If the government faces incentives to fund research projects with a high probability of commercial success, a reduced-form equation may find a correlation between subsidies and R&D. However, this correlation could arise because subsidies stimulate R&D, because firms that do R&D receive subsidies, or because of other factors that cause both R&D and subsidies.

I investigate the Small Business Innovation Research (SBIR) program, which awards more than $1 billion per year in R&D grants to small firms. I examine SBIR's institutional structure through official program documentation, congressional committee hearings, statements from program advocates, and government evaluations to determine whether the program encourages managers to fund research that benefits society but is privately unprofitable. I find evidence that the program does not encourage such behavior. Instead, it appears to reward the funding of commercially viable projects. This evidence motivates an econometric approach for testing whether SBIR increases innovation in firms that win awards. Receiving awards may be endogenous to a firm's R&D activities. An econometric test of the effect of grants on R&D must therefore

[1] Cohen and Noll (1995b) study the feasibility of cooperative R&D agreements. Irwin and Klenow (1996) study the effects of Sematech. Yager and Schmidt (1997) study the institutions of the Advanced Technology Program.

control for the government's decision about which proposals to fund, since those decisions may be correlated with a firm's R&D.

The article proceeds as follows. Section 2 explores the incentives affecting the design and implementation of these government-industry R&D programs. Section 3 provides details about SBIR and demonstrates how these incentives play out in this program. Section 4 describes the dataset. Section 5 presents a least squares regression of employment on the government subsidy. I then develop a three-equation system that incorporates the government's decision on whether to provide a subsidy to a firm and the firm's response to government funding. I use this system of equations to explore the effects of a subsidy on firm employment and on firm-financed R&D spending. Section 6 discusses the empirical results and concludes.

2. Program incentives

■ Government subsidies can increase R&D by supporting research that would not be undertaken without the subsidy. Because government subsidies are generally cheaper than other capital, however, firms will request funds for projects that are privately profitable as well as for projects that benefit society but are privately unprofitable.[2] To increase innovation, government-industry R&D programs, therefore, should not fund the best proposals they receive. Instead, they should fund the best proposals among those that are not likely to receive adequate funds from other sources. An efficient technology program would fund only *marginal* R&D projects that have positive net social benefits but negative net profits for the innovating firm. It would avoid funding *inframarginal* projects, which are expected to be privately profitable and would be pursued without a subsidy.

The government can encourage program managers to fund marginal projects only if it can determine whether managers attempt to comply with this objective.[3] The same problems that managers face in selecting marginal projects also make it difficult for the government to evaluate manager performance. Measuring the total returns to R&D is not simple. It is possible to measure private returns once a project has been funded and completed. For example, indicators of private returns include introductions of new products and changes in sales, employment, patents, and firm value. It is difficult, however, to estimate the social returns to R&D *ex ante* and *ex post,* especially if spillovers are diffuse.[4]

Even if measurement problems could be overcome, conducting an accurate program evaluation would remain difficult. For example, the evaluation must not only measure returns but also determine whether these projects required subsidies. It is often impossible to accurately predict the outcome of R&D, and simply identifying *ex post* private returns would not tell us whether a project needed a subsidy *ex ante.* Thorough evaluations would combine sophisticated engineering and scientific data with complex economic welfare analysis under conditions of uncertainty. Even if these evaluations could be conducted at a reasonable cost, political overseers might have difficulty determining whether the evaluations were competent and unbiased.[5]

---

[2] Yager and Schmidt (1997) discuss how firms may decide which proposals to submit to the Advanced Technology Program (ATP).

[3] The Government Performance and Results Act of 1993 is broadly consistent with the objective of efficient implementation in that it requires federal agencies to measure the performance and results of their programs.

[4] Jaffe (1998) and GAO (1997) discuss problems in devising indicators of research outputs.

[5] Firms would probably be able to game any evaluation system and proposal-selection mechanism to some extent. For example, if spillovers are diffuse, a firm may distort the apparent spillovers in an evaluation. Likewise, firms may exaggerate potential spillovers in their applications if proposal selection explicitly focuses on spillovers.

The technical problems in conducting evaluations affect the incentives that politicians will build into it. Because R&D is so uncertain, any program that attempts to select only marginal projects must take risks and thus will fund some projects that ultimately fail. Government support for commercial R&D, however, is controversial. Neither citizens nor their elected representatives are likely to have the incentive or expertise to understand comprehensive R&D evaluations. Instead, they are likely to rely on less complex information, such as changes in employment or anecdotal evidence about success or failure. Proponents look for "success stories" to prove that a program is successful, while opponents look for "horror stories" to prove that the program is wasteful (e.g., Chait, 1996).[6] The political debate is therefore likely to focus on measurable private returns and anecdotes, largely ignoring the difficult-to-estimate *expected* returns and spillovers.

Distributive politics also discourage elected officials from designing elaborate systems for evaluating R&D programs. Cohen and Noll (1991) argue that politicians face incentives to use technology programs to reward constituents, not to correct market failure. Given these incentives, politicians may not want an accurate evaluation system, since it could identify projects that were selected for constituency reasons.

The above analysis leads to two predictions about the design and implementation of commercial R&D programs. First, program proponents are unlikely to implement a monitoring system that identifies projects that would have been pursued without government subsidies. This prediction is consistent with Jaffe's (1998) observation that government technology programs have never been designed to include comprehensive economic evaluations, making it almost impossible to identify marginal effects. Second, politicians are unlikely to support a program that picks many projects that fail or refuses to support commercially attractive proposals. In other words, politicians will want to see a steady stream of measurable benefits from these programs. These two predictions suggest that in a commercial R&D program, "success" is likely to mean *any* market success, with no distinction between marginal and inframarginal. The desire to "pick winners" induces program managers to support commercially attractive, and thus inframarginal, proposals—a hypothesis I test empirically in this article.

3. The Small Business Innovation Research program

■ The SBIR program requires certain federal agencies to set aside a percentage of their extramural R&D budgets for grants to small business.[7] In fiscal year 1998, the eleven participating agencies awarded more than $1 billion in SBIR grants to small firms.[8] The program consists of three consecutive phases. A firm first submits an application for a Phase 1 award. In Phase 1 the agency can award the firm up to $100,000 to "determine the scientific and technical merit and feasibility of an idea." If the firm wins a Phase 1 award it can apply for a Phase 2 award. In Phase 2 the agency can award the firm up to $750,000 to "further develop the idea." In Phase 3 the firm receives no SBIR funds but is expected to commercialize its resulting product.

[6] Opponents may also cite instances where private returns are large as evidence that the program enriches a few firms at public expense, ignoring the possibility that the innovation is net welfare improving and would not have occurred without the subsidy.

[7] "Extramural R&D" is R&D funded by the federal government through contracts, grants, or cooperative agreements and performed by organizations that are not part of the federal government.

[8] The agencies are the Departments of Agriculture, Commerce, Defense, Education, Energy, Health and Human Services, and Transportation, as well as the National Aeronautics and Space Administration, National Science Foundation, Environmental Protection Agency, and the Nuclear Regulatory Commission.

SBIR aims to "increase private sector commercialization of innovations derived from federal research."[9] Despite this goal, regulatory guidelines do not encourage managers to fund marginal projects. Instead, regulations (*Federal Register,* 1993) instruct the agencies to select proposals on their technical merit and potential for commercial success—the same criteria a private investor might use. The guidelines do not distinguish between marginal and inframarginal projects.

Even if managers attempt to fund marginal research, the program contains no evaluation mechanism to identify such efforts. The U.S. General Accounting Office (GAO) conducts the only official evaluations of the program. GAO relies on surveys of grant recipients and focuses primarily on whether funded projects succeeded in the market. Their reports generally conclude that funded projects tend to succeed in the market, but they do not distinguish between marginal and inframarginal effects (GAO, 1992; GAO, 1995).[10]

In the absence of a mechanism to identify marginal effects, advocates point to *any* funded project that achieved commercial success as evidence of program success. When Congress considered SBIR renewal in 1991, Representative John LaFalce, Chairman of the House Committee on Small Business and a leading SBIR proponent, noted as he opened legislative hearings that

> the public investment in small business innovation, through the SBIR Program, has been productive beyond anybody's expectations. The General Accounting Office study will document that the Federal Government's investment of roughly $1 billion in the SBIR program from 1984–87 has, in addition to fulfilling the government's R&D needs, generated more than $1 billion in commercial spinoff of innovative products and additional developmental funding.
>
> The Small Business Administration study will indicate that roughly one in four SBIR projects result in the development and sale of a new commercial product—a ratio that is remarkably high in the risky R&D business. (U.S. House of Representatives, 1991.)

The "commercial spinoff" clearly was not only commercial success on the margin, and the "remarkably high" success ratio of funded projects may indicate that the program did not take sufficient risks.

The Small Business Administration (SBA) adopts the same anecdotal approach to assessing the program. The SBA's annual reports cite the GAO results and include many "success stories." Bibbens (1996) states that "the SBIR program has been a clear success. The commercialization rate of the SBIR program is the highest of any federal R&D program. . . ." These reports may convey to a program manager that a high commercialization rate is the key indicator of success.

Some program reviews implicitly suggest that efforts to fund R&D on the margin may be punished. GAO (1992), for example, found that the Department of Defense (DoD) had a lower SBIR commercialization rate than did other agencies. While noting that DoD tried harder than other agencies to use SBIR to achieve mission objectives, GAO concluded that the relatively low market success rate of funded projects "raises the question of whether DoD should be placing greater emphasis on private-sector commercialization."

---

[9] Public Law 97-219, July 22, 1982. See Wallsten (1998) for further analysis of SBIR's institutional details.

[10] GAO conducted one survey intended to identify marginal effects. Most respondents indicated that they would not have undertaken their projects without the subsidy (GAO, 1989). This survey is likely to be biased because firms receiving free money are unlikely to report that they did not need it, especially if they hope to receive future subsidies.

□ **Does SBIR fund research on the margin?** The above discussion indicates that SBIR is not designed to encourage managers to fund marginal projects, since the program does not distinguish between inframarginal and marginal market success. However, while the GAO reports and anecdotal success stories do not demonstrate that the subsidy was responsible for a project's success, they do not rule it out, either. Indeed, while the incentive structure may induce SBIR to fund inframarginal research, other factors may lead it to increase innovation. SBIR proponents contend that the agencies discourage small firms from participating in federal R&D, and that only SBIR can increase small business access to federal R&D funds (SBA, 1994). This article does not analyze the assertion that small business is excluded from federal R&D. Even though SBIR's set-aside funding means that the cost of an SBIR dollar is a dollar not spent on other research, the exclusion of small firms from federal R&D would mean that SBIR could stimulate innovative activity if the most cost-effective way to increase innovation is to give more grants to small firms.

To my knowledge, only one other study, Lerner (1999), has seriously examined whether SBIR increases innovative activity. Lerner compared the long-term performance of firms that received SBIR awards in the program's first three years to a matched set of firms that did not. He found no effect of SBIR awards except among firms located in geographic areas with a high degree of venture-capital activity. Awardee firms in these areas increased sales and employment more than other firms did. While Lerner included a control group, the study did not address this endogeneity issue—the possibility that grants and firm growth were correlated because growing firms received grants, not because grants caused firms to grow.

In the remainder of the article, I exploit a unique dataset of firms involved in the SBIR program to test empirically whether the program funds projects on the margin when controlling for firm application and agency proposal selection decisions.

4. Data

■ To test whether SBIR increases innovative activity, I assembled an extensive dataset consisting of firms that won awards, firms that applied for awards but were rejected, and firms that were eligible to apply but did not. The group of awardees consists of 367 firms that each won at least one award from any of the participating federal agencies between fiscal years 1990 and 1992.[11] The group of rejected firms consists of 90 firms that applied either to DoD or to NASA for an SBIR award during this period, but did not receive an award from any agency. The dataset does not include firms rejected by other agencies because only DoD and NASA were willing to identify unsuccessful applicants.[12] DoD simply listed firms that applied for awards, while NASA identified the number of applications each firm submitted each year. The group of eligible firms consists of 22 firms chosen from Compustat that were similar in size and industry classification to firms that did apply.[13] The Appendix contains detailed information on the variables in the dataset.

[11] Although these firms were selected on the basis of having won an award between 1990 and 1992, I also compiled awards data for 1988 and 1989 when it became obvious that firms receive multiple awards each year.

[12] DoD and NASA, however, together grant about 65% of all awards.

[13] The complete dataset contains 513 award winners, 185 rejects, and 79 eligible firms. The statistics presented here are for the firms for which I have complete data. The sample of "eligible" firms is small because these are public firms and can best be compared to the public award winners, of which there are 64 in the dataset.

The industries represented by the firms in the sample are what we would expect for a group of small, high-tech firms. I grouped the firms into nine broad industry categories based on two- and three-digit SICs.[14] The largest number of firms (144) is in the R&D and testing industry, followed by computer and data processing (77), and engineering services (64). Table 1 shows that the 369 firms in the dataset that won awards averaged approximately six Phase 1 and two Phase 2 awards between 1990 and 1992. The 63 firms that won at least *ten* Phase 1 awards account for more than half of all the awards in the sample. Since winning multiple awards is common, and only 10% to 25% of Phase 1 proposals receive funding (GAO, 1995), some firms must submit many applications. Indeed, firms that applied to NASA and won an award from any agency submitted about five Phase 1 applications per year to NASA alone.

Table 2 gives some preliminary evidence of why it is important to control for endogeneity. The table shows that the typical award winners, while small, are not the stereotypical entrepreneurs working out of their garages. These firms averaged more than $4 million in annual sales and employed more than 40 people in 1991, while the unsuccessful applicants were slightly smaller. The table also shows that sales and employment among firms that won awards increased from 1991 to 1993. Sales and employment among rejected firms declined slightly. Finally, although neither group received many patents, the awardees patented more than did the rejects. Early in the sample period, firms that won awards were slightly larger and had more patents than did firms that did not win awards. They also show that firms that won awards grew—both in sales and employment—by more than the rejected firms did.[15] These results do not allow me to rule out either the hypothesis that awards affect firm productivity or that more productive firms win awards.

While these statistics are interesting, their relevance depends on the degree to which the sample is representative of the universe of SBIR firms. I attempt to show representativeness in three ways. First, I compare the percentage of awards from each agency in this sample to the percentage of all awards from each agency in fiscal year 1992. The distribution of awards from agencies represented in the sample is close to the actual distribution of awards, although I slightly oversampled DoD awards and slightly undersampled Health and Human Services (HHS) awards. DoD represents 48% of awards in the sample and 43% of all awards granted. HHS represents 17% of awards in the sample and 22% of all awards. The share of awards in the sample from all other agencies is within one percentage point of the share of all awards from those agencies.

Second, I compare firm-level data on numbers of awards won to data collected by the GAO in a 1988 survey of SBIR award winners. Because my sample covers 1990 through 1992 while the GAO survey covers 1983 through 1987, the comparison is not clean. The longer time span of the GAO sample, which allows more time for firms to win multiple awards, is somewhat offset by the fact that the SBIR program was larger from 1990 to 1992 than from 1983 to 1987. The distribution of multiple award winners is similar, although as expected given the shorter time period, my sample is somewhat skewed toward firms winning fewer awards. Forty-six percent of my sample and 43% of the GAO sample received two or fewer Phase 2 awards. Seven percent of my sample and 18% of the GAO sample received more than five Phase 2 awards.

Finally, I compare firm-level data on sales and employment in the sample to similar data in the GAO survey. Table 3 demonstrates that the distribution of sales and median

[14] I categorized firms that did not fit into these groupings as "other."

[15] All comparisons in this paragraph are significant at better than the 1% level, with one exception. The exception is that I cannot reject the null hypothesis that 1991 and 1993 sales and employment of rejected firms remained constant.

TABLE 1 Multiple Award Winners

Number of Phase 1 Awards	Number of Firms	Averages per Firm			Total across Firms			
		Number of Awards		Total Value	Number		Percent of all Awards	
		Phase 1	Phase 2	($ Millions)	Phase 1	Phase 2	Phase 1	Phase 2
1–4	227	2.0	.8	$0.43	452	174	19.9%	21.9%
5–9	77	6.5	2.5	$1.36	499	191	22.0	24.1
10–19	38	12.6	4.3	$2.56	480	165	21.1	20.8
20+	25	33.7	10.4	$6.20	842	261	37.0	32.9
Any	369[a]	6.2	2.1	$1.23	2,273	793[a]	100	100

[a] The number of firms that won at least one Phase 1 award sums to 367 instead of 369, and Phase 2 awards sum to 791 instead of 793. This anomaly occurs because two firms won one Phase 2 award each in this period for projects that received Phase 1 awards prior to 1990. The final row in the table represents all awardee firms in the dataset.

employment is approximately the same in the GAO survey and in this sample. The available indicators suggest that this sample is representative of all SBIR awardees.

□ **Indicators of innovative activity and productivity.** I want to determine whether winning awards causes firms to undertake additional R&D. There is no perfect way to determine if a small firm, for which there is little available data, increased R&D efforts. I explore the effects of SBIR grants using employment and firm-financed R&D expenditures. Most employees in these small firms are likely to be scientists, engineers, or others who are directly involved in R&D. R&D expenditures are a better indicator of research efforts, but these are available only for the publicly traded firms in the sample.

The program could affect employment in two ways. First, if SBIR grants increase commercial opportunities, firms may hire new workers to expand production. In reality, even if the program is successful we may not observe this effect because small, R&D-intensive firms often sell or license their inventions to others for commercial production. Second, if SBIR increases research activities, the firm may employ additional R&D personnel. Himmelberg and Petersen (1994) note that while spending on R&D by small firms is constrained by cash flow, it varies much less than does spending on physical investments. Indeed, small firms seem to smooth their R&D spending so that it is based on expected permanent cash flow. Smoothing occurs because most R&D expenditures in small firms are for the salaries of scientists and engineers, who cannot costlessly be hired and fired. Thus, if the firm expects an SBIR award to lead to a permanent increase in cash flow, say by commercializing a product that results from the subsidized research, then total R&D investment, and thus employment, may increase.

The second indicator of innovation is firm-financed R&D spending. The effect of a government subsidy on a firm's R&D investment depends on the elasticity of supply of private capital, whether the government funds inframarginal or marginal projects, and the share of the project the government funds. A marginal subsidy to a firm facing an increasing cost of capital and decreasing private marginal returns to investment would increase firm-financed R&D spending and employment. An inframarginal subsidy to the same firm would increase total R&D but decrease firm-financed R&D expenditures. In this case, we would find a negative effect of the grant on firm-financed R&D spending but possibly a positive effect on employment because *total* R&D effort would increase. An inframarginal subsidy to a firm facing a perfectly elastic supply of

TABLE 2 Firm Characteristics

	Awardees	Rejects	Eligible, but Never Applied	All Firms
Employment				
1991				
Sample size	369	90	22	481
Mean	41.28	35.69	49.73	40.62
Standard deviation	(75.79)	(61.13)	(34.44)	(71.81)
Median	15	14	45	16
1993				
Mean	48.76	34.36	45.27	45.91
Standard deviation	(79.70)	(56.58)	(25.99)	(74.32)
Median	20	12	48	18
Sales				
1991				
Sample size[a]	208	53	22	283
Mean	$4,762,361	$4,439,775	$4,337,727	$4,668,937
Standard deviation	($8,470,623)	($7,399,532)	($4,149,197)	($8,004,488)
Median	$1,500,000	$1,653,029	$3,834,500	$1,682,000
1993				
Mean	$6,072,432	$4,329,527	$4,610,091	$5,632,342
Standard deviation	($10,214,000)	($4,902,531)	($4,185,718)	($9,102,694)
Median	$2,197,970	$2,000,000	$3,322,500	$2,244,916
Patents, 1990–1995				
Mean	1.5	.5	2.8	1.4
Standard deviation	(3.16)	(1.55)	(8.48)	(3.39)
Percent with at least one patent	38.1	15.6	40.9	33.9
Percent of sample publicly owned	12.7	7.8	100.0	15.8

[a] I do not have sales information for all firms in the data. Sales statistics presented are thus for a subset of the data.

capital, however, would cause firm-financed R&D to decrease by the amount of the subsidy. In this case we would find that the grant leads to a dollar-for-dollar crowding out of firm-financed R&D. Because total R&D is unaffected, we would find no effect on employment.

□ **Data limitations.** This unique dataset has several advantages. For example, the data contain a large share of all SBIR awards—almost 35% of all awards granted over a three-year period. More important, the detailed firm-level data allow me to control for factors that may allow a firm both to win awards and to do R&D. But the data

TABLE 3 Representativeness of the Sample: Sales and Employment Data

	Percent with Sales/Revenue Within Given Range	
Sales/Revenue Range	This Sample (1991 Sales, Deflated to 1987 Dollars)	GAO Survey (Reported 1987 Gross Revenue)
<$500,000	28.5%	26.5%
$500,000–$999,999	18.5	13.4
$1 million–$4,999,999	30.1	29.0
$5 million–$20 million	18.9	24.8
>$20 million	4.0	4.4
		11.6% no response
Median employment	1991: 15 1993: 20	1988: 20

also have some significant limitations. In addition to award winners, the dataset includes firms that were rejected and eligible firms that never applied. The sample of firms in the latter two groups, however, is neither stratified nor representative.

The rejected firms in the dataset come from only two federal agencies and may not be representative of firms rejected by the other nine agencies. The eligible firms in the dataset, meanwhile, are publicly traded. While similar to publicly traded firms that won awards, they may not represent the overall population of eligible firms, which are largely privately held. Moreover, the sample sizes of the award winners, rejected firms, and eligible firms are not proportional to their population sizes. Award winners make up the majority of firms in the dataset. Firms that were rejected make up a smaller share, and eligible firms an even smaller share. The actual number of firms eligible for the program is far greater than the number that apply, and the number that are rejected is greater than the number that win awards.

The important implication of these issues is that the data do not allow me to project out of sample. For example, the analysis will not allow me to predict what would have happened to the typical firm rejected by agencies other than DoD or NASA had they won awards. Nor can I predict what would happen to the typical eligible firm should it choose to apply. In addition, the time frame studied here is relatively short. Because the data do not, for the most part, extend beyond 1993, it is not possible to determine the long-term effects of SBIR awards. Nonetheless, the data allow me to look at short- and medium-term effects and, more important, to explore econometrically the importance of controlling for the endogeneity of awards.

5. Analysis

■ This section demonstrates empirically the importance of controlling for endogeneity of awards. I first present a least-squares regression of employment on awards, which shows a significant correlation between the two. I next simultaneously estimate a system of equations, which controls for the agencies' selection of which firms win awards and instruments for the endogeneity of awards. This instrumental-variables approach demonstrates that firms with more employees win awards, but that awards do not increase employment. I then use the system of equations to estimate the effect of the grants on firm-financed R&D in the publicly traded firms in the sample (the only firms for which R&D data are available).

□ **Ordinary least squares.** Empirical studies of the effects of R&D grants on firms typically regress some measure of firm productivity on the subsidy. In this section I estimate equation (1):

$$y = \alpha + \beta^*(subsidy) + \delta X + \lambda Z + \varphi G + \epsilon. \quad (1)$$

In equation (1), the dependent variable y is the log of employment in 1993. On the right-hand side, *subsidy* is the total number of SBIR Phase 2 awards the firm won from 1990 through 1992. I use Phase 2 awards because they are about ten times larger than Phase 1 awards, making them more likely to have a measurable effect on the firm. I use *number* of awards instead of dollar amount of awards because each Phase 2 award represents a separate research project, making it more likely to determine whether the project was a marginal one.[16] **X** is a vector of firm-specific variables, including the log of age, log of employment in 1991, the number of patents the firm applied for from 1988 and 1989 (and was ultimately awarded), and dummy variables indicating whether the firm is minority owned, publicly traded, a rejected firm, and a firm that never applied for awards. **Z** is a vector of industry dummy variables, indicating whether the firm's primary SIC code places it into one of the nine broad industry groups. **G** is a vector of four geographic dummy variables, indicating whether the firm is located in the San Francisco Bay Area, Los Angeles, Boston, or Washington, D.C.

The first column of Table 4 shows the results of this regression. The more Phase 2 awards the firm wins between 1990 and 1992, the higher its employment in 1993. This is consistent with most studies of the effects of government subsidies on firm productivity. It is also consistent with Lerner's (1996) findings on the employment effects of SBIR awards and with the findings of the GAO studies, which note that firms that win awards tend to commercialize their innovation. However, this regression ignores the possibility that awards are endogenous.

□ **Structural analysis.** To control for endogeneity I reestimate (1) in a simultaneous system of equations, instrumenting for the endogenous awards variable. The awards process is sequential, and I model it as such. Once the firm chooses to apply, it submits applications to federal agencies for Phase 1 awards. The agencies review the applications and choose which projects to fund. The firm can then apply for a Phase 2 award for each project that received Phase 1 funding. The agencies then decide which projects receive Phase 2 funding. The firm then alters its behavior (i.e., employment or R&D expenditures) in response to the award. Equations (1) through (3) together reflect this process.

$$\#\ Phase\ 1\ awards = \alpha_1 + \delta_1 X_1 + \lambda_1 Z_1 + \varphi_1 G_1 + \phi(budget) + \epsilon_1. \quad (2)$$

$$\#\ Phase\ 2\ awards = \alpha_1 + \delta_2 X_1 + \lambda_2 Z_1 + \varphi_2 G_1 + \phi(budget) + \epsilon_2. \quad (3)$$

$E(\epsilon_j) = 0$ and ϵ_1, ϵ_2, and ϵ in equation (1) are all correlated.

[16] Since each Phase 2 award is worth approximately $500,000 in this period, the choice between number of awards and dollar value of the award is essentially a matter of scaling. The empirical results are essentially identical using dollar values.

TABLE 4 Ordinary Least Squares and Three-Stage Least-Squares Regressions: Awards and Employment

	OLS	Three-Stage Least Squares		
Dependent Variable	log (Employment 1993)	Number Phase 1 Awards	Number Phase 2 Awards	log (Employment 1993)
Constant	.97 (6.39)	−2.96 (1.56)	−1.58 (2.66)	.94 (6.27)
Number of Phase 2 awards, 1990–1992	**.02** **(3.33)**			**.01** **(.94)**
SBIR budget instrument ($100 millions)		11.40 (15.66)	3.91 (17.13)	
log (age)	−.20 (3.39)	.17 (.23)	.20 (.91)	−.19 (3.46)
log (employment_{1991})	.85 (35.85)	.42 (1.56)	.16 (1.86)	.85 (39.99)
$\text{Patents}_{1988-1989}$	−.0007 (.03)	.77 (3.57)	.29 (4.24)	.00 (.23)
Never applied	−.33 (2.47)			−.32 (2.41)
Minority	−.03 (.51)	.08 (.10)	.19 (.70)	
Public	.46 (5.40)	.92 (1.00)	−.22 (.75)	.45 (5.45)
R^2 (481 observations)	.85	.48	.54	.85

Note: Industry and geographic dummy variables are included but not shown. Absolute *t*-statistics in parentheses.

Before explaining each equation in detail, I first describe the important instrumental variable, which is excluded from equation (1) and identifies the endogenous awards variable. An appropriate instrument must be correlated with the endogenous awards variable, but uncorrelated with unobserved factors (e.g., innovation) that affect the dependent variable—in this case, firm employment. An ideal instrument for the number of awards a firm wins would be the funds that are "potentially awardable" to the firm (Lichtenberg, 1988). The federal SBIR budget represents those funds. As described above, SBIR legislation mandates that each agency set aside a percentage (1.25% in this period) of its extramural R&D budget for SBIR grants. The number of awards an agency can grant is tied directly to its SBIR budget, since the legislation also mandates that each agency disburse its entire SBIR budget each year. The SBIR budget should therefore directly affect the number of awards a firm can potentially win. The agencies' R&D budgets, however, bear no relation to a firm's unobserved innovativeness.

The *total* SBIR budget, however, is not an appropriate instrumental variable. First, of course, it exhibits no variation across firms. Second, because a small firm's R&D is typically narrowly focused, not all SBIR funds are truly "potentially awardable" to every firm. The Department of Energy, for example, does not generally fund medical research, while the Department of Health and Human Services does not fund research in high-energy physics. An appropriate instrument should capture the SBIR budget that is potentially awardable to the firm given the type of research it does.

For firms that win awards, I thus define the instrumental variable, *budget,* as the total SBIR budget of all agency-years in which the firm won an award.[17] This amount represents the total SBIR funds available for the type of R&D the firm conducts. That is, for each firm,

$$budget = \sum_{i=SBIR\,Agencies} \sum_{t=1990}^{1992} award_{it} \cdot (SBIR\ budget_{it}),$$

where $award_{it} = 1$ if the firm won an award from agency i in year t, and $SBIR\ budget_{it}$ is agency i's SBIR budget in year t. Likewise, for firms that were rejected, *budget* is defined as the total SBIR budget of the agencies to which the firm applied from 1990 through 1992.[18]

The instrument is more complicated for firms that never applied. For these firms I impute from their four-digit SIC classification a probability of their applying to each agency had they chosen to apply. For each firm that never applied, I calculate the percentage of awardees in the same SIC that received awards from each agency. I use these weights to created a fitted *budget* for the eligible firms. For example, for the eligible firms in SIC classification 8734 (research, development, and testing: testing laboratories), *budget* is defined as follows:

$$budget_{neverapplied} = \sum_{i=SBIR\,Agencies} \sum_{t=1990}^{1992} p(award_{it}) \cdot (SBIR\ budget\ of\ agency_{it}),$$

where $p(award_{it})$ is the percentage of sample firms in SIC 8734 that received awards from agency i in year t.

The variable *budget* thus approximates the SBIR funds potentially available to each firm given the type of research it does, which should be uncorrelated with the firm's unobserved innovativeness. Empirically, the instrument proves to be a strong predictor of the number of awards a firm wins. The instrument, of course, is not perfect. One concern is that because it is based in part on whether a firm won an award from (or applied to) an agency, it may be correlated with the number of awards a firm wins and thus partially endogenous. To test the robustness of the results, I also reestimate the system after separating the instrument into its two components: *count*—the number of agency-years from which the firm won awards (or the predicted number of agency-years for firms that never applied), and *mean budget*—the *budget* variable divided by *count.*

Equations (2) and (3) are the agency's Phase 1 and Phase 2 funding decisions. The agency reviews applications and funds proposals based on their technical merits and potential for commercial success. Although I cannot observe the agency's review of the application, I can observe many factors that I assume indicate a firm's ability to submit good applications—firm "quality" (i.e., firm age and employment, and whether the firm has historically conducted much research, as indicated by its patenting in 1988 and 1989). Higher quality and past R&D experience increase the likelihood that the firm has good ideas to propose and that it knows how to prepare applications. The firm

[17] Lichtenberg (1988) uses a similar instrument to identify the effects of defense procurement contracts to firms.

[18] Ideally, I would include all agencies to which the firm applied, but I have only DoD and NASA application data. However, these agencies account for 65% of awards. Moreover, firms that apply to these agencies differ from firms that apply to, say, HHS and are unlikely to apply elsewhere.

could have obtained this knowledge from prior experience with SBIR, other government procurement actions, or nongovernment financiers. In any case, the more experience a firm has, the more likely it is to prepare a winning application. Equations (2) and (3) include industry dummy variables to control for agency preferences regarding the types of proposals they choose to fund and geographic dummy variable to capture agencies' sensitivity to constituency issues.

Equation (1) is the firm's response to receiving an SBIR award. The dependent variable is the log of employment in 1993, while the included endogenous variable is the number of Phase 2 awards the firm received from 1990 through 1992. The *budget* instrument is excluded to identify the awards variable. I also exclude the minority dummy variable from this equation but include it in the awards equations. I do this because one of SBIR's legislated goals is to include minority-owned firms in federal R&D, meaning that minority ownership could affect a firm's probability of receiving an award. Minority ownership should be unrelated to unobserved innovativeness, however. Other included variables are the same as in equations (2) and (3), with the addition of a dummy variable to indicate whether the firm was one that never applied.[19]

Although these decisions occur sequentially, I estimate them simultaneously. I assume the process is in equilibrium—firms know what criteria agencies use to select awards, and agencies know that the firms have this knowledge. Thus, each actor knows approximately how the other will behave and takes this information into account when making its behavioral decisions. In other words, common factors have causal impacts on all stages of the process—firm quality, for example, will affect proposals, agency selection, and the firm's growth rate.

In addition, many of these common factors are unobserved and thus incorporated in the error terms. For example, some CEOs may be especially adept at raising R&D funds. They might prepare more applications, contact agency program managers and influence the probability of a proposal being funded, and conduct more R&D if they are skilled at obtaining funds from other sources. Some firms may hire consultants or purchase special software to help prepare applications.[20] Software that produces polished applications eases proposal preparation and ensures that the firm does not make a bureaucratic mistake that would force an agency to reject the proposal automatically. Use of such software could thus increase the number of applications a firm submits as well as the probability that each application is funded. Error in the equations could also arise if a firm is founded by a former SBIR official. Such a person would have detailed knowledge of the program that could influence the number of applications she submits, the likelihood of receiving funding for a given application, and her ability to apply for other government funds. Finally, a large corporation may spin off small firms in order to apply for SBIR awards. That the firm was founded in order to win awards could impact the number of applications it submits, the quality of those applications (since the small firm may have its parent's resources at its disposal), and other R&D funds available to the firm.

Because these error terms would be correlated across equations, a simple instrumental-variables approach such as two-stage least squares would be inefficient. Three-stage least squares accounts for this cross-equation error correlation.

□ **Results.** The right three columns of Table 4 provide results of simultaneously estimating equations (1) through (3). Whereas the OLS regression showed a significant

[19] Although I treat it exogenously because of data constraints, the decision to apply is endogenous as well.

[20] See, for example, http://www.win-SBIR.com and http://www.idsonline.com/business/cwnelson.

positive correlation between awards and employment, the 3SLS results do not allow me to reject the hypothesis that awards have no effect on employment. Indeed, the coefficient on the awards variable decreases by more than half relative to the OLS results and becomes statistically insignificant. Equations (2) and (3) yield some interesting results as well. Employment in 1991 and the number of successful patent applications in the two years prior to the awards are both significantly positively correlated with winning awards. That is, larger and more patent-intensive firms win more awards. The *budget* instrument is positive and significant, as well—the larger the SBIR budget, the more awards a firm wins. These results hold whether the analysis uses *budget* as the instrument, *count* and *mean budget* together as instruments, or *mean budget* alone as an instrument.

To summarize, OLS regression shows a link between awards and employment in the years after receiving the awards. 3SLS estimation of a system of equations that controls for project selection and awards endogeneity finds that firms with more employees win awards, but that awards do not appear to affect employment. In the next subsection I apply the structural framework to explore the effects of awards on firm-financed R&D spending.

□ **Firm-financed R&D spending: direct evidence of crowding out.** To analyze directly the effect of awards on R&D spending, I again estimate equations (1) through (3), using firm-financed R&D spending in 1992, rather than employment, as the left-hand-side variable in equation (1).[21] One additional difference from the equations estimated above is that each equation includes the firm's R&D spending in 1990 to control for the firm's baseline spending. I do not normalize by, for example, taking the log of variables as I do above because here we are interested in the effect of each subsidy dollar on the firm's R&D spending. The advantage of this estimation is that it gives us a clearer look into the effects of the SBIR grant on the firm's innovative activities. The disadvantage is that R&D spending is available only for publicly traded firms, substantially reducing the sample size. I estimate the system twice: first with the number of Phase 1 awards as the dependent variable and again with the number of Phase 2 awards as the dependent variable.

Table 5 details the regression results. The point estimates reveal that each Phase 1 award is associated with a $132,000 decrease in firm-financed R&D. Because each Phase 1 award is worth about $50,000, this estimate implies more than double crowding out, which seems implausible. The most likely explanation is that because a large percentage of Phase 1 awards are followed by Phase 2 awards, this estimate is picking up some crowding out associated with a follow-on Phase 2 award. In this sample firms receive, on average, one Phase 2 award for every three Phase 1 awards. If Phase 2 awards also crowd out private R&D, then we should expect the estimate on Phase 1 awards to be biased upward. Indeed, the table indicates that each Phase 2 award is associated with a decrease of approximately $530,000. Since each Phase 2 award is worth about $500,000, the coefficient estimate implies a nearly dollar-for-dollar crowding-out effect of awards. We can also conclude at a 95% confidence level that the true Phase 1 coefficient lies in the interval (−15,224, −248,169) and that the true Phase 2 coefficient lies in the interval (−79,053, −981,837).

Table 6 details the results of estimating a two-equation system that uses the total dollar value of the firm's SBIR awards (i.e., each firm's total SBIR subsidy) as the

[21] I also estimate the system using as the dependent variable R&D in 1993, 1994, the sum of all R&D expenditures from 1990 through 1992, and differenced R&D from 1991 to 1994. The results are all essentially identical to those presented here using R&D in 1992 as the endogenous variable.

TABLE 5 Three-Stage Least-Squares Estimates: Effect of SBIR Awards on Firm-Financed R&D Spending, Publicly Traded Firms Only

Dependent Variable	Number Phase 1 Awards	Number Phase 2 Awards	R&D Spending in 1992	
Constant	3.79 (1.45)	.05 (.11)	3,201,570 (2.19)	2,746,580 (2.04)
Number of Phase 1 awards			**−131,697 (2.22)**	
Number of Phase 2 awards				**−530,445 (2.30)**
SBIR budget instrument ($100 millions)	18.36 (8.20)	4.55 (12.02)		
Age	−.12 (1.71)	−.02 (1.74)	−40,964 (1.13)	−35,644 (1.02)
$Employment_{1991}$	.02 (.94)	.00 (.38)	−794.78 (.09)	−2,487.04 (.29)
$Patents_{1988-1989}$	.73 (1.51)	.16 (1.96)	385,689 (1.59)	373,376 (1.59)
R&D $spending_{1990}$	.00 (2.76)	.00 (3.55)	.99 (4.53)	1.01 (4.89)
Never applied			−1,860,790 (2.47)	−1,898,030 (2.50)
R^2 (81 observations)	.72	.81	.71	.70

Note: I estimated the system twice: once using number of Phase 1 awards and once using the number of Phase 2 awards as the endogenous variable. Because the results for the awards equations were identical in both estimations, I report those columns only once. Industry and geographic dummy variables are included but not shown. Absolute t-statistics in parentheses.

endogenous subsidy variable instead of the number of Phase 1 and Phase 2 awards. The coefficient estimate on the SBIR variable is −.82 and is statistically significant at better than 1%. That is, one SBIR dollar is correlated with a reduction of $0.82 in firm-financed R&D. We can conclude with 95% confidence that the true coefficient lies between −.11 and −1.52. We cannot reject the hypothesis that each dollar in SBIR grants is associated with approximately a $1.00 decrease in firm-financed R&D spending.

The results in this case depend somewhat on the structure of the instrumental variable. Using *count* and *mean budget* together as instruments yields identical results on firm R&D spending, although *count* has more predictive power than *mean budget* does. *Mean budget* has no predictive power in the absence of *count*. Because *mean budget* alone has no predictive power in the first-stage regressions, it is not an appropriate instrument by itself. The results of the R&D regressions may thus be viewed as less robust to the extent that one views *count* as potentially endogenous.

6. Discussion and conclusions

■ My analysis of the SBIR program shows the necessity of dealing with the endogenous nature of federal R&D grants to a firm and the firm's R&D when testing for the effect of federal grants. Using an original firm-level dataset I demonstrate that least-squares regressions reveal a correlation between R&D subsidies and employment. Estimations using simultaneous equation systems and instrumental variables to control

TABLE 6 Three-Stage Least-Squares Estimates: Effect of SBIR Dollars on Firm-Financed R&D Spending, Publicly Traded Firms Only

Dependent Variable	Total Dollar Value of SBIR Awards	R&D Spending in 1992
Constant	69,432 (.23)	2,733,110 (2.02)
Total dollar value of SBIR awards		**−.82** **(2.31)**
SBIR budget instrument ($ thousands)	2.96 (11.58)	
Age	−12,781 (1.58)	
$Employment_{1991}$	736.70 (.37)	−2,195.79 (.25)
$Patents_{1988–1989}$	111,128 (2.00)	381,841 (1.62)
R&D $spending_{1990}$	−.14 (3.25)	1.01 (4.90)
Never applied		−1,806,970 (2.38)
R^2 (81 observations)	.80	.71

Absolute *t*-statistics in parentheses.

for the endogeneity of awards reveal that firms with more employees receive subsidies, but subsidies do not lead to increased employment. Using this approach I also find evidence that SBIR awards crowd out firm-financed R&D spending dollar for dollar among the subset of firms for which R&D expenditures are available.

This investigation clearly shows the importance of controlling for endogeneity when estimating the effects of government subsidies. One interpretation of the empirical results is that the awards have no impact on firm R&D activities. They simply crowd out firm-financed R&D expenditures, have no impact on firm employment, and ultimately have no impact whatsoever on R&D activities. This is not the only interpretation, however. Another possibility is that while the grants did not allow firms to *increase* R&D activity, they instead allowed firms to continue their R&D at a constant level rather than cutting back. That is, while the grants may not have funded additional projects, they may have allowed firms to avoid eliminating ongoing projects.

It is also possible that SBIR grants have an effect over a period beyond the scope of the data used here. This could happen if the subsidy allows the firm to delay refinancing for a longer period of time.[22] Funding by the government could also be a mechanism that helps the firm attract additional private funding. Ultimately, these are both additional empirical questions that could be addressed in future work.

Finally, it is unclear how generalizable the empirical results are to other government programs. SBIR may differ from other programs in important ways. It is a congressionally imposed set-aside program, causing many agencies to dislike it. As a result,

[22] Of course, by definition this would be crowding out if government subsidies took the place of private-sector funding.

they may put little effort into selecting projects on their economic merit rather than simply on their probability of commercial success. Other government-industry technology programs may be different. The Advanced Technology Program (ATP), for example, is similar to SBIR in that it funds research aimed at commercialization, but it differs from SBIR in that it specifically attempts to fund high-spillover research. As a result, the incentives in the ATP may differ from those in SBIR. Nonetheless, any government-industry technology program aimed at producing commercializable products will face incentives similar to SBIR to some degree.

The important point is that government subsidies—in any program in which the government must decide which firms to subsidize—may be endogenous to a firm's R&D activities. Any empirical estimation of the effect of subsidies on firm R&D or productivity must deal with this fact. This article shows that a careful econometric investigation can begin to sort out causality questions by controlling for endogeneity of R&D grants and firm R&D. Future empirical work should be cognizant of these endogeneity issues in order to avoid seriously biasing estimates when testing the effects of government subsidies.

Appendix

■ This Appendix describes the data collection and preparation process. I selected at random the group of firms that won awards from the SBIR subfile maintained in the Federal Research in Progress (FedRIP) database. I derived the group of firms that applied for awards and were rejected from lists of firms supplied by DoD and NASA. I chose at random from the Compustat database the group of firms that were eligible but did not apply. I considered a firm to be "eligible" if it had a primary SIC identical to SICs represented in the awardee group and had fewer than 100 employees.[23] I checked both the reject and eligible firms against SBA's lists of all award winners to ensure that they did not win awards from any agency in any year.

Most data on firm characteristics came from Dun and Bradstreet financial reports. Because the private firms voluntarily provide only sales and employment data, not all data existed for all firms, and occasionally no data were available at all. Data availability and research budget constraints dictated sample size, leaving 513 awardee firms, 185 rejected firms, and 79 eligible firms. Firm-level data include firm sales and employment in 1991 and 1993, geographic location (by Metropolitan Statistical Area), four-digit SIC, whether the firm is minority owned, whether the firm is publicly owned, how old the firm is, and each firm's total patent counts. Patent counts measure the number of patents applied for in a given year and awarded between 1990 and 1995 (inclusive). Thus, patent counts for 1988 and 1989 are the number of patents applied for in those years, but not awarded until sometime between 1990 and 1995. While these numbers capture delays in the U.S. Patent Office, any delays should affect all firms equally and reasonably measure patent-intensiveness. I also compiled data on R&D spending for public firms—the only firms for which these data were available.

For the group of firms that won awards, I tabulated from SBA's *Listing of SBIR Awardees* for the relevant fiscal years the number and dollar amount of Phase 1 and Phase 2 awards each firm won from all agencies from 1988 through 1992. I tabulated them such that each award type, agency, and year has a separate variable. For example, one variable is the number of Phase 1 awards the firm won from DoD, another variable is the number of Phase 1 awards from NASA, and so on for all awarding agencies. This methodology resulted in 220 variables (11 agencies, 2 types of awards, 1 variable for dollar amount and 1 variable for number of awards, and 5 years). The awards data can be aggregated into total awards from all agencies for each firm, or used to analyze awards from any given agency.

References

BIBBENS, T. "Innovation, Commercialization, and SBIR: It's Déjà Vu All Over Again." *The Small Business Advocate*, Vol. 15 (1996), pp. 6–7.

CHAIT, J. "Welfare Queens." *The New Republic*, December 16, 1996, p. 22.

COHEN, L. AND NOLL, R. *The Technology Pork Barrel*. Washington, D.C.: The Brookings Institution, 1991.

——— AND ———. *Research and Development After the Cold War*. Center for Economics Policy Research Publication no. 431, Stanford University, 1995a.

[23] Firms with fewer than 500 employees are eligible. This sample includes only firms with fewer than 100 employees to ensure that they are similar to firms that applied for and won awards.

——— AND ———. "Feasibility of Effective Public-Private R&D Collaboration: The Case of Cooperative R&D Agreements." *International Journal of the Economics of Business*, Vol. 2 (1995b), pp. 223–240.

FEDERAL REGISTER. "Small Business Innovation Research Program Policy Directive." Vol. 58 (January 26, 1993), pp. 6144–6158.

HIMMELBERG, C. AND PETERSEN, B. "R&D and Internal Finance: A Panel Study of Small Firms in High-Tech Industries." *Review of Economics and Statistics*, Vol. 76 (1994), pp. 38–51.

IRWIN, D. AND KLENOW, P. "High-Tech R&D Subsidies: Estimating the Effects of Sematech." *Journal of International Economics*, Vol. 40 (1996), pp. 323–344.

JAFFE, A. "Measurement Issues." In L.M. Branscomb and J.H. Keller, eds., *Investing in Innovation*. Cambridge, Mass.: MIT Press, 1998.

LERNER, J. "The Government as Venture Capitalist: The Long-Run Impact of the SBIR Program." *Journal of Business*, Vol. 72 (1999), pp. 285–318.

LEVY, D. AND TERLECKYI, N. "Effects of Government R&D on Private R&D Investment and Productivity: A Macroeconomic Analysis." *Bell Journal of Economics*, Vol. 14 (1983), pp. 551–561.

LICHTENBERG, F. "The Effect of Government Funding on Private Industrial Research and Development: A Re-Assessment." *Journal of Industrial Economics*, Vol. 36 (1987), pp. 97–104.

———. "The Private R&D Investment Response to Federal Design and Technical Competitions." *American Economic Review*, Vol. 78 (1988), pp. 550–559.

NADIRI, I. "Innovations and Technological Spillovers." NBER Working Paper no. 4423, 1993.

NELSON, R. AND ROMER, P. "Science, Economic Growth, and Public Policy." In B. Smith and C. Barfield, eds., *Technology, R&D, and the Economy*. Washington, D.C.: The Brookings Institution and The American Enterprise Institute, 1996.

ROBSON, M. "Federal Funding and the Level of Private Expenditure on Basic Research." *Southern Economic Journal*, Vol. 60 (1993), pp. 63–71.

STIGLITZ, J. *Economics of the Public Sector*, 2d ed. New York: W.W. Norton and Company, 1988.

U.S. GENERAL ACCOUNTING OFFICE (GAO). *Federal Research: Assessment of Small Business Innovation Research Programs*. GAO/RCED-89-39. Washington, D.C.: U.S. Government Printing Office, 1989.

———. *Federal Research: Small Business Innovation Research Shows Success but Can Be Strengthened*. GAO/RCED-92-37. Washington, D.C.: U.S. Government Printing Office, 1992.

———. *Federal Research: Interim Report on the Small Business Innovation Research Program*. GAO/RCED-95-59. Washington, D.C.: U.S. Government Printing Office, 1995.

———. *Measuring Performance: Strengths and Limitations of Research Indicators*. GAO/RCED-97-91. Washington, D.C.: U.S. Government Printing Office, 1997.

U.S. HOUSE OF REPRESENTATIVES. *The Small Business Innovation Research Program: The First Decade*. Hearings before the Committee on Small Business. Serial No. 102-16. October 1991.

U.S. SMALL BUSINESS ADMINISTRATION; OFFICE OF INNOVATION, RESEARCH AND TECHNOLOGY (SBA). *Small Business Innovation Development Act: Tenth Annual Report*. Washington, D.C.: U.S. Government Printing Office, 1993.

———. *The State of Small Business: A Report of the President*. Washington, D.C.: U.S. Government Printing Office, 1994.

WALLSTEN, S. "Rethinking the Small Business Innovation Research Program." In L.M. Branscomb and J.H. Keller, eds., *Investing in Innovation*. Cambridge, Mass.: MIT Press, 1998.

YAGER, L. AND SCHMIDT, R. *The Advanced Technology Program: A Case Study in Federal Technology Policy*. Washington, D.C.: AEI Press, 1997.

[4]

ELSEVIER

Research Policy 31 (2002) 145–158

research policy

www.elsevier.com/locate/econbase

Public/private technology partnerships: evaluating SBIR-supported research☆

David B. Audretsch[a], Albert N. Link[b,*], John T. Scott[c]

[a] *Institute for Development Strategies, Indiana University, SPEA, Suite 201, Bloomington, IN 47405, USA*
[b] *Department of Economics, University of North Carolina at Greensboro, Greensboro, NC 27412, USA*
[c] *Department of Economics, Dartmouth College, Hanover, NH 03755, USA*

Received 21 October 2000; received in revised form 1 December 2000; accepted 12 December 2000

Abstract

This paper evaluates public support of private-sector research and development (R&D) through the Department of Defense's (DoD's), Small Business Innovation Research (SBIR) Program. Based on alternative evaluation methods applicable to survey data and case studies, we conclude that there is ample evidence that the DoD's SBIR Program is stimulating R&D as well as efforts to commercialize that would not otherwise have taken place. Further, the evidence shows the SBIR R&D does lead to commercialization, and the net social benefits associated with the program's sponsored research are substantial.

Keywords: Small business; Innovation; Public/private partnership; R&D; Program evaluation

1. Introduction

As greater attention is paid to public support of private-sector research and development (R&D) by participants in the innovation process, it becomes imperative for policy makers to offer an economic rationale for their support of public/private technology partnerships as well as to formulate and demonstrate means for evaluating such relationships. While scholars and policy makers have long debated the economic role of government in the innovation process and the importance of market failure as an element of that rationale, Baron (1998) has offered a focused rationale for the Small Business Innovation Research (SBIR) Program: "[T]he rational for SBIR is the same as the general argument for government R&D—positive externalities (meaning) social benefits exceeding private ones."

When market failure is the broad justification for public support of R&D performed in the private sector, the situation is not only one where social benefits exceed private benefits. As a consequence of the divergence of social benefits and costs there must be, absent support from the government, underinvestment in R&D from society's perspective. The fact that social benefits exceed private benefits is not sufficient to establish underinvestment.[1] Furthermore,

☆ An earlier version of this paper was presented at the January 2000 American Economic Association meetings in the session "Government as Venture Capitalist: Evaluating the SBIR Program." We appreciate the comments and suggestions from participants at that session as well as those from two anonymous referees.

* Corresponding author.
E-mail addresses: daudrets@indiana.edu (D.B. Audretsch), al_link@uncg.edu (A.N. Link), john.t.scott@dartmouth.edu (J.T. Scott).

[1] See Baldwin and Scott (1987).

0048-7333/02/$ – see front matter
PII: S0048-7333(00)00158-X

if government wants to justify public support with an argument based on market failure, policy makers must establish that government failure would not prevent improving R&D performance with a proposed policy.[2]

The government has at its disposal mechanisms to address market failure and hence, to overcome private-sector underinvestment in R&D.[3] These mechanisms include creating and maintaining an economic environment conducive for innovation. Examples of such mechanisms include the patent system and the passage of the National Cooperative Research Act of 1984 to encourage research joint ventures; tax incentives to stimulate innovative investments through, for example, the 1981 R&E tax credit and its periodic renewals; and direct public/private technology partnerships to subsidize R&D, with the SBIR Program being one of several such programs.[4]

In addition to its important role in surmounting market failure, government should also be held accountable for its legislated policies. As Link and Scott (1998) point out, the concept of public accountability in the US traces at least to the early 1920s. But, the passage of the Government Performance and Results Act (GPRA) of 1993 emphasized the social importance of performance accountability, especially in the areas of science and technology, and has brought accountability back to center stage.

Our focus in this paper is on the SBIR Program. We do not debate the appropriateness of the government's support of that program but take that as a historical given and turn directly to evaluating the program's results.[5] In Section 2, we briefly describe the SBIR Program, with particular emphasis on the Department of Defense's (DoD's) SBIR Program. In Section 3, we outline our evaluation methodology and present our findings with an emphasis on the social returns associated with SBIR-funded research. Concluding remarks are presented in Section 4.

2. An overview of the SBIR Program

The SBIR Program began at the National Science Foundation in 1977.[6] At that time, the goal of the program was to encourage small businesses—considered by many to be engines of innovation in the US economy—to participate in NSF-sponsored research, especially research that had commercial potential. Because of the early success of the program at NSF, Congress passed the Small Business Innovation Development Act of 1982. The 1982 Act required all government departments and agencies with external research programs of greater than US$ 100 billion to establish their own SBIR Program and to set aside funds equal to 0.2% of their external research budget.[7]

The 1982 Act states the following objectives of the program:

1. To stimulate technological innovation.
2. To use small business to meet Federal research and development needs.
3. To foster and encourage participation by minority and disadvantaged persons in technological innovation.
4. To increase private sector commercialization of innovations derived from federal research and development.

In 1987, the funding percentage increased to 1.25%. The Act was later reauthorized by the Small Business Innovation Research Program Reauthorization Act of 1992, and the funding percentage increased to 1.5%. Since 1997, agencies have set aside 2.5% of their external research budgets for SBIR.

[2] While closure on the economic rationale for government policies supporting innovation may be at hand, Klette et al. (2000) explain that documenting both market failure and successful government programs to overcome it is quite difficult (and when government programs are not successful in overcoming market failure there is government failure). Kealey (1997) even challenges the conventional wisdom that government support for science is useful.

[3] Martin and Scott (2000) associate various policy mechanisms with sector-specific market failures that the mechanisms address.

[4] See Link (1999) for a review of public/private partnerships in the United States.

[5] See Lerner (1999) for a discussion of the appropriateness of the government as a venture capitalist through SBIR Programs. See also Lerner and Kegler (2000) for a review of the literature related to SBIR Programs.

[6] Tibbetts (1999) provides a history of the development of this program.

[7] As a set aside program, the SBIR Program redirects existing research funds rather than appropriating new research funds.

Justification for the reauthorization, as stated in the 1992 Act, was that the program:

> ... has effectively stimulated the commercialization of technology development through federal research and development, benefiting both the public and private sectors of the Nation.

Clearly, given the complete list of stated objectives of the program dating from the 1982 Act, its rationale would be broader than the objective of correcting the market failure of underinvestment in R&D. In addition to addressing underinvestment in R&D, the SBIR Program has as well the goal of promoting diversity per se in the population of firms doing R&D.[8] Here, we focus on the market failure, underinvestment justification for the SBIR Program.

The DoD is the largest federal agency participating in the SBIR Program. Administratively located in the Office of Small and Disadvantaged Business Utilization within the Office of the Secretary of Defense, the SBIR Program funds projects of interest to the Army, Navy, Air Force, Defense Advanced Research Project Agency (DARPA), the Ballistic Missile Defense Office, the Defense Special Weapons Agency, the US Special Operations Command, as well as general DoD research and development.[9]

The DoD's SBIR Program solicits proposals from eligible small businesses twice a year, and the awards are of three types.[10] Phase I awards are relatively small, generally less than US$ 100,000. The purpose of these awards is to help firms assess the feasibility of the research they propose to undertake for the agency in response to the agency's objectives. Phase II awards are larger, averaging about US$ 750,000. These awards are for the firm to undertake and complete its proposed research, ideally leading to a commercializable product or process. So-called Phase III awards do not come from DoD, but rather from the private sector to the researching company to pursue commercialization. Table 1 provides a funding history of DoD's SBIR Program.

Table 1
DoD SBIR budgets and awards, by fiscal year[a]

Fiscal year	Budget (US$ (M))	Phase I awards	Phase II awards
1983	16.70	281	0
1984	42.79	368	115
1985	79.00	513	282
1986	153.00	1031	254
1987	202.00	1264	401
1988	221.80	1056	334
1989	234.40	1021	362
1990	239.26	1140	415
1991	233.53	963	318
1992	241.84	1063	434
1993	384.82	1285	535
1994	276.19	1371	417
1995	445.25	1263	575
1996	453.46	1372	613
1997	543.02	1526	639
1998	553.44	1286	674
1999	541.31	1393	569

[a] Source: http://www.sbirsttr.com/sbirmisc/annrpt.html.

3. A methodology for evaluating the SBIR Program

Our methodology for evaluating DoD's SBIR Program, which can be applied to other programs, includes three elements:

1. A broad-based statistical analysis of SBIR recipients.
2. A case-based investigation of recipients regarding the impacts associated with SBIR awards.
3. A case-based investigation of the social rate of return from SBIR-funded research.

Data limitations associated with the SBIR Program, as well as with most public/private partnerships, and the multi-faceted nature of program evaluation per se, led us to adopt a broad-based evaluation methodology. Although survey-based information on SBIR recipients is used to conduct what some may view as a traditional evaluation exercise (Section 3.1), we are reluctant to draw program evaluation conclusions

[8] Scott (2000b) shows that the private sector is much less likely to provide funding to DoD SBIR-funded projects that are in certain technology areas, in certain geographic regions, or minority-owned. Such projects may well address underinvestment, but they could also promote diversity as an independent goal in itself.

[9] For an overview of the program, see http://www.acq.osd.mil/sadbu/sbir/overview.html.

[10] An eligible independently owned (at least 51% owned by US citizens) and operated company must be for-profit with 500 or fewer employees.

from only that analysis since the respondent sample is limited in size and since there is a limited literature to which these econometric findings could be compared.[11] We have relied on case studies (Sections 3.2 and 3.3) to complement our statistical analysis and to facilitate asking broader evaluation questions. Of course, case studies are not without their limitations including representativeness issues and possible interview biases. However, as a whole, our approaches are complementary and our findings are robust, both of which are important elements of a systematic evaluation effort. We conclude that DoD's SBIR Program is encouraging commercialization from research that would not have been undertaken without SBIR support; and, moreover, it is overcoming reasons for market failure that cause the private sector to underinvest in R&D.

3.1. Innovation and commercialization efforts of SBIR awardees

As noted above, two of the legislated goals of the SBIR Program are "to stimulate technological innovation" and "to increase private sector commercialization of innovations derived from federal research and development." As a first step toward an evaluation of the DoD's SBIR Program, we sought to understand the extent to which innovation and commercialization have been achieved among SBIR awardees. This background analysis in and of itself does not constitute an evaluation of the SBIR Program, but rather it provides some initial insight into the magnitude of the commercialization that occurs, and the characteristics of companies that are commercializing SBIR-funded products.

We rely on data collected by the National Academy of Sciences (NAS) for a representative sample of DoD SBIR awardees.[12] The Academy's mail survey went to a sample of 379 SBIR companies that received a Phase II award, since 1992. A total of 232 surveys were returned partially or totally completed. The 112 completed surveys with all of the information needed for our analysis are analyzed here.

Since, the purpose of this background analysis was to gain a better understanding of the commercialization activity that has occurred, we considered a model to explain cross-company differences in actual sales realized to date (the survey was administered in early 1999) from the technology developed during the DoD-funded Phase II project (ActSales).[13] Thus, our model is

$$\text{ActSales}_i = f(\boldsymbol{X}_i) \qquad (1)$$

where the subscript i indexes each awardee company and the vector $\boldsymbol{X}_i$ contains company-specific characteristics as described in Table 2. The dependent variable in Eq. (1), ActSales, is the actual sales realized to date from the technology developed during the Phase II project, measured in dollars. Although the mean of ActSales is US$ 175,021, that reflects the fact that 78 of the 112 projects reported no sales. For the 34 projects reporting sales, the mean of ActSales is US$ 576,539.

One clear implication of these survey data is that SBIR awardees do commercialize products and services based on their SBIR-sponsored research. That general finding does show the two stated purposes of the SBIR Program noted above—stimulation of socially desirable R&D and innovation that would not occur without the SBIR Program and commercialization of the R&D results—are met. We turn now with an econometric model to details of the commercialization.

Least-squares analysis is not appropriate for estimating Eq. (1) since the dependent variable is truncated at zero.[14] Least-squared results do not explicitly use the information about threshold effects on sales, and thus, such estimation would predict negative sales for some observations. We use the tobit model to estimate Eq. (1); the tobit model can be interpreted as predicting a measure of the value of sales from the SBIR project. If the predicted value is positive, then the company commercializes a product or service and then has sales; otherwise, sales

[11] See Lerner and Kegler (2000) for a review of this literature.

[12] These data are described in Cahill (2000).

[13] See Audretsch et al. (2000a) for a preliminary analysis of *expected* sales.

[14] However, in the present case, the results reported regarding the signs and significance of the explanatory variables and the general conclusions about commercialization do obtain in the corresponding least-squares regressions. The least-squares results were presented in the earlier version of this paper that was presented at the American Economic Association's meetings, and they are available on request from the authors.

Table 2
Characteristic variables in Eq. (1) ($n = 112$)

Variables	Definition	Mean	Range	S.D.
AgeBus	Age of the company measured in years	11.13 years	1–36	7.417
ExpFounder	Equals to one, if the most recent employment of the company founder(s) was in another private company as opposed to a university or government agency, and zero otherwise	0.786	0/1	0.412
Revenues	Company revenues were reported in ranges: less than US$ 100,000; US$ 100,000–499,999; US$ 500,000–999,999; US$ 1,000,000–4,999,999; US$ 5,000,000–19,999,999; and over US$ 20,000,000. The midpoint of each range was used with US$ 50,000 defining the midpoint of the lower bound and US$ 25,000,000 defining the midpoint of the upper bound	US$ 5,547,321	US$ 50,000–25,000,000	7162364
PhaseII	Number of previous Phase II awards	6.01 awards	0–81	12.384
Complete	Equals to one, if the current Phase II award is completed, and zero otherwise	0.366	0/1	0.484
Market	Equals to one, if the company has underway or has completed a marketing plan, and zero otherwise	0.589	0/1	0.494
Active	Number of years since the awarding of the most recent Phase II award	2.82 years	1–6	1.195
Electronics	Equals to one, if the relevant technology area of the award is electronics, and zero otherwise	0.500	0/1	0.502
Computer	Equals to one, if the relevant technology area of the award is computer, information processing, and analysis, and zero otherwise	0.214	0/1	0.412
Materials	Equals to one, if the relevant technology area of the award is materials, and zero otherwise	0.071	0/1	0.259
Mechanical	Equals to one, if the relevant technology area of the award is mechanical performance of vehicles, weapons, and facilities, and zero otherwise	0.045	0/1	0.207
Energy	Equals to one, if the relevant technology area of the award is energy conservation and use, and zero otherwise	0.098	0/1	0.299
Environment	Equals to one, if the relevant technology area of the award is environment and natural resources, and zero otherwise	0.045	0/1	0.207
LifeScience	Equals to one, if the relevant technology area of the award is life sciences, and zero otherwise	0.027	0/1	0.162
FastTrack	Equals to one, if the Phase II award was a Fast Track award, meaning that outside private-sector funding was committed to the project before the SBIR award was made, and zero otherwise	0.161	0/1	0.369
ProbResponse	Probability of response to the original NAS survey (the Probit model predicting probability of response used variables that were available for 109 of the 112 observations)	0.659	0.271–0.998	0.175

Table 3
Estimated tobit results from Eq. (1) (asymptotic t-statistics in parentheses)

Variable	Estimated coefficient	Estimated coefficient
Intercept	−2310466 (−3.161)***	−1406461 (−1.292)
AgeBus	−13045.5 (−0.457)	−11542.7 (−0.419)
ExpFounder	−43146.1 (−0.127)	11598.1 (0.034)
Revenues	0.0532968 (2.250)**	0.0432694 (1.760)*
Phase II	−7401.9 (−0.521)	−6718.9 (−0.476)
Complete	707243.6 (2.117)**	536615.9 (1.583)
Market	1147888 (3.082)***	1171060 (3.213)***
Active	62991.5 (0.414)	29066.1 (0.178)
Computer	654036.6 (1.801)*	597507.1 (1.660)*
Materials	849108.7 (1.773)*	899354.4 (1.884)*
Energy	379947.9 (0.769)	254938.4 (0.519)
Environment	919731.2 (1.710)*	821494.6 (1.555)
LifeScience	124461.1 (0.165)	20594.2 (0.028)
FastTrack	−82285.4 (−0.213)	2868.0 (0.007)
ProbResponse	–	−1078063 (−1.134)
Log likelihood	−540.98	−539.41
χ^2	41.95 (13)***	43.15 (14)***
n d.f.	112	109

* Significant at the 0.10 level.
** Significant at the 0.05 level.
*** Significant at the 0.01 level.

are zero. The results for the tobit estimations are in Table 3.[15]

The tobit results in column (2) of Table 3 lead us to conclude that companies are indeed commercializing products from their SBIR projects. Those that have done so to date are the larger companies in the computer technology area, materials technology area, and environmental technology area that have both completed their Phase II research as well as have formulated a marketing plan.[16] The specification reported in column (3) includes an additional explanatory variable, ProbResponse. If the probability of responding to the NAS survey is associated with actual sales (ActSales), and if that response effect in the error term is correlated with the variables included in $\boldsymbol{X}$, the estimates in column (2) of Table 3 could be biased, hence, the inclusion of the variable ProbResponse.[17]

Using the tobit models in Table 3, we estimated for each project the expected value of sales conditional on the tobit index function being greater than zero (and hence, conditional on the realized sales being greater than zero). Formally, we estimate the conditional mean of the tobit index function in the positive part of its distribution for each observation. The average of these conditional means is US$ 603,231

[15] The intercept reflects an effect for electronics and mechanical technologies. There are just five firms in the mechanical technology area, and a dummy variable for that technology area is a perfect predictor for the occurrence of US$ 0 of sales. However, our a priori knowledge of projects in that area tells us that the probability of sales is not zero; rather, with just five observations, it was happenstance that none of those projects had actual sales. We have included these five cases, along with the 56 firms in the electronics technology area, 44 of which have US$ 0 in actual sales, in the intercept term. In alternative tobit estimations (available upon request from Link or Scott), we deleted these mechanical technology cases from the sample, and the qualitative findings (signs and significance and approximate magnitudes) are unchanged.

[16] As shown in Audretsch et al. (2000a), Fast Track projects have higher *expected* sales, other things held equal in an OLS estimation, but those projects did not begin until 1996, and therefore, by 1999 they had realized less actual sales than the older projects, although the effect is not significant.

[17] There are just 109 observations available for the specification including ProbResponse, because three of the 112 observations did not have all of the variables used in the probit model of response. We also calculated a response hazard rate but its inclusion in Eq. (1) in place of ProbResponse gives results similar to those reported in column (3) of Table 3 (results are not reported here but are available upon request from Link or Scott).

for the specification without the control for response bias and US$ 598,698 for the specification with the control.[18] Estimating the expected sales for each observation given X_i (that is, the probability weighted sum of the outcome of US$ 0 of sales as well as the positive outcomes), we find on average for all of the observations the expected sales are US$ 217,267 for the specification without control for response bias and US$ 223,244 for the specification with the control.[19]

These findings, along with case-based information that in the absence of SBIR support the SBIR firms would not have pursued Phase II research or would have pursued it on a very limited scale, lend support for our preliminary conclusion that the program is meeting two of its stated goals.[20] We reach these conclusions based on our interpretation of the statistically significant tobit coefficients and models.

However, because two program goals are being met, in total or in part, is in our opinion insufficient information upon which to base even a preliminary economic evaluation of the overall program. What must be shown is that companies would not have undertaken the research on their own because of insufficient private returns *and* that SBIR funding overcomes market failures benefiting society with a large social return. In the following two sections, we, therefore, add two more elements of evaluation. We use case studies to show that the SBIR Program encourages new entrepreneurs, actually changing the career paths of some researchers and bringing their talents and ideas into the commercial world. Then, additional case studies are used to demonstrate the large social return from the projects that are made possible by the SBIR Program.

[18] There is little qualitative difference between the results when a response hazard rate is substituted for the probability of response in the specification. Using the response hazard rate, the average of the conditional means is US$ 602,483.

[19] For the specification that replaces the probability of response with the hazard rate, the average value of expected sales is US$ 223,450.

[20] See Link (2000), and Scott (2000a) for case-based information that these companies report that in the absence of SBIR support they either would not have pursued Phase II research, or would have done so on a very limited scale.

3.2. Entrepreneurial behavior and the SBIR Program

To develop information about the extent to which the SBIR Program changes the behavior of knowledge workers and thereby helps to create a science-based entrepreneurial economy, Audretsch et al. (2000b) use case studies along with responses to a survey of a broader sample of firms with SBIR projects. Although the authors emphasize that their results are exploratory and based on a small sample, their results do show that the SBIR Program has influenced the career paths of scientists and engineers by facilitating the startup of new firms. Additionally, the experience of those knowledge workers in commercializing products and services with a small business has had a spillover effect by influencing the career trajectories of their colleagues.

Audretsch et al. report that both their survey and their case studies provide consistent evidence that:

1. A significant number of the firms would not have been started in the absence of the SBIR Program.
2. A significant number of the scientists and engineers would not have become involved in the commercialization process in the absence of the SBIR Program.
3. A significant number of other firms are started because of the demonstration effect by the efforts of scientists to commercialize knowledge.
4. As a result of the demonstration effect of SBIR funded commercialization, a number of other scientists alter their careers to include commercialization efforts.

Especially for scientists and engineers without previous experience with knowledge-based small firms, the SBIR Program appears to encourage the development of knowledge about commercialization possibilities and facilitate commercialization from research-based knowledge.

3.3. Estimating social returns from SBIR-supported projects

3.3.1. Conceptual framework for analysis

It is well known that risk and the closely related difficulties of appropriating return to investments in technology—R&D specifically—will lead

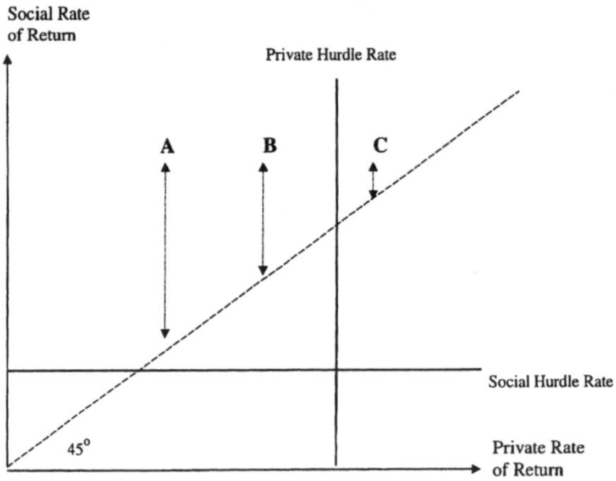

Fig. 1. Gap between social and private rates of return to R&D projects.

to a divergence between private and social benefits. The social rate of return will be greater than the private rate of return as illustrated in Fig. 1.[21] The purpose of this simple heuristic device is to characterize private sector projects with returns not only less than the expected social returns but also less than the private hurdle rate for projects normally undertaken by the firm. In those cases, the divergence of private from social returns can imply the market failure of private underinvestment in socially desirable R&D.

The social rate of return is measured on the vertical axis of Fig. 1 along with society's hurdle rate on investments in R&D. The private rate of return is measured on the horizontal axis along with the private hurdle rate on investments in R&D. A 45° line (dashed line) is imposed on the figure to emphasize that for the projects depicted the social rate of return from an R&D investment exceeds the private rate of return from that same investment. The three illustrative projects discussed below are labeled as project A, B, and C.

[21] A more complete explanation for this figure and for the theoretical model underlying the derivation of social and private returns is in Link and Scott (2001). See also Jaffe (1998) for an application of a similar figure to the Advanced Technology Program.

For project C, the private rate of return exceeds the private hurdle rate, and the social rate of return exceeds the social hurdle rate. The gap (short vertical double arrow line) between the social and private rates of return reflects the spillover benefits to society from the private investment. However, the inability of the private sector to appropriate all benefits from its investment is not so great as to prevent the project from being adequately funded by the private firm. Hence, it is not a candidate for public support.

Consider projects A and B. The gap between social and private returns is larger than in the case of project C; neither project will be adequately funded by the private firm. To address this market failure the government has two alternative policy mechanisms. It can use a tax policy to address the private underinvestment in R&D or it can rely on public/private partnerships as a direct funding mechanism.

If the private return to project B is less than the private hurdle rate because of the risk and uncertainty associated with R&D in general, then tax policy may be the appropriate policy mechanism to overcome this underinvestment. Risk is inherent in a technology-based market, and there will be certain projects for which the rewards from successful innovation are too low for private investments to be justified. Tax policy, such as the R&E tax credit, may in these situations reduce the private marginal cost of R&D sufficiently to provide an incentive for the project to be undertaken privately. For projects like project B, a tax credit may be sufficient to increase the expected return so that the firm views the post-tax credit private return to be sufficient for the project to be funded.

However, for projects like project A, a tax credit may be insufficient to increase the expected return enough to induce the private firm to undertake the project. Such projects could include for example projects expected to yield an innovative product that would be part of a larger system of products. Even if technically successful there might be substantial risk that the product would not interoperate or be compatible with other emerging products, and direct funding rather than a tax credit may be the appropriate policy mechanism to stimulate socially desirable investments in systems.

A priori, it is difficult to generalize about the way that any one firm's under-funded projects will be distributed in the area to the left of the firm's private hurdle rate. However, some generalizations can be made about the portfolio of private sector firms' projects in general. For those R&D projects, like project B, for which the firm will appropriate some returns but for which the overall expected return is slightly too low, a tax credit may be sufficient to increase the expected return to the point that the expected return exceeds the private hurdle rate. Such projects may be of a product or process development nature and are likely to be a part of the firm's ongoing R&D portfolio of projects. For those R&D projects, like project A, for which the firm has little ability to appropriate returns even if the marginal cost of the project is reduced through an R&E tax credit, the firm may not respond to such a tax policy but may respond to a direct funding policy mechanism. Such projects may be of a generic or fundamental technology nature, that is technology from which subsequent market applications are derived and that enable downstream applied R&D to be undertaken successfully. Thus, the economic rationale for public/private partnerships is that such partnerships represent one direct-funding R&D policy appropriate to overcome market failure and that they are more likely to be necessary, compared to fiscal tax incentives, when the R&D is generic or fundamental in nature.

Drawing upon the foregoing arguments, we maintain that a candidate project for SBIR awards is one like project A in Fig. 1. That is, given that the proposed research aligns with the technology mission of DoD, SBIR should fund such projects for which there is a significant potential social benefit but also that are characterized by substantial downside risk such that the firm's expected private return is well below its private hurdle rate.

Case-study information reported by Link (2000) and Scott (2000a) confirms for a small sample of SBIR-supported projects that not only are the firms' private returns less than their private hurdle rates but also that outside investors are unwilling to sponsor fully the research because of both technical and market risk. Hence, at the outset of an SBIR project, not only is a firm's private hurdle rate not expected to be met, neither is the required return for a third party. In conjunction with the evidence presented below that the projects have high social returns, the findings support the belief that the SBIR Program corrects the market failure of underinvestment in socially-valuable research in emerging technologies.

3.3.2. Preliminary estimates of social returns

Link and Scott (2000) interviewed SBIR award recipients for 44 projects in 43 companies for this part of the evaluation study. Each was interviewed toward the end of its Phase II award period. Information collected made possible the calculation of a lower-bound of the prospective expected social rate of return associated with each project and a comparison of that expected social rate of return to both the expected private rate of return to the firm had it pursued the project in the absence of SBIR support and the private rate of return expected by the firm with its SBIR support. The analysis clearly indicates that SBIR is funding projects like project A in Fig. 1, and given such funding the projects have become similar to project C.

The sample of projects studied may not be statistically representative of all projects funded by SBIR. To make such a claim, all would have to be interviewed to obtain information similar to the information collected. This was not done because of resource constraints and the availability of SBIR staff to identify specific individuals to be interviewed and to make initial contacts. Thus, the sample of 44 projects is simply one sample, and the findings are not necessarily

Table 4
Variables for the prospective expected social rate of return calculations

Variable	Definition	Source
d	Duration of the SBIR project	DoD files, verified and updated as necessary during interviews
C	Total cost of the SBIR project	DoD files, verified and updated as necessary during interviews
A	SBIR funding	DoD files, verified and updated as necessary during interviews
r	Private hurdle rate	Interview
z	Duration of the extra period of development beyond Phase II	Interview
F	Additional cost for the extra period of development[a]	Interview
T	Life of the commercialized technology	Interview
v	Proportion of value appropriated	Interview
L	Lower bound for expected annual private return to the SBIR firm	Derived
U	Upper bound for expected annual private return to the SBIR firm	Derived

[a] The additional costs considered were research and development costs and not the costs associated with searching for or negotiating with venture capitalists. See Gans and Stern (2000).

Table 5
Descriptive statistics on variables used in the social rate of return model ($n = 44$)

Variable	Mean	Range	S.D.
d	2.68 years	2–3.5	0.36
C	US$ 1,027,199	US$ 448,000–3,450,000	461901
A	US$ 782,000	US$ 448,000–1,099,966	127371
r	0.33	0.2–0.5[a]	0.08
z	1.30 years	−0.375–5.0[b]	1.07
F	US$ 1,377,341	US$ 0–15,000,000	2972266
T	10.56 years	1–30[c]	7.23
v	0.16	0.0045–0.60	0.16
L	US$ 902,738	US$ 34,842.6–5,300,500	1228850
U	US$ 1,893,001	US$ 258,830–8,666,330	1733581

[a] Eight of the respondents were uncomfortable estimating the private hurdle rate that outside financiers would apply to their projects at their outset. For those, the average value of *r* for the respondents in their region was used in the calculations.

[b] One observation for *z* has a negative value because commercial returns started before the end of Phase II.

[c] One respondent reported that *T* would be several decades, and another reported that *T* would be forever. In both cases, *T* was conservatively entered as the value 30 years. However, because the relevant discount rates are so high, the difference between 30 years and "forever" is not significant. In the integrals in the mathematical model, the term with *T* entered negatively as an exponent would become zero, but with a large value of *T* the term is very small in any case.

representative of all SBIR projects. However, because the findings are strong, and because the anecdotal information obtained from SBIR staff indicates the sample was a reasonable cross-section of funded companies, we are comfortable generalizing to some degree about the findings. In any event, the methodology of Link and Scott (2001) is sufficiently general to be applied to other samples.[22]

[22] A complete description of the sample of companies is available upon request.

Table 4 lists the variables required for the implementation of the model. As noted in the table, data on selected variables were independently available from DoD project files, but all such information was also verified during the interview process and corrected when discrepancies were found. Descriptive statistics for these variables are in Table 5.

Phases I and II values for project duration, total cost, and SBIR funding, were combined into one value to cover both Phases I and II of the project. That is, each project is viewed from the time that Phase I began, and expectations from that point forward are estimated. It is at that point in time that the market failure issues discussed above are especially relevant. Respondents reported the additional period of time beyond the expected completion of Phase II until the research would be commercialized and the additional cost required during that period.

The variable ν, the proportion of value appropriated, deserves some explanation. Firms cannot reasonably expect to appropriate all of the value created by their research and subsequent innovations. First, the innovations will generate consumer surplus that no firm will appropriate, but that society will value. Link and Scott ignore consumer surplus in the calculations of the prospective expected social returns thus motivating the claim that the estimates are lower-bound estimates. Second, some of the profits generated by the innovations will be captured by other firms. Larger firms, for example, will observe the innovation and will successfully introduce imitations. Each respondent as part of the interview process was asked to estimate the proportion of the returns to producers (themselves and other firms) generated by their anticipated innovation that they expect to capture. Then, in an extended conversation, other possible applications of the technology developed during the SBIR project were explored. Each respondent was asked to estimate the multiplier to get from the profit stream generated by the immediate applications of the SBIR project to the stream of profits generated in the broader applications markets—beyond the applications to be made by the SBIR firm—that would reasonably be anticipated. We calculated ν as the product of two proportions. The first is the proportion anticipated by the SBIR firm of the profits from the development and commercialization of the specific applications it planned for the technology being developed. The second is the proportion of profits from all possible applications—including those beyond the applications that the SBIR firm anticipated making itself—taken by the respondent's planned applications of the generic technology being researched and developed in the SBIR project. Thus, ν is the total proportion of the value of the technology appropriable by the researching firm. For example, if an SBIR firm anticipated appropriating 50% of the profits generated by its planned types of applications of its SBIR technology, and if the profits from the types of applications planned by the SBIR firm were expected to be 50% of the profits from all applications of the technology, then the SBIR firm would expect to earn 50% of a profit stream that itself was 50% of the total profit stream attributable to the SBIR generated technology. The firm, therefore, would anticipate appropriating 25% of the profits generated by its SBIR project.

Link and Scott then find the lower bound for the annual private return to an SBIR-sponsored project by solving their investment model for the amount that the private firm must earn to meet its private hurdle rate, or its required rate of return, on the portion of the total investment that the firm must finance.[23] The firm would not invest in the SBIR project on its own unless it expected at least that lower bound for the annual private return on its investment.

To determine the upper bound for the annual private return, the investment model is solved for the expected annual return that if exceeded would imply that the expected rate of return earned by the private firm would be greater than its hurdle rate in the absence of SBIR support, and therefore, SBIR support would not be required for the project.

Sequentially, Link and Scott estimate the average expected annual private return to the firm as the average of the upper and lower bounds, U and L. Knowing the average expected annual private return is $((L + U)/2)$ and knowing the portion of producer surplus that is appropriable, ν, then total producer surplus equals $((L + U)/2\nu)$. This value for total producer surplus is a lower bound for the expected annual social return. Again, it is a lower bound because consumer surplus has not been measured.

The expected private rate of return without SBIR support is the solution for the internal rate of return—the rate of return that just equates the present value of the expected annual private return to the firm to the present value of research and post-research commercialization costs to the firm in the absence of SBIR funding.

Finally, the lower bound on the social rate of return is found by solving the investment model for the

[23] The theoretical mathematical model is available from either Link or Scott. A complete and detailed description of the model, along with an illustrative implementation of it based on project data from the Advanced Technology Program is in Link and Scott (2001).

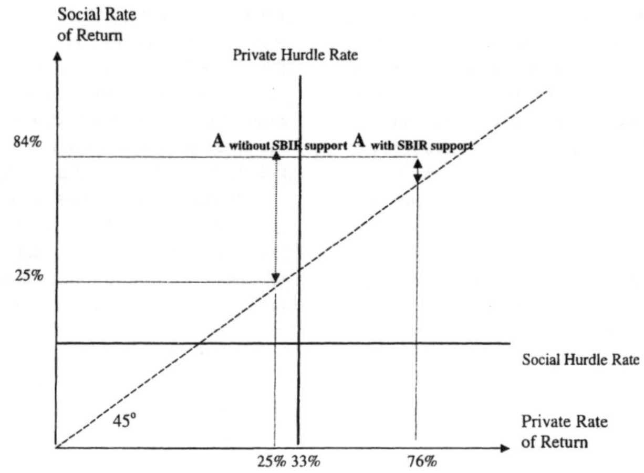

Fig. 2. Gap between social and private rates of return to the average SBIR project, $n = 44$.

internal rate of return with the average expected annual private return $((L + U)/2)$, replaced with the lower bound for the average expected annual social return $((L + U)/2\nu)$.

The investment model was estimated for each of the 44 SBIR-sponsored projects. Mean values of the two resulting important rates of return, averaged across the 44 projects, are shown in Table 6. There are two important points to be seen in Table 6. First, the average of the expected private rate of return in the absence of SBIR support is 25%, clearly less than the average self-reported private hurdle rate of 33% (see Table 5). Thus, in the absence of SBIR support the sample of firms would not have undertaken the research, and in fact each of the sampled firms stated that explicitly during the interviews. Further, the expected social rate of return (lower bound) associated with SBIR funding of these projects is at least 84%, and hence, the projects are expected to be socially valuable.

Table 6
Rates of return for the average SBIR project ($n = 44$)

Variable	Definition
$i_{\text{private}} = 0.25$	Expected private rate of return without SBIR funding
$i_{\text{social}} = 0.84$	Lower bound for expected social rate of return

We cannot conclude that a social rate of return of at least 84% is "good" or "bad," or "better" or "worse" than expected. However, we can compare the estimate of the lower bound of the social rate of return to the opportunity cost of public funds promulgated by the US Office of Management and Budget (OMB). Following the guidelines set forth by OMB (1992) mandating a real discount rate of 7% for constant-dollar benefit-to-cost analyses of proposed investments and regulations, clearly a nominal social rate of return of 84% is above that rate and thus, reflects projects that are socially worthwhile in terms of the OMB standard.[24]

The expected private rate of return with SBIR support for each of the 44 projects can be calculated as the solution for the internal rate of return in the investment model given the actual time series of private investments and private returns. The estimated private rate of return with SBIR support averages 76% for the 44 cases, this value is noticeably above the average private hurdle rate of 33%. However, there is

[24] Link and Scott (1998) discuss the use of this guideline for NIST economic impact assessments.

no way for the SBIR Program to have calculated the optimal level of funding for these 44 projects, or for any projects, unless, as part of the Phase I application all relevant data, including hurdle rates, could have been assessed. In the absence of such information, which in practice would be difficult to obtain because of, if nothing else, self-serving reporting by proposers, the funding implemented by the SBIR Program may be as close to optimal as possible.[25]

Fig. 2 summarizes the estimated values for the average SBIR-sponsored project. Based on the sample of 44 projects, the average gap between the lower-bound social rate of return and the estimated private rate of return without SBIR funding support is 59%.

Two points need to be emphasized, along with our previous comments about the statistical representativeness of the sample of 44 companies. First, the social rates of return estimated for the SBIR projects are very conservative, lower-bound estimates because they do not include consumer surplus in the benefit stream. Second, some might be skeptical about the SBIR awardees' earnest belief that without SBIR funding the projects would not have been undertaken or at least would not have been undertaken to the same extent or with the same speed. With the SBIR Program in place, certainly the pursuit of SBIR funding would perhaps be a path of least resistance. However, if the research would have occurred without the public funding, the estimated upper bound and (hence the average of the upper and lower bounds for the expected private returns) would be too low, and the actual lower bounds for the social rates of return would be even higher than estimated. Further, the gap would remain, although that would not in itself necessarily justify the public funding of the projects.

4. Concluding remarks

Based on our three-part evaluation analysis, and the caveats associated with each, there is ample evidence to support the conclusion that the net economic benefits associated with DoD's SBIR Program are positive. More specifically, our broad-based statistical analysis of SBIR recipients demonstrates that two of the program's objectives—stimulating technological innovation and increasing private sector commercialization of innovations derived from federal research and development—are being met. In addition, the case-based analyses demonstrate that the SBIR Program redirects the efforts of award recipients toward commercial activity that would not otherwise have taken place, and that commercial activity and its attendant spillover effects generate substantial positive net social benefits.

Although our conclusions are robust, it is important to emphasize that our analysis is specific to DoD's SBIR Program. Until studies of other agencies' SBIR programs take place, one should be extremely cautious about generalizing from our DoD SBIR findings to the entire SBIR Program.[26]

References

Audretsch, D.B., Link, A.N., Scott, J.T., 2000a. Statistical analysis of the national academy of sciences survey of small business innovation research awardees: analyzing the influence of the fast track program. In: Wessner, C.W. (Ed.), The Small Business Innovation Research Program. National Academy Press, Washington, DC, pp. 291–306.

Audretsch, D.B., Weigand, J., Weigand, C., 2000b. Does the small business innovation research program foster entrepreneurial behavior? Evidence from Indiana. In: Wessner, C.W. (Ed.), The Small Business Innovation Research Program. National Academy Press, Washington, DC, pp. 160–193.

Baldwin, W.L., Scott, J.T., 1987. Market Structure and Technological Change. Harwood, London.

Baron, J., 1998. DoD SBIR/STTR Program Manager, Comments at the Methodology Workshop on the Assessment of Current SBIR Program Initiatives, Washington, DC, October.

Cahill, P., 2000. Fast track, is it speeding commercialization of the department of defense small business innovation research projects. In: Wessner, C.W. (Ed.), The Small Business Innovation Research Program. National Academy Press, Washington, DC, pp. 43–103.

Gans, J.S., Stern S., 2000. When does funding research by smaller firms bear fruit? Evidence from the SBIR program. NBER Working Paper No. 7877, September.

---

[25] Scott (1998) has proposed using a bidding mechanism that, if applied, would result in the SBIR funding being just sufficient to ensure that the private participants earn just a normal rate of return. The proposal is a novel one, however, it is yet untried. Successful implementation would require additional development in order to reduce it to practice.

[26] As these studies take place, by agency, consideration should be given to having matched samples of firms, that is firms with and without SBIR support, to obtain comparative evaluation estimates of the marginal impact of the program. See Wallsten (2000).

Jaffe, A.B., 1998. The importance of spillovers in the policy mission of the advanced technology program. Journal of Technology Transfer 23, 11–19.

Kealey, T., 1997. The Economic Laws of Scientific Research. MacMillan, London.

Klette, T.J., Møen, J., Griliches, Z., 2000. Do subsidies to commercial R&D reduce market failures? Microeconometric evaluation studies. Research Policy 29, 471–495.

Lerner, J., 1999. The government as venture capitalist: the long-run impact of the SBIR program. Journal of Business 72, 285–318.

Lerner, J., Kegler, C., 2000. Evaluating the small business innovation research program: a literature review. In: Wessner, C.W. (Ed.), The Small Business Innovation Research Program. National Academy Press, Washington, DC, pp. 307–324.

Link, A.N., 1999. Public/private partnerships in the United States. Industry and innovation 6, 191–217.

Link, A.N., 2000. An assessment of the small business innovation research fast track program in southeastern states. In: Wessner, C.W. (Ed.), The Small Business Innovation Research Program. National Academy Press, Washington, DC, pp. 194–210.

Link, A.N., Scott, J.T., 1998. Public Accountability: Evaluating Technology-Based Institutions. Kluwer Academic Publishers, Norwell, MA.

Link, A.N., Scott, J.T., 2000. Estimates of the social returns to small business innovation research projects. In: Wessner, C.W. (Ed.), The Small Business Innovation Research Program. National Academy Press, Washington, DC, pp. 275–290.

Link, A.N., Scott, J.T., 2001. Public/private partnerships: stimulating competition in a dynamic market. International Journal of Industrial Organization 19, 763–794.

Martin, S., Scott, J.T., 2000. The nature of innovation market failure and the design of public support for private innovation. Research Policy 29, 437–447.

Office of Management and Budget, 1992. Economic Analysis of Federal Regulations Under Executive Order 12866, mimeographed.

Scott, J.T., 1998. Financing and leveraging public/private partnerships: the hurdle-lowering auction. STI Review 23, 67–84.

Scott, J.T., 2000a. An assessment of the small business innovation research program in New England: fast-track compared with non-fast track projects. In: Wessner, C.W. (Ed.), The Small Business Innovation Research Program. National Academy Press, Washington, DC, pp. 104–140.

Scott, J.T., 2000b. The directions for technological change: alternative economic majorities and opportunity costs. Review of Industrial Organization 17, 1–17.

Tibbetts, R., 1999. The Small Business Innovation Research Program and NSF SBIR Commercialization Results, mimeograph.

Wallsten, S., 2000. The effects of government-industry R&D programs on private R&D: the case of the small business innovation research program. Rand Journal of Economics 31, 82–100.

[5]

Standing on the Shoulders of Midgets: The U.S. Small Business Innovation Research Program (SBIR)

David B. Audretsch

ABSTRACT. The Small Business Innovation Research (SBIR) program in the United States is one of the most important public policy programs, with a goal of generating innovative hi-technology firms and promoting competitiveness and growth. The purpose of this paper is to describe the SBIR, explain why it has emerged as an important policy instrument, and provide an evaluation of the impact of the SBIR. The findings of this paper suggest that the SBIR has generally accomplished its mission by contributing to the creation of high-technology small firms and enhancing U.S. competitiveness.

1. Introduction

As recently as a decade ago, there were serious concerns about the ability of the United States to withstand competition in the global economy, create jobs and continue to develop. Lester Thurow bemoaned that the United States was "losing the economic race,"[1] because, "Today it's very hard to find an industrial corporation in America that isn't in really serious trouble basically because of trade problems. . . . The systematic erosion of our competitiveness comes from having lower rates of growth of manufacturing productivity year after year, as compared with the rest of the world" (Thurow, 1985, p. 23). W.W. Rostow predicted a revolution in economic policy, concluding that,"The United States is entering a new political era, one in which it will be preoccupied by increased economic competition from abroad and will need better cooperation at home to deal with this challenge."[2]

Final version accepted on January, 2001

Ameritech Chair of Economic Development & Director
Institute for Development Strategies
Indiana University
Bloomington IN 47405
U.S.A.

In the influential study, *Made in America*, directed by the leaders of the MIT Commission on Industrial Productivity, Michael L. Dertouzos, Richard K. Lester, and Robert M. Solow, a team of 23 scholars, spanning a broad range of disciplines and backgrounds, reached the conclusion that for the United States to restore its international competitiveness, it had to adapt the types of policies targeting the leading corporations prevalent in Japan and Germany.

The last decade has seen a reemergence of competitiveness, innovative activity and job generation in the United States. Not only was this economic turnaround largely unanticipated by many scholars and members of the policy community, but what was even more surprising than the resurgence itself was the primary source – small firms. As scholars began the arduous task of documenting the crucial role played by SMEs in the United States as a driving engine of growth, job creation and competitiveness in global markets (Audretsch, 1995), policy makers responded with a bipartisan emphasis on policies to promote SMEs.[3] For example, in his 1993 State of the Union Address to the country, President Bill Clinton proposed, "Because small business has created such a high percentage of all the new jobs in our nation over the last 10 or 15 years, our plan includes the boldest targeted incentives for small business in history. We propose a permanent investment tax creditor the small firms in this country."[4] The Republican response to Clinton was, "We agree with the President that we have to put more people to work, but remember this: 80 to 85 percent of the new jobs in this country are created by small business. So the climate for starting and expanding businesses must be enhanced with tax incentives and deregulation,

Small Business Economics **20**: 129–135, 2003.

rather than imposing higher taxes and more governmental mandates."[5]

In fact, one of the responses of U.S. policy to the competitiveness crisis of the 1970s and early 1980s was to enact the Small Business Innovation Research Program (SBIR). The mandate of the SBIR was to enhance U.S. competitiveness by promoting innovative and high-technology small firms. The purpose of this paper is to explain the link between international competitiveness and the promotion of small firms as a policy response to globalization, as well as to provide an evaluation of the SBIR.

In the second section of this paper we document how the comparative advantage has shifted away from the traditional factors of production towards knowledge-based economic activity. The implications of knowledge for entrepreneurship is explained in the third section. In the fourth section the role of the SBIR in promoting entrepreneurship in the U.S. is evaluated. Finally, in the fifth section, a summary and conclusions are provided.

2. Globalization and the shifting source of comparative advantage

Globalization has triggered a virtual spatial revolution in terms of the geography of production.[6] The (marginal) cost of transforming information across geographic space has been rendered to virtually nothing. Confronted with lower cost competition in foreign locations, producers in the high-cost countries have three options apart from doing nothing and losing global market share: (1) reduce wages and other production costs sufficiently to compete with the low-cost foreign producers, (2) substitute equipment and technology for labor to increase productivity, and (3) shift production out of the high-cost location and into the low-cost location.

Many of the European and American firms that have successfully restructured resorted to the last two alternatives. Substituting capital and technology for labor, along with shifting production to lower-cost locations has resulted in waves of *Corporate Downsizing* throughout Europe and North America. At the same time, it has generally preserved the viability of many of the large corporations. As record levels of both European and American stock indexes indicate, the companies have not generally suffered. For example, between 1979 and 1995 more than 43 million jobs were lost in the United States as a result of corporate downsizing.[7] This includes 24.8 million blue-collar jobs and 18.7 million white-collar jobs. Similarly, the 500 largest U.S. manufacturing corporations cut 4.7 million jobs between 1980 and 1993, or one quarter of their work force.[8] Perhaps most disconcerting, the rate of corporate downsizing has apparently increased over time in the United States, even as the unemployment rate has fallen. During most of the 1980s, about one in 25 workers lost a job. In the 1990s this has risen to one in 20 workers.

Much of the policy debate responding to the twin forces of the telecommunications revolution and increased globalisation has revolved around a trade-off between maintaining higher wages but suffering greater unemployment versus higher levels of employment but at the cost of lower wage rates. There is, however, an alternative. It does not require sacrificing wages to create new jobs, nor does it require fewer jobs to maintain wage levels and the social safety net. This alternative involves shifting economic activity out of the traditional industries where the high-cost countries of Europe and North America have lost the comparative advantage and into those industries where the comparative advantage is compatible with both high wages and high levels of employment – knowledge based economic activity.

Globalisation has rendered the comparative advantage in traditional moderate technology industries incompatible with high wage levels. At the same time, the emerging comparative advantage that is compatible with high wage levels is based on innovative activity. For example, employment has increased by 15 percent in Silicon Valley between 1992 and 1996, even though the mean income is 50 percent greater than in the rest of the country.[9]

The global demand for innovative products in knowledge-based industries is high and growing rapidly; yet the number of workers who can contribute to producing and commercializing new knowledge is limited to just a few areas in the world. Economic activity based on new knowledge generates higher wages and greater employment opportunities reflecting the exploding demand for new and improved products and services. There are many indicators reflecting the shift in the comparative advantage of the high-wage countries towards an increased importance of innovative activity. For example, Kortum and Lerner (1997, p. 1) document an unprecedented jump in patenting in the United States, as evidenced by the explosion in applications for United States patents by American inventors since 1985. Throughout this century, patent applications fluctuated within a band between 40,000–80,000 per year. By contrast, in 1995 there were over 120,000 patent applications. Similarly, Berman et al. (1997) have shown that the demand for less skilled workers has decreased dramatically throughout the OECD, while at the same time, the demand for skilled workers has exploded.

3. Entrepreneurship in a knowledge economy

That SMEs would emerge as becoming more important seems to be contrary to many of the conventional theories of innovation. The starting point for most theories of innovation is the firm. In such theories the firms are exogenous and their performance in generating technological change is endogenous (Arrow, 1962). For example, in the most prevalent model found in the literature of technological change, the model of the knowledge production function, formalised by Zvi Griliches (1979), firms exist exogenously and then engage in the pursuit of new economic knowledge as an input into the process of generating innovative activity. The most decisive input in the knowledge production function is new economic knowledge. Knowledge as an input in a production function is inherently different than the more traditional inputs of labour, capital and land. While the economic value of the traditional inputs is relatively certain, knowledge is intrinsically uncertain and its potential value is asymmetric across economic agents.[10] The most important, although not the only source of new knowledge is considered to be research and development (R&D). Other key factors generating new economic knowledge include a high degree of human capital, a skilled labour force, and a high presence of scientists and engineers.

There is considerable empirical evidence supporting the model of the knowledge production function. This empirical link between knowledge inputs and innovative output apparently becomes stronger as the unit of observation becomes increasingly aggregated. For example, at the unit of observation of countries, the relationship between R&D and patents is very strong. The most innovative countries, such as the united States, Japan and Germany, also tend to undertake high investments in R&D. By contrast, little patent activity is associated with developing countries, which have very low R&D expenditures. Similarly, the link between R&D and innovative output, measured in terms of either patents or new product innovations is also very strong when the unit of observation is the industry. The most innovative industries, such as computers, instruments and pharmaceuticals also tend to be the most R&D intensive. Audretsch (1995) finds a simple correlation coefficient of 0.74 between R&D inputs and innovative output at the level of four-digit standard industrial classification (SIC) industries. However, when the knowledge production function is tested for the unit of observation of the firm, the link between knowledge inputs and innovative output becomes either tenuous and weakly positive in some studies and even non-existent or negative in others. The model of the knowledge production function becomes particularly weak when small firms are included in the sample. This is not surprising, since formal R&D is concentrated among the largest corporations, but a series of studies has clearly documented that small firms account for a disproportional share of new product innovations given their low R&D expenditures.

The breakdown of the knowledge production function at the level of the firm raises the question, *Where do innovative firms with little or no R&D get the knowledge inputs?* This question becomes particularly relevant for small and new firms that undertake little R&D themselves, yet contribute considerable innovative activity in newly emerging industries such as biotechnology and computer software (Audretsch, 1995). One answer that has recently emerged in the economics literature is from other, third-party firms or research institutions, such as universities. Economic knowledge may spill over from the firm conducting the R&D or the research laboratory of a university.

Why should knowledge spill over from the source of origin? At least two major channels or mechanisms for knowledge spillovers have been identified in the literature. Both of these spillover mechanisms revolve around the issue of appropriability of new knowledge. Cohen and Levinthal (1989) suggest that firms develop the capacity to adapt new technology and ideas developed in other firms and are therefore able to appropriate some of the returns accruing to investments in new knowledge made externally.

By contrast, Audretsch (1995) proposes shifting the unit of observation away from exogenously assumed firms to individuals, such as scientists, engineers or other knowledge workers – agents with endowments of new economic knowledge. When the lens is shifted away from the firm to the individual as the relevant unit of observation, the appropriability issue remains, but the question becomes, *How can economic agents with a given endowment of new knowledge best appropriate the returns from that knowledge?* If the scientist or engineer can pursue the new idea within the organisational structure of the firm developing the knowledge and appropriate roughly the expected value of that knowledge, he has no reason to leave the firm. On the other hand, if he places a greater value on his ideas than do the decision-making bureaucracy of the incumbent firm, he may choose to start a new firm to appropriate the value of his knowledge. In the metaphor provided by Albert O. Hirschman (1970), if voice proves to be ineffective within incumbent organisations, and loyalty is sufficiently weak, a knowledge worker may resort to exit the firm or university where the knowledge was created in order to form a new company. In this spillover channel the knowledge production function is actually reversed. The knowledge is exogenous and embodied in a worker. The firm is created endogenously in the worker's effort to appropriate the value of his knowledge through innovative activity.

4. The public policy response

Public policy towards business has gradually shifted from viewing the role of SMEs through the static lens towards the dynamic framework. For example, the United States Congress enacted the Small Business Innovation Research (SBIR) program in the early 1980s as a response to the loss of American competitiveness in global markets. Congress mandated each federal agency with allocating around four percent of its annual budget to funding innovative small firms as a mechanism for restoring American international competitiveness. The SBIR provides a mandate to the major R&D agencies in the United States to allocate a share of the research budget to innovative small firms. Last year the SBIR program amounted to around $1.2 billion. The SBIR consists of three phases. Phase I is oriented towards determining the scientific and technical merit along with the feasibility of a proposed research idea. A Phase I award provides an opportunity for a small business to establish the feasibility and technical merit of a proposed innovation. The duration of the award is six months and can not exceed $70,000. Phase II extends the technological idea and emphasizes commercialization. A Phase II Award is granted to only the most promising of the Phase I projects based on scientific/technical merit, the expected value to the funding agency, company capability and commercial potential. The duration of the award is a maximum of 24 months and generally does not exceed $600,000. Approximately 40 percent of the Phase I Awards continue on to Phase II. Phase III involves additional private funding for the commercial application of a technology. A Phase III Award is for the infusion and use of a product into the commercial market. Private sector investment, in various forms, is typically present in Phase III.

Under the Small Business Research and Development Enhancement Act of 1992, funding in Phase I was increased to $100,000, and in Phase II to $750,000.

The SBIR was an offshoot of the Small Business Investment Company (SBIC) program, which provided more than $3 billion to young firms between 1958 and 1969. During this period this amounted to more than three times the total amount of private venture capital.

The SBIR represents about 60 percent of all public SME finance programs. Taken together, the public SME finance is about two-thirds as large as private venture capital. In 1995, the sum of equity financing provided through and guaranteed by public programs financing SMEs was $2.4 billion, which amounted to more than 60 percent of the total funding disbursed by traditional venture funds in that year. Equally as important, the emphasis on SBIR and most public funds is on early stage finance, which is generally ignored by private venture capital. Some of the most innovative American companies received early stage finance from SBIR, including Apple Computer, Chiron, Compaq and Intel.

Through the Small Business Innovation Research (SBIR) program, the National Institutes of Health (NIH) awarded $266 million in grants to small firms for medical and biopharmaceutical research. It is expected that the SBIR program at NIH will exceed $300 million in 1999.

In addition to the NIH, the United States Department of Defense also uses the SBIR program to fund biotechnology firms. Between 1983 and 1997 there was more than $240 million in SBIR awards for biotechnology companies from the Department of Defense. Phase I accounted for $47 million and Phase II accounted for $194 million (Wessner, 2001).

To evaluate the impact of the SBIR on the commercial activities of SMEs, a large, comprehensive survey was undertaken by the U.S.National Academy's division on Science, Technology, and Economic Policy (STEP) (Wessner, 2001). In addition, case studies were undertaken on the basis of detailed interviews with the founders, owners and employees of over fifty firms (Scott, 2001; Link, 2001; Link and Scott; 2001). All of the case study firms had received SBIR assistance. They are dispersed across the United States and span a broad range of technologies, products and industries. While some are new startups, others have a proven track record of success. These case studies examined the impact of the SBIR in a broad context. In particular, the results from evaluating the SBIR suggested that the benefits of the SBIR extends beyond the impact on the individual recipient firm. The social rate of return, which incorporates this external positive impact, exceeds the positive rate of return. There was no evidence of a negative rate of return associated with the SBIR. There is compelling evidence that the SBIR program has had a positive impact on developing the U.S. biotechnology industry. The benefits have been documented as:

- The survival and growth rates of SBIR recipients have exceeded those of firms not receiving SBIR funding (Lerner and Kegler, 2001)
- The SBIR induces scientists involved in research to change their career path. By applying the scientific knowledge to commercialization, these scientists shift their career trajectories away from basic research towards entrepreneurship (Feldman, 2001 and Audretsch et al., 2001).
- The SBIR awards provide a source of funding for scientists to launch start-up firms that otherwise would not have had access to alternative sources of funding.
- SBIR awards have a powerful demonstration effect. Scientists commercializing research results by starting companies induce colleagues to consider applications and the commercial potential of their own research.

5. Conclusions

Globalization is shifting the comparative advantage in the United States away from being based on traditional inputs of production, such as land, labor and capital, towards knowledge. As the comparative advantage has become increasingly based on new knowledge, public policy has responded in two fundamental ways. The first has been to shift the policy focus away from the traditional triad of policy instruments essentially constraining the freedom of firms to contract – regulation, antitrust in the U.S., and public ownership of business. The policy approach of constraint was sensible as long as the major issue was how to restrain large corporations in possession of considerable market power. That this policy is less relevant in a global economy is reflected by the waves of deregulation and privatisation. Instead, a new policy approach is emerging which focuses on enabling the creation and commercialisation of knowledge. Examples of such policies include encouraging R&D, venture capital and new-firm startups. In particular, the new focus of SME policies is to promote the first type of strategy deployed by SMEs to enhance global competitiveness – innovation. Probably the greatest and most salient shift in SME policy over the last fifteen years has been a shift from trying to preserve SMEs that are confronted with a cost disadvantage due to size inherent scale disadvantages, towards promoting the startup and viability of SMEs involved in the commercialization of knowledge, or knowledge-based SMEs. The SBIR program of the United States is clearly one such policy that has recognized the important role that innovative small firms play in generating competitiveness and growth.

Acknowledgements

I thank Frank Lichtenberg of Columbia University for inspiring the title. I am grateful for the helpful suggestions of Donald Siegel, Chuck Wessner, Al Link, Maryann P. Feldman and John Scott as well as participants at the University of Nottingham September 2000 conference on Entrepreneurship and Technology Policy.

Notes

[1] Lester Thurow, 'Losing the Economic Race', *New York Review of Books*, September 1984, pp. 29–31.
[2] W. W. Restow, 'Here Comes a New Political Chapter in America', International Herald Tribune, 2 January 1987.
[3] For example, U.S. News and World Report (16 August, 1993) reported, 'What do Bill Clinton, George Bush and Bob Dole have in common? All have uttered one of the most enduring homilies in American political discourse: That small businesses create most of the nation's jobs.'
[4] Cited from Davis et al. (1996, p. 298).
[5] Representative Robert Michel, House Minority Leader, in the Republican Response to the 1993 State of the Union Address, cited from Davis et al. (1996, p. 298).
[6] According to *The Economist*, 'The death of distance as a determinant of the cost of communications will probably be the single most important economic force shaping society in the first half of the next century.' 'The Death of Distance', *The Economist*, 30 September, 1995.
[7] 'The Downsizing of America', *New York Times*, 3 March, 1996, p. 1.
[8] See Audretsch (1995).
[9] 'The Valley of Money's Delights', *The Economist*, 29 March, 1997, special section, p. 1.
[10] Arrow (1962) pointed out this is one of the reasons for inherent market failure.

References

Arrow, K., 1962, 'Economic Welfare and the Allocation of Resources for Invention', in R. Nelson (ed.), *The Rate and Direction of Inventive Activity*, Princeton: Princeton University Press.

Audretsch, David B., 1995, *Innovation and Industry Evolution*, Cambridge: MIT Press.

Audretsch, David B. and Maryann P. Feldman, 1996, 'R&D Spillovers and the Geography of Innovation and Production', *American Economic Review* **86**(3), 630–640.

Audretsch, David B. and Paula E. Stephan, 1996, 'Company-Scientist Locational Links: The Case of Biotechnology', *American Economic Review* **86**(3), 641–652.

Audretsch, David B., Albert N. Link and John T. Scott, 2001, 'Statistical Analysis of the National Academy of Sciences Survey of SBIR Awardees: Analyzing the Influence of the Fast Track Program', in Charles Wessner (ed.), *The Small Business Innovation Research Program (SBIR): An Assessment of the Department of Defense Fast Track Initiative*, Washington, D.C.: National Academy Press, pp. 275–290.

Audretsch, David B., Albert N. Link and John T. Scott, 2001b, 'Public/Private Technology Partnerships: Evaluting SBIR Supported Research', *Research Policy*.

Audretsch, David B., Juergen Weigand and Claudia Weigand, 2001, 'Does the Small Business Innovation Research Program Foster Entrepreneurial Behavior? Evidence from Indiana', in Charles Wessner (ed.), *The Small Business Innovation Research Program (SBIR): An Assessment of the Department of Defense Fast Track Initiative*, Washington, D.C.: National Academy Press, pp. 160–193.

Berman, Eli, John Bound and Stephen Machin, 1997, 'Implications of Skill-Biased Technological Change: International Evidence', working paper 6166, National Bureau of Economic Research (NBER), Cambridge, MA.

Cohen, W. and D. Levinthal, 1989, 'Innovation and Learning: The Two Faces of R&D', *Economic Journal* **99**(3), 569–596.

Cooke, Philip and David Wills, 1999, 'Small Firms, Social Capital and the Enhancement of Business Performance through Innovation Programmes', *Small Business Economics* **13**(3), 219–234.

Davis, Steven J., John Haltiwanger and Scott Schuh, 1996, 'Small Business and Job Creation: Dissecting the Myth and Reassessing the Facts', *Small Business Economics* **8**(4), 297–315.

Feldman, Maryann P., 2001, 'Role of the Department of Defense in Building Biotech Expertise', in Charles Wessner (ed.), *The Small Business Innovation Research Program (SBIR): An Assessment of the Department of Defense Fast Track Initiative*, Washington, D.C.: National Academy Press, pp. 251–274.

Gans, Joshua S. and Scott Stern, 2000, 'When Does Funding Research by Smaller Firms Bear Fruit?' Evidence from the SBIR Program, NBER Working Paper #7877, September.

Griliches, Z., 1979, 'Issues in Assessing the Contribution of R&D to Productivity Growth', *Bell Journal of Economics* **10**, 92–116.

Hirschman, A. O., 1970, *Exit, Voice, and Loyalty*, Cambridge, MA: Harvard University Press.

Kortum, S. and J. Lerner, 1997, 'Stronger Protection or Technological Revolution: What is Behind the Recent Surge in Patenting?' working paper 6204, National Bureau of Economic Research (NBER), Cambridge: MA.

Lerner, Josh, 1999, 'The Government as Venture Capitalist: The Long-Run Impact of the SBIR Program', *Journal of Business* **72**(3), 285–318.

Lerner, Joshua and Colin Kegler, 2001, 'Evaluating the Small Business Innovation Research Program: A Literature Review', in Charles Wessner (ed.), *The Small Business*

Innovation Research Program (SBIR): An Assessment of the Department of Defense Fast Track Initiative, Washington, D.C.: National Academy Press, pp. 397–426.

Link, Albert N. and John T. Scott, 2001, 'Estimates of the Social Returns to Small Business Innovation Research Projects', in Charles Wessner (ed.), *The Small Business Innovation Research Program (SBIR): An Assessment of the Department of Defense Fast Track Initiative*, Washington, D.C.: National Academy Press, pp. 275–290.

Link, Albert N., 2001, 'An Assessment of the Small Business Innovation Research Fast Track Program in the Southeastern States', in Charles Wessner (ed.), *The Small Business Innovation Research Program (SBIR): An Assessment of the Department of Defense Fast Track Initiative*, Washington, D.C.: National Academy Press, pp. 194–210.

Scott, John T., 2001, 'An Assessment of the Small Business Innovation Research Program in New England: Fast Track Compared with Non-Fast Track Projects', in Charles Wessner (ed.), *The Small Business Innovation Research Program (SBIR): An Assessment of the Department of Defense Fast Track Initiative*, Washington, D.C.: National Academy Press, pp. 104–140.

Thurow, Lester C., 1985, 'Healing with a Thousand Bandages', *Challenge* **28**, 1–14.

Wessner, Charles W. (ed.), 2001, *The Small Business Innovation Research Program (SBIR): An Assessment of the Department of Defense Fast Track Initiative*, Washington, D.C.: National Academy Press.

Part II
Government's Leveraging of R&D

[6]

Small and Medium-Sized Enterprises and IT Diffusion Policies in Europe

Renata Lèbre La Rovere

ABSTRACT. The main argument to justify an information technology (IT) diffusion policy for small and medium-sized enterprises (SMEs) is that IT could act as a catalyzer to the SMEs' growth, which is important because of the role played by these firms in innovation and regional development. However, due to the heterogeneity of SMEs, present IT diffusion policies in Europe are reaching only a limited number of these firms. The aim of this paper is to discuss the obstacles to the implementation of an IT diffusion policy for SMEs. The main arguments that justify a policy for IT diffusion in SMEs, and the main initiatives taken for IT diffusion in these firms in European countries are summarized. The shortcomings of present policies will be discussed, together with possibilities to increase IT diffusion in SMEs.

Introduction

As the importance of small and medium-sized enterprises (SMEs) to economic development has received increasing attention in the literature, policy-makers have become increasingly aware of the need to devise specific policies for these firms. There are two main findings in the literature that have implications for economic policy concerning SMEs. The first is that the nature of innovation adoption differs according to the size of the firm. The second is that clusters of small firms or industrial districts can be important for regional development. In this context, supporting SMEs can help strengthen innovative activity and regional development, but the policy must be devised to deal with these firms' specific characteristics. An IT diffusion policy for SMEs could help to achieve this aim, by increasing firms' technical capabilities, by reducing obstacles to growth linked to regional isolation and by allowing increased exchange of information between users and producers of innovation.

However, present policies for IT diffusion in European countries reach only a limited number of SMEs. The aim of this paper is to discuss the reasons of this limitation. In the first section we will present the main arguments that justify an IT diffusion policy for SMEs. In the second section we will summarize the main initiatives taken to support IT diffusion in SMEs in Europe. In the third section we will discuss the shortcomings of present policies, based on the distinction between supply and demand-oriented IT diffusion policies. As a conclusion we will discuss possible solutions to overcome present obstacles.

1. Arguments in favor of an IT diffusion policy for SMEs

Until the middle seventies, SMEs had a minor role in the debate about economic development due to the dominance of the mass production paradigm in industry. After this period, this paradigm was increasingly challenged, leading to large firms' fragmentation, unemployment growth and creation of new SMEs (Acs, 1992). The increasing importance of SMEs led to a multiplication of support programmes aimed at those firms, although in most countries a coordinated policy is still lacking (Storey, 1994). Support policies have been justified because of these firms' capacity to innovate and their potential to promote regional development. A brief review of the arguments of the studies on SMEs, innovation and regional devel-

Final version accepted on December 10, 1996

Wirtschafts-und Sozialwissenschaftliche Fakultät
Universität Rostock, Germany
and Instituto de Economia
Universidade Federal do Rio de Janeiro, Brazil
Work supported by CNPq, Brazil

Small Business Economics **11**: 1–9, 1998.

opment is needed to understand why an IT diffusion policy for SMEs is considered necessary.

SMEs, innovation and regional development

As authors started to analyze the potential that SMEs have for innovation, several characteristics of these firms that favor innovation adoption have been identified. Usually SMEs have diversified activities and flexible structures that allow for fast responses to market changes. In addition, SMEs can operate in niches and thus have a high innovation rate (Acs, 1992). Finally, risk-aversion of these firms is lower than that of large firms, and workers are more motivated (Pratten, 1991; OECD, 1995). However, small firms do not necessarily have a greater potential to innovate than large firms. Utterback and Suarez (1993) observe that large firms have clear advantages to innovate since they have greater chances to develop and implement what will become the 'dominant design' in an industry. Furthermore, SMEs have less access to technological information, and thus may be less prone to innovation (OECD, 1995). SMEs also tend to have a larger rate of indebtedness than large firms (Acs and Audrestch, 1993) and thus are more sensible to economic downturns. The main difficulty that faces those enterprises concerning innovation and growth comes from their low ability to obtain financial support.

Therefore, the ability of SMEs to innovate depends on several factors related to the countries' industrial organization and innovation system. As Rothwell and Dodgson (1993) warn, both SMEs and large firms have advantages in innovation adoption, but these advantages differ. While large firms have *material* advantages, due to their greater capability to support R&D, SMEs have *behavioral* advantages that stem from their greater flexibility and ability to adapt to changes in the market. While large firms are important in creating new technologies, small and medium-sized firms are innovative concerning the use of existing technologies (Thwaites and Wynarczyk, 1993).

The behavioral advantages of SMEs in innovation justify an IT diffusion policy for these firms. The use of information technologies can ease communication flows outside the firms, by allowing an increased interaction with users and suppliers. In addition, IT can enhance the flexibility of workers in the firm by improving internal communication flows. Therefore, the innovative activity of SMEs may be enhanced with IT use. Empirical studies show that after IT introduction SMEs usually have a greater ability to produce new products, while large firms tend only to diversify the existing range of products. In the cases where IT adoption is done concurrently to modifications in management, gains are increased (Correa, 1994). Since firms that use one IT application are more prone to use other applications (Wynarczyk et al., 1995) awareness-building efforts concerning IT are cumulative, and thus can lead to a 'virtuous cycle' of productivity growth.

The importance of SMEs to regional development has been highlighted by the literature on industrial districts (clusters) and by the analysis of the links between innovation and space. The analysis of industrial districts suggests that the flexible production paradigm is linked to the organization of firms, especially small firms, in clusters or networks.[1] These clusters contribute to the region's development by preserving the flexibility of firms and at the same time enabling gains associated with scale of production. The analysis of the links between innovation and space goes further, considering networks of firms as 'attractors', i.e. industrial organization forms that can serve as catalyzers to technical progress (Camagni, 1991). The main conclusion that stems from this literature is that in post-fordism, innovation may be closely associated with regional characteristics and the organization of firms in networks.

The empirical evidence of the positive results in performance obtained by clustered SMEs in developed countries is scarce (Schmitz, 1995) and does not provide insights on whether the cluster's experience can be translated to other regions or countries (van Djik, 1995). Neverhteless the fact that clustered firms are a feature of the new industrial paradigm is another justification for increased IT diffusion in SMEs. When those firms are gathered in a cluster or a network, they may extract more benefits from IT use by setting up a 'network of innovators' (Freeman, 1991). An advanced IT infrastructure not only supports present clusters but may encourage the formation of new ones. By encouraging the creation of new clusters, IT diffusion would bring benefits to the region where the firms are located.

Therefore, the main arguments in favor of an IT diffusion policy for SMEs are that IT can increase innovative activity and that IT can support clusters of firms, thus helping regional development. These arguments permeate the policies to support IT diffusion in Europe, as will be discussed in the next section.

2. IT diffusion policies for SMEs

As was observed in section 1, an IT diffusion policy for SMEs is considered important due to these firms' role in innovation and regional development. For instance, the Commission of European Communities states that SMEs 'are the basic operation unit of economic activity in the Community, because of their role in economic activity and regional development and also because of their dynamism, productivity, adaptability and innovation capacity' (CEC, 1990, p. 1). The European Union launched in 1994 an initiative called 'Small and Medium-sized Enterprises Action Programme' (SMEAP 2C) to support SME development. Its main objectives are to 'intensify the priority measures for enterprise policy in order to stimulate growth within the Community'. The main goals of the programme are:

- 'encouraging SMEs, including craft-industry enterprises, to adapt to structural changes and to changes brought about by the internal market;
- stimulate exchange of experiences and transnational cooperation activities;
- promote a better financial environment for enterprises;
- promote a better observation of the economic development of enterprises as part of the dynamic of effective implementation of the internal market' (ECHO, 1995)

The increased access of SMEs to information technologies is considered important to achieve the goals outlined above. The European Union has been trying to stimulate the diffusion of IT by implementing several programmes. Policies to support IT diffusion can be classified as supply-oriented policies and demand-oriented policies. As observed by Gillespie et al. (1995), IT diffusion is better understood as a series of translations: from infrastructure investment to the provision of appropriate services and applications; from such provision to user awareness: from awareness to adoption; from adoption to effective usage; and from effective usage to competitive advantage. While supply policies ensure the first translation, demand policies ensure the others. In other words, supply policies aim to increase the number of firms that are providers of IT, while demand policies stimulate firms to use IT. Supply-oriented policies are focused on the provision of infrastructure, support for R&D, support to co-operation between firms and universities and procurement policies. Demand-oriented policies involve training of workforce, technology centers, consultancy services and provision of technical information.

In the 80's and early 90's most of the European IT diffusion policies, as well as national policies, were supply-oriented, focusing on the improvement of the infrastructure and on the building of technological capabilities. Moreover, policies tended to focus on IT 'national champions', discouraging the entry of new firms in the industry (Soete, 1993). The more recent IT-related programmes represent about one-third of the total EU budget for the period 1994–1998 (BMWi, 1995). As IT is considered important for innovation and regional development, these programmes include support to research and development, regional initiatives and support for cooperation among countries. Most programmes try to encourage SMEs' participation as providers of IT, and some also aim to stimulate IT use in sectors that are dominated by SMEs.

2.1. *Supply-oriented policies*

As the early IT diffusion policies focused in the improvement of the infrastructure, Europe's telecommunications networks were upgraded. In 1996, networks of 18 countries will be 100% capable of offering ISDN, and an optical fiber network will link 19 major European cities by 1998. In addition, an European standard for mobile telephony is already in force and tends to become a global standard (BMWi, 1995). The recent changes brought by globalization have turned the information infrastructure into a strategic asset for regional development (Gillespie et al., 1995). Therefore the issue of an European integrated infrastructure has been receiving the attention of European policy-makers since 1993. In this year the White Paper on Growth, Competitiveness and Employment in Europe reckoned that European telecommunications networks needed investments of 150 billion ECU until 2003. The Bangesmann report, in 1994, recommended that communications use should increase in the EU to guarantee competitiveness of firms, especially SMEs. Increased use should be achieved by enlarging ISDN access, reducing tariffs, developing standards and increasing global connectivity. On July 1994 the Action Plan for 'Europe's Way to the Information Society' was launched. The representatives of IT industries are working with the European Union in this plan to recommend actions towards an integrated infrastructure in four different areas: regulation framework, network infrastructure, social aspects and promotional activities.

However, before the integrated infrastructure is attained some obstacles must be surpassed. The first is that Europe's telecommunications infrastructure is still fragmented, with different digital switching and videotext systems. This is because the building of the infrastructure followed the strategies of national telecommunications operators, which varied across countries. The second is that the integration of the infrastructure depends on the evolution of deregulation, which also varies widely across Europe (EITO, 1995). The third is that European countries have wide differences in IT penetration (see Table I). These differences are linked to the fact that most telecommunications operators until recently had a 'research-oriented' approach to the infrastructure, upgrading it according to their capabilities and disregarding users' needs (Grupp, 1993).

European IT diffusion programmes also aim to stimulate IT diffusion by supporting IT providers, including SMEs. The Telematics Applications programme has allocated 5% of its budget to providers that are SMEs. SME participation as IT providers is also emphasized in 15 different programmes to stimulate IT diffusion in Europe, with aims ranging from improving education to establishing Government networks. In addition, most IT diffusion programs of national Governments in Europe focus on the building of technological capability in the firms (Hana et al., 1995).

TABLE I
IT penetration rates in Europe

Country	IT investment as % of GDP	PCs per white-collar workers
United Kingdom	2.50	0.73
Denmark	2.48	0.73
Germany	2.41	0.79
The Netherlands	2.35	0.85
Finland	2.18	0.52
France	2.09	0.60
Belgium/Luxembourg	2.00	0.77
Ireland	1.71	0.59
Austria	1.67	0.60
Italy	1.37	0.50
Portugal	1.27	0.40
Spain	1.14	0.44
Sweden	1.04	1.13
Greece	0.70	0.40
Western Europe (average)	2.07	0.72

Source: EITO 1995.

However, a good infrastructure and capability to produce IT are necessary but not sufficient conditions for enhanced IT use. Although Europe's share in the total number of telephone lines is 40%, European users represent only 10% of online services use and 10% of Internet use (BMWi, 1995). Therefore, most European users have still to make the translation from adoption to effective usage of IT. The European Council recognized this by stating in 1995 that education and training should be considered priority domains for information and communication technology programmes (DGXIII, 1995). Thus more recent IT diffusion policies, including those for SMEs, are a combination of policies to improve the infrastructure and policies to stimulate use (Correa, 1994; OECD, 1995).

2.2. *Demand-oriented policies*

The available evidence on IT diffusion in SMEs suggests that these firms lag behind large firms in IT usage (Blili and Raymond, 1993; Correa, 1994; DGXIII, 1995; OECD, 1995). Table II shows that even simple IT applications like e-mail have a much lower use in small than in large firms.

The evaluation of supply-oriented programmes such as STAR (Special Telecommunications

TABLE II
Electronic mail use in European Firms (%)

Country	Large firms that use e-mail	Small firms that use e-mail
Sweden	63	29
France	33	23
Benelux	42	19
United Kingdom	49	16
Italy	25	10
Germany	32	7

Source: EITO, 1995.

Action for Regional Development) revealed that the demand for new IT services was lower than expected after the improvement of the IT infrastructure, especially in the case of SMEs. Therefore, supply-oriented policies must be complemented by policies to encourage IT demand. Since small firms have difficulties to assess costs and benefits of IT, demand for IT must be stimulated. Stimulating demand is not easy, especially in the case of firms that constitute a universe as heterogeneous as the SMEs. Empirical studies show that demand for IT is greater in new SMEs than in traditional ones (Computer Industry Report, 1994). In the latter, family-type management hinders adoption of new technologies and orientation to external markets, thus restricting the efforts to increase competitiveness that frequently generate an increase in IT use (Granger, 1995).

Furthermore, IT implementation is different according to the sector in which the firm is located. While in the manufacturing sector IT can support *process innovations*, in the services sector IT can also be *product innovations*. Product innovations have different requirements of investment and skills training compared to process innovations (Antonelli, 1993). Therefore a demand-oriented diffusion policy must take into account sectoral and organizational characteristics of SMEs. It must also cater to SMEs' needs by encouraging exploratory use of new services by a group of selected firms and by reviewing the portfolio of applications periodically (Eurostaf, 1994). In addition, it is essential to analyze user/producer relationships that shape demand for IT applications, to assess how the policy can stimulate future demand (Mansell, 1994). The location of the firms is also important: Gillespie et al. (1995) note that the use of IT applications can vary in different regions of a country.

The challenges involved in defining an IT demand-oriented diffusion policy for SMEs explain the limited scope of concrete initiatives. One of the first examples of an IT demand-oriented policy was the European initiative to support networks of firms in 1988. This initiative was based on information diffusion, creation of European centers of information and building of information networks aimed to improve cooperation between enterprises. These centers favored IT diffusion, stimulating access to information networks and creation of networks of firms. Thanks to those efforts, 25.000 proposals of cooperation between SMEs in the period 1988–1992 were evaluated (CEC, 1990). These results are clearly limited, since the total of SMEs in Europe reaches 13 million. Therefore national Governments in some countries have been setting up IT demonstration centers, aiming to increase awareness of the benefits of use (Hana et al., 1995). The more recent European initiatives to support SMEs as users of IT are found in nine projects included in the IMPACT programme, whose goal is to establish an internal market for information services and to improve the competitiveness of European firms. Eight of these projects are sector-specific, destined to SMEs in the plastics, publishing, footwear, textile, shipbuilding, freight, building and music industries. The other project is focused on creating a network of subcontracting firms. In addition, the 'Telematics Applications' programme that aims to develop new telematic services, recently made a call for proposals to provide services to train staff of SMEs.

Initiatives such as the European IT Work Programme 1994 aim to have feedback from users in projects that support software, technologies for components and subsystems, multimedia systems, open microprocessor systems, high performance computing and networking, technologies for business processes and integration in manufacturing. SMEs would be stimulated to participate, by the concession of exploratory awards to cover the costs of preparing proposals for the programme. User participation would be ensured by networks of excellence, that gather centers of excellence across Europe, and user-supplier collaboration.

Although these projects represent an important step in stimulating IT diffusion in SMEs, they do not reach all European SMEs, being limited to the SMEs that are in networks of excellence or in specific sectors. In the next section we will discuss the difficulties involving the implementation of a policy to stimulate IT diffusion in SMEs.

3. Obstacles to policy implementation

There are several reasons that explain the gap between policy-makers' intentions and concrete results concerning IT diffusion in SMEs. European countries were slow to ensure the first translation of IT diffusion, i.e. the provision of appropriate services in their infrastructure. As observed by Burgelmann (1995), the experience of Minitel and Internet show that the success of new communication services depend on a 'strong central commitment, a coordinated and long-term dedicated financial support, a stable regulatory environment (..) and almost unlimited room for users to experiment whit it'. The deregulation of the telecommunications sector that has been taking place in Europe does not allow such conditions. On the other hand, if Europe achieves an integrated infrastructure and surpasses the obstacles mentioned in section 2.1, there will be possibilities for increased diffusion of new services. The translation from provision of services to awareness is also taking place slowly, as demonstrated by the low rates of use of IT. The European and national initiatives to set up information centers, as well as networks of firms, aim to increase awareness. As these initiatives continue and the integration of the infrastructure is discussed, awareness about the benefits of IT tend to increase.

The translation from awareness to adoption, usage and competitiveness is more complex, especially in the case of SMEs. First, there are several types of SMEs. SMEs can be classified in four different types, according to the market structure where they are located, to the prevalent innovation rate and to their organization. SMEs can be individually in competitive markets with low innovation rates. They can also be individually in highly dynamic industries with high innovation rates. On the other hand, SMEs can be organized as production cooperatives (clusters), or in networks under the dominance of a large firm (Souza, 1993). Each type of SME has specific contributions to economic growth, and will accordingly perceive IT in different ways. The first type perceives IT as a cost reduction tool, while the second perceives IT as a business opportunity. The third and the fourth use IT in their productive processes and in their relations with the other firms, thus improving their interface with the market. Supply-oriented policies benefit mainly the second type of SME, that is operating in a high innovation rate environment and therefore is less risk-averse. The limited range of demand-oriented policies leaves most SMEs outside the reach of IT diffusion policies.

In addition, demand-oriented policies have been limited because defining them is a complex process. A demand-oriented IT policy for SMEs must take into account these firms' specific characteristics: individuality of the firm, lack of formal management system and sectoral/local characteristics (OECD, 1993). Awareness-building of IT potential must be directed towards the firm's owners or managers, which in turn can diffuse this awareness among employees. Information about IT and IT-based management techniques must be provided concurrently and experiments encouraged. Development of IT-based applications that correspond to the firm's needs must also be supported. Finally, IT applications should enable SMEs to have closer relationships with users, which can contribute to the innovation process (Gibbons and Weijers, 1992).

The third problem is defining the level of policy implementation. Policies should not be limited to the regional level, since chances of success vary between countries: Hassink (1996) observes that regional policies have more success in countries where there is a strong tradition of decentralized government, such as Germany. Although most European national Governments have been publishing reports on the 'Information Society', few are making efforts to define policy guidelines and resources that will generate a demand for new IT services (Melody, 1995). In addition, national policies are mostly supply-oriented, and most demand-oriented initatives are limited to the setting up of demonstration centers (Hana et al. 1995). On the other hand, policies defined at a supra-national level, as those of the European Union, also have limits. European policies are based on optimistic assumptions about the 'Information Society' that are failing to materialize. Although these policies emphasize SMEs' participation, most disregard the fact that IT applications for large and for small firms may not be the same (Garnham, 1994).

Fourth, it should be noted that the priority given to SMEs does not necessarily mean a participation of these firms in European projects. As observed by Pyke and Charles (1995), participation in EU projects involves several costs that result from the effort to raise financial resources, the dealing with the bureaucracy and other legal issues. The variety of EU programmes can also be a hindrance for SMEs (DG XIII, 1995). SMEs have few incentives to participate in EU projects since their planning is mostly on a short-term basis. The STAR evaluation, for instance, highlighted the fact that the main beneficiaries of European programmes are large firms. The European Union 'Technology Stimulation Measures for SMEs' programme is trying to address this problem by offering awards to outlines of proposals that can be presented by two SMEs from two Member States. SMEs that have these awards can proceed to prepare a full proposal. As an alternative, SMEs can join research teams already established, by contacting 'Focal Points', i.e., research-related institutions in specific areas. However, these efforts are reaching a limited number of SMEs: in one year of implementation, the programme gave 600 awards and funded 120 projects, reaching 2000 firms (CORDIS, 1996).

Fifth, there is the problem of reconciling macro and microeconomic policy goals. For instance, the deregulation of the telecommunications sector that is taking place in Europe does not benefit SMEs. This is because the elimination of cross subsidies in the pricing schemes of public telecommunication operators increases the price of local calls (Gillespie and Williams, 1989). In addition, the deregulation process has been has been developed by defining access according to geographical range criteria, not affordable prices (Mansell, 1993). Thus special actions for SMEs must be envisaged in order to ensure their access to the integrated infrastructure.

Finally, the heterogeneity of situations leads to mixed results concerning IT implementation. According to the OECD (1995) IT adoption in networks of firms can lead to a more rigid division of roles between SMEs, thus restricting their flexibility. Besides, in SMEs the strategy of IT adoption is not proactive, but a reaction, usually partial, to pressures from clients and providers. Most SMEs introduce innovations in an ad-hoc manner, and only when they perceive clearly that the innovation will bring a business opportunity (Gagnon and Toulose, 1996). Therefore, present efforts of awareness-raising that emphasize only technical advantages of IT may not be successful. In the SMEs' case, it is not sufficient to promote new technologies: information on how these technologies can contribute to organizational improvement must also be provided (STAR, 1993).

4. Summary and conclusions

Although the recent literature on SMEs support the case for a specific IT diffusion policy for those firms, the implementation of this policy has been reaching a limited number of firms in Europe. The challenges to stimulate IT diffusion among those firms are manifold. First, SMEs are extremely heterogeneous, so there are several policy possibilities. Second, the definition of a diffusion policy requires not only an empirical study of SMEs and their IT necessities but also a constant interaction between SMEs, IT suppliers and policy-makers. Third, a successful IT diffusion policy must combine regional, national and European initiatives. Fourth, success in IT diffusion requires a standardization of the information infrastructure, to guarantee full access to information for all firms. Finally, policy-makers must monitor the effects on SMEs of previous efforts to implement IT and of changes in the macroeconomic environment. As observed by the OECD (1995), policy-makers should not view IT as a 'technological medicine' but as a means for SMEs to achieve greater competitiveness.

In view of the above a policy to support IT in SMEs must be flexible enough to cope with the diversity of situations but at the same time have standard procedures of implementation to ease the coordination among the different agencies of support. The policy must identify the weaknesses of SMEs in IT adoption and accordingly provide advice not only technical, but also in the areas of management and finance.

Therefore a diffusion policy must not only focus on the supply of infrastructure and services but also on stimulating the demand for IT. A demand-focused policy must not only include measures that are important to stimulate demand, such as the constitution of information centers and consultant systems for SMEs, but also ensure that there is a link between IT adoption and usage. This supposes policies to support the organizational development of SMEs, the upgrade of management and human resources and the creation of a technical culture that leads the firm to search constant technological updating. Finally, policies to support IT diffusion must target not only the technological capabilities of SMEs but their competitive capabilities as well.

Note

[1] For a discussion of the main authors in the literature, see James and Bhalla (1993). A book by Bagnasco and Sabel (1995) has more recent contributions to this issue.

References

Acs, Z. J., 1992, 'Small Business Economics: a Global Perspective', *Challenge*, November-December.

Acs, Z. J. and D. B. Audretsch, 1993, 'Inovação e Mudança Tecnologica: a Nova Ciencia', in J. P.R. Velloso and L. Martins (eds.), *A Nova Ordem Mundial em Questão*, Ed. Jose Olympio, Rio de Janeiro.

Antonelli, C., 1993, 'The Diffusion of Technological Systems and Productivity Growth: The Case of Information and Communication Technologies', Herausforderung fur die Informationstechnik Internationale Konferenz, Dresden, 15–17 June.

Bagnasco, A. and C. Sabel (eds.), 1995, *Small and Medium-Sized Enterprises*, London: Pinter.

Bianchi, A., J. Cornford, B. Dillon, A. Gillespie, S. O'Siochru and L. Qvortrup, 1995, 'Strategic Options for Information and Communication Technologies in Less Favoured Regions', International Telecommunications Society European Regional Conference, Juan les Pins, France, 1–3 September.

Blili, S. and L. Raymond, 1993, 'Information Technology: Threats and Opportunities for Small and Medium-Sized Enterprises', *International Journal of Information Management* **13**.

Bundesministerium fur Wirtschaft (BMWi), 1995, 'Die Informationgesellschaft: Fakten, Analysen, Trends', November.

Burgelmann, J. C., 1995, 'Issues and Assumptions in Communications Policy and Research in Western Europe: A Critical Analysis', paper for the 1995 IAMCR Conference 'Communication beyond the Nation State', Potorez, Slovenia, June.

Camagni, R. (ed.), 1991, *Innovation Networks: Spatial Perspectives*, London: Bellhaven Press.

Comission of the European Communities (CEC), 1990, 'Enterprise Policy: A New Dimension for Small and Medium-Sized Enterprises', Documents Series 08, Com(90) 528 final.

Community R&D Information Services (CORDIS), database.

Computer Industry Report, 1994, v. 29 n. 22, 30 September.

Correa, C., 1994, Difusión y Políticas de Tecnologias de la Información para Pequeñas y Medianas Empresas en América Latina y el Caribe, mimeo, Buenos Aires, March.

Directorate General XIII (DG XIII), 1995, Task Force Multimedia Educational Software, 'Educational Multimedia: First elements of reflection', I'M Europe database.

Ducatel, K. and J. Millard, 1996, 'Employment and Innovation in Advanced Communication Technologies', *Futures* **28**(2).

European Community Host Organization (ECHO), database.

European Information Technology Observatory (EITO), 1995, *European Information Technology Observatory 95*, Mainz: Eggebrecht – Press KG.

EUROSTAF – Collection Analyses de Secteurs, 1994, 'Le Multimedia:Convergences Strategiques face aux Enjeux des Medias Interactifs', 3eme trimestre.

Freeman, C., 1991, 'Networks of Innovators: A Synthesis of Research Issues', *Research Policy* **20**.

Gagnon, Y. C. and J. M. Toulouse, 1996, 'The Behavior of Business Managers When Adopting New Technologies', *Technological Forecasting and Social Change* **52**.

Garnham, N., 1994, 'Whatever Happened to the Information Society?', in R. Mansell (ed.), *The Management of Information and Communication Technologies: Emerging Patterns of Control*, ASLIB.

Gibbons, M. and T. Weijers, 1992, 'Position Paper on User-Producer Relations', mimeo, Apeldoorn, the Netherlands, June.

Gillespie, A. and H. Williams, 1989, 'A Small Firms' Perspective on the Liberalisation of Telecommunications', in N. Garnham and A. Aksoy (eds), *European Telecommunications Policy Research*, Proceedings of the Communications Policy Research Conference, 22–24 June.

Gillespie, A., R. Richardson and J. Cornford, 1995, 'Information Infrastructures and Territorial Development', paper prepared for the OECD joint ICCP-TDS Workshop on Information Technologies and Territorial Development, Paris, November.

Granger, J. R., 1995, 'Advanced Telecommunication Services and Competitiveness in Small and Medium-Sized Enterprises: the Spanish Case', Fundesco, Madrid.

Grupp, H., 1993, 'Efficiency of Government Intervention in Technical Change in Telecommunications: Ten National Economies Compared', *Technovation* **13**(4).

Hassink, R., 1996, 'Technology Transfer Agencies and Regional Economic Development', *European Planning Studies* **4**(2).

Hana, N., K. Guy and E. Arnold, 1995, *The Diffusion of*

Information Technology: Experience of Industrial Countries and Lessons for Developing Countries, World Bank Discussion Papers 281.

James, J. and A. Bhalla, 1994, 'Flexible Specialization, New Technologies and Future Industrialization in Developing Countries', *Futures*, July/August.

Mansell, R., 1993, *The New Telecommunications: A Political Economy of Network Evolution*, Sage.

Mansell, R., 1994, 'Negotiating the Management of ICTs – Emerging Patterns of Control', in Mansell, op. cit.

Melody, W., 1995, 'Future Telecom Services: New Directions', paper presented at the Annual COST 248 Workshop, 'The Future European Telecommuncations User', Bad Honneff, Germany.

OECD, 1993, *Les Petites et Moyennes Entreprises: Technologie et Competitivité.*

OECD, 1995, *Information Technology (IT) Diffusion Policies for Small and Medium-Sized Enterprises.*

Pratten, C., 1991, *The Competitiveness of Small Firms*, Cambridge University Press.

Pyke, A. and D. Charles, 1995, 'The Impact of International Collaboration in U.K. University-Industry Links', *Industry and Higher Education*, October.

Rothwell, R. and M. Dodgson, 1993, 'Technology-based SMEs: Their Role in Industrial and Economic Change', Inderscience Enterprises.

Schmitz, H., 1995, 'Collective Efficiency:Growth Path for Small-Scale Industry',*Journal of Development Studies* **31**(4), April.

Soete, L., 1993, 'Information Technology Promoting Structural Change', Herausforderung fur die Informationstechnik Internationale Konferenz, Dresden, 15–17 June.

Souza, M. C. A. F., 1993, *Pequenas e Médias Empresas na Reestruturação Industrial,* Ph.D dissertation, UNICAMP, Campinas.

Special Telecommunications Action for Regional Development (STAR), 1993, Community-level Evaluation, Executive Summary, Eubank Press Ltd.

Storey, D. J., 1994, *Understanding the Small Business Sector*, London: Routledge.

Thwaites, A. and P. Wynarckzyk, 1993, 'Innovation and Financial Performance in Small Firms', Centre for Urban and Regional Development Studies (CURDS), Newcastle.

Utterback, J. and F. Suarrez, 1993, 'Innovation, Competition and Industry Structure', *Research Policy* **22**, 1–21

van Djik, M. P., 1995, 'Flexible Specialisation, the New Competition and Industrial Districts', *Small Business Economics* **7**, 15–27.

Wynarckzyk, P., I. Hardill and J. Turner, 1995, 'Technology and Company Performance:Survey of the British Textile and Clothing Filiere', CURDS Working Paper, Newcastle.

[7]

ELSEVIER

Research Policy 28 (1999) 615–624

research policy

www.elsevier.nl/locate/econbase

Innovative output, and a firm's propensity to patent. An exploration of CIS micro data

Erik Brouwer [a], Alfred Kleinknecht [b,*]

[a] SEO, University of Amsterdam, Amsterdam, Netherlands
[b] Department of Economics, Delft University of Technology, Kanaalweg 2 B, NL-2628 EB Delft, Netherlands

Received 15 December 1997; received in revised form 29 April 1998; accepted 4 January 1999

Abstract

We analyse a firm's propensity to patent, using a new indicator of innovative output. We find that a firm's propensity to patent is significantly higher among R&D collaborators, and it varies across sectors and by firm size. With a given innovation output, smaller innovators have a *lower* probability of applying for a patent. Smaller innovators that do apply for patents, however, tend to have higher *numbers* of patent applications than larger firms. The results have obvious consequences for the interpretation of patenting figures as an indicator of innovation. © 1999 Elsevier Science B.V. All rights reserved.

Keywords: Output; Patent; Propensity

1. Introduction

Patents have been and still are frequently used as indicators of innovation, and there is considerable literature about their validity for this purpose (see the survey by Griliches, 1990). One of the arguments that causes doubt refers to differences in what Scherer (1965, 1983) has called the *propensity to patent*, suggested to vary across sectors or across other firm characteristics. In other words, with a given innovation intensity, different firms may have a different patenting intensity, implying that patents may be a problematic innovation indicator.

Apart from soft indications that often derive from case studies or anecdotal evidence, however, there are only few attempts at systematic investigation of the propensity to patent (see Scherer, 1983; Bound et al., 1984). These attempts came down to relating R&D to patent data. Such comparisons have the weakness that any deviation between a firm's R&D input and its patenting intensity can be due to two different reasons that are hard to separate empirically: either to a more or less efficient use of R&D-inputs *or* to differences in the propensity to patent. An additional problem in this context is the undercounting of (often informal and small-scale) R&D in small firms (Kleinknecht, 1987; Pavitt and Patel, 1988). The undercounting of small firm R&D im-

* Corresponding author. Tel.: +31-15-278-3469; Fax: +31-15-278-3956; E-mail: kleinknecht@wtm.tudelft.nl

PII: S0048-7333(99)00003-7

plies, among others, that observations on the relationship between R&D and patenting will be systematically biased by firm size.

To assess the propensity to patent more adequately, a measure of the *results* of the innovation process is required. That is, we need to know whether firms with a given innovation output differ with respect to their actual patenting intensity. One of the few studies that compared patenting to a measure of innovation output is by Acs and Audretsch (1989), and more recently by Arundel and Kabla (1998). While Acs and Audretsch used data on new product announcements in the US, we use an alternative measure of innovation output that is based on a national survey of sales of innovative products in the Netherlands. While Acs and Audretsch had to rely on line-of-business data, our approach (and that by Arundel and Kabla) has the advantage that firm-level data are used. Besides using a direct measure of innovation output in Section 4 of our paper, we also take a look at qualitative answers of respondents about their appreciation of patent protection as a means of appropriation of innovation benefits (Section 3). This information can be compared to earlier findings by Levin et al. (1987).

2. The data and the model

The recently held *Community Innovation Survey* (CIS) produced firm-level indicators of innovative output and qualitative information about a firm's attitude towards patenting. In the Netherlands, the survey also covered a firm's actual patent applications, besides a number of other firm characteristics. We use the Dutch part of the CIS covering the year 1992; moreover, an innovation survey about the year 1988 is available that can be linked, at the firm level, to the 1992 survey which allows for an alternative specification of our model in Section 4. Numbers of firms that responded to both surveys amount to about 1300 manufacturing firms; response rates in the two surveys were 54% and 50.8%. A brief survey among non-respondents indicates that these do not differ systematically from respondents, implying that the database is representative of firms with 10 and more workers in all manufacturing sectors of the country (for detailed documentation see Brouwer and Kleinknecht, 1994).

The innovation output indicator in the 1992 CIS questionnaire is based on differentiation among three types of product:

1. products essentially unchanged during 1990–92;
2. products subject to incremental technological change during 1990–92, and
3. products that were radically changed or newly introduced during 1990–92.

Products in categories (2) and (3) are further subdivided into products 'new to the *firm*' (i.e., already known in the sector) and those 'new to the *sector*' (i.e., not earlier introduced by a competitor). Whereas, the former category is likely to consist of imitative innovations, products 'new to the sector' may be interpreted as 'true' innovations, incorporating new technological knowledge which is likely to be patentable. The firms were asked to report the share in their 1992 sales taken by each of these three types of product. An earlier pre-test of this indicator in five countries was satisfactory (Kleinknecht, 1993); the same holds for the experience in the actual survey (Brouwer and Kleinknecht, 1994) and in subsequent econometric work (see various contributions in Kleinknecht, 1996).

We start from the assumption that our output indicator is indeed a valid measure of a firm's product innovation performance. We interpret differences between a firm's score on the innovation output indicator (products *new to the sector*) and on the patenting indicator as indicating a higher or lower propensity to patent. The interesting question is, of course, whether differences in the propensity to patent among individual firms can be ascribed to such factors as branch of principal activity, firm size, etc. Since patent applications are a discrete count variable, we use a Hurdle Count Data model, trying to identify factors which have an impact on two dependent variables:

1. the probability that a firm will have filed *at least one* patent application at the European Patent Office (EPO), and
2. given that a firm filed EPO patent applications, how *many* patent applications did it file?

Before documenting the Hurdle estimate (in Section 4), we examine additional information about a firm's attitude towards patenting which was also covered by the 1992 CIS. This qualitative information related to a firm's view of the effectiveness of patents and other means for protection of product innovations (e.g., secrecy, time lead on competitors, etc.). In Section 3, we compare the qualitative answers from the CIS to the results reported by Arundel and Kabla (1998). In Section 4, the relationship between innovation output and patent applications is examined.

Before the data are analysed, a note on the structure of the 1992 CIS survey is necessary. The CIS questionnaire started with general questions about the firm (sales, employment, branch, etc.) which had to be filled-in by *all* respondents. Three questions then followed: Had the firm:

1. developed and/or introduced (during 1990–92) any technologically changed products or services?
2. developed and/or introduced (during 1990–92) any technologically changed processes?
3. Or had it plans to do so in the next 3 years?

Thirty-nine percent of firms reported that they had some product innovation activities, 38% reported process innovation activities, and 46% stated that they had plans for innovation activities in the next 3 years (Brouwer and Kleinknecht, 1994, 46).

If at least one of the three questions was answered in the affirmative, the firm had to complete the remainder of the questionnaire. The intention behind this procedure was that non-innovative firms should not be bothered with numerous and irrelevant questions about innovation since this might cause selective non-response. Although the questionnaire also asks for *process* innovation activities, we confine our analysis to *product* innovation, since information about process innovation is confined to a simple 'yes/no'-answer, whereas for product innovation we also have a measure of innovativeness (i.e., shares in sales taken by new products). It should be noted that about 30% of the firms that reported some innovation activities during 1990–92 also had sales in 1992 which were due to products *new to the sector* (introduced during 1990–92), while about 95% had sales of products *new to the firm* (introduced during 1990–92) (Brouwer and Kleinknecht, 1994, 44–45). This difference in percentage is not surprising since the category *new to the firm* consists of imitative innovations, which are more frequent than 'true' innovations. The remaining 5% of firms which had innovation activities during 1990–92 but no sales of innovative products (on whatever definition), may have had process innovations, had failed projects or projects still to be introduced on the market.

3. A firm's appreciation of patent protection

Firms that reported that they had undertaken some innovation activities during 1990–92 were asked to indicate how effective they found patents as a protection of their product innovations against imitators. They could answer on a five-point scale, ranging from 'insignificant' to 'crucial'. Of course, patenting is not the only mechanism enhancing the appropriation of innovation benefits. Table 1a and b gives information about a firm's judgement of the effectiveness of various other protection mechanisms. Insiders will note that the relevant questions in the CIS questionnaire follow (with modifications) the example of the Yale survey on appropriation of innovation benefits (see Levin et al., 1987).

In our CIS database, the relevant questions had to be answered by 1008 out of 2.075 responding firms (i.e., those which had *some* innovation activities during 1990–92). This category includes firms that report that they were busy (during the last 3 years) with the introduction of (1) new processes, (2) products 'new to the firm' (i.e., 'imitators') or (3) products 'new to the sector' (i.e., 'innovators'). The table documents in brackets the answers of the much smaller group of firms that introduced products 'new to the *sector*'; it is remarkable to see that the appreciation of patent protection by the latter hardly deviates from the others.

Table 1a shows that almost half of the firms consider patenting as an insignificant protection against imitators. When looking at percentages of firms which consider a certain type of protection as 'very important' or 'crucial', it turns out that having a time lead on competitors is the most important mechanism for appropriation of innovation benefits. Preventing that qualified people leave the firm (not included in the Levin et al., 1987 survey) is also ranking high, pointing to the importance of human capital. Secrecy is still slightly more important than patent protection. The finding that patenting is not

Table 1
Judgements by innovating firms[a] about the effectiveness of various mechanism for protection of *product (or service) / process* innovations against imitators

Mechanism of protection against imitators	Judgement about effectiveness (percentages)				
	Insignificant	Modest	Moderate	Very important	Crucial
(a) Product (or service)					
Time lead on competitors	20.5 (23.8)[b]	6.1 (6.8)	16.3 (16.3)	37.8 (37.0)	19.4 (16.1)
Keeping qualified people in the firm	17.1 (18.0)	5.5 (5.9)	21.6 (22.2)	39.6 (39.9)	16.1 (14.0)
Secrecy	33.2 (36.6)	13.3 (14.4)	20.8 (21.6)	18.9 (15.5)	13.8 (11.9)
Patent protection	47.0 (50.0)	14.9 (15.0)	12.6 (10.9)	15.1 (15.1)	10.3 (9.0)
Complexity of product or process design	47.6 (52.0)	11.8 (11.8)	19.5 (17.2)	15.1 (13.5)	6.0 (5.5)
Copyright and related laws	61.5 (61.8)	14.6 (14.2)	12.4 (11.2)	8.7 (9.5)	2.8 (3.2)
Certification, normalisation	47.4 (47.8)	16.5 (17.0)	17.6 (17.2)	14.9 (14.0)	3.6 (4.0)
(b) Process					
Time lead on competitors	22.9	6.4	14.8	34.8	21.2
Keeping qualified people in the firm	16.7	3.3	21.7	39.6	18.8
Secrecy	32.4	9.7	16.4	22.7	18.6
Patent protection	55.4	15.6	11.1	11.8	6.2
Complexity of product or process design	45.9	10.0	21.5	15.6	7.0
Copyright and related laws	72.8	13.5	8.2	4.5	1.1
Certification, normalisation	56.8	13.2	14.0	13.9	2.1

Number of observations: 1.008.
Source: CIS-Netherlands, 1992.
[a]Answers by firms which developed or introduced new or improved products, services or processes during the last 3 years.
[b]Figures in brackets cover answers by firms which realised sales of products (or services) *new to the sector* in 1992.

the most important means for appropriation of innovation benefits is consistent with the findings by Levin et al. (1987) even if their questionnaire is not fully comparable with ours.

As expected, the appreciation of patents as a protection for *process* innovation (Table 1b) is somewhat lower than for product innovation. In the case of process innovation, other mechanisms such as 'time lead on competitors', 'secrecy' and 'keeping qualified people in the firm' are more important. The last-named point underpins the importance of (firm specific) human capital and tacit knowledge.

It has often been suggested in the literature that not only technological opportunity and innovation but also the propensity to patent may differ widely by sectors (see e.g., Pavitt and Patel, 1988, 40). In a recent contribution to this journal, Arundel and Kabla (1998) reported, as a measure of the propensity to patent, sales-weighted percentages of innovations for which a patent application was made, using a sample of Europe's largest firms (604 respondents). Their results are not fully comparable to ours, due to conceptual differences and differences in sample selection. While they studied the patenting behaviour in Europe's largest firms (including only a few Dutch firms), we have a representative sample of all manufacturing firms in the Netherlands with 10 or more workers. The sample used by Arundel and Kabla has the advantage of a broad European coverage. However, it misses small- and medium-sized enterprises. Our sample is confined to a small country, but it is representative of firms in all size classes (from 10 workers onwards). Moreover, in their European survey, firms were asked whether they patented innovations, while in our survey we asked information about the general attitude of the management towards patenting. In Table 2, we document their results together with ours, split by sectors. Table 2a covers sectoral propensities to patent *product* innovations; Table 2b gives comparable information about *process* innovations.

It is remarkable to note that in spite of considerable differences in sample selection and in spite of the named conceptual differences, both approaches reveal a fairly consistent picture. Considering the propensity to patent *product* innovations, 'chemicals

Table 2
The propensity to patent product/process innovations: comparing our results to those of Arundel and Kabla (1998)

Manufacturing branches	The propensity to patent product and process innovations (Arundel and Kabla, 1998)	Appreciation of the value of patent protection for product and process innovations in Dutch manufacturing (1992); patent protection is:		
		Insignificant or modestly important	Moderately important	Very important or crucial
(a) Product				
Food, beverages and tobacco	26.1	65.0	10.4	24.6
Textiles and clothing	8.1	64.4	13.5	22.1
Wood, paper and printing	n.a.	69.1	5.8	25.1
Petroleum refining	22.6	46.9	16.9	36.3
Chemicals	57.3	46.9	16.9	36.3
Pharmaceuticals	79.2	46.9	16.9	36.3
Rubber and plastics products	33.7	51.9	11.7	36.4
Class, clay and ceramics	29.3	77.8	19.5	11.8
Basic metals (iron and steel)	14.6	93.8	2.3	9.9
Fabricated metal products	38.8	61.9	14.5	23.6
Mechanical engineering	52.4	51.7	21.9	26.4
Office and computing equipment	56.8	53.77	18.5	27.7
Electrical equipment	43.6	53.77	18.5	27.7
Communication equipment	46.6	53.77	18.5	27.7
Precision instruments	56.4	44.2	19.9	35.8
Automobiles	30.0	67.2	3.2	29.6
Other transport equipment	31.2	n.a.	n.a.	n.a.
Other manufacturing	n.a.	82.6	11.6	5.9
Total manufacturing	35.9	61.9	12.8	25.4
(b) Process				
Food, beverages and tobacco	24.7	58.7	13.9	27.4
Textiles and clothing	8.1	78.5	5.5	16.1
Wood, paper and printing	n.a.	82.0	7.6	10.4
Petroleum refining	29.0	54.4	9.8	35.8
Chemicals	39.0	54.4	9.8	35.8
Pharmaceuticals	45.6	54.4	9.8	35.8
Rubber and plastics products	27.6	62.7	7.4	30.0
Class, clay and ceramics	20.2	74.2	20.4	5.4
Basic metals (iron and steel)	15.1	93.3	3.7	3.1
Fabricated metal products	39.4	70.5	10.7	18.9
Mechanical engineering	16.3	60.3	25.3	14.4
Office equipment	20.9	79.8	6.0	14.1
Electrical equipment	21.5	79.8	6.0	14.1
Communication equipment	22.7	79.8	6.0	14.1
Precision instruments	46.8	76.2	2.5	21.3
Automobiles	17.0	75.1	5.0	19.9
Other transport equipment	10.9	n.a.	n.a.	n.a.
Other manufacturing	n.a.	86.7	0.0	13.3
Total manufacturing	24.8	71.0	11.1	18.0

and pharmaceuticals', 'mechanical engineering', 'office and computing equipment', 'electrical equipment' and 'precision instruments' have an above-average propensity to patent product innovations in both databases. Only in the case of 'automobiles' and 'rubbers and plastics', did our data indicate a slightly above-average appreciation of patenting product innovations, whereas the data by Arundel and Kabla do not show an above-average propensity to patent.

Both studies show that *process* innovations are generally less likely to be patented. Yet, 'chemicals and pharmaceuticals' and 'rubber and plastics' show an above-average propensity to patent in both databases. However, Arundel and Kabla's data indicate that 'fabricated metal products' and 'precision instruments' have an above-average propensity to patent process innovations, while in our data the appreciation of patent protection for processes is close to average in these sectors. The opposite holds for mechanical engineering. In conclusion, we can say that significant inter-sectoral differences in the propensity to patent innovations do exist, and the largest part of these differences is robust to differences in sample selection and to conceptual differences. Of course, the propensity to patent can differ not only by sector but also in other dimensions. In Section 4, we engage into a multivariate analysis of the propensity to patent which will allow to address some more factors.

4. The relationship of innovative output and patents

Before analysing the data, we give a few descriptive figures. Among the 2.078 manufacturing firms included in Table 3a, 2.7% applied for a European patent in 1992 and 12.6% had some sales of innovative products in 1992. An average applicant filed about three patent applications; the maximum number of applications is 81 and the standard deviation is 10.5.

Table 3a shows that 86.1% of the firms are non-innovators according to both measures of innovation. A percentage of 11.2 of the firms had in 1992 sales of innovative products, but had no EPO patent applications. A percentage of 1.4 of the 2.078 firms applied for a European patent in 1992 and had also sales of innovative products. Almost the same percentage (1.3%) had some patent applications, but no sales of innovative products. The latter may mainly consist of applications that have not resulted in commercially viable products. Table 3a suggests that the correlation between innovation output and patenting is far from perfect. While 86.1% of the firms have no patent applications and no sales of innovative products, there remain substantial numbers of firms which apply for European patents but have no sales of innovative products and vice versa.

One could object against Table 3a that we have failed to use an adequate time-lag between patenting and sales of innovative products. It seems reasonable to assume that patent protection of inventions which underlie an innovation will precede the actual market introduction of the innovation. Since we are dealing with 1992 sales of innovative products that have been introduced during the period 1990–92, the patent applications related to the underlying inven-

Table 3
Firms which have some sales of products 'new to the *sector*' in 1992 and firms which applied for one or more European patents in 1992 and 1998

Firms which have:	*No* sales of products new to the sector in 1992	*Some* sales of products new to the sector in 1992	Totals
(a) 1992[a]			
No patent applications in 1992	86.1%	11.2%	97.3%
Some patent applications in 1992	1.3%	1.4%	2.7%
Totals	87.4%	12.6%	100.0%
(b) 1998[b]			
No patent applications in 1988	72.0%	13.4%	85.4%
Some patent applications in 1988	9.5%	5.1%	14.6
Totals	81.5%	18.5%	100.0%

Source: CIS-Netherlands, 1992.
[a]Total number of observations: 2078.
[b]Number of observations: 981.

tions are, in any case, likely to be introduced prior to 1992, many of them even prior to 1990. By linking (at the firm level) the 1992 survey data to those of the survey in 1988, we are able to link 1988 patent applications to 1992 sales of innovative products which is probably more adequate. This has been done in Table 3b. While the use of panel data has the advantage that we can use a more adequate time-lag, Table 3b has the disadvantage that we lose more than half of our observations. In spite of such differences, the figures in Table 3b tell us essentially the same story as those in Table 3a.

Of course, the descriptive information in Table 3a and b does not yet allow to draw strong conclusions about the relationship of patents and innovation output, let alone about factors which influence patenting behaviour. For a better insight into determinants of patenting behaviour, we shall now turn to a multivariate analysis.

When investigating factors that influence a firm's propensity to seek patent protection, we consider the following four factors.

First, as becomes obvious from our above Table 2a and b, there are fairly strong sectoral differences in the propensity to patent. These may have to do with the speed and cost of imitation vs. innovation (Mansfield et al., 1981; Mansfield, 1986; Levin et al., 1987). A greater ease of imitation is likely to strengthen a firm's propensity to patent. We therefore include sector dummies in our estimate.

Second, compared to large corporations, smaller firms are likely to be weaker in terms of market shares, marketing networks, brand names, advertising intensity, etc., and they might compensate this weakness by seeking more intensively for patent protection. This justifies inclusion of firm size as an exogenous variable.

Third, while the first two arguments hint to the famous 'appropriation-of-innovation-benefits' argument, one might also argue that patenting serves as a vehicle for the formalisation of technology exchange arrangements between firms. This would notably apply when firms engage in one or the other joint innovation project. When negotiating about conditions of technological partnerships, previous patenting may facilitate to treat a firm's knowledge as a documented, formalised and, hence, tradable capital asset. We would therefore expect firms that engaged into R&D collaboration agreements to have an above-average rate of patenting. Moreover, the already mentioned 'appropriation-of-innovation-benefits' argument might also apply here: A firm may wish to protect its most precious pieces of knowledge before one engages into a joint project in which the partner gets access to that knowledge.

Finally, we include as control variables a firm's R&D intensity (i.e., its R&D budget as a percentage of sales) and information about whether a firm had introduced new products and the share in total sales taken by such products as a measure of innovative output. We use products *new to the sector* (i.e., not earlier introduced by a competitor), since these are most likely to cover new technological knowledge that is patentable.

Our multivariate analysis should answer a two-fold question:

- what is the impact of factors such as firm size, R&D collaboration and differences between sectors on the *probability* that a firm will have at least one patent application at the EPO?
- what is the impact of the same exogenous variables on actual *numbers* of patent applications, given that a firm has at least one such application?

As numbers of patents are a discrete count variable and many firms have zero patents, we estimate a Hurdle count data model (Mullahy, 1986). Before coming to results, a note needs to be made on time-lags in our model. Our output measure covers shares in a firm's 1992 total sales taken by products introduced during the years 1990–92. As already mentioned above, it is plausible to assume that patent applications tend to precede the market introduction of innovations. Hence, patent applications for inventions underlying innovations introduced during 1990–92 should have taken place prior to 1992. This is an argument in favour of taking patents from earlier years. Unfortunately, our choices for alternative time-lags are extremely limited by the available database.

Besides linking 1992 innovative output to 1992 patent applications, we also have the possibility to link (at the firm level) our 1992 data to an earlier survey from 1988 and use a firm's 1988 patent applications as a proxy for patenting activities related to 1990–92 innovations. One may object against this that we estimate a seemingly strange model, 'explaining' 1988 patent applications by sales of innovative products in 1992. However, given the time-lags between invention, patenting, market introduction and subsequent diffusion, it seems reasonable to use 1992 sales as a control variable when examining factors that influence a firm's 1988 patenting activities. One could argue that a firm's patent applications in 1988 were guided by its perception of future sales potentials.

While this procedure allows to handle a more adequate time-lag, the use of 'panel'-type data has the disadvantage that we lose observations. When linking the two databases, it turned out that the number of patent applicants dropped from 145 to 127. These fairly low numbers forced us to reduce the number of exogenous variables. In our earlier estimates, we experimented with detailed sector dummies. However, in our final estimates we reduced the sector dummies to a simple division between high and low technological opportunity sectors. This division of sectors went along the lines of taxonomy of sector of Pavitt (1984), taking Pavitt's 'supplier-dominated' sectors as low technological opportunity sectors and his categories of 'specialised suppliers', 'scale intensive industries' and 'science-based' sectors in the high technological opportunity category. A look at Table 2a and b suggests that our simplified sector dummy is quite a good approximation of sectoral differences in propensities to patent.

Clearly, time-lags in our two models are less than adequate, although we trust that a firm's patenting behaviour may be fairly constant over time which could mean that our failure to handle adequate time-lags may have less severe consequences. In any case, estimation of models with two different time-lags should give some hints about the robustness of outcomes. Of course, it remains a weakness of our analysis that the data do not allow to link individual innovation cases to corresponding patent applications. However, the data definitely have the advantage that we can link the two types of variables at the firm level in a large sample, whereas in the literature this linking had to be done often at the level of branches or firm size classes. Table 4 summarises our multivariate estimates.

Readers may be concerned about a possibly reversed causality in our estimate: Since patents protect a patent holder's market against imitators, patent protection may enhance sales of innovative products. We therefore constructed an instrumental variable for 1992 sales. However, application of the Hausman test indicated that sales are *not* endogenous. We therefore document the outcomes of the Hurdle model *without* instrumental variables.

It should be noted that, besides experimenting with detailed sector dummies, we also experimented with slope dummies, taking sectors times firm size and sectors times the sales share of innovative products. These estimates gave essentially the same results, and the loglikelihood was not significantly higher than in the model presented here. At first glance, the negative intercepts in Table 4 appear implausible. However, it should be noted that we use a non-linear positive truncated binomial model which does not allow the dependent variable to take a negative value. Simple simulations show that, in spite of the negative intercept, firms can never get a negative number of patent applications.

The above findings can be summarised as follows. We find a systematic relationship between a firm's innovation output (i.e., sales of innovative products) on the one hand, and its actual patenting behaviour, on the other hand. This holds for both the probability that a firm will apply for at least one European patent and for the actual number of EPO patent applications. The correlation between the innovation output indicator and patenting is far from perfect, however.

First, and not surprisingly, there are significant differences among sectors. High technological opportunity sectors tend to have a higher propensity to patent than low technological opportunity sectors.

Second, firms which have R&D collaboration agreements indeed have a higher probability to apply for at least one patent, and they also apply for higher numbers of patents. This corroborates our above arguments about the importance of formalisation of technology exchange arrangements, and about the need to protect knowledge before it is laid open for partners. This finding implies that, when using patents as an innovation indicator, one will overestimate the innovativeness of firms that collaborate on R&D compared to those that do not.

Third, larger firms have a greater probability to seek patent protection. Using the coefficients in Table

Table 4
Count data model estimates of factors which influence: (a) the probability that a firm will have at least one patent application at the EPO (logit part); (b) number of patent applications at the EPO (positive truncated negative binomial regression model) (*t*-values in brackets)

Exogenous variables	Model 1 (1992 patent applications)		Model 2 (1988 patent applications)	
	(a) Probability of at least one patent application	(b) Number of patent applications	(a) Probability of at least one patent application	(b) Number of patent applications
Constant term	−5.40 (−11.1)	−4.47 (−8.1)	−4.43 (8.0)	−1.80 (−3.2)
Firm size (log of number of employees)	0.56 (6.80)	0.62 (7.5)	0.48 (4.9)	0.38 (3.6)
Dummy: firm engaged in R&D collaboration	0.40 (1.9)	0.70 (2.5)	0.54 (2.4)	0.36 (1.3)
Sector dummy: high technological opportunity sectors	0.89 (3.7)	0.68 (2.1)	0.72 (2.9)	0.51 (1.7)
R&D intensity (R&D expenditures as a percentage of sales)	0.09 (5.6)	0.03 (1.7)	0.07 (2.9)	0.009 (0.4)
Control variable: 1992 sales of innovative products[a]	0.94 (4.71)	0.24 (2.5)	0.42 (1.8)	0.26 (3.1)
Number of observations	1320	148	652	127
R-squares (Veall and Zimmermann)	0.31	0.66	0.24	0.44

[a] When explaining whether the firm has at least one EPO patent application (a), we use a dummy for whether the firm had some sales of innovative products; when explaining the number of applications (b), we use logs of actual shares in sales taken by innovative products.

4, simple simulations show that the probability that a firm will have at least one EPO patent increases *more* than proportionately with numbers of workers. Given that a firm has some patent applications, however, the *number* of applications increases *less* than proportionately with firm size. This implies that when taking as an innovation indicator rates of firms that apply for patents, shares of smaller firms that innovate will be *under*estimated. However, within the group of patent applicants, one will *over*estimate the innovation intensity of smaller firms when judging from numbers of patent applications.

The finding that smaller innovators that do apply for patents have relatively higher *numbers* of patents is consistent with the hypothesis that they use the patent system as a compensation for having less market power than their larger counterparts. However, with a given innovation intensity, smaller innovators have a lower probability to apply for patents. This can be taken as an indication that they experience some threshold towards patenting. A possible explanation of our outcome can be found in a recent study by Berger Consultants (1994); they conclude that patenting costs and lack of information about the possibilities of the patent system are likely to have a negative impact on the patenting decision of smaller firms.

In conclusion, we can say that, when using patents as an innovation indicator, the rate of smaller innovators and the innovativeness of firms that do not collaborate with others on R&D will be underestimated. Moreover, patents will understate the innovativeness of firms in certain sectors, notably in low technological opportunity sectors.

Acknowledgements

We use data from the Dutch part of the 1992 *Community Innovation Survey* (CIS-I). The research for the present paper has been supported by the European Patent Office (EPO). Earlier versions of this paper were presented at the June 1998 conference of the International Schumpeter Society at Vienna and at the workshop of the TSER network 'R&D, productivity and innovation' in June 1998 at Paris. We wish to thank Anthony Arundel, Jenny Lanjouw and F.M. Scherer, as well as two anonymous referees of this journal for their written notes on an earlier version of this paper. The usual disclaimers apply.

References

Acs, Z.J., Audretsch, D.B., 1989. Patents as a measure of innovative activity. Kyklos 42, 171–180.

Arundel, A., Kabla, I., 1998. What percentage of innovations is patented? In: Research Policy, 27, 127–141.

Berger Consultants, 1994. The use of patent protection in Europe. A representative survey committed by the European Patent Office, München: EPO-script.

Bound, J., Cummins, C., Griliches, Z., Hall, B.H., Jaffe, A., 1984. Who does R&D and who patents. In: Z. Griliches (Ed.): R&D, patents, and productivity, Chicago: University of Chicago Press, 21–54.

Brouwer, E., Kleinknecht, A., 1994. Innovation in Dutch manufacturing and services, 1992 (in Dutch), Report to the Ministry of Economic Affairs, Den Haag, BTE 27 September 1994.

Griliches, Z., 1990. Patent statistics as economic indicators: a survey. Journal of Economic Literature 28, 1661–1707.

Kleinknecht, A., 1987. Measuring R&D in small firms: How much are we missing? Journal of Industrial Economics 36, 253–256.

Kleinknecht, A., 1993. Testing innovation indicators for postal surveys: Results from a five-country project. In: Kleinknecht, A., Bain, D. (Eds.), New Concepts in Innovation Output Measurement. Macmillan, London, pp. 153–188.

Kleinknecht, A. (Ed.), 1996. Determinants of Innovation: The Message from New Indicators. Macmillan, London.

Levin, R.C., Klevorick, A.K., Nelson, R.H., Winter, S., 1987. Appropriating the returns from industrial research and development. Brookings Papers on Economic Activity 3, 783–831.

Mansfield, E., 1986. Patents and innovation: an empirical study. Management Science 32, 173–181.

Mansfield, E., Schwartz, M., Wagner, S., 1981. Imitation costs and patents: an empirical study. Economic Journal 91, 907–918.

Mullahy, J., 1986. Specification and testing of some modified count data models. Journal of Econometrics 33, 341–365.

Pavitt, K., 1984. Sectoral patterns of technological change. Towards a taxonomy and a theory. Research Policy 13, 343–373.

Pavitt, K., Patel, P., 1988. The international distribution and determinants of technological activities. Oxford Review of Economic Policy 4, 35–55.

Scherer, F.M., 1965. Firm size, market structure, opportunity, and the output of patented inventions. American Economic Review, pp. 1097–1125.

Scherer, F.M., 1983. The propensity to patent. International Journal of Industrial Organization 1, 107–128.

[8]

Profiles of State Technological Transfer Structure and Its Impact on Small Manufacturers

G. Bruce Hartmann
John Masten

ABSTRACT. This paper presents the results of an empirical examination of the methods 10 state agencies use in coordinating technology assistance to small manufacturers. The strategies that agencies use to transfer technology are evaluated in light of 15 economic variables. Where appropriate, regression analysis was used to measure the impact of these variables on the growth rate of small manufacturers. State assistance strategies were determined by mailing a questionnaire to the chief technology transfer officer in each of the states. Survey participants were asked to discuss the mission, goals, degree of goal accomplishment, and size of budget/staff. The study compares their responses and discusses how these agencies might improve the transfer process.

Five economic variables were found to be significantly related to the number of small manufacturers in a state. However when growth rates of the significant variables and small manufacturers were compared the results were not as clear. The states with the fastest growth rates of small manufacturers did focus their technical assistance on small firms. These states tended to place a priority on providing technical assistance to small manufacturers.

1. Introduction

This paper presents the results of an empirical examination of the methods state agencies use in coordinating technological assistance to small manufacturers. Historically, improvements in technological capability have explained 25% of the annual increase in national income. In generating half of all jobs, small firms have contributed significantly to the increases in national income. The ability of small firms to continue to play an important role in the growth of national income is limited by the fact that they only account for 5% of R&D expenditures (Cohen and Noll, 1991; Senchack, 1981). To enhance employment growth and competitiveness, small manufacturers should be a major recipient of publicly supported technological assistance.

Department of Economics
School of Business
Tennessee State University
Nashville, TN 37203
E-mail: harpo.tnstate.edu

Clark and Dobson (1991) noted that neither firms, educational institutions, nor government agencies have responded as rapidly as their foreign counterparts to the technology needs of small manufacturers. Branscomb (1993) suggests that more emphasis needs to be placed on the successful transfer of new technology to industry. Barriers to technology transfer are mostly man made and it requires persistent, vigorous activities to overcome them. To successfully transfer a technology, champions are needed on both sides of the process. For instance, in the study done by Bailey et al. (1990), no champions and no problems were found from a survey of 44 manufacturing firms in Tennessee. When champions were appointed to work directly with the firms, over 60 technology related problems were discovered. The champions were publicly supported agencies and manufacturers who now sought to solve their technical problems.

The number and type of public assistance agencies varies considerably among the states. Some states have technical extension programs dating back to the fifties but these programs have generally been inadequate. More than 80% of the state/federal technical aid programs have been established since 1980. The state/federal technical aid programs usually are carried out by universities, community colleges, and federal agencies located within the state (Masten and Hartmann, 1993).

Improving the technology transfer process has the potential for substantially increasing produc-

Journal of Technology Transfer **25**: 83–88, 2000.

tivity in the manufacturing sector. But there is little awareness of the agencies in some states by small manufacturers (Masten and Hartmann, 1994). In a recent study by AT&T (1993) it was noted that only 4% of small and medium sized manufacturers in the United States are served by government technical assistance programs.

Little is known about how to effectively structure technological assistance to small manufacturers. Many of the technology transfer services provided by states to small manufacturers are similar. There are, however, variations in the level of service and the number of programs. These variations could be associated with a state's economic environment and the public assistance administrative structure. The Clark and Dobson survey (1991) found that the major services provided to manufacturers by the agencies were:

1. review of manufacturing methods and services;
2. productivity and quality assessments;
3. plant layout and operation;
4. acquisition advice;
5. implementing TQM;
6. locating information;
7. networking.

2. Methodology

The purpose of this study is to evaluate the impact of state technological assistance on the growth rate of small manufacturers. At the present time little is known about the effectiveness of state technology transfer administrative organizations and programs. The state has a role in promoting technology transfer to small manufacturers because some of the benefits from improved competitiveness spill-over to society. These spill over benefits include increased employment, productivity, and incomes. The method and intensity of state assistance to small businesses vary among states. The impact of the assistance is difficult to measure since there is wide variation in state economic conditions. In this paper variations in state economic conditions, assistance strategies, and the growth rate of small manufacturers are examined. It is hypothesized that both state economic environment and state assistance strategies affect the rate of growth of small manufacturing businesses.

Fifteen economic variables were identified as having a potential impact on the number of small manufacturers within a state. Stepwise regression analysis was used to determine their significance. Several were found to be significant. Then the states were ranked highest to lowest according to the rate of growth of small manufacturers. This ranking was compared to a similar ranking of the significant economic variables. The rankings were also compared to individual state technical assistance strategies. Information about state assistance strategies was obtained from a questionnaire.

Major state technical assistance strategies are direct assistance, coordinating, and advocating. State assistance to small manufacturers includes decentralized and centralized approaches. A centralized approach refers to the state controlling and providing services to small manufacturers. A decentralized approach refers to the state having a coordinating role with existing service providers. Some state governments only coordinate, some provide direct assistance, while others provide only an advocacy role. Another strategy the state could pursue would be to enhance the overall economic environment. Activities in this strategy could include improving technical educational opportunities and other factors that are associated with a fast growing small business sector. Do these assistance strategies impact the small manufacturing environment? For example, would a centralized role by the state be more associated with a faster growing small manufacturing sector than a decentralized assistance role? Further, will a state that targets small manufacturers for technical assistance have a higher growth rate of small manufacturers. A questionnaire was used to determine state assistance strategies. The result from the questionnaire were compared to the growth rate of small manufacturers within the state between 1980 and 1990. The states selected represented the Southern and Midwestern districts of the Small Business Administration. The selection was predicated on possible logistical assistance in contacting agencies and small manufacturers from SBA. Furthermore, these states represented diverse approaches to technological assistance and economic environments. The survey instrument was designed to elicit the strategies of each state.

The survey instrument contained combinations of open ended, dichotomous, multiple response, rating, and ranking questions. It was sent to 20 states. Ten states responded. The instrument sought to discover the methods used to transfer technical knowledge. Survey participants were asked to discuss their mission, goals, degree of accomplishment, and size of budget/staff. They also estimated the time they spent in advocacy, direct client assistance, and networking. Finally, they listed the top five problems their agency faced in transferring technology to the small manufacturers. The last question was included in the survey to assist states in identifying areas in which they might improve their services to small manufacturers. This information was also used to determine if the state was satisfied with its current assistance strategy.

The growth rate of small manufacturers was examined in the 10 states between 1980 and 1990. Could the growth rate of small manufacturers be explained by the economic environment in each state? To answer this question, 15 variables were regressed against the number of small manufacturers in 1980 and again in 1990. Four of these variables proved to be significant in 1980 and 5 in 1990. The four significant 1980 variables were the number of Ph.D.'s granted, the number employed in manufacturing, percent of engineers in the population, and state and local expenditures on higher education. A priori reasoning would suggest that these variables would be associated with a strong manufacturing sector.

Table 1 lists the 10 states out of the original population of 20 that responded to the questionnaire and thus were included in the study. The states were ranked according the growth rate of small manufacturers between 1980 and 1990. Columns 2 and 3 reveal that there was a wide variation in the growth rate of small manufacturers and the year that the state technical assistance was established. The state rankings would be significantly different if the absolute number of small manufacturers were used in 1990 rather than the grow rate. The year the state office was established was included in the table because of the possibility of experience impacting the growth rate. This variable was used qualitatively and was not included as an economic variable.

Table 1
The growth of small manufacturers and the value of shipments

States ranked by the growth of small manufacturers	% Growth of small business 1980–1990	% Growth in the value of shipments 1980–1990	Year transfer office established
Kentucky	31	81	1990
N. Carolina	29	81	1955
Tennessee	29	65	1963
Indiana	28	56	1989
Alabama	26	64	1986
Georgia	24	75	1956
Arkansas	19	54	1984
Pennsylvania	11	33	1988
Illinois	8	39	1989
Louisiana	8	15	1991

3. Results

Stepwise regression analysis of statewide economic characteristics

Stepwise regression analysis was used to determine if the health of the small manufacturing sector was influenced by the economic environment. Obviously, as suggested from the first hypothesis, there should be a strong relationship. There was a wide variation in the number of small businesses among the states. The 15 selected independent environmental variables were stepwise regressed with the dependent variable, the number of small manufacturers, both in 1980 and 1990.

Stepwise regression is a statistical process in which the independent variables are selected one at a time. The selection process starts with the highest F value and continues selecting until a predetermined cut off F value is reached. The F statistic is the ratio of the explained to the unexplained variation in the dependent variable as each variable is selected. For instance, after the first independent variable is selected, each one of the remaining variables is added to the equation one at time and the variable that produces the highest F statistic is selected. The new equation is now run and a third independent variable is selected. This process is repeated until all variables that produce an F value above the cut off value are added to the equation. When there is a readily explained causation between the dependent variable and a number of independent variables, stepwise regression is an appropriate statistical technique to use to select the strongest relationships. The results of using stepwise regression for 1980 and 1990 are listed in Tables 3 and 4 at the end of this article.

The significant independent variables in 1980 were Total Employment, Manufacturing Employment, Percent of Engineers in the Population, and the Percent Urban Population. In 1990 only Manufacturing Employment continued to be significant; the new significant variables were Population, Ph.D.'s Granted, Percent Engineers in the Population, and Current Expenditures of State and Local Governments for Higher Education. The 98.6% association in 1980 and the 96.9% association in 1990 of the number of small manufacturers with the environmental variables suggests that states might concentrate resources on improving the environmental factors rather than providing or coordinating technical assistance. These relationships support the first hypothesis that the economic environment has a positive impact on the number of small manufacturers.

Table 2 ranks the states that responded to the questionnaire by the percentage growth in the number of small manufacturers between 1980 and 1990. Kentucky experienced the highest rate of growth in small manufacturers while Louisiana had the lowest. The rank order in the growth of the environmental variables is also listed in the table. The correlation between the ranking of the growth rate of the independent variables and the growth rate of small manufacturers ranged between 3% and 57%. The low correlation between growth rates leaves open the possibility that other factors within the state are affecting the growth of small manufacturers. State technological assistance could be one of the other influential variables. Kentucky had the highest growth in the number of small businesses but the relative ranking of the independent variables ranged from 3 to 9. Arkansas 7 on the growth list but had a variation in the ranking of independent variables from 3 to 10. Louisiana, last on the growth list, had rankings from 1 to 10.

Table 3
Stepwise regression: 1980

Step	1	2	3	4
Constant	−966.2	−765.7	−1152.9	−4392.7
Totem	0.00347	0.00234	0.00282	0.00205
t Ratio	19.7	7.36	7.76	4.08
Mfgem		0.0041	0.0049	0.0069
t Ratio		3.88	4.75	5.06
Pengpop			−1.19	−1.53
t Ratio			−2.16	−2.88
Popurban				67
t Ratio				2.05
R Squared	95.57	97.65	98.18	98.58

Table 2
State ranking of significant economic variables and growth rates from 1980 to 1990

States	Engineers employed	Ph.D.'s granted	Population	Manufact. employ.	Expend. high ed.
Kentucky	9	5	7	6	9
N.C.	3	2	2	4	2
Tn.	4	8	3	2	1
Indiana	5	7	6	7	4
Alabama	1	6	4	5	8
Georgia	2	4	1	1	6
Arkansas	8	10	5	3	3
Penn.	7	3	9	10	10
Illinois	6	9	10	9	5
La.	10	1	8	8	7

Table 4
Stepwise regression: 1990

Step	1	2	3	4	5
Constant	−655.61	15.89	1072.59	1497.83	4594.96
Mfgem	0.0158	0.0216	0.0169	0.0158	0.0169
T Ratio	10.28	8.93	7.60	7.65	8.64
Pop(90)		−0.00064	−0.0011	−0.00131	−0.00152
T Ratio		−2.85	−5.23	−6.18	−6.96
Pgranted			5.0	4.4	4.7
T Ratio			3.72	3.5	4.07
Pengpop				1.58	1.52
T Ratio				2.19	2.32
Exphigh					10.1
T Ratio					−2.03
R-Sq	85.45	90.15	94.72	96.00	96.91

4. Analysis of questionnaire responses

Mission and goals

The states were asked to list and rank the five most important long term goals, the five most important short term goals, and the degree of goal accomplishment. The most comprehensive and specific long term goals were given by Kentucky, the state with the highest growth rate. Kentucky's goals concentrated on coordinating technological assistance to manufacturers. Louisiana had just started their agency and had not yet established goals. Several states did list as goals the support of some of the environmental factors. Arkansas wanted to concentrate funding on technical higher education to increase the number of scientists and engineers in the workplace. Other states mentioned policy analysis and other more general goals. Most of the respondents had 3 or fewer goals.

Some of the short term goals included increasing the national competitiveness of their scientists in basic research, providing assistance to a specific number of manufacturers per year, and establishing a technical data base. Arkansas, Pennsylvania, and Tennessee had the most comprehensive listing of short term goals. Arkansas, Georgia, Indiana, and Tennessee were able to rank the degree of short term goal accomplishment. None of the states was completely satisfied with their level of goal completion. Indiana and Georgia were the most satisfied. Interestingly, the goals did not include the role of technology transfer in attracting new manufacturing to the state. There was no particular relationship between goals and the growth of small manufacturers.

Agency characteristics

The agencies had a wide variation in the number of professional and nonprofessional staff. Tennessee had only one professional and one nonprofessional staff member. This is in contrast to Indiana which employed 38 professional and 20 nonprofessional staff. The number of professional staff employed at the state level was related to the degree of direct assistance provided by the state. Those states with larger central offices tended to be more active in securing federal funding to provide services at the state level.

Kentucky's response indicated that they had the greatest influence over other assistance agencies. Most of the respondents indicated that they had little influence over the other agencies. Annual budgets from state sources ranged from $626,000 to $25,000,000. Louisiana had the smallest budget and Pennsylvania had the largest.

Most of the states received all of their funds from state sources. Except for Pennsylvania, the states with the largest budgets and staff spent most of their time in direct assistance. The largest percentage of time for most states was spent in direct assistance, networking was second, and advocacy was third. Kentucky spent only 45% of their time in direct assistance and 50% of their time in networking. Arkansas, Louisiana, and Indiana spent the highest percent of time in direct assistance. Half of the states provided direct assistance to small manufacturers. Except for Tennessee the states provided direct assistance to other state technological assistance agencies.

Most of the states indicated a high level of assistance in the area of networking. Except for Arkansas, the states were satisfied with the level of coordination among agencies. Most states gave a score of 3 on a 5 point scale indicating improvement in coordination among the numerous agencies was possible.

Top five barriers

The states were asked to rank and list the top five problems they faced in improving technology transfer to small manufacturers. Three states listed the number one barrier as lack of resources. Two states indicated that small manufacturers were not aware of the services available to them. Others included identification of resource providers and lack of awareness by small manufacturers of the available services. One state indicated that small businesses are usually the last to use the latest technology and possibly the least willing to try new innovations. Ironically, none of the states listed long or short term objectives that specifically addressed the problem of awareness.

The relationship between state assistance and the growth of small manufacturers is not clear. The size of office, method of transfer, and size of budget were not directly related to growth. Those states with the fastest growth rates did focus at the state level on technical assistance to small manufacturers.

5. Conclusion

Most of the state technology assistance offices have been established since 1980. These offices vary in size and mission. Except for a few, focused assistance to small manufacturers is not a priority. The selected economic variables were highly related to the number of small manufacturers in the state. However, the states with the highest growth rates in the number of small manufacturers during the 1980s did not necessarily have the highest growth rate in engineers employed, Ph.D.s granted, manufacturing employment, population, and expenditures on higher education. The states with the fastest growth rates did focus at the state level on small manufacturing technical assistance needs. The four states with the lowest growth rates all started offices in the 1980s, whereas only two of the fastest growth states established offices during the 1980s.

The implications of this study are:

1. There is no consistent relationship between the growth rate of small manufacturers and traditional economic variables;
2. A state focus on assisting small manufacturers in the technology transfer process probably will increase their growth rate;
3. The selection of a strategy for providing technology assistance is state specific;
4. The state's goals in assisting small manufacturers in the technology transfer process should be a part of their economic development plan.

More research needs to be conducted on the impact of state technological assistance on the health and expansion of small manufacturers. Based on recent growth rates, it is expected that small manufacturers will have an increasing impact on total state manufacturing employment. Technology will play a leading role in determining the health of small manufacturing. Maintaining a leading technological edge will require increased emphasis on technology transfer from public agencies to small manufacturers.

References

AT&T, 1993, *AT&T Small Business Study*, AT&T.

Bailey, T., J. Masten and G.B. Hartmann, 1992, 'The Effectiveness of Publically Supported Technology Transfer in the United States: A Case Study', *Proceedings, International Council of Small Business*.

Branscomb, L.M., 1993, *Empowering Technology*, MIT Press.

Clark, M.K. and E.N. Dodson, 1991, *Increasing Technology Transfer: The Competitiveness of American Manufacturers and State Industrial Extension Programs*, National Governor's Association.

Cohen, L. and R. Noll, *Small Business in the Year 2000*, Washington, D.C., Office of Advocacy, Small Business Administration.

Masten, J. and G.B. Hartmann, 1993, 'The Utilization of Technology Assistance by Small Manufacturers', *Journal of Technology Transfer* **18** (1, 2).

Masten J., G.B. Hartmann and A. Safari, 'Small Business Strategic Planning and Technology Transfer: The Use of Publicly Supported Technology Assistance Agencies', *Journal of Small Business* **33** (3).

Senchack, A.J., 1981, 'Characteristics of Small Manufacturing Firms' R&D Activities', *Journal of Small Business Management*.

[9]

ELSEVIER

International Journal of Industrial Organization
19 (2001) 763–794

International Journal of Industrial Organization

www.elsevier.com/locate/econbase

Public/private partnerships: stimulating competition in a dynamic market

Albert N. Link[a,*], John T. Scott[b]

[a]Department of Economics, University of North Carolina at Greensboro, Greensboro, NC 27412, USA

[b]Department of Economics, Dartmouth College, Hanover, NH 03755, USA

Abstract

This paper sets forth a public/private partnership competition policy that mitigates the appropriability problems associated with innovation that occur in a dynamic market when competitive pressures are present. We illustrate the applicability of our policy proposal using the results from an analysis of research projects jointly funded by the Advanced Technology Program (ATP) and the private partners. Our analysis illustrates that in the absence of ATP funding these projects would not have been undertaken by private-sector firms, and that the social rate of return from the projects is substantial. We also posit a mechanism whereby ATP, or any public agency, can partner with industry to ensure that its public funds are being efficiently allocated.

Keywords: Public/private partnership; Advanced Technology Program; R&D; Technology policy

JEL classification: O31; D62

1. Introduction

An effective competition policy in a dynamic market encourages innovative behavior, given the knowledge that innovation encourages technological advance and technological advance stimulates economic growth and the competitiveness of firms. However, technology-based competition erodes appropriability and in-

*Corresponding author. Tel.: +1-336-334-5146; fax: +1-336-334-4089.

E-mail addresses: al_link@uncg.edu (A.N. Link), john.t.scott@dartmouth.edu (J.T. Scott).

0167-7187/01/$ – see front matter
PII: S0167-7187(00)00093-X

creases risk. Both of these factors can work against the social goals of an effective competition policy because large firms in concentrated markets may be the outcome of a market's adjustments to a regime of rapid innovation. For example, if numerous firms compete through investments in research and development (R&D), each may anticipate many competing innovations in a post-innovation market or each may anticipate a low probability that it will itself innovate. In either case, anticipated appropriable profits from the R&D investments may be too small, and the risk that the post-innovation market profits for the firm would fall below its required rate of return may be too large, to allow socially optimal innovative investments in such competitive markets. Schumpeter (1950) reasoned that the large firms that dominate industrial markets — markets with resources concentrated in the control of a few leading firms — would be able to use the profits from their market power in pre-innovation markets to provide internal finance and insurance funds to cover the risks of innovative investments.

The tradeoff between multiple independent firms competing and the benefits of coordination to reduce risk and appropriability problems is not new to the arena of competition policy, and there are at least two policy responses. One policy approach to deal with the tradeoff was the passage of the National Cooperative Research Act (NCRA) of 1984 (PL-98-462). This Act culminated a five-year effort to ease the antitrust treatment of collaborative research by creating a registration process, later expanded by the National Cooperative Research and Production Act (NCRPA) of 1993 (PL-103-42), under which research joint ventures (RJVs) can disclose the names of their members and their research intentions to the U.S. Department of Justice. RJVs gain two significant benefits from such voluntary filing: if subjected to antitrust action, they are evaluated under a rule of reason that considers a venture's effects on social welfare; and if found to fail a rule-of-reason analysis, they are subject to actual rather than treble damages.[1] However, in certain circumstances cooperative R&D may result in the loss of socially desirable competitive pressures (Scott, 1993). Competitive pressures may be desirable because they would provide socially useful parallel paths for research or desirable diversity, but if such pressures are incompatible with sufficient appropriability of returns, then a second policy might prove useful to allow desirable innovative investment at the same time that competitive pressure exists.

A second policy approach to deal with the tradeoff is the one posited and evaluated in this paper. This policy mechanism involves partial public funding of privately-performed research. The objective of such 'public/private partnership competition policy' is to mitigate the appropriability and risk problems that would inevitably occur in a dynamic market when substantial competitive pressures are present. Such a policy would not only complement the existing RJV-related

[1] Filing with the Department of Justice is distinct from the decision of whether to form an RJV or not. For a review of analyses of the formation decision see Hagedoorn et al. (2000).

policies, but it would also specifically benefit small and medium-sized enterprises (SMEs). A large body of literature suggests that SMEs face significant finance constraints.[2] In particular, SMEs are most likely to face imperfect financial markets where transaction costs preclude access to impacted information. Thus, given the market structures now encouraged by antitrust and competition policy, SMEs may require public funding to achieve socially desirable research.

There is a tradeoff between SMEs, that bring many independent sources of innovative ideas, and large, dominant firms with market power. Large firms with market power may be able to realize advantages of firm size, and their market power may stimulate R&D investment as well. In fact, Schumpeter (1950, p. 87) emphasized that market power might be necessary to provide internal funds for R&D investments:

> *We must ... recognize the ... fact that restrictive practices [monopoly restriction of output to raise price] ... acquire a new significance in the perennial gale of creative destruction, a significance which they would not have in a stationary state or in a state of slow and balanced growth. In either of these cases restrictive strategy would produce no result other than an increase in profits at the expense of buyers except that, in the case of balanced advance, it might still prove to be the easiest and most effective way of collecting the means by which to finance additional investment. But in the process of creative destruction, restrictive practices may do much to steady the ship and to alleviate temporary difficulties.*

There are typically different ways of creating innovative solutions to problems. Given the assumption that numerous firms, perhaps to avoid rent-destroying competition in post-innovation markets, independently pursue different solutions (Scott, 1991; Cohen and Klepper, 1992), society would like competition and independent sources of initiative to survive in the market. Under such conditions, however, it may be difficult to finance R&D. In fact, because of difficulties appropriating sufficient returns, it may be difficult to even justify investing in R&D. The policy we set forth overcomes the appropriability hurdle; it allows for more competitive market structures and independent sources of innovative initiative (with associated appropriability problems as numerous competitors can benefit from the innovative investments of any particular firm), while still generating sufficient R&D investments.

In Section 2, we discuss market failure and the concomitant problem of underinvestment in R&D. In Section 3, we describe the illustrative case analysis for this paper, namely a set of privately-performed projects that received partial

[2] See Lerner (1996); Hall (1992); Hao and Jaffe (1993); Himmelberg and Petersen (1994) and Hubbard (1998).

public funding from the Advanced Technology Program (ATP) at the National Institute of Standards and Technology (NIST). In Section 4, we set forth a conceptual model for estimating the expected private rate of return and the expected social rate of return associated with these research projects, and we implement our model using these ATP project data. Our analysis in this section quantifies the size of the gap between the social and private rate of return. Section 5 concludes the paper with a recapitulation of our quantitative estimates, and then draws on the implications of those estimates to generalize about how policy makers might proceed to formulate a public/private partnership competition policy to effectively stimulate innovative behavior in the private sector. We observe that one implication of the gap between the social and private rate of return (private rate of return with and without public support) is that 'financial engineering' could be used to leverage the public funds that stimulate innovative investment.[3] We posit that if a bidding mechanism were implemented, following Scott (1998), there would be net social cost savings.

2. Market failure and underinvestments in R&D

The purpose of this section is threefold: to provide an overview of the economic concept of market failure, to provide interview information to demonstrate that the projects analyzed in Section 3 would not have been undertaken in the absence of public ATP support, and to describe the elements of market failure that would have brought about such an underinvestment in R&D.

2.1. Underinvestments in R&D

Many point to President Bush's 1990 *U.S. Technology Policy* as the Nation's first formal domestic technology policy statement. Albeit an important initial policy effort, it however failed to articulate a foundation for government's role in technology. Rather, it implicitly assumed that government had a role, and then set forth the general statement (1990, p. 2):

> *The goal of U.S. technology policy is to make the best use of technology in achieving the national goals of improved quality of life for all Americans, continued economic growth, and national security.*

President Clinton took a major step forward from the 1990 policy statement in

[3] According to Scott (1998), financial engineering, a primarily European term, refers to the optimal amount and design of public funding of privately performed investments in technology and innovation carried out by public/private partnerships.

his *Economic Report of the President* (1994) by articulating first principles about why government should be involved in the technological process (1994, p. 191):

> *The goal of technology policy is not to substitute the government's judgment for that of private industry in deciding which potential 'winners' to back. Rather, the point is to correct market failure ...* [4]

Subsequent Executive Office policy statements have echoed this theme; *Science in the National Interest* (1994) and *Science and Technology: Shaping the Twenty-First Century* (1998) are the most recent such examples.

Market failure, as we address it in this paper and of the type which could specifically be termed "technological or innovation market failure," refers to a condition under which the market, including both the R&D-investing producers of a technology and the users of the technology, underinvests from society's standpoint in a particular technology. Such underinvestment occurs because conditions exist that prevent organizations from fully realizing or appropriating the benefits created by their investments. In our explanation of market failure and the reasons for market failure, we essentially reiterate and apply the seminal work of Arrow (1962) in which he identified three sources of market failure related to knowledge-based innovative activity — uncertainty, non-exclusivity, and public goods.

To explain, consider a marketable technology to be produced through an R&D process where conditions prevent full appropriation of benefits. Other firms will realize some of the profits from the innovation, and of course consumers will typically place a higher value on a product than the price paid for it. A firm will then calculate, because of such conditions, that the marginal benefits it can receive from a unit investment in such R&D will be less than could be earned in the absence of the conditions reducing the appropriated benefits of R&D below their potential, namely the full social benefits. Then, the firm may underinvest in R&D relative to what it would have chosen as its investment in the absence of the conditions. Stated alternatively, the firm may determine that its private rate of return is less than its private hurdle rate and therefore will not undertake socially valuable R&D.

2.2. Barriers to technology

Risk and difficulties appropriating returns create barriers to technology, and as a result, there may be an underinvestment in or underutilization of a technology. The premise that markets may fail to undertake socially optimal amounts of R&D has

[4] The conceptual importance of identifying market failure for policy is also emphasized, although without any operational guidance, in Office of Management and Budget (1996).

long been accepted by economists and is now being invoked by policy makers, as the quoted passage from President Clinton above makes clear.[5] Much of the technological market failure literature focuses on underinvestments in the creation or production of technology through R&D. However, the arguments set forth below are generalizable to the purchase and utilization of the technology that results from R&D.

As a starting point to discuss barriers to technology, the concept of risk must be defined. In its most general form, risk measures the probability that actual outcomes will deviate from the expected outcome. So defined, risk can be characterized in terms of the variance of the distribution of possible outcomes centered around the expected outcome.

Our definition of risk, for the purpose of this paper and for the purpose of proffering a public/private partnership competition policy, follows from the following general statement. We use a definition of risk that is focused on the operational concern with the downside outcomes for an investment. The shortfall of the private expected outcomes from society's expected returns reflects appropriability problems. As Arrow (1962) explained, investments in knowledge entail uncertainty of two types — technical and market. The technical and market results from technology may be very poor, or perhaps considerably better than the expected outcome. Thus, a firm is justifiably concerned about the risk that its R&D investment will fail, technically or for any other reason. Or, if technically successful, the R&D investment output may not pass the market test for profitability. Further, the firm's private expected return typically falls short of the expected social return as previously discussed. We elaborate on this concept of downside risk in Section 4 below.

There are several related technological and market factors that will cause private firms to appropriate less return and to face greater risk than society does. These factors underlie what Arrow (1962) identifies as the non-exclusivity and public good characteristics of investments in the creation of knowledge. The private firms' incomplete appropriation of social returns in the context of technical and market risk can make risk in its operational sense unacceptably large for the private firm considering an investment. Operationally, Tassey (1997), for example, defines risk as the probability that a project's rate of return falls below a required, private rate of return or private hurdle rate (as opposed to simply deviating from an expected return). As we illustrate below (both in concept and in terms of the specific ATP-funded projects), for many socially desirable investments, the private firm faces an unacceptably large probability of a rate of return that falls short of its private hurdle rate. Yet, from society's perspective, the probability of a rate of

[5] There is an excellent theoretical and empirical literature that concludes that the private sector will underinvest in R&D because of market failures. A survey of that literature is beyond the scope of this report, but one recent review is in Martin and Scott (1998, 2000).

return that is less than the social hurdle rate is sufficiently small that the project is still worthwhile.

There are a number of factors that can explain why a firm will perceive that its expected private rate of return will fall below its hurdle rate. Individuals will differ not only about a listing of such factors because they are not generally mutually exclusive, but also they will differ about the relative importance of one factor compared to another in whatever taxonomy is agreed upon.

First, high technical risk (that is, outcomes may not be technically sufficient to meet needs) may cause market failure given that when the firm is successful, the private returns fall short of the social returns. The risk of the activity being undertaken is greater than the firm can accept, although if successful there would be very large benefits to society as a whole. Society would like the investment to be made, but from the perspective of the firm, the present value of expected returns is less than the investment cost and is thus less than the amount yielding its acceptable return on investment.

Second, high risk can relate to high commercial or market risk (although technically sufficient, the market may not accept the innovation — reasons can include factors listed subsequently such as imitation or competing substitutes or interoperability issues) as well as to technical risk when the requisite R&D is highly capital intensive. The project may require too much capital for any one firm to feel comfortable with the outlay. The minimum cost of conducting research is thus viewed as excessive relative to the firm's overall R&D budget, which considers the costs of outside financing and the risks of bankruptcy. In this case, the firm will not make the investment, although society would be better off if it had, because the project does not appear to be profitable from the firm's private perspective.

Third, many R&D projects are characterized by a lengthy time interval until a commercial product reaches the market. The time expected to complete the R&D and the time until commercialization of the R&D results are long, and the realization of a cash flow from the R&D investment is in the distant future. If a private firm faces greater risk than society does, and as a result requires a greater rate of return and hence applies a higher discount rate than society does, it will value future returns less than does society. Because the private discount rate exceeds the social discount rate, there may be underinvestment, and the underinvestment increases as the time to market increases because the difference in the rate is compounded and has a bigger effect on returns further into the future.

Fourth, it is not uncommon for the scope of potential markets to be broader than the scope of the individual firm's market strategies so the firm will not perceive or project economic benefits from all potential market applications of the technology. As such, the firm will consider in its investment decisions only those returns that it can appropriate within the boundaries of its market strategies. While the firm may recognize that there are spillover benefits to other markets, and while it could possibly appropriate them, such benefits are ignored or discounted heavily relative to the discount weight that would apply to society. A similar situation arises when the requirements for conducting R&D demand multidisciplinary research teams; unique research facilities not generally available with individual companies; or 'fusing' technologies from heretofore separate, non-interacting parties. The possibility for opportunistic behavior in such thin markets may make it impossible, at reasonable cost, for a single firm to share capital assets even if there were not R&D information sharing difficulties to compound the problem. If society, perhaps through a technology-based public institution, could act as an honest broker to coordinate a cooperative multifirm effort, then the social costs of the multidisciplinary research might be less than the market costs.[6]

Fifth, the evolving nature of markets requires investments in combinations of technologies that, if they existed, would reside in different industries that are not integrated. Because such conditions often transcend the R&D strategy of firms, such investments are not likely to be pursued. That is not only because of the lack of recognition of possible benefit areas or the perceived inability to appropriate whatever results, but also because coordinating multiple players in a timely and efficient manner is cumbersome and costly. Again, as with the multidisciplinary research teams, society may be able to use a technology-based public institution to act as an honest broker and reduce costs below those that the market would face.

Sixth, a situation can exist when the nature of the technology is such that it is difficult to assign intellectual property rights. Knowledge and ideas developed by a firm that invests in technology may spill over to other firms during the R&D phase or after the new technology is introduced into the market. If the information creates value for the firms that benefit from the spillovers, then other things being equal, the innovating firms may underinvest in the technology. Relatedly, when competition in the development of new technology is very intense, each firm, knowing that the probability of being the successful innovator is low, may not anticipate sufficient returns to cover costs. Further, even if the firm innovates, intense competition in application can result because of competing substitute goods, whether patented or not. Especially when the cost of imitation is low, an individual firm will anticipate such competition and may therefore not anticipate returns sufficient to cover the R&D investment costs. Of course, difficulties appropriating returns need not always inhibit R&D investment (Baldwin and Scott, 1987). First-mover advantages associated with customer acceptance and demand as well as increasing returns as markets are penetrated and production expanded can imply that an innovator wins most of the rewards even if it does not 'take all.'

Seventh, industry structure may raise the cost of market entry for applications of

[6] See Leyden and Link (1999) on the role of a federal laboratory as an honest broker.

the technology. The broader market environment in which a new technology will be sold can significantly reduce incentives to invest in its development and commercialization because of what some scholars have called technological lock-in and path dependency.[7] Many technology-based products are part of larger systems of products. Under such industry structures, if a firm is contemplating investing in the development of a new product but perceives a risk that the product, even if technically successful, will not interface with other products in the system, the additional cost of attaining compatibility or interoperability may reduce the expected rate of return to the point that the project is not undertaken. Similarly, multiple sub-markets may evolve, each with its own interface requirements, thereby preventing economies of scale or network externalities from being realized. Again, society, perhaps through a technology-based public institution, may be able to help the market's participants coordinate successful compatibility and interoperability.

Eighth, situations exist where the complexity of a technology makes agreement with respect to product performance between buyer and seller costly. Sharing of the information needed for the exchange and development of technology can render the needed transactions between independent firms in the market prohibitively costly if the incentives for opportunistic behavior are to be reduced to a reasonable level with what Teece (1980) calls obligational contracts. Teece emphasizes that the successful transfer of technology from one firm to another often requires careful teamwork with purposeful interactions between the seller and the buyer of the technology. In such circumstances, both the seller of the technology and the buyer of the technology are exposed to hazards of opportunism. Sellers, for example, may fear that buyers will capture the know-how too cheaply or use it in unexpected ways. Buyers may worry that the sellers will fail to provide the necessary support to make the technology work in the new environment; or they may worry that after learning about the buyer's operations in sufficient detail to transfer the technology successfully, the seller would back away from the transfer and instead enter the buyer's industry as a technologically sophisticated competitor. Once again, if society can use a technology-based public institution to act as an honest broker, the social costs of sharing technology may be less than market costs.

These eight factors that create, individually or in combination, barriers to technology and thus lead to a private underinvestment in R&D are listed in Table 1. While we have discussed these factors individually above, and have listed them

[7] See David (1987) for detailed development of the ideas of path dependency in the context of business strategies and public policy toward innovation and diffusion of new technologies.

Table 1
Factors creating barriers to technology in the TIMA projects

1. High technical risk associated with the underlying R&D	2
2. High capital costs to undertake the underlying R&D	6
3. Long time to complete the R&D and commercialize the resulting technology	1
4. Underlying R&D spills over to multiple markets and is not appropriable	7
5. Market success of the technology depends on technologies in different industries	2
6. Property rights cannot be assigned to the underlying R&D	5
7. Resulting technology must be compatible and interoperable with other technologies	7
8. High risk of opportunistic behavior when sharing information about the technology	0

in the table as if they are discrete phenomena, they are interrelated and overlapping, although in principle any one factor could be sufficient to cause a private firm to underinvest in R&D (see Tassey, 1997).

3. Overcoming market failure through public/private partnerships

3.1. The Advanced Technology Program

The Advanced Technology Program (ATP) was established within the National Institute of Standards and Technology (NIST) through the Omnibus Trade and Competitiveness Act of 1988, and modified by the American Technology Preeminence Act of 1991. The goals of the ATP, as stated in its enabling legislation, are to assist U.S. businesses in creating and applying the generic technology and results necessary to "[c]ommercialize significant new scientific discoveries and technologies rapidly, and refine manufacturing technologies."[8] More specifically:[9]

> *The goal of the ATP is to benefit the U.S. economy by cost-sharing research with industry to foster new, innovative technologies. The ATP invests in risky, challenging technologies that have the potential for a big pay-off for the nation's economy. These technologies create opportunities for new, world-class products, services and industrial processes, benefiting not just the ATP participants, but other companies and industries and ultimately consumers and taxpayers as well. By reducing the early-stage R&D risks for individual*

[8] The term 'generic technology' does not have a generally accepted definition. It is not a National Science Foundation reporting category of R&D spending (Link, 1996). Tassey (1992, pp. 98–99) offers the following definition: "generic technology research is a major step in the sequential evolution of a typical industrial technology. It is the organization of scientific principles into a *functional technical concept*."

[9] http://www.atp.nist.gov/atp/imp_fact.htm.

companies, the ATP enables industry to pursue promising technologies which otherwise would be ignored or developed too slowly to compete in rapidly changing world markets.

ATP received its first appropriation from Congress in FY 1990, and held its first general competition in that same year. Since 1994, ATP has sponsored a number of focused program competitions in addition to its general competitions. According to ATP:[10]

[Focused programs are] multi-year efforts aimed at specific, well-defined technology and business goals. These programs, which involve the parallel development of a suite of interlocking R&D projects, tackle major technology problems with high payoff potential which cannot be solved by an occasional project coming through the general competition.

One such focused program is the Technologies for the Integration of Manufacturing Applications (TIMA) Program. There was a TIMA competition in 1995 from which four research projects were selected to receive ATP support, and in 1997 there was a second TIMA competition from which six research projects were funded.

3.2. The TIMA focused research program

The overall goal of the TIMA focused program is to develop and demonstrate the technologies needed to create affordable, integrable manufacturing systems. Many manufacturing companies need to respond more rapidly to changing markets and evolving opportunities if they are to remain competitive in global markets. Although this need is widely recognized, manufacturers find it difficult to implement the technologies needed for them to become more agile producers. Even highly automated plants and factories struggle to adapt successfully and efficiently and reconfigure production processes to accommodate design changes and new product lines. Customized systems integration efforts are often needed to achieve such changes, but they are not undertaken primarily because of idiosyncrasies in manufacturing software and incompatibilities among software applications.

Typically, factory-floor information systems focus on the operation of production equipment and the control of processes. The systems communicate neither directly nor regularly with administrative information systems, or with design and engineering systems. As a result, upstream information systems are unaware of

[10] http://www.atp.nist.gov/atp/focusprg.htm.

important manufacturing details. Middle-level information systems, known as manufacturing execution systems (MES), bridge this critical information gap.

MES solutions, complex and burdensome as they may be, can be solved by contracting with a large systems provider or integrator. However, once a manufacturer has incurred such a substantial investment it is likely to become dependent on that single vendor and thereby become unaware of, or if aware likely ignore, other vendors that may have more economical or innovative solutions. Because the initial solution involves a re-engineering of the manufacturer's business processes to be compatible with the vendor's requirements, even large manufacturers that can afford the up-front investment cost will by-pass the use of MES technology.

TIMA technologies are expected to benefit a range of companies: companies that employ MES by providing them with a wider range of powerful, integrable applications that will dramatically improve the manufacturer's ability to reconfigure, scale, and adapt their processes; small- and medium-sized manufacturers by making MES more affordable and by providing a direct path toward greater automation through incremental addition of compatible applications; and vendors of MES products by expanding the market, lowering barriers to entry, stimulating innovation, and technical specialization. Consumers may benefit from the adoption of these technologies in at least two ways, a higher quality product and a lower priced product to the extent that greater automation increases competition.

ATP identified a contact person for each of the ten ATP-funded TIMA projects.[11] We interviewed seven individuals corresponding to eight of the ten funded projects.[12] Each of these individuals was asked: *In the absence of ATP funding, would this research have been undertaken?* Three of the eight projects are single participant projects and five are joint ventures. All five of the research joint-venture respondents answered that no, in the absence of ATP funding, the joint venture would not have been formed and that in the absence of the joint venture their companies would not have undertaken the research. For the three

[11] Detailed project descriptions are available from the authors upon request. The ten projects, along with the sponsoring companies, are: Model-Driven Application & Integration Components for MES, sponsored by Vitria Technology, Inc.; An Agent-Based Framework for Integrated Intelligent Planning–Execution, sponsored by IBM Manufacturing Solutions Unit; Advanced Process Control Framework Initiative, sponsored by Honeywell Technology Center; Solutions for MES-Adaptable Replicable Technology (SMART), sponsored by National Industrial Information Infrastructure Protocols Consortium; Virtual Reality Telecollaborative Integrated Manufacturing Environment (VRTIME), sponsored by Searle; Process Integration Using Model-Driven Engines, sponsored by Vitria Technology, Inc.; Agent-Enhanced Manufacturing System Initiative, sponsored by Advanced Micro Devices; EECOMS: Extended Enterprise Coalition for Integrated Collaborative Manufacturing Systems, sponsored by IBM Corporation, CIIMPLEX; Distributed Factory System Framework; sponsored by Consilium, Inc.; ANTS Scheduling and Execution System, sponsored by Denab Robotics, Inc.

[12] One respondent was the contact person for two projects. Vitria Technology, Inc. chose not to participate in this study without explanation. Vitria Technology is involved in two research projects.

single participant projects, one respondent reported that the research would not have been undertaken without ATP funding. One reported that maybe the research would have been undertaken but it would have been at a reduced level. One reported that maybe the research would have been undertaken but it would have been at a slower pace.

Respondents were asked to describe their understanding of the characteristics of technical risk (the project being technically successful) and market risk (the project being commercially successful) associated with the research project, and how that risk affected the fact that their company could not have undertaken the research at all or at the same level, scope, and speed in the absence of ATP funding. Of course, risks inherent in the project are viewed in the context of the respondent's firm, and we cannot be certain to what degree the perceived risk reflects risk inherent in the project and to what degree it reflects the capabilities of the firm. Table 1 is used as a summary device for generalizing from the responses given during the interviews and tallying the responses indicating the various barriers. Again, these barriers to technology are listed as if they are discrete phenomenon, when in reality they are interrelated. More than eight responses are recorded in Table 1; there were numerous occasions where we inferred from the respondent's discussion that there was more than one barrier to technology, that is, more than one reason for their underinvestment in R&D.

The tally in Table 1 captures our interpretation of the discussions with the project leaders of the TIMA projects. As such, the tally may underestimate the magnitude of the importance of some of the barriers to technology that can lead to market failure. Given this caveat, at least two interesting patterns emerge from Table 1. One, the TIMA research projects are characterized by both technical and market risk. Technical risk is explicitly cited only twice, although it may be evident in the cases where there are concerns about the large capital cost needed to undertake the research and in the cases of concern about interoperability with other technologies. Market risk is primarily evident in the need for the resulting technology (generic technology to be applied to software development) to interface with users' information technology. The respondents discussed with us each aspect of risk, and they emphasized the interfacing issues and interoperability issues. We concluded those issues were the major sources of risk contributing to market failure.

4. Toward a public/private partnership competition policy

4.1. The conceptual model

Jaffe (1998, p. 18) argues that for ATP to be effective in achieving its statutory objectives, it "must try to determine which projects proposed to it will generate large spillovers ..."[13] His arguments that lead to this conclusion assume that ATP should only select those projects that would not be funded by the private sector in the absence of ATP funding, or if funded would be funded at a considerably lower level so that only partial results would have been realized and these results would have taken longer to occur. Jaffe points to a number of ATP-sponsored studies and an independent study by the General Accounting Office. These studies conclude that ATP grantees are of the opinion that their research would not have taken place in the absence of ATP funding or would have been funded at a considerably lower level and hence would have taken place only partially and would have taken considerably longer to complete. However, Jaffe's argument also sets general parameters for a broader public/private partnership competition policy, as we discuss in the concluding section of this paper.

Fig. 1 illustrates Jaffe's conclusion.[14] The social rate of return is measured on the vertical axis along with society's hurdle rate on investments in R&D; the private rate of return is measured on the horizontal axis along with the private hurdle rate on R&D. A 45-degree line (dashed) is imposed on the figure under the assumption that the social rate of return from an R&D investment will at least equal the private rate of return from that same investment. Two separate research

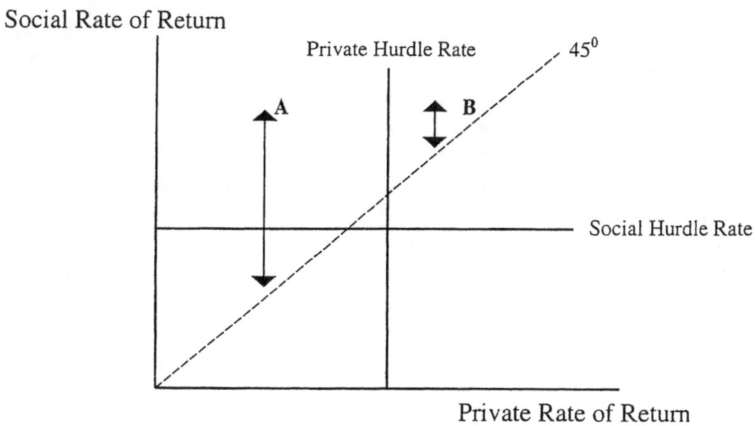

Fig. 1. Spillover gap between social and private rates of return to R&D.

[13] Jaffe does not quantify what a 'large' spillover is. Neither he nor we can define such a concept in the absence of a benchmark. Rather, we present below an estimate of the size of the spillover gap that characterizes the TIMA projects.

[14] See Link (1998, 1999) for a more detailed discussion of public/private partnerships and related public policy relative to this diagram.

projects are labeled Project A and Project B. Each, for our purposes, is shown with the same expected social rate of return.

For project A, the private rate of return is less than the private hurdle rate because of barriers to technology. As such, the private firm will not choose to invest in project A, although the social benefits from undertaking the project would be great. The vertical distance marked by project A is, in the Jaffe sense, the spillover gap; it results from the additional value society would receive above what the private firm would receive if project A were undertaken. Project A is precisely the type of project in which the public should invest, namely one in which the private sector would not invest because of market failure and one from which society would greatly benefit. In contrast, project B yields the same social rate of return as project A, but most of that return is capturable by the innovator, and the private rate of return is greater than the private hurdle rate. Hence, project B is one in which the private sector has an incentive to invest on its own or, alternatively stated, there is no economic justification for public funds being allocated to support project B.

Referring back to the interview information suggesting that the research projects would not have been undertaken or would have been undertaken at a reduced level or pace, we conclude that the TIMA research projects have similar characteristics to those of Project A in Fig. 1 in that each of the respondents views the expected private rate of return absent ATP funds to be less than his company's private hurdle rate. Hence, these TIMA projects are valid candidates for ATP support in the Jaffe sense, and in the broad sense these are the types of projects that a public/private partnership competition policy could affect.

Fig. 1 is the conceptual basis for understanding the result that public funding has on projects subject to market failure, as discussed above. And, Fig. 1 is useful for motivating the formulation of the competition policy that we set forth in Section V below. However, for completeness, it is also important as groundwork to illustrate what happens to the entire distribution of the rate of return as ATP (or any public agency) adds funds to a project like Project A in Fig. 1. This illustration emphasizes the concept of risk that underlies the above discussion of market failure.

ATP support of a private research project, whether it be a project with a single participant or a joint venture, reduces the risk that characterizes the project. In terms of Fig. 1, ATP's support shifts project A to the right so that the firm's private rate of return is then greater than its private hurdle rate.

In Fig. 2, we alternatively illustrate that reduction in risk in terms of a rightward shift in the distribution of the rate of return for the private firms. We illustrate the rate of return in Fig. 2 in order to illustrate our previous conceptualization of operational risk. The rightward shift of the distribution, and the concept of reducing the probability of returns lower than acceptable to the private investors, applies equally well to the absolute level of net return (absolute return minus private investment) expected from the project.

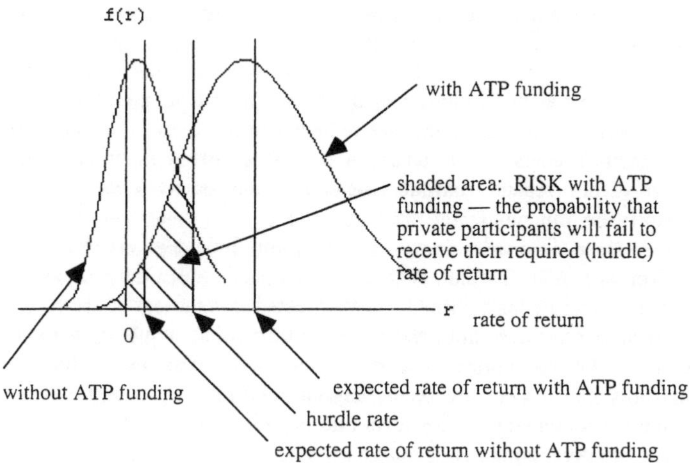

Fig. 2. Private risk reduction resulting from ATP funding.

For each distribution in Fig. 2 — without ATP funding (left distribution) and with ATP funding (right distribution) — the expected rate of return is shown.[15] As drawn, with ATP funding the expected private rate of return and the variance in the private rate of return from the research project will increase. One can generalize that this will always be the case.[16]

Consider the left distribution — the distribution of the rate of return for the private firm without ATP funding. As drawn, the private hurdle rate is to the right of the expected rate of return without ATP funding, meaning that the private firm will not undertake this research because the firm will not receive its required rate of return. The risk of the project equals the area under this without-ATP distribution that is to the left of the private hurdle rate. For those used to thinking of the variance of the distribution as the measure of risk, the downside risk —

[15] Note that the expected rate of return does not necessarily correspond to the greatest frequency or probability density because the distribution of rates of return need not be symmetric.

[16] The expected private rate of return with ATP support is: $r=$[return $-$ (total project cost $-$ ATP funding)]/[total project cost $-$ ATP funding]. Let $Z=$(total project cost $-$ ATP funding). Then, $r=$ (return $-Z)/Z=$[(return$/Z)-1$]. The variance of r is: $[(1/Z)^2\text{Var(return)}]$, and it is a general proposition that as ATP funding increases (and hence Z decreases) the variance in the private rate of return increases (since $(1/Z)$ gets larger). It is also a general proposition that the expected private rate of return $=E[(\text{return})/Z)-1]$ must increase for the same reason. Further, neither the expected social rate of return nor the variance in the social rate of return change at all. The social cost is the same and the social return is the same.

which is the probability of a rate of return less than the hurdle rate — may seem unusual. Variance measures the possibility that outcomes can differ from the expected outcome, while the downside risk measures the probability of an outcome departing to the downside of the hurdle rate. Note that the technical risk and the market risk for the project are reflected in the variance of the distribution — the technical goals may exceed or fall short of expectations and market acceptance of the project's technical outcomes may do the same. The downside risk refers to the outcomes that fall short of the hurdle rate.

Consider the right distribution — the distribution of the rate of return for the private firm with ATP funding. With ATP funding, the private firm will expect a return greater than its hurdle rate — the expected private rate of return with ATP funding is drawn to the right of the private hurdle rate.[17] While ATP funding will not increase the probability that the research will be successful, assuming hypothetically that it were undertaken absent ATP funding, it will reduce private risk by increasing the expected private rate of return because the expected rate of return will be based on a smaller private outlay.[18] Hence, ATP funding leverages the private firm's investment as illustrated by a greater expected return and a greater variance in the distribution as explained above.

The shaded area in Fig. 2 is what we call the downside risk of the project — that is, it is the probability that the project will yield a rate of return less than the private hurdle rate even with ATP funding. Hence, the amount of downside risk with ATP funding is visually less than the downside risk associated with the research project in the without-ATP funding case.

Although we will conclude that ATP funding reduces risk, as defined operationally in terms of reducing the probability of a rate of return below the private hurdle rate, we emphasize that our analysis below is in no way wed to any particular measure of risk or any particular model of capital asset pricing with associated systematic and non-systematic risk. Instead, our treatment encompasses any and all such models because the relevant risk, however it is perceived by private firms, is captured in the private hurdle rate, and the distributions of returns are otherwise represented by their expected values. In describing the effect of ATP funding on the distribution of private rates of return, we are describing an underlying reality that would be reflected in the private hurdle rate — as determined by some model — and in the expected value of the returns. Thus, Fig. 3 re-specifies Fig. 1 and shows specifically the implications of ATP funding reducing downside risk.

[17] ATP funding need not affect the firm's private hurdle rate; that rate is set by corporate policy in most cases. Conceivably, because the operational measure of risk falls, the hurdle rate might fall as well in the presence of ATP funding, and the simulative effect of ATP funding would hold a fortiori.

[18] To capture the idea of limited liability for investors, we bound that return below by zero. Thus, the rate of return can be quite negative when the return falls below the amount invested, but because the return is bounded below at zero, the rate of return is bounded below by (−100%).

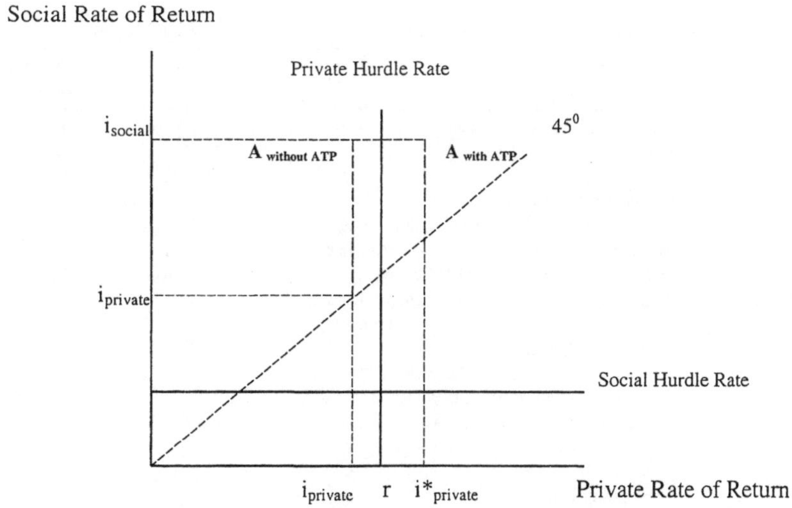

Fig. 3. Spillover gap between social and private rates of return to ATP-funded TIMA research.

4.2. Quantifying the conceptual model

Based on our extensive interviews with contact individuals in the TIMA projects, we collected quantitative information on the variables noted in the first two columns of Table 2.

We relied on these project-specific data to calculate the critical values shown above in Fig. 3. In particular, we calculate the expected social rate of return associated with the TIMA projects, i_{social}; the private rate of return without ATP funding, i_{private}; and the private rate of return with ATP funding, i^*_{private}. Given our understanding developed in the interviews along with the derived estimates for these critical values we conclude:

– The private rate of return to TIMA projects is less than the private hurdle rate, hence these projects would not have been pursued without ATP support, as initially inferred from interviews.

– There are spillover benefits to society associated with these TIMA projects as evidenced by the size of the spillover gap, $(i_{\text{social}} - i_{\text{private}})$.

For our calculation of the expected social rate of return from TIMA research projects, we calculate a lower bound for the expected social rate of return in two different ways. One, we calculate the expected social rate of return for each of the eight TIMA projects studied, and then we average across the rates; and two, we first average the eight project-specific data for each relevant variable, and then

Table 2
Descriptive statistics on values used in the estimation of the social rate of return ($n=8$)

Variable	Definition	Mean	Standard deviation
d	Duration of the ATP project	2.51 years	0.578
C	Total cost of the ATP project	\$12,749,125	1.16e+07
A	ATP funding	\$6,421,875	5.48e+06
r	Private hurdle rate	0.25	0.063
z	Duration of the extra period of development	1.03 years	0.619
F	Additional cost for the extra period of development	\$1,644,784	1.35e+06
T	Life of the commercialized technology	7.69 years	3.96
m	Proportion of producer surplus appropriated by the participants in their particular application of the generic technology	0.675	0.152
k	Proportion of total applications of the generic technology developed by the project addressed by the participants' applications	0.216	0.118
v	Proportion of producer surplus appropriated	0.135	0.071
L	Lower bound for expected annual private return to investing firm	\$5,250,439	5.59e+06
U	Upper bound for expected annual private return to the investing firm	\$9,885,615	1.05e+07

second estimate one expected social rate of return based on average data for each variable. We will use the latter calculation in our exposition, but the results of both methods are presented. Neither of these methods precisely characterizes the expected social rate of return to the TIMA research program, but each approximates what might be called an expected social rate of return for an average TIMA research project.

The calculation of the expected social rates of return using each of our two methods relies on information collected during the telephone interviews and information published by ATP about each project. Interview respondents were assured that their individual responses would remain confidential; hence only the average values of interview data and derived metrics are reported herein. The mean values and standard deviations for all of these variables are in the last two columns of Table 2.

Data related to project duration, d, total project cost, C, and the level of ATP funding, A, are reported in ATP's Project Brief for each project provided to us by ATP. Data on the next six variables — r, z, F, T, m, and k — were obtained through the interview process.

Respondents were asked the following question about their private hurdle rate, r: *For projects like this one, what is your company's hurdle rate or minimum expected rate of return over the life of the project?*

Regarding the duration of the extra period of development, z, and the additional research cost for the extra period of development, F, respondents were asked:

Approximately, how soon after the project's completion will the technology be commercialized? And, *Approximately, what is the level of the additional investment expenditures expected to be made by your company during this period?*

Once commercialized, the life of the technology, T, was determined from responses to the question: *Approximately, what is the projected life of the technology being developed?*

The variable v represents the proportion of producer surplus appropriated by the project team; larger (smaller) values of v are interpreted to mean that less (more) spillover of technical knowledge is anticipated and thus appropriability problems are less (more) critical. From an analytical perspective, this is a critical value for our analysis. We calculated v as the product of two information data collected from the interviews. It equals the product of the proportion (m) of total profits anticipated from the development and commercialization of the specific software being developed, and the proportion (k) of all possible applications of the generic technology addressed by the specific software being researched in the ATP project.[19] Thus v, the product of these two interview elements, is the total proportion of the value of the technology appropriable by the researching firms. Loosely speaking, and focusing on the mean values in Table 2, the funded firms expect to appropriate about 70 percent (m) of the market value of the software they are developing in the 20 percent (k) (by value) of the applications areas they are focusing on. Roughly, the funded firms expect to capture about 14 percent of the profit potential of the generic technology being developed, and 86 percent of the profit potential associated with the generic technology being developed will spill over to imitators.

The variables L and U are discussed just below.

Given the published data in the Project Briefs and interview data on each of the variables noted in Table 2, the first step in the calculation of the expected social rate of return is to calculate the variable L, the lower bound for the annual expected private return from a TIMA research project. L is a derived variable. It was derived for each of the eight projects using project-specific data as reported in Table 2. It was also calculated separately using the averaged values of each relevant variable, as discussed in our exposition below of the 'average' TIMA project.

Eq. (1) consists of three general terms. Each term represents the present value for a particular flow that is realized over a particular time period. The first term in the equation represents the present value of the negative cash flows that result to the firm from the cost of conducting the project, $C-A$, from its start to its expected completion, $t=0$ to d. The second term is the present value of the future negative cash flows from the additional cost, F, of taking the generic technology

[19] Seven of the eight participants offered an estimate of m and k. For the eighth project we imputed a value of m and k equal to the average of the reporting seven. Thus, eight observations are used for the calculation of each mean in Table 2.

from the ATP project, at $t=d$, and commercializing it, at $t=d+z$. Finally, the third term is the present value of the expected net cash flows from the project, L, after it has been commercialized, at $t=d+z$, over its estimated life, to $t=d+z+T$. Note that the discount rate in Eq. (1) is the firm's hurdle rate, r. Therefore, the value for L that solves Eq. (1) is the value for which the private firm just earns its hurdle rate of return on the portion of the total investment that it must finance. The firm would not invest in the ATP project unless it expected at least L for the average annual private return so that its hurdle rate would exactly be met. Thus, L is a lower bound estimate.

$$-\int_0^d \left(\frac{C-A}{d}\right) e^{-rt}\, dt - \int_d^{d+z} \left(\frac{F}{z}\right) e^{-rt} dt + \int_{d+z}^{d+z+T} L\, e^{-rt}\, dt = 0$$

$$\Rightarrow -\left(\frac{C-A}{d}\right)\left(\frac{-1}{r}\right) e^{-rt}\big|_0^d - \left(\frac{F}{z}\right)\left(\frac{-1}{r}\right) e^{-rt}\big|_d^{d+z} + (L)\left(\frac{-1}{r}\right) e^{-rt}\big|_{d+z}^{d+z+T} = 0$$

$$\Rightarrow \left(\frac{C-A}{dr}\right)(e^{-rd}-1) + \left(\frac{F}{zr}\right)(e^{-r(d+z)} - e^{-rd}) - \left(\frac{L}{r}\right)(e^{-r(d+z+T)} - e^{-r(d+z)}) = 0 \tag{1}$$

Given specific values for the variables d, C, A, r, z, F, and T, Eq. (1) is solved for the unknown variable, L. To illustrate using the mean values for these seven variables from Table 2, Eq. (1) solves for L equal to $3,883,680. Again, this derived value of L is the lower bound for the estimate of the expected annual private return using the mean values of the relevant variables in Table 2. Since we will illustrate the solution of the three equations that follow also using the mean values of the relevant variables in Table 2, we will refer for ease of exposition to the solutions as those for the 'average' TIMA project.

The second step in the calculation of the estimated social rate of return is to calculate the variable U, the upper bound for the expected annual private return for each TIMA project. U was derived for each of the eight projects, by solving Eq. (2) using values for the variables in Table 2. Note that the first term in Eq. (2) is the total negative cash flow from the cost of conducting the research project, C. Thus, U is an upper bound for the annual private return because a return greater than U would imply that the firm would earn a rate of return in excess of its hurdle rate in the absence of ATP funding, and therefore ATP funding would not be required for the project. Note that all reference to ATP is purposively absent in Eq. (2) because we are calculating the expected annual private return that would result in the private firm just meeting its hurdle rate in the absence of ATP involvement.

$$-\int_0^d \left(\frac{C}{d}\right) e^{-rt}\, dt - \int_d^{d+z} \left(\frac{F}{z}\right) e^{-rt}\, dt + \int_{d+z}^{d+z+T} U\, e^{-rt}\, dt = 0$$

$$\Rightarrow -\left(\frac{C}{d}\right)\left(\frac{-1}{r}\right) e^{-rt}\Big|_0^d - \left(\frac{F}{z}\right)\left(\frac{-1}{r}\right) e^{-rt}\Big|_d^{d+z} + (U)\left(\frac{-1}{r}\right) e^{-rt}\Big|_{d+z}^{d+z+T} = 0$$

$$\Rightarrow \left(\frac{C}{d}\right)(e^{-rd}-1) + \left(\frac{F}{zr}\right)(e^{-r(d+z)} - e^{-rd}) - \left(\frac{U}{r}\right)(e^{-r(d+z+T)} - e^{-r(d+z)}) = 0. \tag{2}$$

To illustrate the solution of Eq. (2), values for the variables d, C, r, z, F, and T are given. Thus, Eq. (2) solves for U. To illustrate using the mean values for these given variables in Table 2, the derived value of U is \$7,267,910. This value represents the upper bound of the annual private return for the 'average' TIMA project.

Using the mean values (other than L and U) in Table 2, an estimate of the average expected annual private return to the firm is calculated as $[(L+U)/2]$, the mean of the upper and lower bounds on the average expected annual private return for the 'average' TIMA project.[20] The average expected annual private return to the participating firm or firms equals v times the average expected annual return that will be captured by all producers using the technology (producer surplus). Knowing the average expected annual private return is $[(L+U)/2)]$ and knowing the portion of producer surplus that is appropriable, v, then total producer surplus equals $[(L+U)/2v]$and hence this value is a lower bound for the average expected annual social return. It is a lower bound because consumer surplus has not been measured.

The expected private rate of return without ATP funding is the solution to i in Eq. (3), given solution values for L and U from Eqs. (1) and (2). The solution value of i in Eq. (3), represents the rate of return that just equates the present value of the expected annual private return to the firm to the present value of research and post-research commercialization costs to the firm in the absence of ATP funding.

$$-\int_0^d \left(\frac{C}{d}\right) e^{-it}\,dt - \int_d^{d+z} \left(\frac{F}{z}\right) e^{-it}\,dt + \int_{d+z}^{d+z+T} \left(\frac{L+U}{2}\right) e^{-it}\,dt = 0$$

$$\Rightarrow \left(\frac{C}{di}\right)(e^{-id} - 1) + \left(\frac{F}{zi}\right)(e^{-i(d+z)} - e^{-id}) - \left(\frac{L+U}{2i}\right)(e^{-i(d+z+T)} - e^{-i(d+z)}) = 0 \qquad (3)$$

The expected private rate of return without ATP funding, i, was estimated for each of the eight TIMA projects by solving Eq. (3) given the values for d, C, z, F, T, L, and U. To illustrate the solution of Eq. (3) using the mean values of the relevant variables in Table 2 along with the values derived from Eqs. (1) and (2)

[20] We want an estimate of the average expected annual return, and the simple average of the lower and upper bounds is the natural measure. Alternatively, one could use both L and U and obtain an upper and lower bound on each solution to Eqs. (3), (4) and (5) below. One could then think of their average as a better estimate. However, our procedure is more direct and more simply exposited, and because we have presented the values for L and U, readers interested in re-estimating Eqs. (3), (4) and (5) can do so. As we show below, our results are in general quite insensitive to the choice of computing disaggregated metrics and then aggregating across individual results, versus aggregating variables and computing a single aggregated metric.

of L and U for the 'average' project, the expected private rate of return for the 'average' TIMA project is 0.20 or 20 percent.

Finally, the lower bound on the social rate of return is found by solving Eq. (4) for i, given values for the other variables. Note that Eq. (4) is identical to Eq. (3) with the exception that the average expected annual private return, $[(L+U)/2]$, is replaced with the lower bound for the average expected annual social return, $[(L+U)/2v]$.

$$-\int_0^d \left(\frac{C}{d}\right) e^{-it}\, dt - \int_d^{d+z} \left(\frac{F}{z}\right) e^{-it}\, dt + \int_{d+z}^{d+z+T} \left(\frac{L+U}{2v}\right) e^{-it}\, dt = 0$$

$$\Rightarrow \left(\frac{C}{di}\right)(e^{-id} - 1) + \left(\frac{F}{zi}\right)(e^{-i(d+z)} - e^{-id}) - \left(\frac{L+U}{2iv}\right)(e^{-i(d+z+T)} - e^{-i(d+z)}) = 0 \quad (4)$$

As with each of Eqs. (1) through (3), Eq. (4) was estimated for each of the eight TIMA projects. Again, to illustrate the solution of Eq. (4) using the average values of d, C, z, F, T, and v from Table 2 with the derived values of L and U for the 'average' project, the expected social rate of return for the 'average' TIMA project is derived to be at least 0.63 or 63 percent.

To summarize, we have extracted an expected social rate of return from our model. We first designed a set of questions that allowed us to gather the additional information, to supplement the information in the Project Briefs, needed to derive an estimate of the private benefit stream anticipated by the participants. Second, we gathered information revealing the benefits created by the project that would be captured by producers that are not participants in the project. We then had an estimate of the total producers' benefits from the project to compare with the project's costs. Thus, we use the private participant's expected benefits and the project's costs to compute the private expected rate of return. We use the total expected benefits, to all producers whether they are participants or not, to compare to the total costs for the project to compute the lower bound on the social rate of return. It is a lower bound, because we have not measured the consumer surplus created by the project for the ultimate consumers of the final goods that are made using the project's technology.

We can summarize the findings from our analysis for the 'average' TIMA research project, based on the mean values in Table 2 (excepting those for L and U) and the derived values from Eqs. (1) through (4). There are two important points to be made. First, the average expected private rate of return in the absence of ATP funding is 20 percent, clearly less than the average private hurdle rate of 25 percent since the estimate of the upper bound for the average cash flows would just allow the hurdle rate to be reached. Thus, in the absence of ATP funding the TIMA firms would not have undertaken this research, and in fact they expressed this fact explicitly as noted earlier. Second, the expected social rate of return associated with ATP's funding of the TIMA projects is at least 63 percent, and hence the projects are expected to be socially valuable.

As discussed above, we estimated Eqs. (1) through (4) for each of the eight projects. Doing so, we derived eight values for L and eight values for U (the averages for these values are shown in Table 2 and are somewhat higher than the single values derived for the 'average' project and reported in the text), and eight lower bound estimates of the expected social rate of return — one set of rates for each TIMA project. For each of the eight projects, just as for the 'average' project, the private rate of return without ATP funding of course falls short of the private hurdle rate, yet because of the spillovers the projects are all socially valuable. The average of the eight expected private rates of return absent ATP funding is 19 percent (0.19 with a standard deviation of 0.06); the average expected social rate of return is 72 percent (0.72 with a standard deviation of 0.22). These derived rates of return are observably close to our 'average' project's private and social rates of return of 20 percent and 63 percent respectively. Thus, we are comfortable using these figures to characterize the 'average' TIMA project.

Our broad-brush approach to estimating the various rates of return does not try to build in more detail for the model than the data can support. For example, instead of trying to develop the details of an unknown diffusion pattern for the generic technology, we estimated the average expected annual return over the lifetime of the technology. Our methodology gives a reasonable fix on the average expected annual private returns to the ATP project participants because we can solve for what they must be expecting as lower and upper bounds, and then we average the two expectations. What we do not know with certainty is the multiplier to apply to the participants' average annual expected returns to estimate returns to all producers. Our multiplier is based on the interview responses that estimate the variables m and k, and hence v. We do not have a good estimate of the lag from $d + z$, when the ATP project's technology is commercialized, until the technology is imitated not only by producers copying the particular application of the generic technology but also by producers developing new applications of the generic technology. Thus, because there will be some lag before imitation and because there will be some development costs for producers using the generic technology, the net average expected annual earnings to all producers using the generic technology will be somewhat less than $[(L + U)/2v]$ unless diffusion is instantaneous and costless. For the particular information technology that we are studying, the diffusion may well be relatively quick and additional development costs may well be minimal; however, the possibility that it is not the case can be modeled by lowering our multiplier $(1/v)$ to reflect the fact that if diffusion is not rapid, then the ATP participants' average annual returns are actually a larger proportion of the average social returns than the proportion v. To estimate the sensitivity of our results to the speed of the diffusion of the technology, we varied v in the following experiment. If v is increased by 10 percent, the estimated social rate of return for the 'average' TIMA project falls from 63 percent to 61 percent; if v is increased by 50 percent, the estimated social rate of return falls to 54 percent; and if v is increased by 100 percent, the estimated social rate of return falls to 47 percent. Thus, our conclusion that ATP's TIMA projects are socially valuable is robust with respect to the parameter v when it is increased to capture slower diffusion rates (and any development costs to be netted out) for the applications of the generic technology.[21]

It is important to re-emphasize that the profit potential of the generic technology being developed that is appropriated by the TIMA firms represents producer surplus. Our analysis does not, as we have previously stated, attempt to capture consumer surplus, and of course consumer surplus would not be appropriated by the researching firms. Thus, when we refer to a lower bound estimate of the expected social rate of return we are explicitly acknowledging that there are social benefits to the consumer surplus generated from the TIMA research-based software, but we are just not capturing them. Thus, our social rate of return estimate for the 'average' TIMA project is clearly a lower-bound estimate.

Finally, we explicitly note that our model in Eqs. (1) through (4) is a continuous time model, as is appropriate. However, if the model is approximated as a discrete time model, we calculated that the estimated social rate of return for the 'average' TIMA project is again 63 percent, given the rounding of times to the nearest whole year and given appropriate placement (a particular choice of beginning or end of years that results in replicating the solution found with the continuous model) of the discrete cash flows. As would be expected, however, in the discrete model, the estimation is not insensitive to the placement of the cash flows, and for that reason we present and solve the more accurate continuous time model.

5. Formulating a public/private partnership competition policy for a dynamic market

The case-based analyses presented in this paper can be summarized as follows. The TIMA projects under study would not have been undertaken in the absence of ATP's public funding. Because of technical and market risk and because of appropriability issues — key elements of market failure — firms perceive that their expected rate of return had they pursued this research in the absence of ATP's support would have been less than their required rate of return. ATP funding of these projects reduces risk to the private firm. And, as a result of ATP funding, these projects are being researched and the expected social rate of return from them is estimated to be at least 63 percent.

Of course, one cannot conclude that a social rate of return of at least 63 percent is 'good' or 'bad,' or 'better' or 'worse' than expected. Those are non-axiomatic

[21] We thank Jeanne Powell of the ATP for raising this interesting issue.

conclusions. However, one can compare our estimate of the lower bound of the social rate of return to the opportunity cost of public funds. Following the guidelines set forth by the Office of Management and Budget (1992) to use a real discount of 7 percent for constant-dollar benefit-to-cost analyses of proposed investments and regulations, then clearly a nominal social rate of return of 63 percent is above that rate and thus is socially worthwhile.[22]

Consider Fig. 4. Project A characterizes the 'average' TIMA project. Shown is our estimated private rate of return absent ATP funding of 20 percent, and our estimated lower bound on the social rate of return of 63 percent. Hence the spillover gap associated with the average TIMA project is at least 43 percent.

In comparison to these estimates, we calculated the private rate of return with ATP funding to be 33 percent. This rate comes from the solution to Eq. (5) for i, given the mean values of d, C, A, z, F, and T from Table 2 and the derived values of L and U from Eqs. (1) and (2).

$$-\int_0^d \left(\frac{C-A}{d}\right) e^{-it}\,dt - \int_d^{d+z} \left(\frac{F}{z}\right) e^{-it}\,dt + \int_{d+z}^{d+z+T} \left(\frac{L+U}{2}\right) e^{-it}\,dt = 0$$

$$\Rightarrow \left(\frac{C-A}{di}\right)(e^{-id}-1) + \left(\frac{F}{zi}\right)(e^{-i(d+z)} - e^{-id}) - \left(\frac{L+U}{2i}\right)(e^{-i(d+z+T)} - e^{-i(d+z)}) = 0 \qquad (5)$$

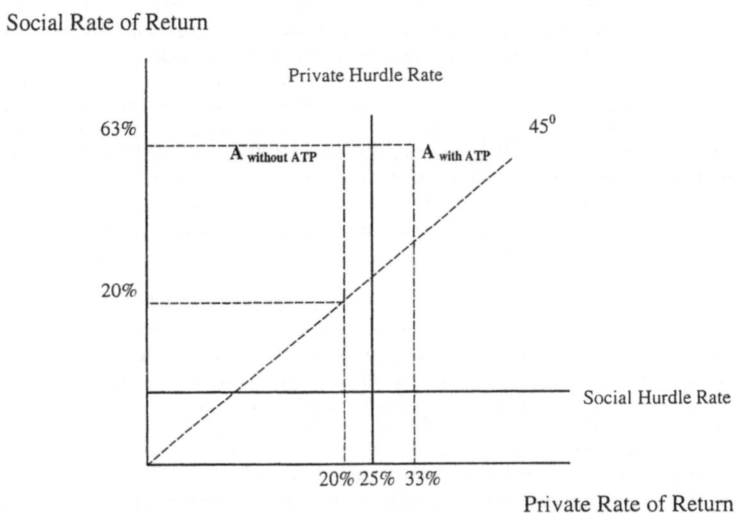

Fig. 4. Spillover gap between social and private rates of return to ATP-funded TIMA research.

[22] Link and Scott (1998) discuss the use of this guideline for NIST economic impact assessments.

The estimated rate of return with ATP funding of 33 percent is perhaps reasonably close to the private hurdle rate of 25 percent.[23,24] Had ATP funding been greater (less), the estimated rate of return with ATP funding would have been greater (less). However, there is no way for ATP to have calculated the optimal level of funding for these TIMA projects unless, as part of the focused program proposal, all relevant data, including hurdle rates, could have been assessed.[25] In the absence of such information, which in practice would be difficult to obtain because of, if nothing else, self-serving reporting by proposers, the funding scheme that ATP has implemented appears to be close to optimal.[26]

In principle, a bidding mechanism could be used by ATP to ensure that the

[23] Our estimation procedure has not forced the closeness of these rates. Although greater ATP funding (an increase in A) has no effect on our estimate of U, it does of course require that our estimate of L and hence of $[(L + U)/2]$ be lower. However, the initial private investment $(C - A)$ falls as ATP funding increases, and that effect dominates causing our estimate of the private rate of return to increase as ATP funding increases. For example, if ATP funding for the average TIMA project were increased by 50 percent, our estimate of the private rate of return with ATP funding would have risen from 33 percent to 42 percent.

[24] The expected private rate of return with ATP funding was calculated for each of the eight projects. The mean of these eight rates is 0.35 or 35 percent, with a standard deviation of 0.094.

[25] The reader will note that it is mathematically possible to solve Eq. (5) for the level of ATP funding, A^*, that would equate the estimated private rate of return with ATP funding to the hurdle rate of 25 percent. Such an exercise may lack policy relevance in the sense that the calculation has the benefit of hindsight and information collected after ATP funding was allocated and furthermore we do not know the true weights for our estimated upper and lower bounds. We assume the equal weights of 0.5 for each, but although our basic conclusions about the relation between the social and private rates of return are not sensitive to that assumption, the calculation of A^* is dependent on whether the actual expected annual returns are closer to the upper or the lower bound. We believe it is safer for the ATP funding to be on the 'high side' in any case. In a separate study (Hall et al., 1998), we have found that greater ATP funding dramatically reduced the probability of termination of ATP information technology projects. The theoretical reason is clear; having more ATP funds shifts the distribution of the project to the right and lowers the probability that the private participants in the project, who are constantly monitoring the progress of the project, would want to terminate the project because the probability of an unacceptably low rate of return was too high.

[26] The above findings should be interpreted in light of three particular limitations of this paper. First, the conclusions presented here are based in part on interview data collected from a key participant in each TIMA project. As such, there is obviously some uncertainty built into the expressed estimates, although they are the best point estimates available. Perhaps more important, this interview information may be time dependent, meaning that it was collected at this particular early point in the progress of the research project. As such, the estimates presented herein are not definitive estimates to be expected at the completion of each project, but rather characterizations of the expected social benefits at this point in time. Second, we have referred to our lower bound estimate of 63 percent — although this caveat refers to any of the calculated social rates of return — as a metric to characterize the 'average' TIMA project. And third, our analyses have not specified any diffusion paths. We refrained from going beyond the limits of our data and did not attempt to model much less quantify when in time rivals will imitate the participants' applications of the generic technology, or when in time others will commercialize other aspects of the generic technology. Still, with these limitations in mind, it appears at this point in the progress of the TIMA projects that they are socially worthwhile.

private participants in TIMA research projects earn just a normal rate of return. Such a bidding mechanism is what Scott (1998) refers to as a hurdle-lowering auction.

Motivated by the case analysis of the ATP-funded TIMA projects, a broad public/private partnership competition policy should address three broad questions:

1. How can the public get the best private partner for each partnership?
2. How can society determine the optimal level of public funding for the partnership, not too much yet enough to overcome the underinvestment resulting from market failure?
3. How can society avoid the potential for opportunistic behavior to which both the government and the private sector's partners are exposed?

Below we list several premises that we argue should motivate the formulation of a public/private competition policy, along with the implementation implications of those premises.

Premise 1: The private sector knows more than the government about the investment characteristics of the technology projects — or at least has the resources to make the best estimate of the streams of returns and the risk.

Implication 1: A competition policy should include a mechanism for setting up a public/private partnership that provides the incentive for private parties to determine who is best suited to be the private partner in a public/private partnership.

Premise 2: The government desires to overcome the project's underinvestment resulting from market failure and to do so at the least cost to the public.

Implication 2: A competition policy should include a mechanism that gives the selected private partner for the public/private partnership an incentive to undertake the desired level of investment while providing a proportion of the project's funding that is consistent with a normal expected private rate of return given the appropriability and risk characteristics of the project.

Premise 3: All parties related to the public/private partnership want to overcome the potential for opportunistic behavior by the other party.

Implication 3: Policy should include a mechanism that provides both the public- and the private-sector partners an incentive to participate in the project in a way that maximizes the total value of the project's expected outcome rather than the value to the individual partner that could of course use opportunistic behavior to benefit at the expense of the overall project's results.

General characteristics of the mechanism design: What are the general characteristics of the optimal mechanism design for public/private partnerships that will achieve the desired incentives for the private sector to choose the best private partner, for the private partner(s) to carry out the desired amount of investment at the least cost to the public, and for avoiding opportunistic behavior by either the public or the private partner?

Consideration of the questions suggests that the optimal mechanism design for what we advocate and call a public/private competition policy would have the private parties use a contingent valuation method to bid for the right to be the private partner. In particular, the bidding could be a hybrid bidding mechanism that combines an up-front bid, a periodic payment bid, and finally a royalty bid: private firms would bid for the right to be the private partner in the public/private partnership project that the government would fund.[27] Alternatively, instead of bids being accepted directly from the companies that will be performing the R&D, private venture capital companies that would manage the public investments might bid for the rights to manage the projects.

As a simple example of how a bidding process would deliver the public funding to a public/private partnership such as any of the TIMA projects discussed in this paper, consider the following. Suppose that from society's perspective, an R&D investment project would cost $100 now and generate the expectation of $130 in one year and nothing thereafter. Suppose further that the threshold rate of return justifying public funding — society's hurdle rate — is 10 percent. Thus, the R&D project yields a social rate of return of 30 percent, which exceeds society's hurdle rate of 10 percent, and of course the net present value of $[(130/(1.1))-100]$ is greater than zero. Suppose that from a private perspective the project costs $100 and, because of incomplete appropriation of returns, yields the expectation of just $105 in one year. Suppose further that given the private risk the private hurdle rate is 15 percent. Thus, the private sector would not undertake the project which has an internal rate of return of 5 percent, which is less than the hurdle rate of 15 percent; and, of course, net present value is then negative.

In the context of the foregoing example, which conceptually describes each of the TIMA projects, the bidding process would work as follows. The government announces that it will 'buy' the R&D project, paying the $100 investment cost.[28] The government then opens the bidding for the right to be the private partner in the public/private partnership. Private firms will bid the amount $X such that $X(1.15)=$105, implying that $X=$91.30. The cost to the public of the project would then be $8.70. With great uncertainty about the future returns, the use of royalty bidding rather than the up-front bidding can yield more to the government. Also, private firms with better capabilities for doing the project would be expected to bid higher than those firms that are less well suited to the project.

Suppose the ATP, or the government in general, wanted to use a public/private partnership to develop a project. It would announce that it would provide an

[27] There is a large literature describing bidding mechanisms in great detail. McAfee and McMillan (1987) provide an excellent review, and they set out the general hybrid mechanism with the up-front bid as well as the royalty bid. Hansen (1985) and Samuelson (1986) provide analyses of the royalty bidding and bidding for the up-front fee and the royalty rate simultaneously.

[28] Martin and Scott (2000) provide detailed discussion of the circumstances in which market failure and underinvestment would be expected to occur; the discussion is needed to inform the identification of projects that would be funded.

up-front payment of \$$F$ to support the R&D investment project to be conducted by the winning bidder in an auction to determine the private partner(s) for the public/private partnership (and when there are several partners as in an RJV we are viewing the RJV as the partner). Further, the government pledges to provide a periodic flow of funds \$$C$ throughout the project's life to support the flow costs of the R&D project. The fixed cost \$$F$, and the flow cost \$$C$, correspond to the typical abstraction of the structure of costs for R&D investment projects.[29] Bidders vie for the right to be the private partner in the project by submitting a three-part bid — first a bid for how much the private firm will pay the government up-front, and second a bid on the periodic flow payment during the life of the R&D project, and third a bid on the royalty rate that it would pay the government on the innovation produced by the public/private partnership and licensed (perhaps exclusively) to the private partner.

As McAfee and McMillan (1987) make clear, in the context of the appropriate combinations of assumptions about the characteristics of the asset being auctioned and the participants in the auction and their beliefs about the value of the asset, there are nontrivial choices to be made about the exact nature of the auction. Apart from the usual choices for auctions in general, there would be choices specific to the new institutional use of auctions to determine the private partner for the public/private partnership. For example, institutional arrangements must be designed to ensure that the government's payments of \$$F$ and \$$C$ go solely for the purchase of R&D investments; the private partner's profits from the R&D investment project will come after the innovation is introduced. However, for this paper, full details of the ideal auction in different circumstances will not be developed. Instead, the paper presents the basic idea and observes that the three-part bidding mechanism proposed has the potential for leveraging public funding optimally. Should the Advanced Technology Program have undertaken such a hurdle-lowering auction ex ante, in terms of Fig. 4, the expected rate of return to the TIMA projects with ATP funding (actually estimated to be 33 percent in our model) would not have exceeded the project participant's private hurdle rate (25 percent). Hence, society would have been better off in terms of a more optimal allocation of R&D resources.

Finally, two things must be emphasized. First, the high social rates of return estimated and reported for the TIMA projects are very conservative, lower-bound estimates because they do not include consumer surplus in the benefit stream. The profits that will be generated by the technology are obviously a small proper subset of the social benefits that the technology will generate, but the estimation method measures only the return in the form of profits to the innovator and to other producers of the technology. Second, one might be skeptical about the TIMA respondents' earnest belief that the projects would not have been undertaken, or at

[29] See Lee and Wilde (1980).

least would not have been undertaken to the same extent or at the same speed, without ATP funding. If the research would have occurred without public funding, the estimated upper bound and hence the average of the upper and lower bounds for expected private returns would be too low, and the actual lower bounds for the social rates of return would be even higher than we have estimated. Further, the gap between the social and private rates of return would remain, although that would not in itself justify the public funding of the projects.

Acknowledgements

We would like to thank David B. Audretsch, William L. Baldwin, and an anonymous referee for helpful comments and suggestions.

References

Arrow, K.J., 1962. Economic welfare and the allocation of resources for invention. In: Universities — National Bureau Committee for Economic Research, The Rate and Direction of Inventive Activity. Princeton University Press, Princeton, pp. 609–625.

Baldwin, W.L., Scott, J.T., 1987. Market Structure and Technological Change. Harwood Academic Publishers, London.

Cohen, W.M., Klepper, S., 1992. The tradeoff between firm size and diversity in the pursuit of technological progress. Small Business Economics 4, 1–14.

David, P.A., 1987. Some new standards for the economics of standardization in the information age. In: Dasgupta, P., Stoneman, P. (Eds.), Economic Policy and Technological Performance. Cambridge University Press, Cambridge, England, pp. 206–239.

Economic Report of the President, 1994. Government Printing Office, Washington, DC.

Executive Office of the President, 1990. U.S. technology policy. Office of Science and Technology Policy, Washington, DC.

Hagedoorn, J., Link, A.N., Vonortas, N., 2000. Research partnerships. Research Policy 29, 567–586.

Hall, B.H., 1992. Investment and research and development: Does the source of financing matter? University of California at Berkeley Working Paper No. 92–194.

Hall, B.H., Link, A.N., Scott, J.T., 1998. Universities as partners in ATP-funded research projects. Preliminary final report submitted to the Advanced Technology Program.

Hansen, R.G., 1985. Auctions with contingent payments. American Economic Review 74, 862–865.

Hao, K.Y., Jaffe, A.B., 1993. Effect of liquidity on firms' R&D spending. Economics of Innovation and New Technology 2, 275–282.

Himmelberg, C.P., Petersen, B.C., 1994. R&D and internal finance: A panel study of small firms in high-tech industries. Review of Economics and Statistics 76, 38–51.

Hubbard, R.G., 1998. Capital-market imperfections and investment. Journal of Economic Literature 36, 193–225.

Jaffe, A.B., 1998. The importance of 'spillovers' in the policy mission of the Advanced Technology Program. Journal of Technology Transfer 23, 11–19.

Lee, T., Wilde, L.L., 1980. Market structure and innovation: a reformulation. Quarterly Journal of Economics 94, 429–436.

Lerner, J., 1996. The government as venture capitalist: The long-run impact of the SBIR program. NBER Working Paper 5753.

Leyden, D.P., Link, A.N., 1999. Federal laboratories as research partners. International Journal of Industrial Organization 17, 572–592.

Link, A.N., 1996. On the classification of R&D. Research Policy 25, 379–401.

Link, A.N., 1998. Public/private partnerships as a tool to support industrial R&D: Experiences in the United States. Final report prepared for the working group on technology and innovation policy, division of science and technology at OECD.

Link, A.N., 1999. Public/private partnerships in the United States. Industry and Innovation 6, 191–216.

Link, A.N., Scott, J.T., 1998. Public Accountability: Evaluating Technology-based Institutions. Kluwer Academic Publishers, Norwell, Massachusetts.

McAfee, R.P., McMillan, J., 1987. Auctions and bidding. Journal of Economic Literature 25, 699–738.

Martin, S., Scott, J.T., 1998. Financing and leveraging public/private partnerships. Final report prepared for the working group on technology and innovation policy, division of science and technology at OECD.

Martin, S., Scott, J.T., 2000. The nature of innovation market failure and the design of public support for private innovation. Research Policy 29, 437–447.

Office of Management and Budget, 1992. Circular No. A-94: Guidelines and discount rates for benefit-cost analysis of federal programs. Washington, DC.

Office of Management and Budget, 1996. Economic analysis of federal regulations under Executive Order 12866. Washington, DC.

Office of Science and Technology Policy, 1994. Science in the national interest. Executive Office of the President, Washington, DC.

Office of Science and Technology Policy, 1998. Science and technology: Shaping the twenty-first century. Executive Office of the President, Washington, DC.

Samuelson, W.F., 1986. Bidding for contracts. Management Science 32, 1533–1550.

Schumpeter, J.A., 1950. Capitalism, Socialism and Democracy. Harper & Row, New York.

Scott, J.T., 1991. Research diversity induced by rivalry. In: Acs, Z.J., Audretsch, D. (Eds.), Innovation and Technological Change. Harvester-Wheatsheaf, London, pp. 132–151.

Scott, J.T., 1993. Purposive Diversification and Economic Performance.. Cambridge University Press, Cambridge, England.

Scott, J.T., 1998. Financing and leveraging public/private partnerships: The hurdle-lowering auction. STI Review 23, 67–84.

Tassey, G., 1992. Technology Infrastructure and Competitive Position. Kluwer Academic Publishers, Norwell, Massachusetts.

Tassey, G., 1997. The Economics of R&D Policy. Quorum Books, Westport, Connecticut.

Teece, D.J., 1980. Economies of scope and the scope of the enterprise. Journal of Economic Behavior and Organization, September 1, 223–247.

[10]

Public Sector Support for Innovating SMEs

David North
David Smallbone
Ian Vickers

ABSTRACT. Innovation support for SMEs is considered from the point of view of the "needs" of SMEs in the context of a specific regional economy – London's Lee Valley region. As identified in the literature and, more specifically, in a survey of 100 innovating SMEs in the study region, attention is focused on the innovation support needs that relate to the size characteristics of SMEs. The difficulties of financing innovation are shown to stem from the attitudes of SME owner managers as well as from supply-side failures. The paper examines four different innovation support instruments, taken as constituting key elements of the support "system" within the region. Drawing on the findings of interviews with client businesses, the paper considers the strengths and weaknesses of each policy instrument. Conclusions relate to the overall impact of innovation support provision, in particular the need for greater integration in the design and implementation of policy instruments.

1. Introduction

In recent years, there has been a growing recognition amongst policy-makers of the importance of innovation to the competitiveness of national and regional economies and of the particular role of small and medium sized enterprises (SMEs) in this process. In the case of the U.K., a series of Government White Papers on Competitiveness during the 1990s emphasised the important contribution of SMEs to the development of a more competitive economy, and recognised that "entrepreneurship and innovation are central to the creative process in the economy and to promoting growth, increasing productivity and creating jobs" (DTI, 1998, para. 2.1). This has resulted in an increasing interest in the kinds of policy instruments needed to stimulate innovation in SMEs and to help overcome the barriers to innovation. There are now a number of different public and quasi public support instruments aimed at assisting SMEs with their innovation related needs, ranging from some nationally run policy initiatives to other, more locally specific initiatives. The purpose of this paper is to examine the appropriateness of these initiatives in relation to the needs of SMEs in a specific regional economy – London's Lee Valley region.

Centre for Enterprise & Economic Development Research
Middlesex University Business School
London NW4 4BT
U.K.
E-mail: d.north@mdx.ac.uk, d.smallbone@mdx.ac.uk, i.vickers@mdx.ac.uk

2. SMEs and innovation support

In line with recent policy developments, the study adopts a broad interpretation of what constitutes innovation in an SME context. Following Porter's (1990) emphasis on the importance of the link between innovation and competitiveness, the definition used here includes the development and/or adoption of new products and processes, incremental improvements to products and processes, and new approaches to marketing and/or new forms of distribution. An emphasis on the diffusion and adoption of new knowledge and technology and the role of more incremental innovations in supporting competitiveness (which are less dependent on a formal research and development [R&D] function) is particularly relevant in the context of SMEs (Rothwell and Zegveld, 1982; Bessant et al., 1993, 1994). Such an understanding also corresponds to an increased interest, on the part of many academics, in learning processes, both within firms and between firms and the role of external sources of knowledge, stimulus and support (e.g. see Asheim, 1996; Lundvall and Borrás, 1997; Cooke and Morgan, 1998).

Given this approach to defining innovation, the justification for providing innovation support which is specific to SMEs rests on the recogni-

Small Business Economics **16**: 303–317, 2001.

tion that there are certain size-related characteristics of SMEs that have important implications both for their support needs and for the way in which those needs are addressed. First, SMEs are constrained by their limited internal resources, particularly with respect to finance, management resources and their knowledge-base compared with larger firms. Limited financial resources clearly make it harder for smaller firms to support and to reap the benefits of innovative activity, particularly in terms of R&D and the marketing of new product innovations. In this respect the difficulties that small technology-based firms face in gaining external finance to support innovation has been the subject of considerable attention in the U.K. in recent years (CBI, 1997; Bank of England, 1996; HM Treasury, 1998).

Furthermore, limited human resources can influence a firm's propensity and ability to be aware of, and respond to, opportunities and threats presented by the external environment. This includes scanning for new knowledge relevant to the business that in a large firm would be typically undertaken by dedicated staff, e.g. R&D staff in the case of technical knowledge (Cohen and Levinthal, 1990; Seaton and Cordey-Hayes, 1993). The limited internal resource base of smaller firms is a key justification for interventions designed to provide external support for innovation in SMEs. With respect to financial support this applies particularly at the start-up stage or in very young businesses where the opportunities to resource innovation and business development internally are more limited than in established firms. The limited internal knowledge-base and scanning ability of smaller firms is a key rationale for the publicly subsidised provision of technology support by intermediary organisations in terms of information, advice and consultancy (e.g. see Hassink, 1996, 1997; Seaton and Cordey-Hayes, 1993).

Secondly, a typical characteristic of small firms is their distinctive organisational culture, stemming from the combination of ownership and management that typifies the majority of SMEs. Hence there is a need to understand the role of the individual traits of the owner-manager in the way in which the business is managed and developed, which in turn affects management behaviour, attitudes to risk (and hence innovation), and the nature and extent of external financing that is sought. These attitudinal and behavioural characteristics all have potential implications for the nature and extent of the support needs, as well as effective delivery of external support to small companies. The behavioural characteristics of small firms that stem from the combination of ownership and management can result in a reluctance, or even a resistance, to taking in external help for a variety of reasons. These include: doubts about the value for money on the part of the business owner or manager; a scepticism about generalist advice, particularly where this is offered by advisers who lack a detailed sectoral knowledge; and a preference for autonomy which small business owner-managers may perceive is threatened by the use of external advice. This may result in a greater use of informal rather than formal channels of support in cases where the professional management resources (i.e. where managers lack formal management training or management qualifications) are limited (Smallbone, 1997).

Thirdly, SMEs have less ability than larger companies to shape and influence their external environment, e.g. in their relationships with customers, suppliers, sources of finance and the labour market. This means that the smaller firm is typically faced with a more uncertain external environment than a larger firm. As a consequence, competitiveness often relies on the firm's flexibility, responsiveness to customers and adaptability to external changes, which are key attributes that need to be considered and enhanced when designing support programmes and initiatives targeted at SMEs.

A further issue which arises when considering the "support needs" of SMEs is the distinction between "expressed needs" and "latent needs" or potential. The distinction refers to the difference that can exist between what a business owner or manager's stated wants or expressed needs are, compared with what might potentially emerge from a systematic audit of the strengths and weaknesses of the firm in terms of resources and competence. Differences between "expressed" and "latent" needs/potential often stem from the difficulties that many SME owners-managers have in diagnosing the external support requirements of their business, particularly in cases where they have limited formal professional management training. For example, what may be expressed as a need for marketing support may conceal a deeper latent need for a root and branch strategic review of the business, with possible implications for the core business definition. Differences between expressed and latent support needs have important implications for policy, particularly where the objective is to elicit the full potential of the business, or to influence the nature of change, rather than the support offered being simply "client"- or "market"-led.

The distinctiveness of SMEs affects their support needs and how such support is delivered if it is to be effective. In addition, ongoing relationships with external providers of support are more likely to provide a basis for competence-building and learning than more one-off and piecemeal interventions (e.g. see Hassink, 1996, 1997). At the same time, it must be recognised that it is neither realistic nor desirable that all of the support needs of firms (expressed or latent) can or should be addressed through public policy. In this respect, the aim should be to focus those public resources that are available on the needs of the economy at the regional (or national) level rather than simply on the expressed needs of individual firms themselves. Thus, the nature of innovation support policy may be shaped by regional (or national) priorities set within a strategy for innovation (Morgan and Nauwelaers, 1999; Thomas, 2000). In producing such strategies, attention should be paid to the strengths and weaknesses of the innovative capability of the economy, and the priorities for raising it, taking into account the wider socio-economic benefits. This more strategic rationale for providing support for the innovation support needs of SMEs might include policy measures aimed at stimulating the creation of new, innovative businesses and raising the level of innovative activity in existing enterprises, rather than solely meeting the expressed needs of SMEs themselves.

3. Aims and methodology

This paper is concerned with assessing the appropriateness of the public and quasi public policy instruments that are available to support innovative activity in SMEs within a chosen case study area, namely the Lee Valley area of North East London and adjoining parts of the outer metropolitan area. Using the results of a survey of a sample of innovative firms in the study region, the paper considers the extent to which the existing policy instruments are addressing the needs of firms and whether there are some gaps in provision which public support needs to address. The paper focuses on 4 of the main types of innovation support policy initiative which apply within the study area and will consider the underlying rationale of each one, and the kind of SME that each is targeted at. In addition, the main strengths and weaknesses of the instruments will be identified, particularly with respect to their overall impact.

The evidence on which the paper is based is drawn from a European Commission Targeted Socio-Economic Research (TSER) Programme project entitled "SME Policy and the Regional Dimension of Innovation" (known as SMEPOL).[1] The main objective of the SMEPOL project has been to undertake a comparative analysis of existing policies and programmes to encourage and support innovation in SMEs with a view to identifying policy lessons and establishing best practice.

The authors have been responsible for the U.K. national study which has focused on London's Lee Valley area and adjacent parts of the outer metropolitan area (OMA) in the counties of Hertfordshire and Essex. One of the main reasons for the choice of the Lee Valley area is that it has been experiencing industrial decline for some years and has qualified for support from EU Structural Funds (Objective 2 status) over the 1994–99 period. It is an area where there has been concern over the lack of innovation in SMEs resulting in several policy initiatives designed to encourage innovation and the adoption of new technologies. Part of the OMA was selected for inclusion to provide a contrast with the Lee Valley area since it has been a growing and dynamic economy with a high proportion of innovative SMEs. In fact, the latest Community Innovation Survey (Craggs and Jones, 1998) demonstrates that SMEs in Hertfordshire have a higher propensity to innovate than those in London, especially in manufacturing.[2]

This paper draws upon a number of surveys that have been conducted of SME owner/managers within the study area. First, a telephone survey was conducted in 1998 of 100 SMEs that, on the basis of answers to a series of filter questions, were shown to be "innovative" in that they had all introduced some changes to their business in terms of product, process or marketing innovations in the period 1993–98 The main aims of this survey were to (i) identify the main barriers which firms were facing in making different types of innovation; (ii) investigate the extent to which external resources and assistance are being used to support innovation in SMEs; and (iii) identify the extent to which SMEs have innovation support needs. This survey is also used as a "control group" in comparison with the surveys of clients of particular agencies or participants in particular programmes. The survey sample was focussed on three broad sector groups: (i) food processing; (ii) engineering (including computer manufacturing, machinery equipment manufacture, electrical and instrument engineering); and (iii) "business services" (including computer, telecommunications and other business services). A sector focus is justified by previous research which demonstrates considerable variation between sectors in the nature and extent of innovation in SMEs (North et al., 1997). The aim was to select sector groups that are potentially important to the economic base of the study region, whilst capturing the variety of types of innovation in SMEs by including sectors with different technology bases.

Secondly, telephone and some face-to-face surveys were conducted in 1998–99 with a sample of SMEs in the study area which had been recipients of various types of innovation support. They comprised: (i) 40 firms which had received funding under the Government's Smart Award Scheme; (ii) 70 firms which had received assistance from the Innovation and Technology Counsellors provided by Business Link; (iii) 62 firms that had been clients of the Lee Valley Centre, which is a university based technology support centre; and (iv) 14 firms which had been assisted by the Lee Valley Business Innovation Centre. These client surveys provided insights into the types of businesses that were being targeted, the type of assistance received and the extent to which it addressed the needs of the firms. Finally, face to face interviews were conducted with managers and support staff involved with the various initiatives; these helped identify problems and strategic issues affecting the delivery of innovation support in the region from the point of view of practitioners.

4. The key innovation support instruments

Figure 1 presents an overview of the innovation support system within the study area, identifying the main programmes and agencies involved. Much of the funding for innovation support comes from U.K. central government or EU level programmes, although some programmes apply everywhere throughout the U.K. whilst others apply specifically to firms in designated assisted areas. Figure 1 attempts to show the linkages between SMEs and the various programmes and agencies. In most instances, the experiences of SME owner managers will be at the level of dealing with the advisers and consultants within the regional level agencies themselves, rather than with those responsible for the management of the programmes.

The U.K. Government's Smart Scheme is a national programme which applies to eligible SMEs wherever they are located. Originally launched in 1986, Smart focuses on supporting the development of new technology-based products that are "new" within a sectoral or national context. It is based, therefore, on a narrower view of innovation than that used in other innovation support policies targeted at SMEs. Its rationale is also based upon the view that there is a gap in the venture capital market as a result of the reluctance of venture capitalists to make investments of less than about £250,000 (Bank of England, 1996). Smart takes the form of an annual competition for financial grants run by the Department of Trade and Industry (DTI) and aims to stimulate highly innovative but marketable technology in small firms. During the period of this study the award offered two kinds of grant: first, *feasibility studies* into innovative technology for firms with up to 50 employees; and secondly *development projects* up to pre-production prototype stage of new products and processes, involving a significant technological advance, for firms of up to 250 employees.

The second example of innovation support

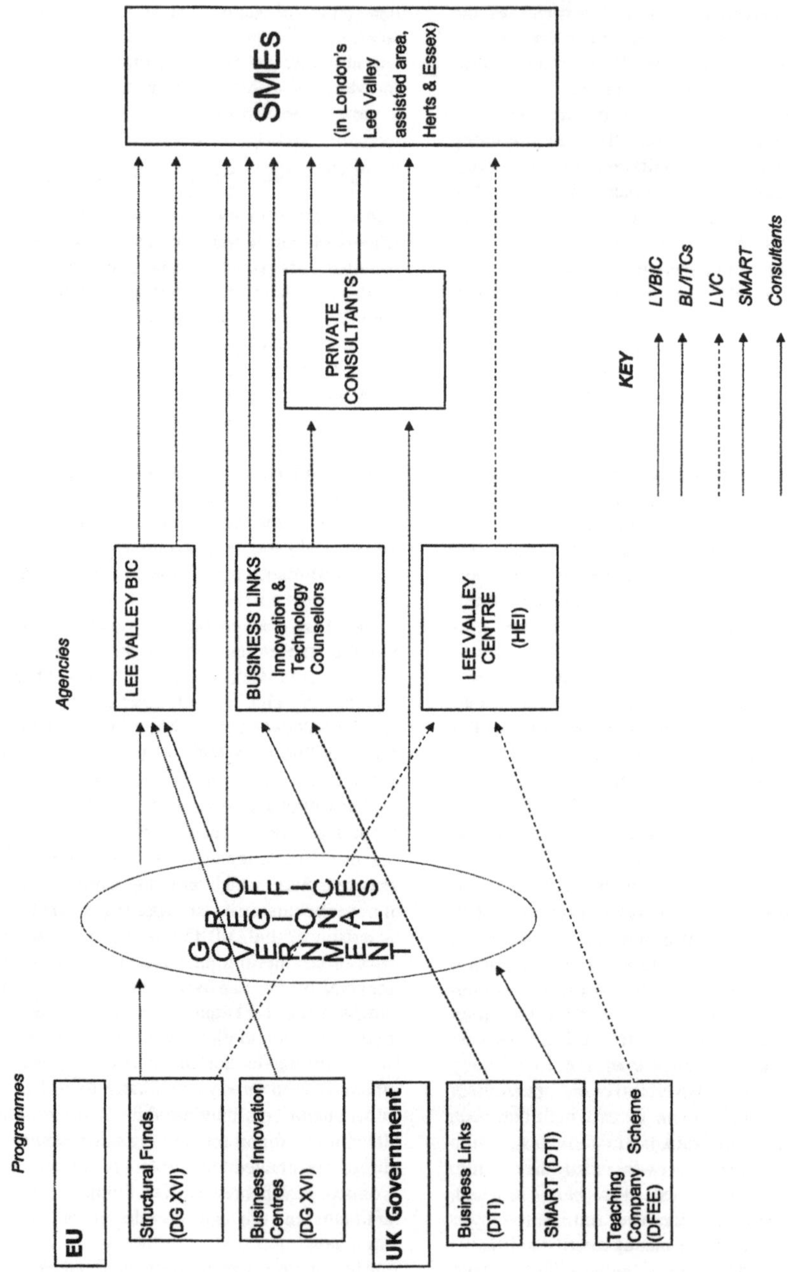

Figure 1. The innovation support system within the study region showing the main channels of assistance.

again originates with U.K. central government and has a national coverage, although in the form of a network of Business Link (BL) companies set up by the Department of Trade and Industry in 1993. Within the study region there are 3 such BL companies. BL is based on "the one-stop-shop concept" that encourages firms to access a wide range of business support services (including personal business advice, diagnostic and consultancy services, marketing training and advice, exporting and international trade advice, and financial advice and information) through a single channel. In addition, most BLs employ some specialist advisers including Innovation and Technology Counsellors (ITCs) whose role is to assist firms with new product development and the adoption of new technologies. A characteristic of BL is that it aims to offer broadly-based innovation support and encourages innovations that are "new to the firm" as well as those that are "new to the industry". Moreover, it takes a fairly holistic view of the development of businesses and attempts to provide an integrated range of services to client firms. Thus technical advice is unlikely to be given in isolation from advice on other aspects of the business.

The EU funding in the form of Structural Funds has been pivotal in initiating two regional level innovation support initiatives, namely the Lee Valley Centre (LVC) and the Lee Valley Business and Innovation Centre (LVBIC). As such, these initiatives have a role in raising the innovative capability of businesses within the Lee Valley region. The LVC was set up by Middlesex University in 1995, as part of the programme for the regeneration of the Lee Valley subregion using EU Structural Funds. It provides an example of a Higher Education Institution (HEI) becoming involved in assisting SMEs with innovation, design and technology transfer. Thus the stated aim of the Lee Valley Centre (LVC) has been "through the promotion of innovation, new technology, good design and effective communication, to help businesses expand and become more competitive and profitable and to stimulate industrial and commercial growth in the Lee Valley region". During the period 1995–1998 the Centre provided its services under four semi-autonomous functions: a design management centre, a telematics support centre, a technology transfer centre, and a Teaching Company Scheme centre.

The LVBIC is one of a Europe-wide network of 140 Business Innovation Centres (BICs) (of which 10 are in England) created since 1984 by the EC Regional Directorate, DGXVI, with a view to stimulating the creation and expansion of innovative SMEs in Less Favoured Regions of the European Union. BICs are therefore essentially regional/local scale initiatives for innovation support, although they represent part of a wider strategy for stimulating innovation in disadvantaged areas of the European Union. The support provided by BICs generally takes the form of services to support new product development, intellectual property protection, business and market planning, and provision of venture capital. The emphasis of the BICs is on the commercialisation of innovative ideas. In this respect they aim to provide a comprehensive package of support for innovative new ventures and existing projects. In emphasising the commercial application of innovation, BICs also aim to address the weakness that has been identified in small technology based firms of an over emphasis on technical development at the expense of marketing and general management skills (Monck et al., 1988; Oakey, 1991; Berry and Taggart, 1997).

The LVBIC was launched in November 1995 by a partnership led by London Borough of Enfield and including Middlesex University, Ford Motor Company, and North London Chamber of Commerce. It is located on a newly designated science park, The Lee Valley Innova Science and Business Park, which includes incubation units for innovative business start-ups. The distinctive characteristic of the LVBIC is its focus on the provision of an ongoing package of intensive support for selected innovative individuals and enterprises, with a particular emphasis on more technological opportunities and on product innovations.

The initiatives examined are not the only forms of innovation support available to SMEs in the study region, but they are the main ones. Perhaps the most important government scheme not directly evaluated for this study is the Teaching Company Scheme, which involves placing a graduate in the appropriate specialism in a company for a period of two years in order to support the transfer of knowledge and skills.[3]

Before looking at the effectiveness of each of the four policy support instruments, the next section of the paper will draw upon the results of the control group survey of 100 firms (described above) to gain some insights into the support needs of SMEs within the study region.[4]

5. Evidence on the innovation support needs of SMEs

One way of identifying "support needs" is to ask managers about the main barriers to innovation. With regards to product/service innovation, the majority of firms experienced constraints, although one constraint was identified more frequently than the others. Thus whilst three quarters of the firms actively involved in developing new products identified at least one barrier, finance proved to be the most commonly identified barrier (49% of firms), followed by time constraints (14%) and shortages of skilled labour (10%). However, there were significant sectoral variations (significant at the 0.01 level) in the extent to which finance was perceived as a constraint: 74% of engineering firms that expressed an interest in new product development, compared with 42% of business service and 32% of food processing firms. A similar sectoral pattern applies to those firms that have actively sought external finance to support product/service innovations. New product development in engineering is likely to be more costly in terms of materials and equipment than in other sectors and therefore more likely to be beyond the internal resources of the firm.

Similarly, finance was the main constraint identified by SME managers as affecting process innovations and, once again, this most commonly applied to engineering firms. Three quarters of those firms that had identified a need to raise the firm's level of technology and technological competence experienced barriers, with finance being identified by over half (59%). In the majority of these cases the difficulties resulted from raising finance internally (45%) rather than externally (14%). Financial constraints, both internal and external, were significantly more likely to be identified by engineering firms than by firms in other sectors (78% compared with 42% in business services and 16% in food processing) (0.0001 level), reflecting the more technically specialised and capital intensive character of engineering. Other barriers mentioned included: "a shortage of management time" (20%), "a lack of in-house expertise" (10%), or that "the volume of production was insufficient" to warrant technological upgrading (10%).

Although finance was identified as the main barrier to making both product/service and process innovations, a key finding was that the problem is not simply that firms looking for external finance cannot obtain it, but that most small firms do not seek it. For instance, although half of the firms actively interested in developing new products/services reported a shortage of finance as one of the main barriers to achieving it, less than one in five had previously sought external finance to support product/service innovation, and only slightly more had sought external finance for process innovation. Hence the financial constraints that SMEs report do not result solely from supply side failure but, as noted in section 2, from the conservative attitudes of many SME owners and managers to external finance. The problem is not simply that firms looking for external finance cannot obtain it, but rather that most small firms do not seek it either because they believe they stand little chance of getting it, or because they perceive it as too expensive, or because they simply wish to minimise their exposure to debt. In other words, the financial constraints on innovation that managers emphasised in the survey result from the reliance on finance that is internally sourced (i.e. retained profits and/or increased personal equity on the part of the owner) rather than necessarily because firms had been refused finance by banks or other providers. This suggests that one of the tasks of innovation support should be to encourage attitudinal change on the part of owner-managers with respect to external finance, where this is seen as necessary for the growth and development of the business. This would involve increasing their awareness of forms of financing (e.g. asset-based) that may reduce their perceived exposure to risk.

When the extent to which SMEs actually use external advice and consultancy to support the innovation process is examined, the key finding was that less than half of innovating firms had made use of such assistance over the 1993–98 period. Forty four per cent of those firms engaging in product or service innovation had made some use of external advice or consultancy, compared with 27% of those undertaking process innovation and 25% of those introducing new marketing methods (Table I). External advice and consultancy was more likely to have been utilised by engineering firms (64% of firms with innovative products), compared with 33% in food processing and 32% in business services (difference significant at the 0.05 level). This finding is congruent with previous research showing that external assistance is more commonly used in technology-based sectors, where firms are more likely to need to draw on specialist technical expertise outside of the firm (Smallbone et al., 1993; North et al., 1997).

It would appear that the majority of firms obtain external advice from private rather than public sector sources (Table I); just under half of the firms using external advice and consultancy used private consultants. In contrast, less than a fifth used Business Link, the most frequently used public sector source of assistance.

External support was found to be used most frequently by firms in the 10–49 size band, both in relation to finance and external advice and consultancy. For firms in this group sustained innovation can place heavy demands on the organisation's internal resource base, in terms of both finance and management, particularly if they are seeking to make the transition from owner-management to team management, which makes them a potential target for innovation support measures.

6. The effectiveness of innovation support in the study region

Having considered the support needs of SMEs and the extent to which they are seeking external assistance, attention is now focused on the kind of support that is currently available to SMEs within the study region, concentrating upon the four contrasting examples previously described: (i) the national Smart Award Scheme; (ii) the Innovation and Technology Counsellor support provided by Business Link; (iii) the regional level Lee Valley Centre; and (iv) the regional level Lee Valley Business and Innovation Centre.

6.1. *The Smart Scheme*

Given the strict nationally applied criteria that applicants for Smart have to meet, its appropriateness to the needs of firms in particular regional economies is likely to depend upon the regional sectoral and business structure. Within the study region the take-up of the Smart Scheme is much higher in the outer metropolitan area (especially in Hertfordshire) than in London, and especially in the Lee Valley area. As mentioned earlier,

TABLE I
Sources of external advice and consultancy by area of innovation

	Product	Process	Marketing	No. of firms using each source at least once	Percentage of those firms (n = 60) using external assistance
Private consultants	15	9	7	27	45%
Business link	7	1	4	10	17%
Suppliers	4	3	2	9	15%
Higher education	4	1	0	4	7%
Sector bodies	3	0	0	3	5%
Business associations	1	0	0	1	2%
Innovation centres	0	0	1	1	2%
Customer	0	0	0	0	–
Other	6	2	3	5	28%
Innovating firms using at least one form of assistance to support innovation	28 (44%)	15 (27%)	17 (25%)	60 (60%)	
Total no of innovating firms	63 (100%)	55 (100%)	69 (100%)	100 (100%)	

evidence indicates that SMEs in Hertfordshire have a higher propensity to innovate and are more likely to be in sectors with a strong R&D element than those in London. This is reflected in the pattern of demand for the Smart Scheme. A low demand and dearth of good quality Smart applications are found in areas like the Lee Valley, where the more technology-based industries are under-represented in the industrial structure.

In view of the fact that the Smart Scheme had been in existence for more than ten years at the time of the survey, a high level of awareness might reasonably be expected. Just over a third (37%) of firms in the control group survey were aware of the Scheme, but this rose to nearly two thirds (63%) in the case of the engineering sector where firms are more likely to be involved in technical development and therefore eligible for Smart.

From the survey of 40 client firms in the study region it is clear that Smart winners are drawn from a narrow segment of the SME population (see Smallbone et al., 2000b for a more detailed account). They are predominantly from manufacturing sectors, especially medical and optical instruments and various machinery and equipment manufacturing sectors, although a few computer service firms are also represented. The majority are young (a median age of 6 years) and very small firms (a median size of 4 employees). In line with the aims of the Smart Scheme, these firms are highly R&D oriented, both in terms of the proportion of staff time allocated and in terms of the proportion of current income dedicated to R&D.

The Smart Scheme is clearly fulfilling an innovation support need amongst the relatively narrow band of firms that qualify for it. Over half (57%) of the client firms identified finance as the main barrier to product innovation prior to obtaining a Smart award. The results also show a high level of additionality. According to their owners three quarters of the new businesses founded in the year that the Smart Award was received would not have started without the grant. In the case of existing businesses, 35% of the projects would not have gone ahead, 52% would have been delayed and a further 7% would have proceeded on a smaller scale. However, the Scheme appears less successful in its aim of helping innovative firms gain access to sources of external finance. In this respect there is little evidence of additionality in terms of levering additional private funding into the innovative projects of Smart winners. Only a third of firms had sought external finance after the award was obtained, and typically they raised just 10% (median) of the total project costs from private sources (from banks, venture capital companies, and informal risk capitalists). A further problem relates to the lack of support for commercialisation once the Smart project is completed; in this respect the successful achievement of commercial objectives is much less common (40% of surveyed clients) than the achievement of technical/design objectives (88% of clients).

Overall, it would seem that the Smart Scheme has been of limited value to the innovation support needs of both existing SMEs in the study region and for the majority of start-ups. Its focus, until recently, on more novel technology-based innovations and its strict criteria mean that very few firms have qualified to receive support, especially in the Lee Valley area. At the same time, the types of innovation projects supported by Smart can potentially contribute to the development of activities that are new to the region, although the very small numbers involved hitherto limits the overall impact. The low take-up of both Smart and the Teaching Company Scheme (which also tends to be focussed on more technology-based firms) within the Greater London region as a whole was identified as a cause for concern during interviews with regional government officers responsible for administering the Scheme.

There have been a number of recent developments aimed at addressing some of the identified shortcomings of Smart. First, Smart winners' clubs have been encouraged in a number of regions including London with the intention of helping to address some of the problems commonly experienced by innovative small firms (particularly with respect to commercialisation) and facilitating mutual support and learning between award winners. Secondly, following earlier criticisms of the overly stringent conditions of the Scheme, in 1999 it was expanded in England to include three further elements. These new elements provide smaller grants to support: (i) *Micro Projects* – low cost technology based innovations in businesses with fewer than 10 people; (ii) *Technology*

Reviews – expert reviews to help SMEs benchmark themselves against best practice technology in the sector; and (iii) Technology Studies – for SMEs to help identify technological opportunities which may lead to innovative products and processes. These new elements are aimed at helping small firms find new applications for existing technology, as opposed to the more novel developments typically supported by the Scheme. This is likely to have the effect of making the Scheme more appropriate to a wider range of SMEs, including the kind of "technology follower" SMEs that tend to predominate in the Lee Valley region.

6.2. *Business Link Innovation and Technology Counsellors*

In view of the high level of national publicity and media coverage that the Business Link (BL) initiative received during the 1990s, together with the wide range of business support services that can be accessed via BL, it is hardly surprising that 84% of control group respondents had heard of this support agency, a far higher level of awareness than with the other support measures. There was little variation in awareness levels between either size or sector groups. However, despite this high level of awareness overall, less than a fifth (17%) of the control group had used assistance from BL in relation to making their innovations. Thus, on this evidence, it appears to be a minority of innovating SMEs which avail themselves of the kind of services provided by BL's Innovation and Technology Counsellors (ITCs).

From the survey of 70 BL client firms, it appears that BL is providing assistance to innovating firms drawn from a wide range of different sectors, although the evidence indicates that the kind of assistance provided by ITCs may be more relevant to firms in the manufacturing (and particularly in engineering) rather than the service sectors. The majority of clients appeared to be well established, having been in existence for at least 10 years, indicating that it is not just young firms that are in need of innovation support.

A key feature of BL business support services is that they are targeted at firms that can demonstrate growth potential. The survey evidence indicates that this is being achieved since the BL clients had a better growth performance than did the innovative firms in the control survey; for example, the BL client firms achieved a median net employment growth of 38% over the 1993–98 period compared with a 24% increase in the control group. The implication is that there are likely to be firms that are trying to innovate but which are not qualifying for BL support because they are not growing sufficiently. There may be a tendency for BL advisers to select firms which can already demonstrate growth rather than those that could achieve it with appropriate assistance.

Given the broad approach to innovation that underpins the kind of support provided by BL, it is not surprising to find that the client firms have been involved in making a number of changes, particularly in terms of new market development and opening up new geographical markets. This is likely to reflect a particular strength of BL which is the provision of assistance with market development and the adoption of new marketing methods. The holistic approach to innovation adopted by BL means that firms that may be seeking help with product development or particular technical problems may end up receiving assistance with the marketing of their products and services. In fact, half (49%) of the surveyed firms have received assistance with marketing. BL is playing a key role, therefore, in providing the marketing support that innovating SMEs require if they are to exploit the full potential of their innovations.

Part of the rationale for BL is that it helps SMEs access the support available from private sector consultants and advisers. The evidence from BL clients suggests that this is being achieved, since three quarters of the firms had been directed to other sources of specialist expertise by the BL advisers. This proved to be less likely in the case of very small firms (64%) than medium firms (82%), perhaps indicating that BL advisers are more likely to be able to satisfy the more generic requirements of very small firms themselves.

In terms of the benefits of BL assistance to its client firms, the survey evidence indicates that BL does play an important role in helping certain types of SME overcome barriers to innovation. Whereas just under a third (31%) of the managers interviewed considered that the assistance received from BL had enabled them to overcome particular difficulties relating to making innovations, this applied to a higher proportion (40%) of the very small firms (i.e. in the 1–9 size band). This further suggests that the more generalist advice that BL advisers are able to provide is likely to be of more assistance to very small firms than to larger firms, where more specialist help may be required. Interestingly, the majority (88%) of the BL clients consider they would have gone ahead with their innovations in some form or another without BL assistance. Whilst this might appear to suggest that BL assistance has low additionality, its main value appears to be in terms of helping firms reap the full commercial benefits from their innovations.

6.3. *The Lee Valley Centre*

Compared to a national programme like BL with all its media promotion, the level of awareness of regional level innovation support agencies is much lower, with under a quarter (23%) of the control group respondents having heard of the Lee Valley Centre (LVC), and only 2% of them actually having received assistance from the Centre. As with other sources of innovation support, awareness levels tended to be higher amongst engineering firms (31%) than amongst business service (18%) and food firms (16%).

Many of the difficulties experienced by the LVC appear to relate to the problems of targeting SMEs and identifying their support needs. At the time of its formation, there was no systematic attempt by the LVC managers to carry out an audit of the needs of SMEs in the region. It was assumed that once established there would be a steady growth in demand for the services offered. In the event, some of the projects have clearly met the needs of SMEs in the region, whilst others have not. The most successful project has been for the telematics support centre, there being a clear demand from a wide range of SMEs for basic ICT training courses including Internet access and Webpage design. The design management centre has also met with an increasing demand for its services, particularly since having developed a "menu" of clearly identifiable services that it could offer businesses. However, the Teaching Company Scheme initiative and the technology transfer function have proved less successful, resulting in both functions being closed down in early 1998. In the case of the former, SMEs have been reluctant to make the commitment to taking on a graduate for a period of two years to assist with the transfer of technical and management skills. In the case of the latter, similar problems were experienced in engaging the interest of local SMEs in the ambitious range of manufacturing support and technology transfer services offered.

The survey of 62 LVC client businesses shows that the Centre attracts firms from a wide range of manufacturing and service sectors, including software, technical consultancy, wholesaling, food, clothing, and furniture. Not surprisingly, they consisted mainly of firms that were interested in improving their level of technology, with two thirds having a plan for upgrading their technological base. However, only a small minority (7%) considered themselves to be abreast of the latest technology and/or leaders in their field. This suggests that the kinds of services provided by the LVC are more likely to be relevant to "technology followers" rather than to "technology leaders".

According to the respondents, the most commonly identified barriers to improving the technology of the firm were "insufficient internal financial resources" (cited by 35% of clients) and "difficulty in obtaining external finance" (35%), followed by "lack of in-house expertise" (15%) and "not enough time" (11%). These findings are broadly in line with the barriers to improving technology experienced by firms in the control group (section 4). It indicates that the kind of support provided by the LVC will be worthwhile only if firms are also able to overcome these financial constraints to innovation and technical change. It also demonstrates the importance of raising awareness within local businesses of sources of external finance, changing their attitudes towards using external finance, and enabling them to access the finance available.

The LVC example also illustrates some of the well-documented impediments to collaboration between HEIs and SMEs which stem from a mismatch between the organisational structure and culture of universities and the needs and culture of SME owner managers (e.g. Charles and Howells, 1992; Hassink, 1996). First, the possibilities for collaborating with SMEs are affected by the extent to which the knowledge resources available correspond to the needs of businesses. In this respect, some of the LVC projects have received a number of enquiries from companies seeking expertise in disciplines not covered by the parent university. And secondly, a number of potential projects have not progressed because university specialists do not have the time nor the motivation to prioritise helping local SMEs. Loss of academic staff from the engineering departments, together with the more immediate pressures to cater for students, have meant that staff have been unable in some cases to provide the rapid response required by businesses.

6.4. *The Lee Valley Business and Innovation Centre*

As with the LVC, awareness of the Lee Valley Business Innovation Centre (LVBIC) amongst the control group was relatively low, with only a quarter of the respondents being aware of the Centre and none of them having received any assistance. Similarly, awareness levels were higher in the case of engineering firms (31% of them), compared with business service firms (18%) and food firms (22%). According to the LVBIC staff, the fact that they are not generally well-known is an advantage, given the large number of referrals and the limited resources available for providing such intensive support. The majority of LVBIC's clients (65%) are referred by Business Link, with a further 25% from banks and professional advisors. Of these referrals, 80% tend to be rejected straight away because they are judged to be insufficiently innovative and/or commercially non-viable.

To assess the effectiveness of the LVBIC, 14 face-to-face interviews were conducted with LVBIC clients during March 1999.[5] The interview findings indicate that the support provided by the LVBIC does address the needs of embryonic and start-up businesses, particularly given the prior negative experiences recounted by most of these clients with respect to the weak support for innovative individuals and small businesses in the U.K. The particular strength of the LVBIC appears to be the ongoing nature of a package of affordable support for selected innovative projects. A key component of the support package provided by LVBIC is the provision of finance in the form of a short-term loan or as equity, with the BIC's contribution typically being around 10–15%. LVBIC does not charge fees but aims to gain a return through taking out an equity in the innovating company or via a royalty return on licence. As well as providing much-needed financial support in the early stages of product innovation, the LVBIC has also helped some of its clients lever in funds from other sources, including obtaining bank loans under the DTI's Loan Guarantee Scheme (LGS), since commercial lenders feel more comfortable in supporting an innovatory project when the risk is shared with other partners. It has also helped some of its clients prepare successful applications for Smart awards, and is valued as an entry point for accessing other networks, including other sources of specialist expertise (i.e. universities) and other government and EU support programmes. Some clients, particularly those located in the incubation business units within the Centre, also valued the close proximity and contact with other innovative businesses. In this respect the Centre appears to offer an environment conducive to mutual support and learning.

Finally, a number of clients drew attention to a possible important area of weakness in the support service, notably an inability to support larger projects financially, particularly with respect to commercialisation, and the lack of access to sources of venture capital. With regard to the wider U.K. climate for innovative small enterprises, the key problem identified by all clients is insufficient financial support for innovation. In this respect the support that the LVBIC is able to provide is also limited, although at the time of the study efforts were being made to establish a hedge fund drawing on the European Regional Development Fund.

7. Conclusions

The four examples of public/quasi public sector support for innovation in SMEs in the study region differ in terms of their origins, their rationale, and the kinds of SMEs targeted. Although they do not constitute a fully integrated innovation support system, on balance they tend to complement rather than duplicate each other. In terms of the numbers of firms using public sector innovation support, it is Business Link that has the highest level of use and appears to be relevant to the needs of the largest proportion of firms across a wide range of sectors. In comparison, the Smart Scheme is targeted at a relatively small number of the most innovative firms in the technology-based sectors, and the Lee Valley Centre has experienced difficulties in targeting firms and in developing services that SME owner managers see as relevant to their needs. The LVBIC is playing a distinctive role in helping to incubate new businesses based on innovative products and services, rather than meeting the innovation support needs of existing firms.

Overall, the public support for innovation that exists within the study region has to be seen as rather selective, with many SMEs lying outside the target groups of the policy measures. This leads to the identification of three groups of firms that appear to have some innovation potential but which have been inadequately catered for by the available innovation support instruments:

(i) *Firms that are engaged in product/service innovation but which fall below the stringent criteria of the Smart Scheme*: Because of the focus of the Smart Scheme on cultivating more innovative applications of technology, there are large numbers of SMEs within the study region which are engaged in product and service innovation but do not qualify for financial assistance because they are either not in R&D intensive sectors or their innovations are of a more adaptive/incremental nature. Previously, some of these firms would have qualified for a Regional Innovation Grant which applied to SMEs within the Lee Valley Assisted Area, but this has since been withdrawn. As the various surveys have shown, finance emerges as the main barrier to making innovations, showing the need for some form of financial assistance with innovation with a lower "innovation" threshold. It is possible that new elements introduced to the Smart suite of assistance in England in 1999 will help to address this problem.

(ii) *Firms that are growth orientated and innovating but find growth difficult to achieve:* Whilst Business Link has been successful in implementing the national policy objective of supporting SMEs with growth potential, its targeting criteria may be somewhat conservative by focusing mainly on firms that are actually growing (not least because they are easier to identify) than on firms that are growth orientated but as yet not achieving it. The control group survey has shown that there are innovating firms in the SME population in the study region which are growth seeking but currently underachieving. There is likely to be potential added value in helping these firms to improve their performance.

(iii) *New firms:* Although there has been a recent recognition of the need to improve the availability and quality of information and advice for new business start-ups at a national level (particularly those with high growth potential), through most of the 1990s the emphasis of Business Link has been on support for established firms. Since new businesses represent one of the means for introducing new types of economic activity into a region, there would seem to be a case for a more concerted effort to encourage and support innovative start-ups, particularly those that are knowledge based and focused on the Lee Valley area. In the past, the Smart Scheme has been too narrowly focused to meet the support needs of all new firms of this type and the number of businesses assisted by Smart and the LVBIC combined is very small. In this regard, it should be noted that in December 1998, the U.K. Government announced a new programme of support for start-ups with high growth potential that would appear to address this gap.

The fact that problems in accessing finance to support innovation (including marketing/commercialisation) is a theme that emerges from both the control group survey and the evaluations of the individual policy initiatives indicates that this is a key deficiency to be addressed. The policy response here is likely to vary according to the type of innovative activity involved, ranging from encouraging and facilitating a bank loan to finance the acquisition of new equipment, to finding an equity partner to support the commercialisation of a new product. The overall conclusion here relates to the need for greater "holism" in the design of support instruments and their integration within the wider support system, including private sector sources of venture capital. The U.K. Government's commitment to the creation of new regional venture capital funds will clearly have an important potential role to play in this respect (DTI, 1998).

As identified in the second section of this paper, there is another rationale for an innovation support policy besides that of meeting the needs of the firms themselves. This has to do with the wider benefits associated with strengthening the innovative capability of firms within a regional economy. However, the study leads to the conclusion that the overall impact of the existing public policy measures on the innovative capability of SMEs in London's Lee Valley region is partial for various reasons. First, the small scale and fragmented nature of all but the Business Link initiative limits their overall effect. Secondly, the penetration of most of these support initiatives, as measured by the number of firms assisted, is typically small with only Business Link being used by a significant minority of firms and then mainly for non-specialist support. Thirdly, the survey of innovative SMEs revealed that there is a low level of awareness of the support available, again with the exception of Business Link services. This suggests that a more strategic and co-ordinated approach to providing innovation support is needed if the aim is to strengthen the innovative capability of SMEs in the region. In this respect, it remains to be seen how effective the current attempt to produce an Innovation and Knowledge Transfer Strategy for the London region will be in responding to this challenge (LDP/GOL, 1999).

Acknowledgements

The authors would like to thank the anonymous referee for his helpful comments on an earlier draft of this paper.

Notes

[1] This involved seven partner countries: Austria, Denmark, Italy, the Netherlands, Norway, Spain, and the U.K.

[2] Data supplied to the authors by the Department of Trade and Industry.

[3] The Teaching Company Scheme has been subject to a number of in-depth evaluations in recent years, e.g. Senker and Senker (1994) and TCS (1997).

[4] A fuller analysis can be found in Smallbone et al. (2000a).

[5] The interviewees were taken from a list of 21 provided by the Centre's Managing Director. Without a comprehensive list of clients it was not possible to determine the sample's representativeness of the total client base. The desire for confidentiality was a key issue constraining the authors' access to information on the LVBIC, both on the part of the Director of LVBIC and senior staff in Directorate Generale XVI which oversees the BIC network as a whole.

References

Asheim, B. T., 1996, 'Industrial Districts as "Learning Regions": A Condition for Prosperity?', *European Planning Studies* **4**(4), 379–400.

Bank of England, 1996, *The Financing of Technology-Based Small Firms*, London: Bank of England.

Berry, M. J. and J. H. Taggart, 1998, 'Combining Technology and Corporate Strategy in Small High Tech Firms', *Research Policy* **26**, 883–895.

Bessant, J., J. Burnell, R. Harding and S. Webb, 1993, 'Continuous Improvement in British Manufacturing', *Technovation* **13**(4), 241–254.

Bessant, J., S. Caffyn, J. Gilbert, R. Harding and S. Webb, 1994, 'Rediscovering Continuous Improvement', *Technovation* **14**(1), 17–29.

CBI, 1997, *Tech Stars: Breaking the Growth Barriers for Technology-Based SMEs*, London: Confederation of British Industry.

Charles, D. and J. Howells, 1992, *Technology Transfer in Europe: Public and Private Networks*, London: Belhaven Press.

Cohen, W. M. and D. A. Levinthal, 1990, 'Absorptive Capacity: A New Perspective on Learning and Innovation', *Administrative Science Quarterly* **35**, 128–152.

Cooke, P. and K. Morgan, 1998, *The Associational Economy*, Oxford: Oxford University Press.

Craggs, A. and P. Jones, 1998, *U.K. Results from the Community Innovation Survey*, Economic Trends, No. 539, October 1998.

DTI, 1998, *Our Competitive Future: Building the Knowledge Driven Economy*, Department of Trade and Industry Government White Paper, London: HMSO.

Hassink, R., 1996, 'Technology Transfer Agencies and Regional Economic Development', *European Planning Studies* **4**(2), 167–184.

Hassink, R., 1997, 'Technology Transfer Infrastructures: Some Lessons from Experiences in Europe, the U.S. and Japan', *European Planning Studies* **5**(3), 351–370.

HM Treasury, 1998, *Financing of High Technology Business: A report to the Paymaster General*, London: HM Treasury.

LDP/GOL, 1999, *London: The Knowledge Capital. An Innovation and Knowledge Transfer Strategy for London*, London: The London Development Partnership and Government Office for London.

Lundvall, B.-Å. and S. Borrás, 1997, *The Globalising*

Learning Economy: Implications for Innovation Policy. Luxembourg: Office for Official Publications of the European Communities.

Monck, C. S. P., R. B. Porter, P. R. Quintas, D. J. Storey, and P. Wynarczyk, 1988, *Science Parks and the Growth of High Technology Firms*, Croom Helm, Beckenham, Kent.

Morgan, K. and C. Nauwelaers (eds.), 1999, *Regional Innovation Strategies: The Challenge for Less Favoured Regions*, London: Stationery Office.

North, D., D. Smallbone and R. Baldock, 1997, *Innovation and New Technology in Small Rural Firms*, Rural Research Report No. 33, London and Salisbury: Rural Development Commission.

Oakey, R. P., 1991, 'High Technology Small Firms: Their Potential for Rapid Industrial Growth', *International Small Business Journal* **9**(4), 30–42.

Porter, M., 1990, *The Competitive Advantage of Nations*, London: Macmillan.

Rothwell, R. and W. Zegveld, 1982, *Innovation and the Small and Medium Sized Firm*, London: Frances Pinter.

Seaton, R. A. F. and M. Cordey-Hayes, 1993, 'The Development and Application of Interactive Models of Industrial Technology Transfer', *Technovation* **13**(1), 45–53.

Senker, P. and J. Senker, 1994, 'Transferring Technology and Expertise from Universities to Industry: Britain's Teaching Company Scheme', *New Technology, Work and Employment* **9**(2).

Smallbone, D., 1997, 'Selective Targeting in SME Policy: Criteria and Implementation Issues', in D. Deakins, P. Jennings and C. Mason (eds.), *Small Firms: Entrepreneurship in the 1990s*, London: Paul Chapman Publishing.

Smallbone, D., R. Leigh and D. North, 1993, 'The Use of External Assistance by Mature SMEs in the U.K.: Some Policy Implications', *Entrepreneurship and Regional Development* **5**(3), 279–295.

Smallbone, D., D. North, I. Vickers and I. McCarthy, 2000a, *SME Policy and the Regional Dimension of Innovation: U.K. National Report*, London: Centre for Enterprise and Economic Development Research, Middlesex University.

Smallbone, D., D. North, I. Vickers, and I. McCarthy, 2000b, 'Policy Support for R&D in SMEs: The U.K. Governments Smart Award Scheme', Chapter 11 in W. During, R. Oakey and M. Kipling (eds.), *New Technology-Based Firms at the Turn of the Century*, Oxford: Pergamon, pp. 143–159.

TCS, 1997, *Quinquennial Report*, Faringdon, Oxon: The Teaching Company Directorate.

Thomas, K., 2000, 'Creating Regional Cultures of Innovation? The Regional Innovation Strategies in England and Scotland', *Regional Studies* **34**(2), 190–198.

[11]

ELSEVIER

Research Policy 30 (2001) 611–624

research policy

www.elsevier.nl/locate/econbase

The relative effectiveness of patents and secrecy for appropriation

Anthony Arundel *

Maastricht Economic Research Institute on Innovation and Technology (MERIT), Senior Research Fellow, Tongersestraat 49, P.O. Box 616, 6200 Maastricht, Netherlands

Received 20 May 1999; received in revised form 6 September 1999; accepted 10 February 2000

Abstract

Changes in patent law, global competition, and the rise of several new technologies suggest that patents could be of greater value than secrecy for many innovative firms, particularly small firms that are unable to use market power or other factors to appropriate their investments in innovation. Data from the 1993 European Community Innovation Survey (CIS) for up to 2849 R&D-performing firms is analysed to determine the relative importance of secrecy vs. patents. This particular combination of appropriation methods is interesting because secrecy can act as a mutually-exclusive alternative to patents. The results show that a higher percentage of firms in all size classes rate secrecy as more valuable than patents. However, ordered logit regressions show that the probability that a firm rates secrecy as more valuable than patents declines with an increase in firm size for product innovations, while there is no relationship for process innovations. The firm's R&D intensity has no effect on the relative value of secrecy vs. patents. There is weak evidence to show that participation in cooperative R&D increases the value of patents over secrecy for product innovations. The results are relevant to ongoing debates over changes to the patent system and the need for policies to encourage small firms to patent more. © 2001 Elsevier Science B.V. All rights reserved.

Keywords: Appropriation; Innovation surveys; Patents; Secrecy

1. Introduction

Both anecdotal and empirical research suggest that patents have increased in importance since the early 1980s. The evidence for this ranges from a marked increase in the number of patent applications in the United States, from approximately 60,000 in 1984 to 120,000 in 1995 (Kortum and Lerner, 1999), to the decision of several large firms, such as Texas Instruments and IBM, to aggressively pursue licensing revenue from their patented inventions (Thurow, 1997). These changes are partly thought to be driven by the establishment in 1982 of a new Federal Court of Appeals, which strengthened the ability of firms to enforce their patent rights in the United States. A second possible explanation is an increase in the economic importance of proprietary knowledge, due to a shift from competition based on price towards competition based on technical innovation. A third explanation is the rise of new technologies, such as biotechnology and information technology, where a large number of small firms have been active. These small firms could rely on patents to signal expertise,

* Tel.: +31-43-388-3702; fax: +31-43-321-6518.

E-mail address: a.arundel@merit.unimaas.nl (A. Arundel).

PII: S0048-7333(00)00100-1

either to attract research partners or investment (Mazzoleni and Nelson, 1998). All of these factors, taken together, are thought by some observers to have created a 'pro-patent era'.

A shift in business strategies towards patenting could be marked by either an increase in patent propensity rates (the percentage of innovations that are patented) or a more intensive exploitation of patents, or both. Whether or not either of these outcomes has occurred is subject to debate. Both Thurow (1997) and Granstrand (1999) use case studies and other data sources to argue that firms are exploiting their patents more intensively. In contrast, Kortum and Lerner's (1999) analysis of patent statistics shows that the increase in patenting in the United States is more likely to be due to an increase in the generation of patentable inventions than to an increase in patent propensities, although their results do not address the issue of whether or not patents are exploited more intensively.

The European policy-making community tends to accept the 'pro-patent' view, as illustrated by a recent report on patent policy to the European Commission (ETAN, 1999) and the discussion of patenting issues in major European policy documents such as the *Green Paper on Innovation* and the *First Action Plan for Innovation*. The possible rise of a 'pro-patent era' has created some concern, based on the belief that the presumed decline in the global competitiveness of European firms is partly due to their failure to aggressively use patents to protect their investments in innovation. The response of policy-makers within the European Commission, as shown by the policy prescriptions in the *Green Paper on Innovation* and the *First Action Plan for Innovation*, is to encourage European firms, particularly Small- and Medium-sized Enterprises (SMEs), to patent more. Another thrust of European policy is to strengthen the patent system by lowering application costs and by considering other changes, such as the introduction of a grace period, that would permit the release of information about an invention without forgoing the right to apply for a patent at a later date.

These policy proposals to encourage patenting do not always consider why firms might choose alternative appropriation methods to patents. Patents have several disadvantages as an appropriation mechanism, which partly explains why the majority of innovations in the United States and Europe in the early 1990s were not patented (Arundel and Kabla, 1998). One important drawback to patents is the requirement to fully disclose the invention, since the disclosed information can release valuable information to competitors on potentially profitable research areas or how to invent around the patent. The importance of disclosure as a reason not to patent has been established in both theoretical models of patent behaviour (Horstmann et al., 1985; Scotchmer and Green, 1990; Harter, 1993) and in survey-based research (Arundel and Kabla, 1998; Cohen et al., 1998). Firms that do not wish to disclose information can forgo patenting and use secrecy to protect their investment in the invention.

Given that patents and secrecy are mutually exclusive in this scenario, an interesting question is which method is more effective as an appropriation mechanism, and under what conditions? I would argue that there is a widespread although subtle bias on the part of some policy-makers, economists, and jurists in favour of patents, which are seen as a 'gold standard' for effective appropriation. As an example, Friedman et al. (1991) comment that some of the legal decisions in the United States on trade secrecy suggest that 'no rational person with a patentable invention would fail to seek a patent'. Although Friedman et al. then give three situations in which secrecy would be preferred to patents, they agree with the underlying assumption that patents are best. Similarly, some of the theoretical literature assumes, either implicitly or explicitly, that all inventions are patented (Tandon, 1982; Scotchmer, 1991), leaving no room at all for secrecy. The theoretical models of patenting and disclosure also start with the assumption that firms will patent all of their inventions *unless* disclosure is of sufficient concern to impel the firm not to patent, although a more balanced perspective is provided by Takalo (1998).

This preference for patents is called into question by the empirical evidence based on innovation surveys in the United States (Levin et al., 1987; Rausch, 1995; Cohen et al., 1998), Europe (Arundel et al., 1995; Harabi, 1995), and Australia (McLennan, 1995). These surveys consistently show that manufacturing firms give secrecy a higher average rating as an appropriation method for both product and process innovations. The single exception is Japan, where firms give a higher rating to patents for product innovations, although secrecy is more valuable for process innovations (Cohen et al., 1998).

The survey results for process innovations can be explained by the ability of firms to keep process innovations hidden from their competitors for long periods of time. In contrast, the role of secrecy is not as evident for product innovations, since once the product is on the market it can be reverse-engineered by competitors. This creates a puzzle. One explanation is that secrecy could be used during the pre-market development phase. This would give the firm time to refine its invention and build up a lead-time advantage over competitors.

These possible uses of secrecy for product innovations show that patents and secrecy are not always mutually exclusive appropriation methods, as assumed by many of the theoretical models cited above. A firm could use secrecy to protect an invention during a development phase and then rely on patents or other appropriation methods when the invention is on the market. Of course, firms can also use secrecy for some of their innovations and patents for others. The ability of firms to strategically use both patents and secrecy complicates an evaluation of their relative effectiveness. It also suggests that a clear understanding of appropriation strategies requires better information about the conditions that impel firms to prefer secrecy to patents — or the *relative* value of secrecy vs. patents to appropriation.

One factor of policy significance, given the desire in the European Union to encourage SMEs to patent more, is the relationship between firm size and the effectiveness of secrecy and patents. One possibility is that small firms could find patents to be more effective than secrecy. In addition to the value of patents for small high-technology firms, as noted above, small firms frequently lack the manufacturing capacity or marketing networks to be able to rapidly recoup their investment in innovation through the sales of their own products. Consequently, these firms could be forced to rely on patents to create a breathing space from competition while they build up their manufacturing or marketing capabilities. In contrast, large firms could be less reliant on patents, both because they can use their marketing strength to create lead-time advantages and because large markets increase the financial returns from investment in process innovation (Cohen and Klepper, 1992; Klepper, 1996), where patents are of less value than secrecy.[1]

There are also several plausible reasons why small firms could find patents of *less* value than large firms, leading them instead to rely more on secrecy. In addition to patent application costs, which could be a greater barrier to small than large firms, small firms could find it exceedingly difficult to protect their patents from infringement, which would increase the relative value to them of secrecy compared to patents. This difficulty could be expressed, relative to large firms, as either a higher level of concern over legal costs or as a lower appreciation of the effectiveness of patents as an appropriation mechanism. Another possibility is that many small firms, with the exception of those that are pursuing a high-technology strategy, could be less likely than large firms to develop patentable innovations. Instead, many of their innovations could be based on minor incremental improvements that are not worth patenting.

Policies to encourage firms to patent more, particularly high-technology SMEs, could improve the competitiveness of these firms if patents are relatively more effective, under real conditions, than secrecy. In contrast, these policies could be less effective than expected if firms find secrecy to be more valuable than patents. Clearly, the design of patent policy would benefit from empirical evidence on the relative importance of secrecy and patents and a better understanding of the factors that shape the decision to keep an invention secret or to seek patent protection.

The purpose of this paper is to provide some empirical evidence of relevance to these issues, using

[1] The greater manufacturing and marketing capabilities of large firms imply that they should also patent proportionately fewer of their innovations than small firms. Some evidence on this issue is available from the PACE survey of Europe's largest industrial firms, although many of these 'firms' are in fact divisions with sales in the range of SMEs. The PACE results show that the percentage of both product and process innovations that are patented *increases* with sales as a measure of firm size (Arundel and Kabla, 1998).

the results of the 1993 Community Innovation Survey (CIS) in Europe. The empirical analyses focus on three questions. First, what is the relative importance of secrecy vs. patents for European manufacturing firms? Second, do small firms find patents of greater value than secrecy compared to large firms? Third, what factors other than size influence the relative importance of secrecy vs. patents? The results also provide some information of relevance to the debate over a 'pro-patent' era, although it is not possible to determine, using a cross-sectional survey at one point in time, if the value of patents has been changing over time.

The most important questions for European policy concern the relative importance of patents and secrecy for Europe's most innovative firms. For this reason, all analyses are limited to a subset of R&D-performing firms. Of course, firms that do not perform R&D can also develop patentable innovations, but many firms in this group probably only develop very minor innovations that do not meet the novelty requirements for a patent. These firms are, therefore, excluded in order to maximise the percentage of firms that have the option of using patents, in addition to the option of using secrecy. Furthermore, a few of the analyses are limited to a subset of highly innovative firms with R&D intensities (RDINTENS) above 10%.

The measure of the relative importance of patents vs. secrecy is at the firm level and therefore differs from the published results of other surveys which compare average differences in the importance of patenting and secrecy. For example, the most commonly used approach is to calculate the percentage of firms that give a rating of '4' or '5' on a five-point subjective scale. These are subject to inter-rater differences in the interpretation of the importance scale. Using relative scores, as described below, avoids these problems.

The results show that firms of all sizes find secrecy to be relatively more important than patents, but small firms find secrecy to be of greater importance than larger firms. This relationship with firm size, however, is less marked for small firms with RDINTENS above 10%. Several additional factors, such as the types of information sources used by the firm, also influence the relative importance of secrecy vs. patents.

2. Data sources and methodology

The 1993 CIS contains information on the value of both secrecy and patents for innovative manufacturing firms in Norway plus six EU countries: Germany, Luxembourg, the Netherlands, Belgium, Denmark, and Ireland. For further details on the CIS methodology, see Archibugi et al. (1994). The combined crude response rate for these seven countries is 36.1%, although the response rate varies from a low of 22% for Germany to a high of 79% for Luxembourg. The response rate exceeds 50% in Luxembourg, Norway, Denmark, and the Netherlands. The response bias in the 1993 CIS is in the direction of more innovative firms. This means that the response rate for R&D-performing firms is probably higher than the overall response rate. This bias will also favour the value of patents over secrecy, assuming that R&D-performing firms are more likely to develop patentable innovations.

A firm is defined in the CIS as innovative if it introduced a product or process innovation between 1990 and 1992. This definition of an innovator includes firms that only make minor improvements or introduce new products and processes developed by other firms. Many of these 'minor' innovators might not develop patentable innovations. To avoid this problem, all analyses are limited to a subset of firms that report that they perform R&D on a continuous basis. A maximum of 2849 R&D-performing firms are available for analysis, of which 1021 firms have fewer than 100 employees and 1689 firms have fewer than 250 employees.[2]

[2] Firms with more than 100 employees and with a reported R&D intensity of over 100% (less than 0.5% of the sample) are also excluded from the analyses because such high R&D intensities are unlikely to occur in firms of this size, suggesting either errors in data coding, or that the firm is a legally independent unit that provides R&D services to a much larger parent firm. In the latter case, an accurate analysis would require linking the R&D unit to the parent firm, which is not possible with the CIS data. Firms with less than 100 employees and R&D intensities over 100% are included because many of them will be new technology-based firms where R&D expenditures can exceed revenue for many years.

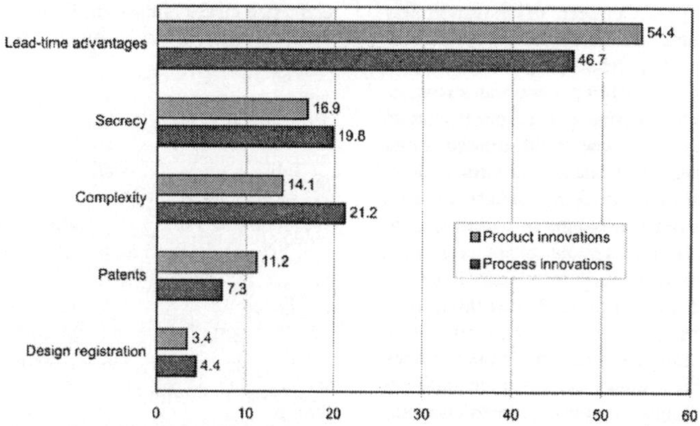

Fig. 1. Percentage of 2849 R & D-performing firms that give their highest rating to each appropriation method.

The relevant CIS question asks each respondent to 'evaluate the effectiveness of the following methods for maintaining and increasing the competitiveness of product [or process] innovations introduced during 1990–1992'. The methods include patents, secrecy, design registration, complexity of product design, and lead-time advantages over competitors. Five ordinal response categories are provided, ranging from 'insignificant' to 'crucial'.

One option for the presentation of descriptive results is to use the mean score for patents and secrecy, while another option is to use the percentage of respondents who reply that an appropriation method is 'very significant' or 'crucial'. However, both of these methods suffer from inter-respondent variation in what is meant by each of the subjective response categories. For example, a response of 'very significant' for one firm could be equal to a response of 'slightly significant' for another firm.[3] This can create significant amounts of noise in the data that can mask possible differences in appropriation methods by a variable of interest, such as firm size or sector of activity. As an example, inter-rater differences in the interpretation of a subjective scale could have masked sectoral differences in the importance of secrecy in analyses using data from the American Yale survey of the early 1980s (Cockburn, 1992; Cockburn and Grilliches, 1988).[4]

The method used here avoids these problems by relying on the internal consistency of each respondent's answers. For example, assume that the respon-

[3] Inter-rater differences can be checked by including different questions about the same problem in the survey or by empirical tests of what respondents mean by their ratings. Although these techniques are widely used in health surveys (Streiner and Norman, 1995), it is not possible to use these techniques for the CIS.

[4] Cockburn (1992) evaluates five possible causes for statistical noise in the Yale survey: coding errors, scoring and scale construction, respondent heterogeneity in the interpretation of the rating scales, firm vs. sector conditions, and the wording of the question. Many of these problems either do not apply to the 1993 CIS or the analytical method used here avoids them. First, coding errors are unlikely to be a serious problem in either survey. Second, the Yale survey asks the respondents to evaluate the importance of each appropriation method for their 'line of business', while the CIS asks for an evaluation for the respondent's own firm. The CIS version thereby avoids problems in confusing conditions for the firm and for the sector. Another effect of this change is that greater heterogeneity is expected in the CIS than in the Yale survey, since micro differences at the firm level are common, whereas the Yale survey should produce greater agreement among the respondents about the importance of each appropriation method for their line of business. Third, the CIS question is simpler than the Yale version, although there is still room for improvement. Fourth, the remaining two problems with scoring and inter-rater differences in the interpretation of the rating scale are avoided by the method of analysis based on relative differences in the importance of patents and secrecy.

dent for firm A gives a score of 3 (moderately significant) to patents and a score of 4 (very significant) to secrecy, while the respondent for firm B gives a score of 5 (crucial) to patents and a score of 4 to secrecy. Both respondents give an equal score to secrecy, but we do not know if the importance of secrecy is objectively equivalent, since respondents A and B could be using different subjective scales. In contrast, we can safely assume that each respondent gives internally consistent responses, so that the *relative* importance of secrecy vs. patents is robust. Using this method, we can conclude that the respondent for firm A rates secrecy as more important than patents while the respondent for firm B rates secrecy as less important than patents. A further advantage of this method for comparing two appropriation methods is that it is unaffected by the trend for large firms to give all appropriation techniques higher scores than small firms (Arundel, 1997).

This method is used to compare the relative importance of different appropriation methods, to determine the relative importance of secrecy vs. patents, and to serve as the dependent variable for a series of regression analyses, as discussed below.

3. Descriptive results

In order to place the relative importance of patents and secrecy in perspective, Fig. 1 gives the percentage of firms that gave their highest score to each of the five appropriation methods.[5] Lead-time advantages received the highest score from 54.4% of the respondent firms for product innovations and from 46.7% of firms for process innovations. Both patents and secrecy are given the highest rating by a notably lower percentage of firms. There is little variation in the rankings by firm size (results not shown). For example, lead-time advantages consistently receive the highest ranking for product innovations across all eight size classes shown in Table 1 below, followed by complexity, secrecy, patents, and design registration.[6]

3.1. Firm size, secrecy, and patents

Table 1 gives the relative importance of secrecy vs. patents for eight size classes. The rows in each subsection of Table 1 for products and processes add to 100%, aside from rounding-off errors. For example, all 183 firms with less than 19 employees are accounted for in the results for product innovations: 17.5% report that patents are more important than secrecy, 38.3% give equal ratings to patents and secrecy, and 44.3% give a higher score to secrecy than patents.

The results for product innovations show that a higher percentage of firms, in all size classes, rate secrecy as more important than patents. Furthermore, secrecy is relatively more important than patents for small firms than for large firms. The trend for the relative importance of secrecy to decline with firm size is statistically significant ($p = 0.001$). Most of the trend is due to a decline in the percentage of firms that give equal scores to patents and secrecy. The results differ for process innovations. The percentage of firms that give a higher score to patents than to secrecy remains comparatively stable, between 9.7% and 12.4%, with the exception of the largest size class. The trend by size class is also not statistically significant.

The results given in Table 1 are for all R&D-performing firms, including firms that spend very little on R&D. Since small, highly innovative firms have attracted a great deal of policy interest, the analyses were repeated after limiting the firms to those with RDINTENS above 10%, as shown in Table 2. The number of size classes has been reduced from eight to five to account for the smaller

[5] Tied maximum scores are divided equally between the relevant methods. For example, assume that the highest score given by firm A is 4, and that it is given to both patents and secrecy. Both of these methods would then be allotted 0.5 firms each. The percentages sum to 100% of the total firms for product and processes separately.

[6] The high importance given to lead-time advantages raises an interesting possibility in respect to patents that would be worth investigating in the future. Firms could forgo patenting if it reduces the ability of the firm to pursue lead-time advantages, e.g., if the time required to prepare patent applications distracts staff from more important tasks. This could be one explanation for Scherer et al., (1965) finding that the number of patent applications is closely related to the number of in-house patent lawyers.

Table 1
Relative importance of patents and secrecy for all R&D-performing firms (standard errors in parentheses)

Employees	N	Product innovations			Process innovations		
		Patents more important	Equal importance	Secrecy more important	Patents more important	Equal importance	Secrecy more important
< 19	183	17.5 (2.8)	38.3 (3.6)	44.3 (3.7)	10.4 (2.3)	40.4 (3.6)	49.2 (3.7)
20–49	386	17.6 (1.9)	23.6 (2.2)	58.8 (2.5)	12.4 (1.7)	27.5 (2.3)	60.1 (2.5)
50–99	452	23.0 (2.0)	28.5 (2.1)	48.5 (2.4)	11.1 (1.5)	37.4 (2.3)	51.5 (2.4)
100–249	668	20.7 (1.6)	28.0 (1.7)	51.3 (1.9)	11.8 (1.3)	35.9 (1.9)	52.2 (1.9)
250–499	479	20.5 (1.8)	30.1 (2.1)	49.5 (2.3)	12.3 (1.5)	29.6 (2.1)	58.0 (2.3)
500–999	319	24.5 (2.4)	24.8 (2.4)	50.8 (2.8)	9.7 (1.7)	23.2 (2.4)	67.1 (2.6)
1000–1999	186	23.7 (3.1)	33.9 (3.5)	42.5 (3.6)	10.8 (2.3)	30.6 (3.4)	58.6 (3.6)
> 2000	176	30.7 (3.5)	26.1 (3.3)	43.2 (3.7)	19.9 (3.0)	23.3 (3.2)	56.8 (3.7)
Significance of the trend by size				$p < 0.0001$			ns

number of firms. The trend is no longer statistically significant for either product or process innovations, although for all size classes more firms find secrecy to be more important than patents than the reverse. A comparison of Tables 1 and 2 indicates that small R&D-intensive firms find patents to be more important than the average for all small R&D-performing firms. For example, 33.3% of R&D-intensive firms with a workforce between 100 and 249 employees find patent more important than secrecy, compared to 20.7% of all R&D-performing firms in this size class.

3.2. Other influences on the relative importance of secrecy vs. patents

The results of Tables 1 and 2 establish that R&D-performing firms rank secrecy as a more effective appropriation mechanism than patents for both product and process innovations. Furthermore, Table 1 shows that a higher percentage of small than large R&D-performing firms give a higher rating to secrecy for product innovations, although there is no notable difference by firm size for the most R&D-intensive firms. However, before concluding that it will be an uphill battle to encourage more patenting among SMEs, it is important to address other possible influences on the relative importance of these two methods and to check for confounding factors. Other than firm size, three groups of factors come to mind that could influence the relative importance of secrecy vs. patents: the firm's own innovation strategies, the influence of different types of information sources on the firm's innovative activities, and technology effects linked to the firm's sector of activity. Sections 3.3–3.5 describe each factor and the con-

Table 2
Relative importance of patents and secrecy by firm size for R&D-intensive firms (standard errors in parentheses)

Employees	N	Product innovations			Process innovations		
		Patents more important	Equal importance	Secrecy more important	Patents more important	Equal importance	Secrecy more important
< 19	59	22.0 (5.4)	33.9 (6.2)	44.1 (6.6)	15.3 (4.7)	39.0 (6.4)	45.8 (6.5)
20–99	92	23.9 (4.5)	27.2 (4.7)	48.9 (5.2)	16.3 (3.9)	25.0 (4.5)	58.7 (5.2)
100–249	57	33.3 (6.3)	19.3 (5.3)	47.4 (6.7)	12.3 (4.4)	36.8 (6.4)	50.9 (6.7)
250–999	53	24.5 (6.0)	24.5 (6.0)	50.9 (6.9)	9.4 (4.1)	24.5 (6.0)	66.0 (6.6)
> 1000	40	32.5 (7.5)	30.0 (7.3)	37.5 (7.8)	22.5 (6.7)	17.5 (6.1)	60.0 (7.8)
Significance of trend by size				ns			ns

struction of variables to capture their effect in the regressions. The influence of these factors on the relative importance of secrecy vs. patents is explored using regression models.

3.3. Innovative strategies

Two measures of a firm's innovative strategy are constructed using the CIS data. First, an important characteristic of a firm's innovative strategy is its RDINTENS. This could be positively related to the value of patents by increasing the proportion of inventions that are patentable, or by increasing the value of competing on the basis of innovation relative to price. A positive effect for RDINTENS is also suggested by the difference between Tables 1 and 2, particularly for small firms, as noted above.

Second, the percentage of R&D spent on product (RDPROD) and process (RDPROC) innovation could influence the relative importance of secrecy and patents. One would expect firms that focus on product innovation to rate patents more highly than secrecy, and vice-versa for firms that focus on process innovation.

3.4. Information sources

Firms can acquire the information that they need to innovate from both internal sources, such as from their own in-house R&D activities, and from external sources, such as the technical literature, patent databases, customers, their parent firm when applicable, and through cooperative research with other firms or publicly-funded research institutions. Firms that extensively use external sources that require the sharing of valuable information could find patents of great value. An example is cooperative research, where patents could play an important role in clarifying ownership of the results. A recent study that linked patent data to the CIS results for the Netherlands found that firms that participate in collaborative R&D are more likely to apply for a patent than firms that do not participate in collaborative R&D (Brouwer and Kleinknecht, 1999). Alternatively, firms that focus strongly on internal information sources could give greater emphasis to secrecy. Another factor is related to ownership. Subsidiary firms could rely on their parent firm for major innovations and therefore find patents of less value than secrecy.

The CIS contains relevant data for three aspects of information sources that are relevant to the trade-off between secrecy and patents. First, the CIS contains one question, in a yes-or-no format, that asks if the firm had any cooperative R&D arrangements with other enterprises or institutions in 1992 (COOPRD). Second, another group of questions are used to identify subsidiaries. Third, a series of questions ask about the value to the firm's innovative activities of both internal sources within the firm and 11 external information sources. A dummy variable is constructed using this information (INTERNAL). Firms are classified as focusing on in-house information sources if the rating for sources within the firm is equal to or higher than eight external sources that require sharing information.[7] Three publicly available information sources are not included in the construction of this variable because they are less likely to require the sharing of information and therefore should not affect the relative value of secrecy and patents to the inventive firm: patent disclosures; conferences, meetings or journals, and fairs or exhibitions.[8]

3.5. Technology or sector effects

Innovation surveys, including the CIS, have shown that there are large differences in the effectiveness of patents by sector of activity, with patents most useful for firms that market products that are expensive to develop but relatively cheap to imitate, such as chemical compounds and mechanical equipment. These differences can be partly captured by includ-

[7] The eight external information sources are suppliers of materials and components, suppliers of equipment, clients or customers, competitors, consultancy firms, universities, government laboratories, and technical institutes.

[8] Of course, information could be shared with other firms, including competitors, at a conference or through attendance at a trade fair. However, the type of information that is shared (or revealed) is likely to be less important than the type of information shared with other external sources such as suppliers or customers, where firms could have built up a long-term relationship.

ing sector dummies in a regression model. The drawback to including sector dummies is that it does not provide any information on what aspects of each sector influence the effectiveness of patents. In effect, sector dummies do not help to explain *why* there are differences by sector. An alternative approach is to try to identify the factors, operating at either the firm or sector level, that influence the effectiveness of different appropriation methods.

An important factor that should vary by sector consists of the disadvantages of patents. These include the disclosure requirement and limits to the effectiveness of patents in preventing imitation. Both can act at the level of an entire sector as a result of their link to other factors such as technological opportunity. For instance, the ability of firms to 'invent around a patent' will partly depend on the number of different technical solutions to a problem and the rate of technical progress. Similarly, many low-technology sectors with few technological opportunities, such as basic metals, should also find patents ineffective in preventing imitation since many technologies are based on general engineering principles that are not patentable.

Unfortunately, the CIS does not contain data on the disadvantages of patents. Instead, this information is obtained from 610 respondents to the PACE survey of Europe's largest industrial firms who answered a question on the reasons why they did not apply for a patent for one or more innovations. An average measure of the disadvantages of patents by sector is calculated from the percentage of firms that rate at least one of two drawbacks to patents as 'extremely important': "limits to the effectiveness of patents in preventing imitation" or "the amount of information disclosed in a patent application" (PATPROB). Separate variables for each of these disadvantages were also included in the regression models, but did not provide robust results. This is probably because these two disadvantages are linked — a patent can be ineffective in preventing imitation because it releases information, although there are other factors that can limit effectiveness, such as a high density of alternative technologies. Firms in sectors where these two disadvantages of patents are strong should prefer secrecy over patents. Sector averages for patent disadvantages are only calculated for manufacturing sectors with more than five PACE respondents, which required excluding the printing and publishing sector (NACE 22) from the regression analyses given below.[9]

3.6. Regression results

The dependent variable for product innovations in the regressions consists of six categories of the difference between the respondent scores for secrecy and patents. The highest value of the dependent variable is given when the respondent rates secrecy as three or more categories higher than patents. This occurs, for example, when the respondent gives secrecy a score of 5 (crucial) compared to a score of 2 (slightly important) for patents. The lowest value of the dependent variable occurs when secrecy is given a score that is two or more categories lower than the score for patents. The dependent variable for process innovations consists of five categories. Although the scoring system appears to be based on an interval variable, it is important to emphasise that it is only possible to assume an ordinal scale of measurement. For this reason, the ordered logit is the appropriate regression model to use since it makes no assumptions about the distance between two values of the dependent variable. Details of the model are given in Appendix A.

The results of four regression models, two each for product and process innovations, are given in Table 3. A positive coefficient for an independent variable increases the probability that firms find *secrecy* to be more effective than patents, while a negative coefficient shows the reverse. Models 1 and 3 include sector dummies, while models 2 and 4 include the sector-level measure of the disadvantages of patents (PATPROB).

The measure of firm size used in these regressions is not the number of employees, as in Tables 1 and 2, but the log of the absolute amount that the firm spends on R&D (LRDEXP). This measure of firm size improves the fit of the model, although using the log of the number of employees provides roughly

[9] In total, results are available in PACE for 17 sectors with an average of 36 respondents per sector, after combining three related sectors, textiles (NACE 17), clothing (NACE 18), and leather goods (NACE 19). These three sectors are also combined in the CIS analyses. The PACE methodology is given in Arundel et al. (1995).

Table 3
Ordered logit results for finding secrecy to be more important than patents to "maintain and increase the competitiveness of innovations"

Variable	Product innovations				Process innovations			
	1		2		3		4	
	β	p	β	p	β	p	β	p
Constant	1.583	0.000	1.202	0.000	0.363	0.024	0.180	0.288
Firm size								
LRDEXP	−0.084	0.007	−0.101	0.001	−0.062	0.059	−0.049	0.114
Innovation strategy								
RDINTENS	0.003	0.977	0.001	0.990	−0.045	0.565	−0.050	0.602
RDPROD	−0.143	0.120	−0.180	0.043				
RDPROC					0.236	0.028	0.238	0.020
Information sources (both dummy variables)								
INTERNAL	0.154	0.002	0.150	0.002	0.038	0.447	0.041	0.400
COOPRD	−0.089	0.078	−0.096	0.053	0.049	0.338	0.039	0.441
Sector level variables								
PATPROB			0.855	0.002			0.771	0.007
Sector dummies	Yes				Yes			
Model χ^2	122.3	0.0000	97.7	0.0000	273.3	0.0000	238.1	0.0000

All models include 1995 firms and control for the country of location, with Denmark as the reference country. The reference sector for models 1 and 3, which include sector dummies, is food and beverages. The estimated threshold values (μ) (not shown in the table) are statistically significant ($p < 0.0000$) in all models and increase monotonically. The number of firms is less than in Tables 1 and 2 because of the exclusion of firms with one or more missing values for the independent variables. Results are similar when the dependent variable consists of three categories, as used in Tables 1 and 2, or when an ordered probit model is used.

similar results. R&D spending probably works better than the number of employees because it is a more accurate measure of the innovative size of the firm. For product innovations, the relative importance of secrecy declines with firm size, which means that secrecy is more important to small than to large firms. In contrast, firm size is only of borderline significance for process innovations. Both of these findings corroborate the descriptive results of Tables 1 and 2.

The firm's RDINTENS has no effect on the relative importance of secrecy vs. patents, which contrasts with the earlier comparison of Tables 1 and 2. This result suggests that what matters is not the firm's RDINTENS, but the absolute amount spent on R&D. For example, a very large firm with relatively high levels of R&D spending but a low RDINTENS will follow a similar strategy on patents and secrecy (other factors being equal) than a medium-sized firm with the same amount of spending on R&D. A comparable finding was noted in analyses of the PACE survey data, which found that the absolute amount spent on R&D had a much stronger influence on a range of innovation strategies than RDINTENS (Arundel et al., 1995). The explanation for this probably lies in the organisation of innovation. An R&D lab of a given size probably follows the same strategies as those of its peers, regardless of the total number of employees within the firm.

The other variable for the firm's innovation strategy is the percentage of all RDPROD or RDPROC, depending on the regression. RDPROD is negative, as expected, but not significant in both models for product innovations. In contrast, the equivalent variable RDPROC is more robust for process innovations, indicating that firms that spend a high percentage of their R&D expenditures on process innovations are more likely to use secrecy to protect them.

The two variables for information sources only influence the results for product innovations. Firms that focus on in-house information (INTERNAL) are more likely to give secrecy a higher rating than patents. In contrast, participation in cooperative R&D (COOPRD) reduces the probability that a firm will give a higher rating to secrecy than to patents, although the results are of borderline significance. This corroborates the results of Brouwer and Kleinknecht (1999) and provides further support, albeit indirect, for the theory that patents help to clarify the ownership of the intellectual output of cooperative R&D alliances. The firm's status as a subsidiary or as an independent firm had no effect in any of the models and is therefore not included in the regressions shown in Table 3.

The variable for the average disadvantage of patents (PATPROB), by sector, is strongly significant and positive for both product and process innovations. This supports the theoretical literature, which finds that firms will use secrecy when disclosure is a serious disadvantage to patenting. Another implication is that this sector-level variable is capturing some of the effect that is normally covered by sector dummies. However, it is not entirely clear if PATPROB is capturing the same effect for different sectors, since sectors where patents have a high rating for disadvantages include several low technology sectors such as food and beverages and several sectors where patents are known to be of value, such as chemicals and petroleum products.

The use of the relative importance of secrecy and patents as the dependent variable in the regressions has a considerable influence on the interpretation of the value of patents and secrecy. For example, the ordered logit analyses for product innovations were repeated using the responses for secrecy alone as the dependent variable. These range through five response categories from 'insignificant' (1) to 'crucial' (5). In this regression, the value of secrecy *increases* significantly with firm size, in contrast to the decline in the relative value of secrecy compared to patents with firm size, as shown in Table 3. Similar results were found for patents in an earlier study (Arundel, 1997). These results are due to the fact that large firms find all appropriation methods to be of greater value than small firms. The failure to look at the relative value of secrecy vs. patents would erroneously suggest that secrecy is less important to small firms than to large firms.

4. Conclusions

The results of these analyses show that a higher percentage of R&D-performing firms in all size classes find secrecy to be a more effective means of appropriation than patents. This is also true when the analysis is limited to firms with RDINTENS above 10%. Furthermore, small firms, on average, do *not* rely more on patents than on secrecy in comparison with large firms. Instead, small firms are less likely than large firms to find patents to be of greater value than secrecy for product innovations, although there is little difference by firm size for process innovations. These results are also confirmed by regression analyses that include the effect of other factors. There is one exception by firm size. As shown in Table 2, small R&D-intensive firms do not differ very much from large firms in their preferences for either secrecy or patents, although a higher percentage of all R&D-intensive firms find secrecy to be more important than patents than the opposite.

These results, showing that secrecy is an effective alternative to patenting, suggest that we should also see a negative relationship between the strength of secrecy as an appropriation method and the patent propensity rate. Relevant information on this issue is available from an analysis of the PACE survey (Arundel and Kabla, 1998). Firms that found secrecy to be an effective appropriation method had a lower propensity to patent product innovations than firms that found secrecy to be ineffective.

The CIS results, based on thousands of firms across the full spectrum of manufacturing sectors, have some relevance to the debate over the existence of a 'pro-patent' era. Between 1990 and 1992, patents were less important than secrecy to most European firms in this sample, including R&D-intensive firms. This indicates that there are marked limitations to the value of patents as an appropriation method. The timing of the survey is also well within the 'pro-patent' era, as shown by patent applications in the United States, which had already increased by about 60% between 1984 and 1992 (Kortum and Lerner, 1999). It is possible that there could have either been a lag in the adoption of pro-patent policies by European firms compared to American firms, or patents could have markedly increased in importance since 1993. A possible lag is supported by slightly lower patent propensity rates for European than American firms in the early 1990s (Arundel and Kabla, 1998). However, Cohen et al.'s (1998) recent comparison of the results of two American innovation surveys, in the early 1980s and then in the early 1990s, shows that secrecy *increased* in importance relative to patents over this time period. One possible explanation is an increase in concern over the ease of 'inventing around' a patent. The paradoxical result is that we could be witnessing a rise in the importance of both patents and secrecy, with the latter increasing in importance more quickly to make up for the inadequacies of the former.

The finding that firms that engage in cooperative R&D are more likely to find patents of greater value than secrecy does, however, provide some support for a pro-patent era, given an increase in cooperative R&D over the last decade (Granstrand, 1999).

The results are relevant to three aspects of European innovation programmes: to encourage patenting among all firms, to encourage patenting by high-technology firms, and possible improvements in patent legislation.

First, the results suggest that policies to encourage overall patenting will not automatically influence either the patent policies of firms or the competitiveness of European firms. This is because patents are a relatively unimportant appropriation method in many different sectors, with more firms finding secrecy of greater value than patents than the opposite. The regression results for PATPROB, or the disadvantages of patents at a sector level, indicate that a major reason for why firms prefer secrecy is concern over disclosure or other limits to the effectiveness of patents in preventing imitation. This result is in line with theoretical models on the effect of disclosure as a reason not to patent, but the strong relative importance of secrecy compared to patents in this study suggests that future theoretical models should explore the opposite problem: under what conditions will a firm decide not to keep its invention secret?

Second, a major area of European policy concern is the competitiveness of small high-technology start-up firms. In this respect, the similarity between the relative importance of patents and secrecy among small R&D-intensive firms and large R&D-intensive firms suggests that the former do not need additional government support for patenting. The appropriation strategies of these small high-technology firms are already similar to that of their larger peers.

Third, the results suggest one problem with current patent legislation that deserves closer attention. The fact that most small firms value secrecy more highly than large firms, even after controlling for factors such as the sector of activity and RDINTENS, is of concern. This difference is unlikely to be due to smaller firms having few patentable innovations, because the analyses have intentionally excluded firms that do not perform R&D and are, therefore, less likely to develop patentable inventions. One explanation of the difference is that small firms lack the financial reserves to protect their patents from infringement. If true, this would point to the need for policies to provide a more level playing field for infringement suits. This possibility was investigated by looking for interaction effects between the measure of the disadvantages of patents (PATPROB) and firm size. These analyses suggested that patent disadvantages *were* a more serious problem for small than for large firms, but the results were not robust, probably because this data was only available at the sector level.

The results also raise several questions. One of the main justifications for a patent system is to disseminate knowledge through disclosure. This could be socially beneficial by reducing the amount that firms invest in R&D. The widespread use of secrecy by firms could arguably be both economically inefficient and retard the rate of technical progress, suggesting that patent policies should be designed to encourage firms to patent rather than using secrecy. However, repetitive R&D is not necessarily an entire loss, since some R&D is required for imitation and such R&D could also lead to useful discoveries, since no R&D programme is likely to be identical. Nor is secrecy likely to substantially slow down the rate of technical change, at least for product innovations, which can be reverse-engineered. How these factors play out in terms of imitation lags and costs, or the effect of patents in forcing competitors to explore new, unprotected avenues of research, could be a fruitful avenue for further research.

Finally, there are at least two important limitations to this study. The measure of the value of secrecy and patents in the CIS is a composite of all of the firm's innovations. Their relative importance could differ for the firm's most economically important innovation, or for innovations that include the greatest technical advances. Second, with the exception of Germany, all of the data are from the smaller European economies. Conditions could differ in the larger European economies, as well as in other countries outside of Europe.

Acknowledgements

The analysis of the CIS-1 data was funded under a 1995 EIMS research grant from DG XIII of the European Commission, while the remainder of the study was funded by EIMS grant 98-190. All opinions and errors in this article are the responsibility of the author only and cannot be attributable to the European Commission or to any of its staff or agencies. I would also like to thank Guy Gellatly of the Micro-economic Analysis Division of Statistics Canada and two anonymous reviewers for their comments.

Appenidx A

The ordered logit model can be used to examine the impact of a range of exogeneous variables on a dependent variable which takes a finite set of ordered values $(1,2,\ldots,n)$ (Liao, 1994). The method of estimation is maximum likelihood. The model assumes that the dependent variable y is generated by a continuous latent variable y^* whose values are unobserved. The model assumes that there are a set of ordered values $(r1, r2,\ldots,rn-1)$ and a variable y^* such that:

$$y = 1 \text{ if } y^* < r1$$

$$y = k \text{ if } rk-1 < y^* < rk \qquad \text{for } 1 < k < n$$

$$y = n \text{ if } rn-1 < y^* \tag{1}$$

The unobserved variable y^* is modelled as a linear function of the (N,k) vector of exogeneous variables X:

$$y_i^* = \beta X_i + \varepsilon_i \qquad i = 1,\ldots,N \tag{2}$$

where ε_i has a distribution function f derived from the logistic cumulative distribution function:

$$F(x) = 1/(1+\mathrm{e}^{-x}) \tag{3}$$

Given the characteristics X_i of individual i, the probability that y_i is found in category k is:

$$\text{Prob}(Y_i = 1/X_i) = F(r1 - \beta X_i)$$

$$\text{Prob}(Y_i = k/X_i)$$

$$= F(rk - \beta X_i) - F((rk - 1) - \beta X_i)$$

$$\text{Prob}(Y_i = n/X_i) = 1 - F((rn - 1) - \beta X_i) \quad (4)$$

The scoring of the dependent variables in the regressions given in Table 3 are as follows: For product innovations: 0 = secrecy two or more categories less than patents, 1 = secrecy one category less than patents, 2 = secrecy and patents given equal scores, 3 = secrecy given a score one category higher than patents, 4 = secrecy given a score two categories higher than patents, and 5 = secrecy given a score three or more categories higher than patents. For process innovations: 0 = secrecy less than patents, 1 = secrecy equal to patents, 2 = secrecy one category higher than patents, 3 = secrecy two categories higher than patents, and 4 = secrecy three or more categories higher than patents.

References

Archibugi, D., Cohendet, P., Kristensen, A., Schaffer, K.A., 1994. Evaluation of the Community Information Survey (CIS) phase I. DG XIII of the European Commission, EIMS Publication 11, Luxembourg.

Arundel, A., 1997. Enterprise strategies and barriers to innovation. In: Arundel, A., Garrelfs, R. (Eds.), Innovation Measurement and Policies. European Commission, Office for Official Publications of the European Communities, EUR 17019, Luxembourg, pp. 101–108.

Arundel, A., van de Paal, G., Soete, L., 1995. Innovation Strategies of Europe's Largest Industrial Firms: Results of the PACE Survey for Information Sources, Public Research, Protection of Innovations and Government Programmes. Directorate General XIII, European Commission, EIMS Publication 23.

Arundel, A., Kabla, I., 1998. What percentage of innovations are patented? Empirical estimates for European firms. Research Policy 27, 127–141.

Brouwer, E., Kleinknecht, A., 1999. Innovative output and a firm's propensity to patent. An exploration of CIS micro data. Research Policy 28, 615–624.

Cockburn, I., Grilliches, Z., 1988. Industry effects and appropriability measures in the stock market's valuation of R&D and patents. AEA Papers and Proceedings 78, 419–423.

Cockburn, I., 1992. Measuring appropriability: what can we learn from the Yale survey? INSEE International seminar on Technological Appropriation, Paris, 9–10 June.

Cohen, W.M., Nelson, R.R., Walsh, J., 1998. Appropriability conditions and why firms patent and why they do not in the American manufacturing sector, Mimeo. Carnegie Mellon University, Pittsburgh.

Cohen, W.M., Klepper, S., 1992. The tradeoff between firm size and diversity in pursuit of technological progress. Small Business Economics 4, 1–14.

ETAN (European Technology Assessment Network), 1999. Strategic Dimensions of Intellectual Property Rights in the context of Science and Technology Policy, ETAN Working Paper, DG XII, European Commission, Luxembourg.

Friedman, D.D., Landes, W.M., Posner, R.A., 1991. Some economics of trade secret law. Journal of Economic Perspectives 5, 61–72.

Granstrand, O., 1999. The Economics and Management of Intellectual Property. Edward Elgar, Cheltenham.

Harabi, N., 1995. Appropriability of technical innovations: an empirical analysis. Research Policy 24, 981–992.

Harter, J.F.R., 1993. The propensity to patent with differentiated products. Southern Economic Journal 61, 195–200.

Horstman, I., MacDonald, G.M., Slivinski, A., 1985. Patents as information transfer mechanisms: to patent or (maybe) not to patent. Journal of Political Economy 93, 837–858.

Klepper, S., 1996. Entry, exit, growth and innovation over the product life cycle. American Economic Review 86, 562–583.

Kortum, S., Lerner, J., 1999. What is behind the recent surge in patenting? Research Policy 28, 1–22.

Levin, R.C., Klevorick, A.K., Nelson, R.R., Winter, S.G., 1987. Appropriating the returns from industrial research and development. Brookings Papers on Economic Activity 3, 242–279.

Liao, T.F., 1994. Interpreting Probability Models: Logit, Probit, and Other Generalized Linear Models. Sage Publications, London.

Mazzoleni, R., Nelson, R.R., 1998. The benefits and costs of strong patent protection: a contribution to the current debate. Research Policy 27, 273–284.

McLennan, W., 1995. Innovation in Australian Manufacturing 1994. Australian Bureau of Statistics, Canberra, catalogue no. 8116.0.

Rausch, L., 1995. Estimates from the Manufacturer's Innovation Survey Pilot. National Science Foundation, Washington, DC, April, Mimeo.

Scherer, F.M., 1965. Firm size, market structure, opportunity, and the output of patented inventions. American Economic Review, 1097–1125.

Scotchmer, S., Green, J., 1990. Novelty and disclosure in patent law. RAND Journal of Economics 21, 131–146.

Scotchmer, S., 1991. Standing on the shoulders of giants: cumulative research and the patent law. Journal of Economic Perspectives 5, 29–41.

Streiner, D.L., Norman, G.R., 1995. Health Measurement Scales: A Practical Guide to Their Development and Use. Oxford Univ. Press.

Takalo, T., 1998. Innovation and imitation under imperfect patent protection. Journal of Economics 67, 229–241.

Thurow, L.C., 1997. Needed: a system of intellectual property rights. Harvard Business Review, 95–103, September/October.

[12]

Leveraging Research and Development: Assessing the Impact of the U.S. Advanced Technology Program

Maryann P. Feldman
Maryellen R. Kelley

ABSTRACT. This paper examines the factors that affect a firm's chances of winning an award from the Advanced Technology Program (ATP) and the subsequent impact of the award on a firm's success in raising additional funds for its research and development (R&D) activities. Analysis of data from a survey of 1998 ATP applicants shows that proposals with higher ratings by technical and business/economic experts have a greater chance of winning an award. Further, the projects and firms selected by ATP are more willing to share their research findings with other firms, and tend to be those that open up new pathways for innovation through combining technical areas or by forming new R&D partnerships. Most of the non-winners have not proceeded with any aspect of the R&D project proposed to ATP and, of those that have, most did so at a smaller scale. Furthermore, the ATP award has prestige value for the winning firms; the halo effect from the award increases the success of these firms in attracting additional funding from other sources. Our conclusion is that the ATP is leveraging activities that have a strong potential for broad-based economic benefit.

1. Introduction

Through a number of different agencies, the U.S. federal government provides research funding to universities and colleges, government-owned laboratories, and for-profit enterprises. Support for basic research at universities is widely accepted and the relationship to technical progress is well established. A common question regarding the public funding of research conducted by private firms is whether the public interest, with regard to technical advance and economic growth, are being promoted. After all, a private firm is expected to fund its own research when it perceives a technical opportunity with reasonable profit expectations.

In some cases, however, private firms may not pursue promising technical opportunities for the following reasons:

- R&D scientific and technical frontiers are risky and the chances of failure are high.[1]
- An individual firm may not have the capabilities required to develop the technology. Complex new technologies may require collaboration and information sharing; however, the cost of establishing research and development partnerships and making them work productively may provide disincentives to undertaking the project.
- Private incentives may not be sufficient to induce a firm to undertake the project in the face of difficulties in appropriating the resulting benefits, i.e. (the resulting knowledge may flow to others who may benefit from the R&D without sharing the cost).[2]

Public-private R&D partnerships provide a policy instrument that may alleviate these concerns and encourage private-sector investment. A government R&D partnership may provide a catalyst for private firms to undertake research which will have broad-based knowledge benefits for other firms and other industries. Government programs may provide a neutral forum for competitors to work together on mutually beneficial research (NRC 1996: footnote 19).[3] In addition,

Maryann P. Feldman
Johns Hopkins University
3400 North Charles Street
Baltimore, MD 21218
U.S.A.

and

Maryellen R. Kelley
Pamet Hill Associates
4545 Connecticut Ave., NW
Washington, D.C. 20008
U.S.A.

Small Business Economics **20**: 153–165, 2003.

government funding may reduce the cost to the firm of establishing working relationships with R&D collaborators and provide an incentive for firms to undertake such joint efforts. R&D collaborations provide the opportunity for a firm to acquire knowledge and complementary technical expertise that it would otherwise be difficult to obtain. Moreover, these cooperative arrangements facilitate the diffusion of economically relevant knowledge within the private sector, thus further leveraging the public investment. Finally, government R&D funds may reduce the firm's scientific and technical risk sufficiently to bring the project within an acceptable rate of return for private-sector investment (Griliches, 1992).[4]

Recently, governments around the world have encouraged public-private partnerships (NRC, 1996).[5] The long-term goal of these programs is to achieve greater productivity growth in an economy through the creation of knowledge that may become incorporated in industrial processes, products and services (Kelley, 1997). The logic for public investment is that, in the long run, the economic benefits to consumers, other firms and the national economy will exceed the private returns realized by the firm that initially received the research award. In the United States, the Advanced Technology Program (ATP) provides an example of a government R&D program that attempts to encourage broadly applicable technical advance. Since 1990, the ATP has provided selective cost-sharing awards for early stage industry-defined research projects with broad-based economic potential (Hill, 1998).[6]

1.1. *Rationale for public-private R&D partnering*

The mission of the ATP thus depends upon its success in selecting projects with attributes that indicate a high potential to generate substantial economic benefits. Of course, there is no way of guaranteeing, in advance, whether an R&D project actually will result in successful commercialization or if the knowledge created will be valued by other organizations. The ATP's review and selection process aims to identify high quality research projects proposed by industry that have clearly delineated pathways through which the projects may achieve broad-based economic benefits. High ratings on the technical and economic dimensions by the ATP's reviewers are a pre-requisite for winning an award.

In addition to the assessment of the proposal itself, certain behaviors evident in a firm's R&D strategy may provide an indication of how readily the knowledge generated by the government funded R&D project may diffuse to and benefit other organizations.[7] In particular, a firm that is connected to other businesses for assistance in planning and carrying out its R&D, or has extensive linkages to the university research community, may offer more opportunity for other organizations to learn about the technical advances from the funded project (e.g. Doz, 1996; Hamel, 1991; Inkpen, 1995; Khanna et al., 1998; Powell et al., 1996).[8] Moreover, some firms attempt to guard access to their research results, while others see an advantage in sharing knowledge and are more willing to disseminate their research findings to other firms.[9] Clearly, when a firm openly shares information, the results of its research become known more quickly.

Since the ATP wants to encourage the diffusion of new knowledge by industry, it may favor projects where firms are attempting to forge new ties in new collaborative R&D efforts, or projects proposed by firms that are embedded in networks, or of company R&D and commercial linkages and firms that are more willing to share the knowledge that their R&D creates. The first research question we address in this paper concerns how these differences in R&D strategies of firms are related to winning an ATP award.

- Compared to non-winners, do firms that win an ATP award have R&D strategies with greater potential for diffusing knowledge and incorporating technical advance into commercial applications?

To the extent that the ATP is selecting projects in firms with R&D strategies that are more accessible to other researchers and firms, the more likely it is that the knowledge generated by the project will be taken up and used. Projects with such attributes are more likely to yield economic benefits, all other things being equal, that extend beyond the firms directly engaged in the project.

1.2. *Leveraging private sector R&D investment*

Government funding may also spur additional private investment in R&D activities.[10] For those outside the firm, the value of R&D activities in the early stages of technology development is difficult to assess since there is greater risk and uncertainty.[11] As a result, private investors typically focus on later stage projects. In order to secure private investment, a technology must be at a proof of concept phase. Public funding is a means for supporting early stage technology development and moving the project to a point where private investors are more willing to fund.[12] Finally, at any given time, there are a large number of investment opportunities and it may be difficult for private investors to accurately evaluate potential investments. When information about a company's R&D activity comes from a credible source, such as a government agency with a reputation for scientific integrity as well as programmatic expectations of economic significance, the investment decisions of other organizations may be favorably affected. In light of a large number of potential investment opportunities, the announcement of the ATP award may provide information that certifies an awardee's R&D activities from the perspective of the investment community (Lerner, 1999).[13] Hence, our second research question focuses on how ATP's funding makes a difference to firms in carrying out the type of high-risk, potentially high payout R&D proposed to the program. We address this question in two parts.

- Do non-winners continue with the proposed research project without ATP funding?
- Does the ATP award induce additional investment by other organizations in the R&D project?

The next section provides an introduction to the ATP and its mission. Section 3 presents the data that we will use to answer our three research questions. Section 4 describes how we measure the potential of the project to provide broad-based economic benefits and addresses the first research question. Section 5 considers if non-awarded firms and projects proceeded and we conclude that the majority of the non-awardees did not pursue the research. Section 6 considers the third question on the ability of the ATP award to induce additional funding. We conclude that ATP appears to provide a certification function that increases the amount of funds that firms are able to raise subsequently.

2. Selecting R&D projects with broad-based benefits: Background on the advanced technology program

ATP depends on the initiative of industry to define the goals for the proposed research projects.[14] All applicants are provided a guide to use in preparing proposals. The guide includes information on the evaluation criteria that the ATP employs in selecting awardees. To merit funding, the project must have both scientific and economic potential:

> The research must be challenging, with high technical risk . . . aimed at overcoming an important problem(s) or exploiting a promising opportunity . . . [and] must have a strong potential for advancing the state of the art and contributing significantly to the U.S. scientific and technical knowledge base.
>
> The proposed technology must have a strong potential to generate substantial benefits to the nation that extend significantly beyond the direct returns to the proposing organization(s). The proposal must explain why ATP support is needed and what difference ATP funding is expected to make. The pathways to economic benefit . . . includ[e] the proposer's plan for getting the technology into commercial use, as well as additional routes that might be taken to achieve broader diffusion of the technology.
>
> (*Advanced Technology Program Proposal Preparation Kit*, 1999, pp. 7–8).

Independent technical experts and specialists in the business planning evaluate every proposal. The selection process involves an assessment of the project's contribution to technical advance, its economic potential, and the need for public funding. In any given year, fewer than 20 percent of proposed projects actually receive funding and the overall average is 12 percent for the competitions conducted between 1990 and 1999.

3. Study of the 1998 ATP applicants: Data description

To address our research questions about the ATP, we focus on the competition held during the summer of 1998. A total of 502 proposals were submitted. Because ATP guidelines encourage joint projects, 822 organizations applied in this competition. Our primary interest is the 741 for-profit enterprises that applied to ATP in 1998.[15] This group of firms constitutes our sampling frame. The award winners were announced in October 1998.

ATP funded 80 proposals that involved 161 organizations, including 147 for-profit enterprises. Table I shows the distribution of firms by award status and firm size. About 20 percent of all applicant firms in 1998 received an ATP award. Small firms, employing fewer than 500 employees, constituted nearly 70 percent of all applicants. Although firms in this size category were only 52 percent of ATP award recipients, they accounted for 83 percent of all awards that were made to individual companies in that year. There were also 27 joint venture (JV) projects funded in 1998. These JV projects include 65 percent of all awardees and 87 percent of the firms with more than 500 employees.

To collect data on the 1998 applicants, we conducted a survey that asked about the applicant's perceptions of the process, about their firm's R&D strategies, and their experience with other funding sources in the year following their ATP application.[16] All interviews were completed over a six-month period (June–December 1999). Our respondent was the person identified as the technical lead for the proposed project. If this person was no longer with the company, we asked to speak to the individual who was most knowledgeable about the proposal and the company's R&D activities in the area identified in the proposal.[17] The telephone questionnaire required 20–30 minutes to complete.[18]

Our sample consisted of 100 percent of the winning firms and a simple random sample of 50 percent of the non-winners. Thus, we contacted 297 non-winners and 147 awardees one year after the 1998 ATP selection process. We completed interviews for 119 winners for an 81 percent response rate (119/147). For the non-winners, we discovered that within one year there were 49 cases that we could not interview, either because the company was no longer in business (23 cases) or because the person responsible for preparing the ATP proposal was no longer employed at the company and the company was not pursuing any aspect of the R&D proposed to ATP (26 cases). We adjusted our response rate accordingly. We completed interviews for 122 non-winners, for a 50 percent response rate (122/(297-49)). This yields an overall effective response rate of 61 percent.

The survey results were matched with company and project-specific data from other sources. First, we used independent sources such as the CorpTech Database and Hoovers Online Company and Industry Network to verify survey responses concerning employment, financing, and the founding date of the company.[19] Second, we matched each record with information from ATP administrative records on the technology area of the proposal, the results of the ATP proposal review process, and the firm's prior history of applications and awards.

4. Does the ATP favor firms that have greater potential for knowledge diffusion?

Our first research question requires considering projects and firm characteristics associated with greater economic potential. A project is expected to have greater economic potential if the firm embarks on a new technical problem or forms a new R&D collaboration. Although collaborations are widely recognized as an important R&D strategy (Doz, 1996; Hamel, 1991; Inkpen, 1995; Khanna et al., 1998; Powell et al., 1996), estab-

TABLE I
Distribution of applicants by award status and firm size

	Award status		
	Award winners	Non-winners	All applicants
Percent of firms ≥ 500 employees	48%	27%	31%
Percent of firms < 500 employees	52%	73%	69%
Number of firms	147	594	741
Percent of all applicants	20%	80%	100%

lishing a new relationship is difficult and costly for firms (Harrigan, 1988). Table II shows the percent of projects proposed to ATP in 1998 that involved two or more organizations as research partners, and the percent of new partnerships and new technical areas proposed by ATP award winners compared to non-winners. Seventy-nine percent of the 1998 ATP applicants in our sample included other organizations in their proposal. There is no difference between award-winners and non-winners in their propensity to have partners. However, the percent of award-winners identifying a new collaborator as their most important research partner, is significantly higher than occurs among non-winners (59 percent versus 42 percent). Another indication that award winners are more likely to open new pathways to innovation is their significantly greater tendency to propose a project in a technical area new to the firm (47 percent versus 19 percent).

One firm's R&D activities may have the potential to augment other firms' innovative capabilities but only if research results and knowledge are shared. Competitors' reluctance to collaborate and share research results may be likely. However, for certain complex technical areas, sharing information may be essential for the firm's technical advance. Moreover, an award winning firm's willingness to diffuse research results affords a greater opportunity for other firms to benefit from government-funded research.

TABLE II
Indicators of the creation of new pathways to innovation

New partnerships	
Percent of 1998 applicants who included other organizations in the ATP proposal	79%
If Yes, was this a new partnership?	
Award winners	59%
Non winners	42%
All applicants*	48%
χ^2	5.502[a]
New technical area	
Percent of applicants proposing a project that was not part of the company's R&D plan (in a technical area new to the company)	
Award winners	47%
Non winners	19%
All applicants*	28%
χ^2	21.418[b]

Notes:
* Percentages are weighted to reflect the proportion of winners and non-winners in the overall population.
[a] χ^2 is statistically significant at $p < 0.05$.
[b] χ^2 is statistically significant at $p < 0.01$.

As indicated in Table III, we used three statements about the company's approach to construct a measure of the firm's tendency towards either openness or secrecy in conducting its own R&D. The possible scores ranged from 0 to 3. We expected to find that most firms have a tendency to secrecy (Nelson, 1990; Liebeskind, 1997). Firms that answered yes to two of the three questions were considered to exhibit a tendency towards openness in their communications with other firms about their research. Overall, we find that only 23% of all applicants exhibited a tendency towards openness. However, compared to non-winning applicants, a significantly greater percentage of firms can be characterized as having a tendency towards openness in their research communications (31% compared to 19%).

The potential for contributing to scientific and technical advance is expected to be enhanced by a firm's linkages to universities and other institutions. For example, universities are an important source of new knowledge that may be applied to a broad range of industrial problems (Mowery and Rosenberg, 1989). In carrying out R&D a firm may draw upon its connections to university faculty, graduates, laboratory facilities, and intellectual property to augment its internal capabilities. These relationships are also potential pathways for reciprocal knowledge flows from the firm to the university-based research community and then perhaps on to other firms. We expect that the more connections that a firm has to university resources, the greater the potential for knowledge flows. As indicated in Table IV, we measured twelve different ways a firm may be linked to universities.[20] Included are universities as R&D partners in general and as resources for the ATP project specifically. The number of linkages serves as an index of the strength and diversity of connections to the university-based research community.

The ATP mandate stresses commercial and economic benefits as program goals; however, the program does not fund product development or

TABLE III
Tendency towards openness or secrecy

	Award status		
	Award winners	Non-winners	All applicants*
Percent of firms scoring 2 or 3 on scale measuring *Tendency to openness in attitudes about revealing information on own research to other firms*†	31%	19%	23%

Notes:
$\chi^2 = 4.636$. Statistically significant at $p < 0.05$ with 1 degree of freedom.
† Values of this scale range from 0 to 3, where 0 indicates a strong tendency towards secrecy and 3 means a high degree of openness, i.e., a willingness to share information and to do little to impede other firms from learning about the results of its internal research program.
* Percentage are weighted to reflect the proportion of winners and non-winners in the overall population.

Questionnaire Items in Tendency to openness in attitudes about revealing information on own research to other firms

To what extent do you intend to make your research results available to other firms and industries?	1 = almost always or sometimes 0 = rarely or never
Do you think that keeping your company's R&D knowledge from spreading to other firms is important to your firm's long run success?	1 = no; 0 = yes
Would you ever consider *not* engaging in new R&D activity because you believe another firm may benefit from it?	1 = no; 0 = yes

market research. As a consequence, the potential of a technology to attract non-ATP resources for commercialization is critical. This could take the form of additional private capital attracted by the ATP awarded company, or might include resources provided by other companies who take up the technology and incorporate it into their commercial efforts. Connections to other firms, in the planning and development of the project, and more generally, as sources of financial and technical support, not only may be important to carry out the research but also may facilitate the eventual commercialization of the technology and the overall eventual impact of the project on the economy. We asked about applicant's connections to other firms in preparing the proposal for ATP, as potential collaborators on the project, and more generally, in providing technical and financial resources to the applicant in the two years prior to the application. We used the 19 questions shown in Table V to construct a business linkage index. A higher number of links to other businesses signals connection to a diversity of resources that are particularly important for companies seeking to commercialize their technologies.

Table VI shows the results of our comparisons of award-winners to all other applicants on these two indices. We find that there is no significant difference among firms in the extent of their linkages to universities. However, award-winners have a significantly greater number of linkages to other businesses.

The results of our survey show that the ATP tended to make awards to those companies who proposed projects involving the formation of important new partnerships. In addition, we find that award-winning projects are more likely to involve new areas of research for the firm. When compared to unsuccessful applicants, the firms receiving funding from ATP were more likely to have extensive linkages to other businesses and to profess a greater willingness to share information about their research findings with other firms. This indicates that the knowledge and technologies developed by the award-winning projects were more likely to be diffused quickly through the extensive linkages of the firms and their greater willingness to make their research results available to other firms and their greater tolerance of knowledge leakages. Surprisingly, we find no

TABLE IV
Questionnaire items in university linkages index

For ATP project and proposal:

1. Did your company first learn about ATP from someone at a university?
2. Did a university help you identify the research partner you consider to be the most important for the project you proposed to ATP?
3. In preparing the technical plan portion of your proposal, did you get assistance from someone at a university?
4. In preparing the business plan portion of your proposal, did you get assistance from someone at a university?
5. [If technical lead on the ATP project has been employed with the company less than 5 years], was this person previously employed at a university?

Other ties to university resources:

6. Does your company have any contracts or licensing agreements for intellectual property at universities?

In the two years prior to your ATP application have you used assistance from a university program

7. to address a technical problem?
8. to prepare a business or marketing plan?
9. to recruit R&D employees?
10. In the two years prior to your ATP application have you formed an alliance with a university to address your needs for equipment and facilities?
11. In the two years prior to your ATP application have any of your R&D personnel attended training or technical programs sponsored by a university?
12. In the two years prior to your ATP application, for your R&D or technology development activities, has your company received funds from a university program?

Note: University Linkages Index = Σ Number of Connections (Number of 'yes' answers to these questions).

statistical difference between winners and non-winners in the extent of their linkages to universities.[21] That suggests that the ATP is attracting firms with similar connections to the resources available from universities in our nation's innovation system. Given the types of projects that were proposed to the ATP, perhaps it is not surprising that we find that all applicants had similar links to universities.

In sum, the following attributes increase the likelihood of winning an ATP award:

- Riskier research projects that are new to the firm;
- Research projects that form new R&D partnerships between organizations;

TABLE V
Questionnaire items in business linkages index

For ATP project and proposal:

1. Did your company first learn about ATP from someone at another company, a consulting firm, or a venture capital firm?

In preparing the technical plan portion of your proposal, did you get assistance from

2. someone at another company?
3. a consulting firm?

In preparing the business plan portion of your proposal, did you get assistance from

4. someone at another company?
5. a consulting firm?
6. [If technical lead on the ATP project has been employed with the company less than 5 years], was this person previously employed at another company?
7. Did someone at a venture capital firm help you identify the research partner you consider to be the most important for the project you proposed to ATP?

Other business ties:

In the two years prior to your ATP application have you had assistance in addressing a technical problem from

8. another company?
9. a private consulting firm?
10. a private venture capital firm?

In the two years prior to your ATP application have you had assistance in preparing a business or marketing plan from

11. a private consulting firm?
12. a private venture capital firm?

In the two years prior to your ATP application, has your company received financing for your R&D or technology development activities from

13. another company?
14. a private venture capital fund?
15. an individual (angel) investor?

In the two years prior to your ATP application, to address your needs for equipment and facilities, has your company used

16. an alliance with another company?
17. secured bank financing?
18. private investor or angel financing?
19. venture capital financing?

Note: Business to Business Linkages Index = Σ Number of Connections (Number of 'yes' answers to these questions).

TABLE VI
Resource linkages to universities and to other businesses by ATP award status

	Award status		
	Award winners	Non-winners	All applicants*
Mean number of business-university linkages	5.4	5.5	5.5
Standard deviation	1.6	1.4	1.5
Mean number of business-to-business linkages**	4.5	3.7	4.0
Standard deviation	3.8	3.1	3.5

Notes:
* Means for all applicants are weighted to reflect the proportions of winners and non-winners in the overall population.
** T-statistic for difference between means is significant at < 0.05 level.

- Firms that demonstrate a tendency towards openness in communications about research with other firms and institutions; and,
- Firms that have a more extensive set of business-to-business linkages.

These attributes distinguish award-winners from other applicants.

5. One year later: What happened to non-winners and the R&D projects they proposed to the ATP?

Our second research question addresses the concern that firms may proceed with projects in the absence of government funding. During the review and selection process, the ATP attempts to determine whether or not the R&D activity that is being proposed is likely to be carried forward at the same speed and scale by the company without any assistance from the ATP. Hence, if ATP attracts proposals that companies would proceed with anyway at the same funding level, with the same timeframe and goals, we would expect to see a high proportion of non-winning projects proceeding during this period of time. However, as Table VII indicates, when we interviewed firms one year later, this was not the case.

If a project is believed to be of central importance to a firm that has already funded this type of R&D from its own resources, the ATP is likely to reject the proposal. Similarly, if a company has access to other funding sources (within or outside the firm), the ATP may choose to award its scarce resources to firms that lack access to such funding sources.

More than three-fifths of the non-winners (62 percent) have not proceeded with any aspect of the R&D project that they proposed to ATP. This number includes the non-winners that we discovered had gone out of business in the previous year. Also included are the projects where the persons responsible for preparing the proposal no longer worked for the company and there was no one whom we could identify who knew about the proposal or any continuation of that work in the same technical area.[22] In addition to these 49 cases, there were another 66 non-winners who indicated that, in the previous year, their company had not proceeded with any aspect of the project proposed to ATP.

Nearly 37 percent of the non-awardees began work on the proposed project at some level of effort. However, in most instances (77 percent), the project was pursued at a smaller scale than the company had proposed to the ATP. Only five percent of the firms that received no funding from ATP were proceeding at the same scale as they originally had proposed the previous year, while less than four percent of the non-winners were proceeding with a larger scale effort than had been proposed to ATP.

These results suggest that, for the most part, ATP is attracting applicants that need support in order to proceed with their R&D plans. A small minority of non-winners pursued its R&D plans at the same or an even greater scale than originally proposed to ATP. It seems reasonable to assume that the few non-winners who proceeded at the

TABLE VII
The extent to which non-winners pursue the ATP proposed R&D project without ATP funding

Did not proceed with the project, at any scale	62%
Began project on a *much smaller scale* than proposed to ATP	17%
Began project on a *somewhat smaller scale* than proposed to ATP	12%
Began project at about the *same scale* as proposed to ATP	5%
Began project a *somewhat larger scale* than proposed to ATP	3%
Began project on a *much larger scale* than proposed to ATP	1%

Note: Three respondents were unable or refused to answer this question.

same or greater scale than they had proposed to the ATP were rejected because the ATP saw no need to fund these projects: the projects were promising but the need for public funding was not apparent.

6. Halo effect: Does the ATP award increase R&D funding from other sources?

When an enterprise receives an R&D contract from a government agency the award may be a signal to non-government sources of funding, such as banks, venture capital firms and other potential investors, that the firm has a potential future stream of revenue from a reliable customer (the U.S. government). The ATP awards differ from the usual government R&D contracts to industry, and hence the economic information conveyed by the award is quite different. There is no promise of follow-on funding by the agency. The ATP seeks to fund R&D projects that have commercial applications, and there is no procurement connection – the ATP does not purchase the technology it helps to develop through its awards. In light of the unique features of the program, we assess the potential influence of an ATP award on the subsequent behavior of the investment community to arise from the information the award signifies about both the technical quality and economic potential of the project and firm.

For an ATP award to influence the subsequent decision of other sources of research funding implies that the announcement of the award itself provides valuable and credible information to the investment community. The host agency of the ATP is the National Institute of Standards and Technology (NIST), which is widely recognized to be an important and reliable source of technical information and expertise in a number of areas.

6.1. *Pursuing R&D funding from other sources: The impact of the ATP award*

The highly selective and competitive nature of the ATP award may signal potential investors about the quality of the firms' R&D project. As a result, other investors such as angels investors or venture capitalists, other businesses, as well as state governments and other federal agencies may invest in the firm. We expect the ATP awards to confer a halo effect to winning firms and projects, boosting the chances of a firm's success in attracting additional funding for its R&D activities.

To test for this effect, we considered the group of applicants that pursued other funding sources in the year following the ATP application. Table VIII compares the percentages of firms in the two award status categories that sought additional funding from other sources and the percentages that actually succeeded in attracting funding from these sources. Although a minority of all applicants sought funding from other sources, the percentage of ATP award winners seeking additional funding was much lower. On average, however, the ATP awardees who sought funding from other sources were much more successful in attracting funds than non-winners. We asked firms about the amount of funding that they received for the ATP-proposed research project from non-ATP sources, including private individuals, venture capitalists, other firms, state and local government programs, and other federal government agencies.

6.2. *The ATP halo effect*

Winning an ATP award significantly increases the firm's success in attracting additional funds from other sources for R&D activities. Our findings

TABLE VIII
Percentage of firms that applied to other funding sources to support their R&D activities and their success rates by ATP award status

	Award status		
	Award winners	Non-winners	All applicants
Percent of all firms in award category that applied to other funding sources	25%	47%	38%
Percent of those firms seeking funding that succeeded in attracting additional investment for their R&D activities	73%	33%	44%
$\chi^2 = 12.592^a$			

Notes:
* Percentages are weighted to reflect the proportion of winners and non-winners in the overall population.
[a] χ^2 is statistically significant at $p < 0.01$.

support the inference that the ATP award confers a halo effect on winners making them more likely to attract other funding. Through their investment decisions, other funding sources, both private and public, demonstrate that they believe that the ATP award provides valuable additional information about the quality of the project. Thus, our results confirm that the ATP award appears to send a market signal that certifies that the firm and the technology are promising.

7. Conclusion: The potential of the ATP for small business

Growth theorists describe two roles for government R&D policy (Jovanovic, 2000). The first role is in funding the supply of R&D. Most research to date has focused on this topic. The second role for government is providing incentives for private sector R&D and generating a set of behavioral responses that would be expected to stimulate R&D, promote innovation and economic growth. We know the most about the supply of R&D resources. Less is known about the ways in which government programs provide incentives for activities that generate innovation and economic growth.

In this paper, we have shown how a government program is leveraging private sector R&D activities by selecting projects with greater potential for diffusing knowledge and for contributing to the development of new technologies for commercial use. In essence, the mission of the Advanced Technology Program is to support private sector R&D projects that offer potential for contributing to technical advance and for realizing economic value. Our findings suggest that the operation of the program is consistent with its mandate and objectives. While the program is selecting projects that are rated highly by independent technical and business experts, it awards technically risky, new projects proposed by firms that have strong external linkages to other institutions and exhibit a willingness to diffuse knowledge based on their research findings.

We conclude that ATP is selecting projects and firms that have greater potential for increasing the circulation of new knowledge and for having the business connections necessary to realize economic benefits from its activities. The extent of a firm's linkages to other enterprises and to the resources of universities indicate that a firm is well-positioned to both tap the capabilities of other actors in our nation's innovation system and be a conduit for carrying the knowledge generated by an ATP project to use by other firms. These types of connections are important to the overall operation of the entire U.S. innovation system. We also find that the ATP supports the formation of new linkages and the initiation of R&D projects that bridge different technical areas, opening up new pathways to innovation.

We provide evidence that the investment community values the ATP award. Among firms that seek additional funding, we find that ATP award winners are more successful than non-winners. Since few R&D projects proposed to ATP actually proceed at a comparable level, we conclude that the ATP is stimulating additional investment in risky R&D projects that would otherwise not be funded by the firms themselves or other funding sources. For small businesses that naturally have a tendency to engage in high-risk research projects, form collaborative relationships in order to augment their internal resources, and require additional funding and legitimacy, the ATP program may provide an important vehicle for their growth and development.

Acknowledgements

We thank Wesley Cohen, Bronwyn Hall, Frank Lichtenberg, Jeanne Powell, Rosalie Ruegg, Andrew Wang, and Charles Wessner for their detailed comments on earlier drafts of this paper. We are also grateful for the many helpful comments we received from participants at the workshop sponsored by the National Academy of Science and the University of Nottingham in September 2000.

Notes

[1] Link et al. (forthcoming) note that the traditional distinctions between basic and applied research have become meaningless in the commercialization of complex technologies. Firms may identify research areas that require additional fundamental research.

[2] The profit incentive that motivates innovative activity by an individual firm also discourages information sharing and collaborative R&D activities between companies.

[3] An example of productive and profitable research undertaken by public-private partnerships is SEMATECH.

[4] The general argument for government funding of certain types of R&D activities, as articulated by Griliches (1992), is that such subsidies provide an incentive to firms to undertake high-risk R&D projects in which the public rate of return exceeds the private rate of return. This includes, for example, the case where an industry as a whole may benefit from the development of an enabling technology. Private firms typically use some pre-determined benchmark rate of return known as a hurdle rate. The project will only be acceptable if the expected rate of return is above that benchmark. By reducing the cost of the project to the firm, government funding will increase the expected rate of return and may make private companies willing to pursue them.

[5] See National Research Council (1996, pp. 12–40) for a review of international programs that encourage collaboration. For an introduction and review of the objectives of recent U.S. R&D programs, see Kelley (1997).

[6] For a discussion of the legislation and policy issues that led to the establishment of the ATP, see Hill (1998).

[7] Only when scientific and technical advances become known and are incorporated into new products and processes can ATP expect to see any economic benefits realized from the funded project.

[8] Studies of the strategies that managers use to learn about other organization's products, technologies, and business practices emphasize the importance of both formal linkages between firms and the informal network ties among engineers, scientists, and managers employed in different organizations. In general, this literature indicates that cooperation among firms is an important pathway for learning about technical advances in other organizations. More specifically, linkages that involve the sharing of resources, whether technical or financial, become important conduits for knowledge transfer among firms.

[9] Paradoxically, in order to gain direct and early access to the knowledge and technologies being developed in other organizations, a firm has to be willing to give some of its own accumulated knowledge and technologies to others (Liebeskind, 1996).

[10] Previous research suggests that the ways in which government provides funding to the private sector is important. David et al. (2000) indicate that the private sector is less likely to increase its own R&D spending when government R&D funding is provided through contract R&D programs in which there is potential for follow-on funding and technology procurement.

[11] See the remarks of David Morgenthaler of Morgenthaler Venture Capital (forthcoming).

[12] See Feldman et al. (2000) for an analysis of in-depth case studies that found this relationship. These case studies were used to design this statistical study that we report on here.

[13] Lerner (1999) suggests that the Small Business Innovation Research (SBIR) program award provides a certification function that informs private investors of a notable investment opportunity.

[14] The ATP is fundamentally different from other government R&D programs as the focus is industry-defined. This contrasts with the SBIR as a case in point.

[15] Private firms must lead ATP projects but there is frequent collaboration with not-for-profit organizations. In 1998 alone, there were 38 Universities, 29 non-profit organizations, seven government labs and seven foreign-owned U.S. subsidiaries included in projects proposed to the ATP.

[16] The survey instrument was reviewed and approved by the Office of Management and Budget and the JHU Committee on the Use of Human Subjects. OMB granted approval (no. 0693-0027) for Johns Hopkins University to conduct the survey on March 24, 1999.

[17] Prior to calling our respondents to conduct the interviews, we followed standard survey method procedures, sending a letter to all potential respondents in the selected sample to explain the purpose of the survey, identifying ATP as the sponsor of the study, and asking for cooperation. This letter also contained a statement of confidentiality and assurance that responses to any of the survey questions would remain

anonymous, and would not be publicly released in any form that would identify a specific individual or company. The letter included a list of questions that the respondent might find helpful to have in advance of the telephone interview. In addition, our mailing included an introductory letter from the ATP. The Schaefer Center for Survey Research at the University of Baltimore conducted the telephone interviews.

[18] A copy of the questionnaire is available from the authors upon request.

[19] See http://www.corptech.com/ and http://www.hoovers.com/.

[20] See Feldman and Kelley (2001) for further discussion.

[21] The average applicant scored 5.5 on the university linkage index with no statistical difference between award winners and non-winners.

[22] In most respects, the projects for firms that either went out of business or where the Principal Investigator (P.I.) was no longer employed at the firm were similar to the other non-awardees. However, we found one important difference. When compared to the evaluation of their business plans, the defunct firms were rated as having stronger, on average, technical plans. This suggests that they were relatively weaker on the business front.

References

Advanced Technology Program Proposal Preparation Kit, 1999, U.S. Department of Commerce Advanced Technology Program National Institute of Standards and Technology.

Cohen, Linda R. and Roger G. Noll, 1991, *The Technology Pork Barrel*, Washington, DC: The Brookings Institution.

Cohen, Wesley M. and Daniel A. Levinthal, 1990, 'Absorptive Capacity: A New Perspective on Learning and Innovation', *Administrative Science Quarterly* **35**, 128–152.

David Paul A., Bronwyn H. Hall and Andrew A. Toole, 2000, 'Is Public R&D a Complement or Substitute for Private R&D? A Review of the Econometric Evidence', *Research Policy* **29**(4–5), 497–529.

Doz, Y., 1996, 'The Evolution of Cooperation in Strategic Alliances: Initial Conditions or Learning Processes?' *Strategic Management Journal* **17**, 55–83.

Feldman, Maryann P. and Maryellen R. Kelley, 2001, 'The Case for Government R&D Additionality: An Investigation of the Advanced Technology Program Selection Process'. Prepared for the U.S. Department of Commerce Advanced Technology Program, National Institute of Standards and Technology. NISTIR 6577. January.

Feldman, Maryann P., Maryellen R. Kelley, Joshua Schaff and Gabriel Farkas, 2000, 'Reinforcing Interactions between the Advanced Technology Program and State Technology Programs. Volume II: How The States Assist High-Tech Start-up Companies', Prepared for the U.S. Department of Commerce Advanced Technology Program, National Institute of Standards and Technology. NISTIR 6523. June.

Flamm, Kenneth, 1988, *Creating the Computer: Government, Industry, and High Technology*, Washington, DC: Brookings Institution.

Gittelman, Michelle and Bruce Kogut, 2000, 'Why Do Firms Publish Their Research (By Their Own Scientists?): Publication and Patenting by U.S. Biotechnology Firms'. Paper presented at the Western Economic Association meeting in Vancouver, BC, July 1.

Griliches, Z., 1992, 'The Search for R&D Spillovers', *Scandinavian Journal of Economics* **94** (Supplement), 29–47.

Guellec, D. and B. von Pottelsbughe, 2000, 'The Impact of Public R&D Expenditures in Business R&D'. Paper presented to NBER Summer Institute, July 25.

Hagedoorn, John, Albert N. Link and Nicholas S. Vonortas, 2000, 'Research Partnerships', *Research Policy* **29**(4–5), 567–586

Hall, Bronwyn H., Albert N. Link and John T. Scott, forthcoming, *Universities as Research Partners in ATP-Funded Projects*. U.S. Department of Commerce Technology Administration, National Institute for Standards and Technology.

Hamel, G., 1991, 'Competition for Competence and Inter-Partner Learning Within International Strategic Alliances', *Strategic Management Journal* **12**, 83–103.

Harrigan, Kathryn R., 1988, 'Joint Ventures and Competitive Strategy', *Strategic Management Journal* **9**, 141–158.

Hill, Christopher T., 1998, 'The Advanced Technology Program: Opportunities for Advancement', in Lewis Branscomb and James Keller (eds.), *Investing in Innovation: Creating a Research and Innovation Policy that Works*, Cambridge: The MIT Press.

Inkpen, A. C., 1995, *The Management of International Joint Ventures: An Organizational Learning Perspective*, London: Routledge.

Jaffe, Adam B., 1996, *Economic Analysis of Research Spillovers: Implications for the Advanced Technology Program*, U.S. Department of Commerce Technology Administration, National Institute for Standards and Technology, GCR 97-708.

Jovanovic, Boyan, 2000, 'Growth Theory', NBER Working Paper 7468.

Kelley, Maryellen R., 1997, 'From Mission to Commercial Orientation: Perils and Possibilities for Federal Industrial Technology Policy', *Economic Development Quarterly* **11**(4), 313–328.

Khanna, T., R. Gulati and N. Nohria, 1998, 'The Dynamics of Learning Alliances: Competition, Cooperation, and Relative Scope', *Strategic Management Journal* **19**(3), 193–210.

Klette, Tor Jakob, Jarle Moen and Zvi Griliches, 2000, 'Do Subsidies to Commercial R&D Reduce Market Failures? Microeconomic Evaluation Studies', *Research Policy* **29**(4–5), 471–495.

Lerner, Josh, 1999, 'The Government as Venture Capitalist: The Long Run Impact of the SBIR Program', *Journal of Business* **72**(3), 285–318.

Liebeskind, Julia P., 1996, 'Knowledge, Strategy and the Theory of the Firm', *Strategic Management Journal* **17**, 93–107.

Liebeskind, Julia P., 1997, 'Keeping Organizational Secrets: Protective Institutional Mechanisms and their Costs', *Industrial and Corporate Change* **6**(3), 623–663.

Link, Albert N., David B. Audretsch, Maryann P. Feldman,

John E. Jankowski, Donald Siegel, Paula Stephan, Gregory Tassey and Charles Wessner, forthcoming, 'The Economics of Science and Technology', *The Journal of Technology Transfer*.

Morgenthaler, David, forthcoming, 'The Venture Capital Perspective', in Charles W. Wessner (ed.), *The Advanced Technology Program: Assessing Outcomes*, Washington, DC: National Academy Press.

Mowery, David and Nathan Rosenberg, 1989, *Technology and the Pursuit of Economic Growth*, Cambridge: Cambridge University Press.

Narayanan, V. K., G. E. Pinches, J. M. Kelm and D. M. Lander, 2000, 'The Influence of Voluntarily Disclosed Qualitative Information', *Strategic Management Journal* **21**, 707–722.

National Research Council, 1996, *Conflict and Cooperation in National Competition for High-Technology Industry*, Washington, DC: National Academy Press.

National Science Board, 1998, *Science & Engineering Indicators – 1998*, Washington, DC: U.S. Government Printing Office.

Nelson, Richard R., 1990, 'Capitalism as an Engine of Progress', *Research Policy* **19**(3), 193–214.

Nelson, Richard R., 1992, 'What Is "Commercial" and What Is 'Public' about Technology, and What Should Be?', in Nathan Rosenberg, Ralph Landau and David Mowery (eds.), *Technology and the Wealth of Nations*, Stanford: Stanford University Press, pp. 57–71.

Powell, Walter W., Kenneth W. Koput, and Laurel Smith-Doerr, 1996, 'Interorganizational Collaboration and the Locus of Innovation: Networks of Learning in Biotechnology', *Administrative Science Quarterly* **41**, 116–145.

Sako, Mari, 1992, *Prices, Quality and Trust: Inter-firm Relations in Britain & Japan*, New York: Cambridge University Press.

Sako, Mari, 1994, 'Supplier Relationships and Innovation', in Mark Dodgson and Roy Rothwell (eds.), *The Handbook of Industrial Innovation*, Cheltenham, UK and Brookfield, MA: Elgar.

Teece, David J., 1992, 'Competition, Cooperation and Innovation: Organizational Arrangements for Regimes of Rapid Technological Progress', *Journal of Economic Behavior and Organization* **18**, 1–25.

Trajtenberg, Manuel, forthcoming, 'R&D Policy in Israel: An Overview and Reassessment', in M. P. Feldman and A. N. Link (eds.), *Technology Policy for the Knowledge-Based Economy*, Boston, MA: Kluwer Academic Publishers, pp. 409–454.

von Hippel, Eric, 1988, *The Sources of Innovation*, New York: Oxford University Press.

Wallsten, Scott J., 1998, 'Rethinking the Small Business Innovation Research Program', in Lewis Branscomb and James Keller (eds.), *Investing in Innovation: Creating a Research and Innovation Policy that Works*, Cambridge: The MIT Press.

[13]

ELSEVIER

Journal of Business Venturing 18 (2003) 81–104

JOURNAL of BUSINESS VENTURING

Public versus private venture capital: seeding or crowding out? A pan-European analysis

Benoît Leleux[a,*], Bernard Surlemont[b]

[a]*IMD, International Institute for Management Development, 23 chemin de Bellerive, P.O. Box 915, Lausanne CH-1001, Switzerland*
[b]*University of Liège, Liège, Belgium*

Received 1 June 1999; received in revised form 1 February 2001; accepted 1 February 2001

Abstract

This paper analyzes the relationship between public and private sources of venture capital in Europe and the development of the industry, controlling for characteristics of the legal systems, in 15 European countries over the period 1990–1996. Large public participation is correlated with smaller VC industries, but analyses do not support the view that public venture capitalists are acting to seed the industry or that are they crowd-out private funds. On the contrary, public involvement seems to cause greater amounts of money to be invested in the industry as a whole. We argue that the effects of public intervention, whatever the motives, are real and probably result from demonstrating/sanctioning the social merit of venture capital and from signaling an enduring commitment to it.

Keywords: Venture capital; Signaling; Public interventions; Seed funds; Entrepreneurship support

1. Executive summary

Despite the very significant involvement of public money in the venture capital industry in Europe, the role played by government intervention and its effects has not been clearly addressed so far in the literature. In particular, there is little evidence of the effects of publicly sponsored or managed venture capital funds on the development of active national venture capital industries, or the mechanisms through which public involvement could generate

* Corresponding author. Tel.: +41-21-618-0335; fax: +41-21-618-0707.
E-mail address: leleux@imd.ch (B. Leleux).

0883-9026/03/$ – see front matter © 2003 Elsevier Science Inc. All rights reserved.
PII: S0883-9026(01)00078-7

additional wealth. This is particularly striking given the relatively long-standing commitment of many governments to the cause of regional development through equity investments and support. Such paucity of facts may be traced to difficulties in obtaining information about the industry as a whole in Europe, and its public or quasi-public players in particular. Recent developments have made such information more readily available, and brought the questions back into the political and economic debate. This study constitutes the first comprehensive analysis of the role played by public investors in the development of European national venture capital industries, both conceptually and empirically. The results provide a first useful guide to designing new policies toward private equity investments to shape more effective markets for the future.

Legislatures and governments play a number of roles in energizing private equity markets. First, they define both the legal and fiscal environments in which investors operate, providing various degrees of protection and/or taxation to different constituencies, flows (capital gains, dividends, etc.) and structures (partnerships, sole proprietorships, etc.). Witness, for example, the numerous philosophies and approaches to bankruptcies. Some countries, such as the US, organize the bankruptcy process to facilitate reorganization and the pursuit of activities under the direction of the incumbent management team. Others, such as France or Belgium, support the claims of creditors and owners, favoring an orderly liquidation under the supervision of a court-appointed trustee, a sale-of-assets route to bankruptcy. Second, they sometimes intervene directly in the venture capital process by funding and managing public venture funds, targeting, for example, regional development or industry restructuring. Third, they provide incentives or impediments to private equity investments by regulated private companies such as banks, pension funds or insurance companies.

If and when public entities opt for a direct involvement in venture capital funds, they tend to do so in ways that could be counterproductive. First, public fund managers are often civil servants and government employees, and as such may not have the experience nor the drive necessary to select and support entrepreneurial companies. Second, the incentive structures these managers face often differ markedly from the traditional private fund arrangement, where partners share in the profits through a predefined formula (referred to in the industry as the "carry"), a performance-linked bonus. A fee-based incentive package, common in public institutions, creates different incentives than the profit-based incentives of private VCs. Finally, and more damaging to the industry as a whole, if public funds forego some expected returns for their policy objectives, financing projects at below-market rates, they may end up attracting the best projects, leaving only "lemons" for private VC firms to fund, making the entry of new, independent private equity funds more difficult. These arguments can be used to posit a number of negative effects for direct public intervention in the venture capital process: (1) capital can be misallocated, and (2) additional entry barriers are created against an industry the government actually claims to be supporting. More generally, the three questions of interest are whether direct public interventions (1) tend to "seed" underdeveloped VC industries; (2) prevent the emergence of active VC markets by "crowding-out" private funds; or (3) facilitate their development by signaling and certifying their social value.

The empirical evidence in this study, the first pan-European analysis of its kind, does not support the seeding nor the crowding-out hypothesis, nor does it support the proposition that public investments are detrimental to the industry as a whole. Countries with large proportions of public venture capital are shown to have relatively smaller markets overall. Such a relationship does not point to causality, i.e., is the lack of private equity funds forcing state intervention in the VC industry or is the latter preventing the emergence of a vibrant VC industry? To be able to infer causality, it is necessary to move beyond the static analysis into the dynamics of the relationship between the variables, i.e., how they jointly evolve over time. Granger causality studies, detailed in the appendices, are used to infer such direct causal links. When such dynamic causality links are statistically supported, we will indicate that one variable "Granger causes" another, or that the two variables are "causally related in a Granger sense" to reflect the nature of the causality relationship tested, to provide a strong distinction with the simple correlation measures. The analyses of the 8-year, 15-country panel dataset support the view that (1) public intervention in the venture capital industry tends to be the consequence of the development of the industry, not its cause. This does not support the seeding hypothesis. Public sector participation tends to cause, in the Granger sense, larger amounts to be raised for venture capital investments overall. This evidence does not support the crowding-out hypothesis and we conjecture it may be related to signaling legitimacy by national authorities. Finally, consistent with their charters, public venture capitalists tend to invest more often in later-stage deals in industries with large human resource needs. If the stated rationale for public intervention in the VC industry is to correct the apparent inability of the private sector to develop a market for funds targeting early-stage, high-risk ventures, then the results of the study do not support such a statement. On the other hand, the effects of such interventions, irrespective of their rationale, appear very real and benefit the industry as a whole.

2. Introduction

Despite the very significant involvement of public money in the venture capital industry in Europe, the role played by government intervention and its effects has not been clearly addressed so far in the venture capital literature. In particular, there is little evidence about the effects of publicly sponsored or managed venture capital funds on the development of active and dynamic national venture capital industries, nor the mechanisms through which public involvement could generate additional wealth. This is particularly striking given the relatively long-standing commitment of many governments to the cause of selective regional development through equity investments and support. Part of the reason for this state may be traced to difficulties in obtaining information about the industry as a whole in Europe, and its public or quasi-public players in particular. Recent developments have made such information more readily available, and brought the questions right back to the forefront of the political and economic debate. This study constitutes the first comprehensive attempt at understanding the role played by public investors in the development of European national venture capital industries, from both a conceptual and an empirical standpoint. The results of the study provide a first guide to designing new policies toward private equity investments, which hopefully will help shape more effective markets for the future.

A number of arguments can be made with respect to the role played by legislatures and governments in energizing private equity markets. First of all, they determine both the legal and fiscal environments in which investors operate, providing various degrees of protection to different constituencies and taxing various flows (capital gains, dividends, etc.) and structures (partnerships, sole proprietorships, limited liability companies, etc.) differently. Witness, for example, the numerous philosophies and approaches to bankruptcies. Some countries, such as the US, organize the bankruptcy process to facilitate reorganization and the pursuit of activities under the direction of the incumbent management team. Others, such as France or Belgium, support the claims of creditors and owners, favoring an orderly liquidation under the supervision of a court-appointed trustee, a sale-of-assets route to bankruptcy (Warren, 1987). Second, they can also intervene directly in the venture capital process by funding and managing public venture funds with various policy objectives, e.g., regional development or industry restructuring. Third, they can favor or hinder the development of VC funds by highly regulated, private companies such as banks, or insurers, by providing the incentives or impediments to private equity investments.

If and when public entities opt for a direct involvement in the setting up and management of venture capital funds, they tend to do so in ways that are likely to be counterproductive for a number of reasons. First of all, public fund managers are often civil servants and government employees, and as such may not have the experience nor the drive necessary to select and support entrepreneurial companies. Second, the incentive structures these managers face often differ markedly from the traditional private fund arrangement, where partners share in the profit through performance-linked bonuses. Public sector managers, on the other hand, often face fee-based incentive packages, with very different incentive effects. Finally, and more damaging to the industry as a whole, if public funds forego some expected returns for their policy objectives, financing projects at below-market rates, they may end up attracting the best projects, leaving only "lemons" for private VC firms to fund, making the entry of new, independent private equity funds more difficult. These arguments can be used to posit a number of negative effects for direct public intervention in the venture capital process: (1) capital can be misallocated, and (2) additional entry barriers are created against an industry the government actually claims to be supporting. More generally, the three questions of interest are then whether direct public interventions tend to "seed" underdeveloped VC industries, prevent the emergence of active VC markets by "crowding-out" private funds, or facilitate their development by signaling and certifying their social value.

The empirical evidence in this study, the first pan-European analysis of its kind, does not support the seeding nor the crowding-out hypothesis, nor does it support the proposition that public investments are detrimental to the industry as a whole. Countries with large proportions of public venture capital are shown to have relatively smaller markets overall. Such a relationship does not point to causality though, i.e., is the lack of private equity funds forcing state intervention in the VC industry or is the latter preventing the emergence of a vibrant VC industry? To be able to infer causality, it is necessary to move beyond the static analysis into the dynamics of the relationship between the variables, i.e., how they jointly evolve over time. Granger causality studies, detailed in the appendices, are used to infer such direct causal links. When such dynamic causality links are statistically supported, we will indicate that one variable "Granger causes" another, or that the two variables are "causally related in a Granger sense" to reflect the nature of the causality relationship tested, to provide a strong distinction with the simple correlation measures. The analyses on the 8-year, 15-country panel dataset support the view that (1) public intervention in the venture capital industry tends to be the consequence of the development of the industry, not its cause. This evidence does not support the seeding hypothesis. Public sector participation tends to cause, in the Granger sense, larger amounts to be raised for venture capital investments overall. This evidence does not support the crowding-out hypothesis. We conjecture that state or public intervention may be seen or interpreted as signaling the government's support of such venture-financing activities and a certification of the "public value" of such endeavors. Finally, consistent with their charters, public venture capitalists tend to invest more often in later-stage deals in industries with large human resource needs.

3. Legal systems and European venture capital markets

Why are there such large differences among the size, breadth and valuation of various capital markets around the world? La Porta et al. (1996) conjecture that differences in the nature and effectiveness of financial systems around the world can be traced, in part, to differences in investor protections against expropriation by insiders, as reflected by legal rules and the quality of their enforcement. The authors show that these investor protection rules vary systematically by legal origin, distinguishing four major legal systems: English, French, German and Scandinavian. Comparing legal rules across 49 countries, they show that these rules differ in content as well as in the history of their adoption. In the area of protection against expropriation by insiders, common law countries, i.e., the English system, protect both shareholders and creditors the most, French civil law countries the least, the latter also having the lowest quality of enforcement. Poor investor protection is also shown to be correlated with more highly concentrated ownership of shares. In a follow-up article, La Porta et al. (1997) assess the ability of firms in different legal environments to raise external finance through either debt or equity. On the same set of 49 countries, they find evidence to support the assertion that the legal environment affects the size and breadth of capital markets across countries: those with poorer investor protection tend to have smaller and narrower capital markets.

Legal systems are both a reflection of their environments as well as a very potent influence on them, so that the underlying causality will always remain questionable. The "legal environment" line of research is slowly gaining recognition though. Shleifer and Vishny (1997) examine the agency problem between entrepreneurs and investors and the cross-country differences in the legal solutions adopted. Rajan and Zingales (1995) examine the capital structure decision in the perspective of contract enforcement possibilities, while Modigliani and Perotti (1996) focus on the protection of minority interests and the development of capital markets. King and Levine (1993) and Rajan and Zingales (1996) push the rationale a step further to the relationship between financial markets and growth and development, supporting the view that larger markets generate more GNP growth for their countries.

European venture capital markets and their main actors have previously been investigated from a number of perspectives. Sapienza et al. (1996) survey the venture capitalists' perceptions of the value added in their various roles and interactions with investee companies. Manigart (1994) analyzes the sectorial and environmental forces facilitating the creation of venture capital companies in three major European countries, stressing the importance of industry density over factors such as the creation of tax-transparent entities, state guarantees against losses or the establishment of secondary stock markets, which did not seem to effect founding rates over the period considered. Roure et al. (1990) focus on venture capitalists' strategies, Ooghe et al. (1991) on the industry's growth patterns, while Birley et al. (1996) examine the venture capitalists' investment decision-making process across Europe.

A potentially contentious area as yet unexplored is the role played by public and quasi-public funds in the development of the venture capital industries as a whole. In particular, no attempt as been made so far to ascertain their ability to generate the dynamic VC market they often claim to be seeding. In this paper, we empirically test three nonexclusive propositions:

Proposition 1: Public venture capitalists substitute for insufficient private venture capital funds and "seed" the development of the venture capital industry in their respective countries.

Proposition 2: Public venture capital sends a strong and credible signal to investors of the commitment and support of political forces towards entrepreneurial finance activities, and as such helps bootstrap the industry's development.

Proposition 3: Public venture capitalists exhibit mixed objective functions, where financial performance is put on the same footing as regional development, job creation or maintenance, and improvement of self-perception objectives.

We thus hypothesize three major objectives to public venture capital programs: (1) VC industry launch; (2) VC industry development support; (3) regional employment support and other politically or socially motivated goals.

4. Database, research hypotheses and methodologies

The analysis combines the statistical evidence compiled by the European Venture Capital Association (EVCA) over the period 1990–1996 with information on European legal systems from La Porta et al. (1996, 1997) and other general macroeconomic variables from Leleux (1998) to investigate the direct (through public venture capital firms) and indirect (through the characteristics of the respective legal system as far as the protection awarded investors, directors and shareholders) public involvement in the venture capital process.

The data include the cumulative funds raised to date in each of the 15 countries considered (namely Austria, Belgium, Denmark, Finland, France, Germany, Ireland, Italy, the Netherlands, Norway, Portugal, Spain, Spain, Switzerland and UK), the breakdown of VC funds raised by type of investors, the distribution of VC investments by fund types (independent, captive, semi-captive and public funds), stage of investment (seed, startup, expansion, replacement and buyout capital), and industry, and syndication statistics for the years 1990–1996.

The core proposition is that public involvement through direct investments is in general nonoptimal and even detrimental to the development of venture capital industries. Public intervention is more likely to appear in legal environments where private venture capital funds have a difficult time taking root because of insufficient legal protection or excessive taxation of investors: the perceived need for state involvement may be a direct result of the state's very inability to lay down the ground rules for private equity investments. If such is the case, we can conjecture that this seemingly inefficient arrangement survives because it provides private benefits to its organizers, namely political organizations which see in these state-dependent venture capitalists an additional lever to distribute favors to their respective constituencies.

More specifically, the research hypotheses to be examined by this paper are as follows:

Hypothesis 1: Countries offering relatively poor or unreliable investor protections (French and German civil law countries in particular) develop smaller VC industries, where the major fund suppliers will be tied to the public sector, either directly (public sector investors) or indirectly through regulations (banks, pension funds, insurance companies).

These state-related investors have the ability to influence the very political process behind the legal interpretation or even implementation of the law and are better able than others to avoid the potentially costly limitations of the legal system. This hypothesis is the extension to the venture capital markets of the findings in La Porta et al. (1997).

Hypothesis 2a: Strong public sector involvements in the venture capital industry lead to relatively smaller venture capital industries.

Even though public interventions may be warranted by the unfavorable legal climate, such interventions do not eliminate the roots of the problem, and hence will do nothing to support the industry. The alternative is Hypothesis 2b below: if public intervention in venture capital markets is a response not to the lack of private equity but to the political desire to get involved in an already developing or well-developed vote-sensitive private equity sector, then the size of the venture capital market causes, in the Granger sense, public sector participation Granger and Newbold (1974). Hypothesis 2b actually combines two different postulates: first, public VCs *seed* the industry (a small industry size causes active involvement by public VCs) and second, public VCs are *late entrants* into the industry (a large industry size causes active public VC involvement).

Hypothesis 2b: Public sector intervention in the venture capital industry is a consequence, not the cause, of venture capital market development.

Interventions by public entities could also "certify" politically and socially the role played by the industry and signal a desire to support private equity investments, leading to larger amounts actually being raised and invested in the industry from both public and private sources. This is outlined in Hypothesis 2c:

Hypothesis 2c: Public sector participation in the venture capital industry causes, in the Granger sense, larger amounts of money to be raised for venture capital investments overall.

If state-supported venture capital funds are also used to provide support for politically sensitive projects, these investments can be conjectured to target the demonstrably larger employment generators, such as firms with large, established employee counts, not the small entrepreneurial teams with unknown prospects, which may be light years away from generating any sizeable employment or tax money. To the extent that the individuals making investment decisions within state-related venture capital funds tend to be themselves products from the states' civil services, they could also show a preference for assets in place rather than growth opportunities. A more likely investee in that case would then be a larger, established firm in an employment-generating industry, as reflected in Hypotheses 3 and 4:

Hypothesis 3: Captive and public sector VC firms will tend to favor later-stage deals.

Hypothesis 4: Captive and public sector VC firms are more likely to invest in industries that utilize large amounts of human resources (industrial products and services, chemicals and materials).

Syndication is often seen as a rational response to information asymmetries by a portfolio optimizer, offering the VC fund the ability to diversify further its investments with a limited capital base. Public VC funds operate under different constraints; in particular, availability of funds is often less of a problem than in independent funds, and their objective function may include nonfinancial items such as political and employment goals. With a mixed objective function, portfolio diversification may be less critical. Furthermore, public venture capital tends to operate with broader fund foci, providing them with internal diversification benefits across industries and technologies. These three factors (larger capital basis, mixed objective function and broader fund focus) alleviate the pressure on the need to syndicate. This is reflected in Hypothesis 5 below:

Hypothesis 5: Captive and public sector VC firms syndicate a lower fraction of their investments than independent operators.

Multiple regressions with control variables are used to investigate the explanatory power of a number of independent variables, single-handedly and in combination. A difficulty arises though when testing for causality in the Granger sense for Hypotheses 2a-c above. Granger causality tests are designed to handle pure time series: applying the tests to panel data generates a number of problems due to the characteristics of these datasets Granger and Newbold (1988). In particular, when dynamic models are estimated using time series of cross-sectional data, the usual least squares methods do not lead to consistent estimators for the parameters of the two most commonly used models for panel data (i.e., fixed effects and error components models). This inconsistency results from the fact that the disturbance terms are serially correlated in these models, which causes the lagged endogenous variable to be correlated with those disturbances (Sevestre and Trognon, 1992). To overcome the specification problem, a solution inspired by Gordon (1988) and Mehra (1995) is used. The presence and direction of causality between, e.g., CFR, the cumulative venture capital funds raised up to a given year, and PSPCT, the percentage of investments done by public venture capitalists, can tested using Eqs. (1) and (2):

$$\mathrm{CFR}_t = \alpha_{1,0} + \sum \alpha_{1,s}\mathrm{CFR}_{t-s} + \sum \beta_{1,s}\mathrm{PSPCT}_{t-s} + \text{control variables} \tag{1}$$

$$\mathrm{PSPCT}_t = \alpha_{2,0} + \sum \alpha_{2,s}\mathrm{PSPCT}_{t-s} + \sum \beta_{2,s}\mathrm{CFR}_{t-s} + \text{control variables} \tag{2}$$

where the control variables account for the items previously shown in the literature to impact the size of the venture market or capture missing elements in the regressions. The tests are run using lags up to 6 years, with the most appropriate lag determined by the form of the equation with the lowest Akaike Information Criterion (AIC). For each relationship postulated in the hypothesis, different specifications are tested to optimize the form of the relation: absolute values, logarithms or growth rates, with appropriate theoretical justifications.

For Eqs. (1) and (2), the decision rule is as follows:

- PSPCT does not cause CFR in the sense of Granger if all $\beta_{1,s}=0$. In other words, there is no seeding or additive effect taking place.
- CFR does not cause PSPCT in the sense of Granger if all $\beta_{2,s}=0$. In this case, there is no crowding-out in the industry.

Other decision rules can be inferred from the regression coefficients as well. For example, Hypothesis 2a implies that $\beta_{1,s}<0$, Hypothesis 2b that $\beta_{2,s}\neq 0$, while $\beta_{2,s}<0$ suggests seeding and $\beta_{2,s}>0$ is more supportive of late entry by public VCs. Alternatively, if private venture capital funds somehow crowd-out public venture capital, for instance if their value added far exceeds the cost advantage of public VCs for the entrepreneur, then $\alpha_{2,s}$ and $\beta_{2,s}$ are expected to be negative.

The *F* tests used are derived from the Wald test. These specifications suppose that all variables considered are stationary (e.g., here CFR and PSPCT). That assumption in the context of dynamic linear models has been shown by Sevestre and Trognon (1992) to be not too stringent. On the other hand, that specification offers the advantage of being able to use standard *F* statistics to test the hypotheses, so that no recourse has to be made to Dickey–Fuller tables, as indicated by Koenker (1982) and Judge et al. (1985). Furthermore, the specification can easily be changed to accommodate a linear trend. In that case, the specification becomes:

$$\mathrm{CFR}_t = \alpha_{1,0} + \sum \alpha_{1,s}\mathrm{CFR}_{t-s} + \sum \beta_{1,s}\mathrm{PSPCT}_{t-s} + \text{control variables} + \lambda_1\mathrm{TIME} \tag{3}$$

$$\mathrm{PSPCT}_t = \alpha_{2,0} + \sum \alpha_{2,s}\mathrm{PSPCT}_{t-s} + \sum \beta_{2,s}\mathrm{CFR}_{t-s} + \text{control variables} + \lambda_2\mathrm{TIME}. \quad (4)$$

To overcome potential heteroskedasticity problems, t tests robust to violations of the homoskedasticity assumptions are used. All standard deviations are also estimated using the White (1980) correction for heteroskedasticity. The specification issues are explained in greater details in the appendices.

5. Descriptive statistics

Fifteen European countries are represented in the sample: Austria, Belgium, Denmark, Finland, France, Germany, Ireland, Italy, the Netherlands, Norway, Portugal, Spain, Spain, Switzerland and UK. Venture capital industry data and general macroeconomics information for each of these countries were collected for the years 1990–1996 inclusive.

Table 1 illustrates the evolution of the total cumulative funds raised by country over the period investigated. It highlights the preponderance of UK funds: at the end of 1996, funds in UK (population 57.8 million) had raised in excess of ECU 25.6 billion for venture capital investments, whereas at the same time France (population 58.2 million) stood at ECU 11.7 billion, Germany (population 81.8 million) at ECU 5.1 billion and Italy (population 56.7 million) at ECU 4.8 billion.

Fig. 1 investigates the major sources of venture capital funds by country over the period 1990–1996. Again, vast disparities can be highlighted. In UK, most of the funds raised came from pension funds (40.3% of the total raised) and banks (21.5%), whereas for France, the

Table 1
Cumulative funds raised in Europe, 1990–1996 (in Euro 1000)

	1990	1991	1992	1993	1994	1995	1996
Austria	45,164	45,164	45,813	47,871	47,951	50,756	74,831
Belgium	657,988	749,355	842,230	964,844	1,101,040	1,268,995	1,429,094
Denmark	243,591	268,822	282,278	303,792	415,520	457,575	456,182
Finland	180,339	190,558	167,421	155,060	206,798	277,797	331,109
France	5,346,311	6,528,282	7,500,859	8,570,026	9,670,061	10,590,148	11,696,952
Germany	2,177,208	3,007,792	3,885,773	4,267,187	4,275,140	4,714,364	5,168,330
Ireland	252,292	291,900	340,989	353,293	539,849	539,631	574,174
Italy	1,286,532	1,510,780	4,020,334	3,703,518	3,948,962	3,824,373	4,883,744
The Netherlands	2,214,769	1,508,812	1,560,028	1,701,854	1,846,893	1,907,032	3,271,028
Norway	143,153	251,453	262,821	318,544	353,203	403,919	538,823
Portugal	134,446	146,516	167,762	245,309	348,613	437,744	474,525
Spain	694,356	1,006,449	1,157,597	1,232,440	1,207,419	1,323,140	1,395,800
Sweden	642,381	750,470	1,067,976	1,010,174	1,133,941	1,654,673	1,862,157
Switzerland	338,379	359,171	393,727	479,062	564,849	642,819	789,025
UK	14,960,652	16,271,778	16,715,819	17,038,249	20,882,025	21,517,399	25,652,582

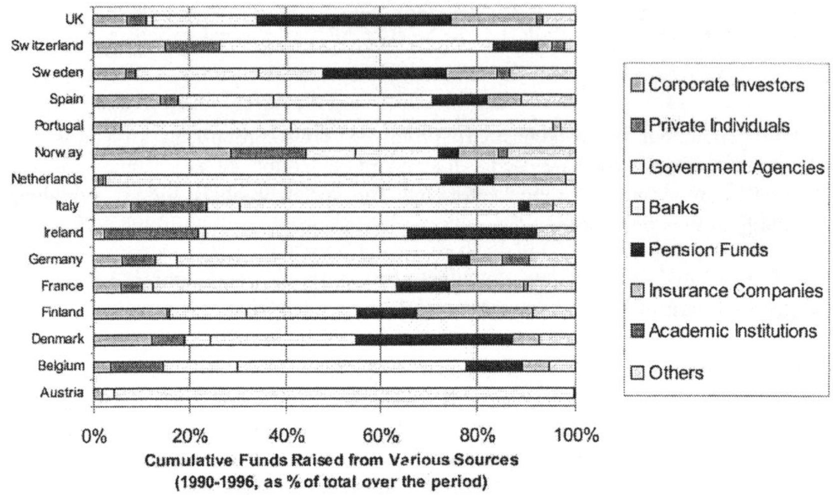

Fig. 1. Distribution by sources of VC funds raised, 1990–1996.

two largest sources of funds were banks (50.5%) and insurance companies (15.3%). Fig. 2 highlights the large variations by country in terms of the preferred stages of investment for the venture capital funds. As a whole, European VCs favor later-stage financings, with seeds and start-ups collecting only a very small percentage of the funds invested.

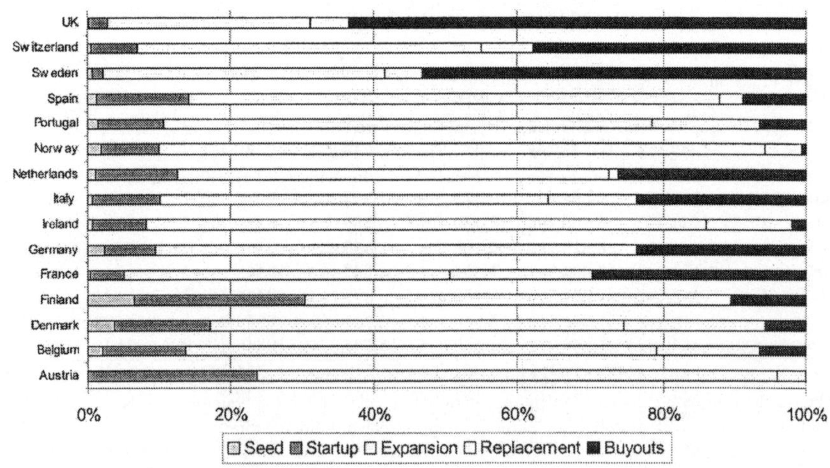

Fig. 2. Distribution of VC investments by stages, 1990–1996.

6. Legal systems and venture capital development

The first hypothesis investigates the correlation between the nature of the legal system, in particular the level of protection afforded investors, and the development of national venture capital industries. The legal systems are classified according to the system developed by La Porta et al. (1997).

Where La Porta et al. (1997) document the better access to equity finance common law countries offer over civil law countries, Table 2 supports a similar finding for venture capital equity. Cumulative funds raised as of 1996 are higher for common law countries (ECU 313.8 per capita) compared to civil law countries (between ECU 66.6 and 125.8 per capita depending on the grouping). Comparisons of means between common and civil law

Table 2
Domestic venture capital markets in Europe and legal systems

Country	Cum VC funds/pop (ECU)	VC executives/ 1 m pop	GDP growth	Rule of law	Directors' rights	Shareholder rights
Ireland	166.91	7.27	4.25	7.80	3	0
UK	460.78	15.93	2.27	8.57	4	0
English origin average	313.85	11.60	3.26	8.19	3.50	0.00
Belgium	145.01	9.33	2.46	10.00	0	0
France	217.46	11.78	2.54	8.98	2	0
Greece	4.06	3.75	2.46	6.18	1	1
Italy	85.47	2.99	2.82	8.33	0	0
The Netherlands	229.92	9.63	2.55	10.00	2	0
Portugal	47.77	7.35	3.52	8.68	2	0
Spain	37.29	6.44	3.27	7.80	2	0
French origin average	109.57	7.32	2.80	8.57	1.29	0.14
Austria	9.97	0.79	2.74	10.00	2	0
Germany	65.93	4.34	2.60	9.23	1	0
Switzerland	123.94	5.81	1.18	10.00	1	0
German origin average	66.61	3.65	2.17	9.74	1.33	0.00
Denmark	87.83	5.01	2.09	10.00	3	0
Finland	69.15	12.74	2.40	10.00	2	0
Norway	122.73	11.24	3.43	10.00	3	0
Sweden	223.82	9.13	1.79	10.00	2	0
Scandinavian origin average	125.88	9.53	2.43	10.00	2.50	0.00

The table classifies countries on the basis of the origin of their legal system, as in La Porta et al. (1997). Cum VC Funds/Pop is the cumulative amount of venture capital raised (in ECU) in the country at of the end of 1996 per inhabitant; VC Executives/1 m Pop is the number of equivalent full-time venture capital executives per million inhabitants; GDP growth is the average annual percentage growth of per capita domestic product for the period 1970–1993; Rule of law is an assessment of the law and order tradition in the country; the last two variables are indices of the respective rights of directors and shareholders.

countries for the cumulative venture capital funds raised per capita are significant at the 1% level (F test = 9.024). The numbers are similarly statistically different (although with slightly lower F test) using an alternative specification for the relative size of the VC markets, namely cumulative funds raised standardized by the country's Gross Domestic Product (GDP). The second measure of venture capital industry development is the number of full-time equivalent venture capital executives in activity at the end of 1996, expressed either as a raw number or per million inhabitants. Common law countries (English origin in Table 2) average 11.6 VC executives per million inhabitants, whereas French civil law countries average only 7.32 and German law countries 3.65 VC executives/million inhabitants. As previously reported in La Porta et al. (1997), antidirector rights (the ability for shareholders to remove relatively expeditiously inefficient managers) are strongest in common law countries and weakest in French civil law countries. No differences emerge in terms of shareholder rights. These results support the first part of Hypothesis 1: countries offering relatively poor investor protections (French and German civil law countries in particular) develop smaller VC industries.

To investigate the second part of the hypothesis, namely that most venture capitals in countries with weak investor protection laws are tied to the public sector, either directly (public sector investors) or indirectly through regulation (banks, pension funds, insurance companies), we analyze the distribution of sources of funds over the period 1990–1996 by country and legal system. Group 1 countries (German law system) source, on average, 75.9% of their funds from the public or publicly regulated industries. The number for Group 2 countries (French civil law) is 64.1%, for Group 3 countries (English common law) 62.4%, while Group 4 countries (Scandinavian law grouping) average 55.9%. The F test for difference of means between the various legal groups is significant at the 1% level (F test = 38.44). Extending the proposition beyond the nature and content of the legal systems to the quality of their enforcement, following the "Rule of Law" definition and measurement of La Porta et al. (1996), did not provide additional insights: Rule of Law by itself does not explain the percentage of VC funds coming from the public and publicly regulated sources. It is the very nature of the legal systems in terms of investor protection, more than the quality of the enforcement of these rules, that seems to influence VC market size.

7. Public venture capital and industry development

The second set of hypotheses investigates the relationship between public interventions in the venture capital industries with the development of these markets, focusing in particular on the dynamics of the relationship and the causality linkages.

Hypothesis 2a posits that countries with strong public sector involvement in the VC industry will tend to have smaller venture capital markets. Fig. 3 illustrates that relationship: countries with large public VC firms (percentage of total VC funds invested over the years 1990–1996 accounted for by public investors) appear to be associated with smaller cumulative funds raised, as posited by Hypothesis 2a.

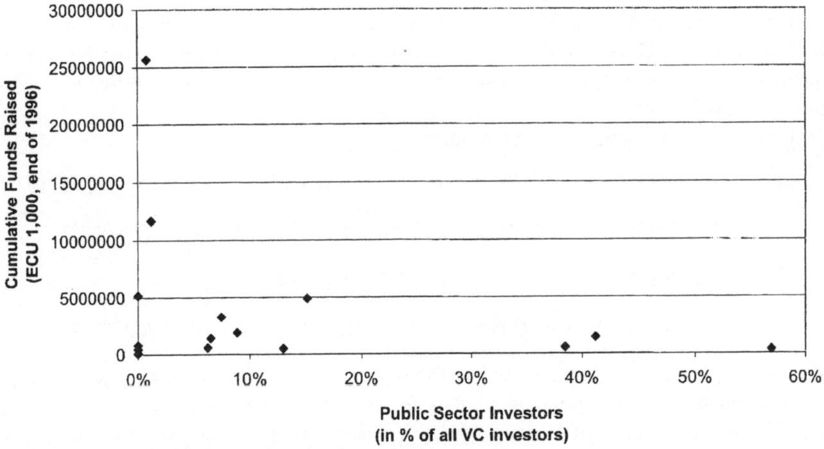

Public sector investors include all government agencies specifically set up to provide risk equity,whatever the rational for the intervention (job creaton, selective regional development, focused sectorial assistance, etc.). The x-axis represents the percentage of public sector investors in the pool of VC money raised over the period 1990-1996 in the countries covered by the study. Cumulative funds raised (in thousands of Euros) refer to the cumulative amounts of venture capital funds raised by the venture capital industry in each country over the period 1990-1996 as per EVCA databases.

Fig. 3. Cumulative funds raised and public investors, 1990–1996.

The exponential nature of the relationship is interesting, possibly indicating that even small amounts of public/state involvement in the VC industry are associated with negative effects. In other words, the possible disruptive elements brought about by state involvement may become active at a very early stage. Points in the lower left corner of the exhibit (countries with no public sector interventions and minuscule VC markets) limit the quantitative analyses of the relationship.

The existence of correlation does not point, however, to the direction of causality; in other words, it is not possible to infer whether the lack of private equity funds induces state intervention in the VC industry or if the latter prevents the emergence of a large VC industry. To ascertain such causality, Granger causality tests are used on the 15-country, 8-year panel data sample, using a process described in detail in the Database, Research Hypotheses and Methodologies section utilizing OLS methods and robust estimators jointly accounting for both heteroskedasticity and autocorrelations in the residuals (heteroskedasticity- and autocorrelation-consistent, HAC, covariances]. The procedure is also described in Appendix A. The AIC is then used to determine the most appropriate lag in the causality model. Information criteria are used in general to provide a measure of information that strikes a balance between goodness of fit and parsimonious specification of the model, with the "best" model being the one afflicted with the smallest information criterion. The AIC is widely used in time series analysis to determine the appropriate length of the distributed lag. Lutkepohl (1991, Chapter 4), for example, presents a number of results regarding consistent lag order selection in VAR models. The information criterion approach is explained in more details in Appendix B. A better description of the Granger causality analyses can be found in Appendix C, as well as the Wald test used to investigate the joined significance of lagged regression coefficients.

Hypothesis 2a is tested using the model:

$$\begin{aligned}\mathrm{CFR} = {} & \alpha_0 + \alpha_1^*\mathrm{CFR1} + \alpha_2^*\mathrm{CFR2} + \alpha_3^*\mathrm{CFR3} + \alpha_4^*\mathrm{CFR4} + \alpha_5^*\mathrm{CFR5} + \alpha_6^*\mathrm{CFR6} \\ & + \beta_1^*\mathrm{PSPCT1} + \beta_2^*\mathrm{PSPCT2} + \beta_3^*\mathrm{PSPCT3} + \beta_4^*\mathrm{PSPCT4} + \beta_5^*\mathrm{PSPCT5} \\ & + \beta_6^*\mathrm{PSPCT6} + \lambda_1^*\mathrm{YEAR} + C(15)^*\mathrm{CO} + C(1)^*\mathrm{RULELAW} \\ & + C(2)^*\mathrm{ANTIDIRECTOR} + C(3)^*\mathrm{CREDITOR} + C(4)^*\mathrm{LAWTYPE}\end{aligned}$$

where the α, β, and λ coefficients are per the regression models above, *C*(.) is the regression coefficient for the control variables, CFR is the cumulative VC fund raised in each country, PSPCT is the percentage of venture capital investments done by public sector firms, the numbers attached to the variables refer to the respective lags (e.g., CFR3 is the cumulative VC fund raised in a country as of 3 years earlier), YEAR is a control for annual variations, CO is a country control to account for national idiosyncrasies not included otherwise in the regression, RULELAW, ANTIDIRECTOR and CREDITOR are, respectively, the rule of law, antidirectors' right and creditors' right indices from La Porta et al. (1997) and LAWTYPE is the simple typology of legal systems proposed by La Porta et al. (1996). The fact that the same number of lags is not available in all years in the panel data does not create a major problem since the most significant specification is always the one-period lagged one.

The PSPCT coefficients are never jointly different from zero, so Hypothesis 2a — that heavy state intervention causes smaller VC industries — is not supported. In other words, there is no evidence to support the hypothesis that public venture capitalists chase-out or "crowd-out" private venture capitalists from the industry.

Hypothesis 2b postulates that the (relative) development of venture capital markets causes public sector intervention. Hypothesis 2b is tested using the model:

$$\begin{aligned}\mathrm{CFR_{GDP}} = {} & \alpha_0 + \alpha_1^*\mathrm{CFR_{GDP}1} + \alpha_2^*\mathrm{CFR_{GDP}2} + \alpha_3^*\mathrm{CFR_{GDP}3} + \alpha_4^*\mathrm{CFR_{GDP}4} \\ & + \alpha_5^*\mathrm{CFR_{GDP}5} + \alpha_6^*\mathrm{CFR_{GDP}6} + \beta_1^*\mathrm{PCPCA1} + \beta_2^*\mathrm{PCPCA2} + \beta_3^*\mathrm{PCPCA3} \\ & + \beta_4^*\mathrm{PCPCA4} + \beta_5^*\mathrm{PCPCA5} + \beta_6^*\mathrm{PCPCA6} + \lambda_1^*\mathrm{YEAR} + C(1)^*\mathrm{CO} \\ & + C(2)^*\mathrm{RULELAW} + C(3)^*\mathrm{ANTIDIRECTOR} + C(4)^*\mathrm{CREDITOR} \\ & + C(5)^*\mathrm{LAWTYPE}\end{aligned}$$

where the α, β, and λ coefficients are per the regression models above, *C*(.) is the regression coefficient on the control variables, CFR_GDP is the cumulative VC fund raised, standardized by the country's GDP, PCPCA is the percentage of public and captive funds in the total invested in the VC industry, the numbers attached to the variables refer to the respective lags

Table 3
Test of Hypothesis 2b: CFR_GDP Granger causes PCPCA

Variable	Coefficient	Standard error	t statistic	Probability
C	− 13.12078	30.13969	− 0.435332	.6643
PCPCA1	0.285028	0.099386	2.867884	.0051
CFR_GDP1	13.05877	5.561952	2.347875	.0210
YEAR	0.006842	0.015117	0.452569	.6519
CO	0.004046	0.006070	0.666553	.5067
RULELAW	− 0.001391	0.020422	− 0.068116	.9458
ANTIDIRECTOR	− 0.070891	0.032390	− 2.188650	.0311
CREDITOR	− 0.123398	0.026585	− 4.641563	.0000
LAWTYPE	0.032343	0.031443	1.028614	.3063
R^2	.494340			
Adjusted R^2	.450842			
Standard error of regression	0.210705			
Sum squared residuals	4.128888			
Log likelihood	18.82348			
Durbin–Watson statistics	1.698063			
Mean dependent variable	0.437508			
S.D.-dependent variable	0.284332			
AIC	− 0.192617			
SC	.038998			
F statistic	11.36475			
Prob (F statistic)	0.000000			

Included observations: 102. Excluded observations: two after adjusting endpoints. Newey–West HAC standard errors and covariance (lag truncation = 4).
The regressions coefficients are defined as follows: C(.) is the regression coefficient for the control variables, CFR_GDP is the cumulative VC fund raised, standardized by the country's GDP, PCPCA is the percentage of public and captive funds in the total invested in the VC industry, the numbers attached to the variables refer to the respective lags (e.g., CFR_GDP1 refers to the standardized cumulative funds raised as defined above as of 1 year earlier, or lag 1), YEAR is a control for annual variations, CO is a country control to account for national idiosyncrasies not included otherwise in the regression, RULELAW, ANTIDIRECTOR and CREDITOR are, respectively, the rule of law, antidirectors' right and creditors' right indices from La Porta et al. (1997) and LAWTYPE is the simple typology of legal systems proposed by La Porta et al (1996).

and the control variables as above. The results of the lag − 1 specification, which offers the smallest AIC, are presented in Table 3. The model specification is highly significant (R^2 adjusted=.45; F statistic = 11.36) and the coefficient on the lag − 1 CFR_GDP variable is positive and significant. In other words, the relative size of the venture capital markets, expressed as the cumulative funds raised relative to the country's GDP, causes, in the Granger sense, a larger participation by public and captive venture capitalists into the industry. The best-fitting lags are for 1- and 2-year periods, implying that public sector intervention is relatively rapid after the industry reaches a critical mass. The results of the causality analyses do not support the initial seeding hypothesis: to the contrary, public venture capitalist appears to come late to the market, more as a response to an already primed industry growth.

Hypothesis 2c paints a more positive image of the public sector involvement in the venture capital industry. Even though the evidence presented above suggests a "reactive" involvement by the public sector (and hence not supporting the seeding hypothesis), it may still contribute to the development of the VC markets overall. Hypothesis 2c is tested using the model:

$$\begin{aligned}\text{VCSTOT} = {} & \alpha_0 + \alpha_1^*\text{VCSTOT1} + \alpha_2^*\text{VCSTOT2} + \alpha_3^*\text{VCSTOT3} + \alpha_4^*\text{VCSTOT4} \\ & + \alpha_5^*\text{VCSTOT5} + \alpha_6^*\text{VCSTOT6} + \beta_1^*\text{PUB}_{\text{INV}}1 + \beta_2^*\text{PUB}_{\text{INV}}2 \\ & + \beta_3^*\text{PUB}_{\text{INV}}3 + \beta_4^*\text{PUB}_{\text{INV}}4 + \beta_5^*\text{PUB}_{\text{INV}}5 + \beta_6^*\text{PUB}_{\text{INV}}6 + \lambda_1^*\text{YEAR} \\ & + C(1)^*\text{CO} + C(2)^*\text{RULELAW} + C(3)^*\text{ANTIDIRECTOR} \\ & + C(4)^*\text{CREDITOR} + C(5)^*\text{LAWTYPE}\end{aligned}$$

where the α, β, and λ coefficients are per the regression models above, $C(.)$ is the regression coefficient, VCSTOT is the total new funds for venture capital investments (=new commitments), PUB_INV is the percentage of public investors in the VC pool, the numbers attached to the variables refer to the respective lags and the control variables are as above. The results of the lag − 1 specification, which offers the smallest AIC, is presented in Table 4. The model is again very significant and supports, at the 5% degree of confidence, the existence of causality between PUB_INV, the percentage of public funds in the industry and the total amounts invested in the industry, with a lag of 1 year. This would tend to support the view that public investments in venture capital may signal or legitimize support for venture capital investments as a whole, opening the floodgate of private investments in the sector. This can be also interpreted as another version of the virtuous circle, whereby money attracts more money into the pool of investment.

Hypothesis 3 posits a link between public sector VC firms and later-stage deals. To test the hypothesis, we ran a simple regression with the percentage of investment made into late-stage deals, defined as expansion, replacement and buy-out financing, as dependent variable and the percentage of VC funds invested by public investors. The explanatory power of the simple model is relatively high, with an adjusted R^2 of 18.1% and an F test of 4.086, significant at the 6% level. The regression coefficient attached to the percentage of public venture capitalists in the industry is negative and significant. In other words, as hypothesized, public venture capitalists tend to be associated with later-stage deals in general. An examination of the residuals indicates one significant outlier, UK, where insignificant public sector involvement in the VC industry leads to an industry where 97.2% of the investments over the period 1990–1996 went to later-stage deals.

Hypothesis 4 posits that public sector venture capital funds are more likely to invest in industries that utilize large amounts of human resources (industrial products and services, chemicals and materials) than any other sector. Again, the argument for such a relationship is that public venture capitalists tend to operate under "mixed constraints" objective functions, i.e., they try to optimize a variety of sometimes conflicting factors such as job creation or maintenance as well as investor value creation. Statistical tests did not highlight any significant relationship between the public and/or captive status of VC investors and a preference for high employment industries. Hypothesis 4 is not supported by the empirical data in this study.

Table 4
Test of Hypothesis 2b: PUB_INV Granger causes VCSTOT

Variable	Coefficient	Standard error	*t* statistic	Probability
C	− 52,615,374	541,20,886	− 0.972182	.3334
VCSTOT1	.574518	.170564	3.368352	.0011
PUB_INV1	451,965.3	200,722.2	2.251696	.0266
YEAR	26,564.65	27,155.86	0.978229	.3304
CO	12,879.74	11,492.52	1.120706	.2652
RULELAW	− 75,410.03	57,294.04	− 1.316193	.1913
ANTIDIRECTOR	113,559.7	71,882.37	1.579799	.1175
CREDITOR	137,982.8	101,825.6	1.355089	.1786
LAWTYPE	− 78,059.08	45,784.58	− 1.704921	.0915
R^2	.530380			
Adjusted R^2	.490833			
Standard error of regression	458,068.2			
Sum squared residuals	1.99E + 13			
Log likelihood	− 1498.479			
Durbin–Watson statistic	2.377172			
Mean dependent variable	340,988.6			
S.D.-dependent variable	641,948.5			
AIC	28.98999			
SC	29.21883			
F statistic	13.41142			
Prob (*F* statistic)	.000000			

Included observations: 104 after adjusting endpoints. Newey–West HAC standard errors and covariance (lag truncation = 4).

The regressions coefficients are defined as follows: *C*(·) is the regression coefficient for the control variables, VCSTOT is the total new funds for venture capital investments (= new commitments), PUB_INV is the percentage of public investors in the VC pool, the numbers attached to the variables refer to the respective lags (e.g., PUB_INV1 refers to the percentage of public investors in the VC pool as of 1 year earlier, or lag 1), YEAR is a control for annual variations, CO is a country control to account for national idiosyncrasies not included otherwise in the regression, RULELAW, ANTIDIRECTOR and CREDITOR are, respectively, the rule of law, antidirectors' right and creditors' right indices from La Porta et al. (1997) and LAWTYPE is the simple typology of legal systems proposed by La Porta et al (1996).

A final hypothesis tested is whether or not captive and public sector VC firms syndicate investments as much as independents firms. Econometric analyses did not support the existence of a statistical relationship between the two variables. The empirical evidence is thus not supportive of Hypothesis 5.

8. Conclusions and discussions

The paper analyzes the relationship between public and private sources of venture capital in Europe and the development of the venture capital industry as a whole, controlling for characteristics of the legal systems in each country. Public and quasi-public venture capital firms are significant players in many European venture capital markets, fulfilling the political goals of facilitating access to growth equity and generating or restructuring economic activity in selected regions.

This paper investigates the rationales and effects of public intervention in the venture capital markets across Europe. The primary objective is to provide evidence supporting either the beneficial role played by public intervention in developing national venture capital industries (the seeding hypothesis) or the negative effect it may have on the formation of dynamic VC markets (the crowding-out hypothesis). The central proposition is that by offering capital at marginal (below market) rates of returns to entrepreneurs, public venture capitalists may appropriate the best projects, leaving sub-optimal ones to be financed by private venture capitalists, hence delaying or even preventing the formation of active private funds.

The empirical evidence presented, drawn from 15 European countries over the period 1990–1996, supports a negative correlation between the magnitude of public participation and the size of the venture capital industry. Panel data analyses do not support the view that public venture capitalists seed the industry and get it off the ground: their involvement appears to be in response to the industry development itself. On the other hand, the crowding-out hypothesis is not supported either: public involvement seems to cause, in the Granger sense, greater amounts of money to be invested in the industry as a whole. The results seem to indicate that if the rationale for public intervention is to create from the ground-up markets for risk capital that private investors were not able to generate, then the rationale is not supported: even though the effects of such public intervention are very real and beneficial for the industry as a whole, they do not constitute seeding of a young industry. In other words, referring to the initial propositions, we found evidence to support Proposition 2 (industry support) but not Proposition 1 (seeding). No evidence could be found to support Proposition 3, namely that public venture capitalists would exhibit mixed objective functions. The latter finding may be blurred by country-specific factors, such as the nature of public intervention in venture capital companies (fund provision, management selection, actual management by public servants, etc.) or the behavior of existing independent funds with respect to syndication.

A possible limitation of this type of study is the nature and quality of the data available. In particular, EVCA data are far from comprehensive and remain in essence at a macro level. Venture capitalists, public and private, differ in their abilities to identify superior concepts and to help them succeed: a micro dimension to the study would obviously be a worthy initiative to identify the factors that actually explain the behavior and performance of individual venture capitalists within larger groupings. At this point though, limitations in data prevent such analyses.

The policy implications of this study are important. First of all, the absence of support for the crowding-out hypothesis should provide comfort to those standing for the active role public entities can play in developing venture capital markets. Even though there is no indication here that the potential agency problems associated with public interventions do not exist, the signaling and certification effects of public involvement may overwhelm the negative implications of such direct public interventions. Second, the evidence does support the view that public venture capitalists tend to be latecomers to the industry and have very little to do with its initiation, at least in the past 10 years in Western Europe. Governments of countries with weak venture equity markets, such as Greece or Austria, may consider the evidence to become pro-active in this sector, setting up fiscal and legal environments more conducive to the emergence of active private equity markets and "putting their money where their mouth is" by investing directly.

A number of important questions have not been addressed directly by this study and would warrant further investigations. First, no attempt was made to identify "micro" issues in public intervention in venture capital funds, which may take many different forms and intensities, from simply providing funds to selecting managers to actually setting up captive funds managed by public servants. Second, the study utilized mainly aggregated data and did not investigate actual behavioral information collected from venture capitalists in the different countries. It may prove very instructive to complement this investigation with actual surveys of venture capitalists on the ground to obtain first-hand confirmation of the actual objectives pursued. At this point, the objectives could only be stated and tested as propositions. Third, we voluntarily limited ourselves to analyzing only one channel (direct or indirect VC fund involvement) through which governments and public entities could boost their respective venture capital industries, while documenting another (the general legal environment and its "investor friendliness"). Clearly, governments have at their disposal a larger palette of administrative, legislative and taxation tools to improve the general attractiveness of the industry. Analyzing the portfolio of activities engaged in by governments may also highlight the relative efficacy of different measures, in particular those targeting entrepreneurs (ease of installation, support of fundamental research, clear technology licensing rules out of governmental research laboratories, etc.) versus those affecting investors. Finally, it may be worth extending the study to non-EVCA member countries and newcomers to the venture capital industry, which provides the unique ability to test the hypotheses *de novo*.

Appendix A

A.1. HAC covariances

When the form of heteroskedasticity is not known, it may not be possible to obtain efficient estimates of the parameters using weighted least squares. OLS provides consistent parameter estimates in the presence of heteroskedasticity, but the usual OLS standard errors will be incorrect and should not be used for inference. Before we describe the techniques for HAC covariance estimation, note that using the White heteroskedasticity-consistent or the Newey–West HAC covariance estimates does not change the point estimates of the parameters, only the estimated standard errors. There is also nothing to prevent us from combining various methods of accounting for heteroskedasticity and serial correlation. For example, weighted least squares estimation might be accompanied by White or Newey–West covariance matrix estimates.

A.2. Heteroskedasticity-consistent covariances (White)

White (1980) derived a heteroskedasticity-consistent covariance matrix estimator, which provides correct estimates of the coefficient covariances in the presence of heteroskedasticity of unknown form. The White covariance estimators are shown in place of the standard OLS formula.

A.3. HAC covariances (Newey–West)

The White covariance matrix assumes that the residuals of the estimated equation are serially uncorrelated. Newey and West (1987) relax that assumption, proposing a general covariance matrix estimator that is consistent in the presence of both heteroskedasticity and autocorrelation of unknown form. Such estimates are also presented.

Appendix B

B.1. AIC

As part of the output for most regression procedures, we report various information criteria that can be used as guides in model selection (Grasa, 1989). The notion of an information criterion is to provide a measure of information that strikes a balance between this measure of goodness of fit and parsimonious specification of the model. The various information criteria, which include the AIC, the Schwarz criterion (SC) and the Hannan–Quinn criterion (HQ), only differ in how they strike that balance. The "best" model is the one afflicted with the smallest information criterion. The information criterion has been widely used in time series analysis to determine the appropriate length of the distributed lag. Lutkepohl (1991, Chapter 4), for example, presents a number of results regarding consistent lag order selection in VAR models.

Appendix C

C.1. Estimating Granger causality

In order to illustrate the process by which the direction and lag of causality are estimated, let us consider two fundamental variables in the study: CFR, or cumulative funds raised by the venture capital industry in a country, and PSPCT, the percentage of the funds provided by public sector sources.

C.2. Step 1: test causality in one direction: CFR causes PSPCT

In this case, we test the following relationship up to lag 6:

$$\begin{aligned}\mathrm{PSPCT}_t = {} & \alpha_0 + \alpha_1\mathrm{PSPCT}_{t-1} + \alpha_2\mathrm{PSPCT}_{t-2} + \alpha_3\mathrm{PSPCT}_{t-3} + \alpha_4\mathrm{PSPCT}_{t-4} \\ & + \alpha_5\mathrm{PSPCT}_{t-5} + \alpha_6\mathrm{PSPCT}_{t-6} + \beta_1\mathrm{CFR}_{t-1} + \beta_2\mathrm{CFR}_{t-2} + \beta_3\mathrm{CFR}_{t-3} \\ & + \beta_4\mathrm{CFR}_{t-4} + \beta_5\mathrm{CFR}_{t-5} + \beta_6\mathrm{CFR}_{t-6} + \text{control variables}\end{aligned}$$

We then use the Wald test (see Appendix D below): H_0: $\beta_1=\beta_2=\beta_3=\beta_4=\beta_5=\beta_6=0$ for the hypothesis that all the β_i coefficients are not jointly equal to zero.

If the $\beta_1\ldots\beta_6\neq 0$, then the variable associated with these regressors (here CFR) causes PSPCT in the sense of Granger. If all $\beta_1\ldots\beta_6=0$, then it is necessary to redo the regression but stopping with a lag 5, testing again that all $\beta_1\ldots\beta_5\neq 0$. The process is then repeated (decreasing the lag by one increment each time) until all $\beta_i=0$. In most cases here, only lag 1 provides significant results.

C.3. Step 2: test causality in the other direction: PSPCT causes CFR

In this case, we test the following relationship up to lag 6:

$$\begin{aligned}\mathrm{CFR}_t = {} & \alpha_0 + \alpha_1\mathrm{CFR}_{t-1} + \alpha_2\mathrm{CFR}_{t-2} + \alpha_3\mathrm{CFR}_{t-3} + \alpha_4\mathrm{CFR}_{t-4} + \alpha_5\mathrm{CFR}_{t-5} \\ & + \alpha_6\mathrm{CFR}_{t-6} + \beta_1\mathrm{PSPCT}_{t-1} + \beta_2\mathrm{PSPCT}_{t-2} + \beta_3\mathrm{PSPCT}_{t-3} + \beta_4\mathrm{PSPCT}_{t-4} \\ & + \beta_5\mathrm{PSPCT}_{t-5} + \beta_6\mathrm{PSPCT}_{t-6} + \text{control variables}\end{aligned}$$

We then again test that all the β_i coefficients are not jointly equal to zero, using again the Wald test, and keep on decrementing the lags until all $\beta_i=0$.

C.4. Step 3: criterion to decide on the direction of causality

The β_i coefficients jointly different from zero determine the direction of causality as well as the most appropriate lag of the relationship. If causality is established at two different levels of lag, it becomes necessary to use the AIC to determine the best model (in this case the one with the smallest AIC score).

Three situations can then obtain in Granger causality tests: (1) one direction of causality is supported; (2) no direction of causality is supported; and (3) causality is supported in both directions, also known as simultaneity.

Appendix D

D.1. Wald test

The Wald test computes the test statistics by estimating the unrestricted regression without imposing the coefficient restrictions specified by the null hypothesis. The Wald statistic measures how close the unrestricted estimates come to satisfying the restrictions under the null hypothesis. If the restrictions are in fact true, then the unrestricted estimates should come close to satisfying the restrictions. The Wald test can be applied to equations estimated by least squares, two-stage least squares, nonlinear least squares, binary, ordered, censored, truncated and count models, and systems estimators. The F statistic compares the residual sum of squares computed with and without the restrictions imposed. If the restrictions are valid, there should be little difference in the two residual sum of squares and the F value should be small.

References

Birley, S., Leleux, B., Muzyka, D., 1996. Trade-offs in the investment decisions of European venture capitalists. J. Bus. Venturing 11, 273–287.

Gordon, R.J., 1988. The role of wages in the inflation process. Am. Econ. Rev. 78, 76–83.

Granger, C.W.J., Newbold, P., 1974. Spurious regressions in econometrics. J. Econometrics 14, 111–120.

Granger, C.W.J., Newbold, P., 1988. Some recent developments in a concept of causality. J. Econometrics 39, 199–211.

Grasa, A.A., 1989. Econometric Model Selection: A New Approach. Kluwer Academic Publishing, New York.

Judge, G., Hill, R.C., Griffiths, W.E., Lutkepohl, H., Lee, T.-C., 1985. The Theory and Practice of Econometrics. Wiley, New York.

King, R., Levine, R., 1993. Finance and growth: shumpeter might be right. Q. J. Economics 108, 717–737 (August).

Koenker, R., 1982. Robust methods in econometrics. Econ. Rev. 1, 213–255.

La Porta, R., Lopez-de-Silanes, F., Shleifea, A., Vishny, R.W., 1996. Law and finance. NBER Working Paper 5661.

La Porta, R., Lopez-De-Silanes, F., Shleifea, A., Vishny, R.W., 1997. Legal determinants of external finance. J. Finance 72 (3), 1131–1150.

Leleux, B.F., 1998. Legal systems, state interventions and venture capital: some European evidence. Babson College Working Paper Series, Babson Park, MA.

Lutkepohl, H., 1991. Introduction to multiple time series analysis Springer-Verlag, Berlin.

Manigart, S., 1994. The founding rate of venture capital firms in three European countries (1970–1990). J. Bus. Venturing 9, 525–541.

Mehra, Y.P., 1995. Wage growth and the inflation process: an empirical approach. In: Bhaskara, R. (Ed.), Co-Integration for the Applied Economist. Macmillan, London.

Modigliani, F., Perotti, E., 1996. Protection of Minority Interest and the Development of Security Markets. MIT, Cambridge.

Newey, W., West, K., 1987. A simple positive semi-definite, heteroskedasticity- and autocorrelation-consistent covariance matrix. Econometrica 55, 703–708.

Ooghe, H., Manigart, S., Fassiw, Y., 1991. Growth patterns of the European venture capital industry. J. Bus. Venturing 6 (6), 381–404.

Rajan, R., Zingales, L., 1995. What do we know from capital structure: some evidence from the international data. J. Finance 50, 1421–1460.

Rajan, R., Zingales, L., 1996. Financial dependence and growth. NBER Working Paper 5758.

Roure, J., Keeley, R., Van der Heydew, T., 1990. European venture capital: strategies and challenges in the 90s. Eur. Manage. J. 8 (2), 243–252.

Sapienza, H.J., Manigart, S., Vermeir, N., 1996. Venture capitalist governance and value added in four countries. J. Bus. Venturing 11, 439–469.

Sevestre, P., Trognon, A., 1992. Dynamic linear models. In: Sevestre, P. (Ed.), The Econometrics of Panel Data. Kluwer Academic Publishing, New York.

Shleifer, A., Vishny, R.W., 1997. A survey of corporate governance. J. Finance 52, 737–783.

Warren, E., 1987. Bankruptcy policy. Univ. Chicago Law Rev. 54, 775–814.

White, H., 1980. A heteroscedasticity-consistent covariance matrix estimator and a direct test for heteroscedasticity. Econometrica 48, 817–833.

Part III
Government's Infrastructure Policies

[14]

Changes in Brazilian Public R&D Institutions Management: The National Institute of Technology Case-Study

Anne-Marie Maculan
Deborah Moraes Zouain

ABSTRACT. Based on the experience of the National Institute of Technology (INT), this paper sets out to analyze some of the organizational changes adopted by this R&D institution to fit the new demands of valorizing research results. First, are discussed the elements of change that allow to consider the emergence of a new paradigm for administering the research system in Brazil. Second, are presented the changes introduced to the INT. The hypothesis is that these changes are meant to introduce new criteria for selecting projects and to evaluate performance in order to create a dynamic interaction with economic agents. The capacity to adopt new orientations seems to be a necessary condition for INT to remain within the Brazilian research system.

1. Introduction

Brazil's Science and Technology (S&T) policy has been radically reformulated as a result of the creation of a federal ministry for Science and Technology in 1985, the new orientation for economic policy and the opening of the domestic market in 1990 (see Maculan, 1995). It became necessary to promote a more intense and effective diffusion of research activities as an instrument of modernizing the industrial base and strengthening the competitiveness of Brazilian companies. This political reorientation means breaking with the old pattern of industrialization and technological development. It calls for changing the way public R&D institutions are organized, redefining the objectives, the financing mechanisms and the evaluating criteria of research quality. The need to demonstrate the economic worth of the research carried out and the applicability of the knowledge that is generated, as well as establishing contractual relations with the companies, amounts to a real challenge for these institutions (universities, CNPq[1] institutes and federal or state R&D centers[2]).

We need to better understand what the true dimensions and impacts of these changes are. This paper sets out to analyze some of the organizational changes adopted by Brazilian R&D institutions to fit the new demands of valorizing research results based on the experience of the National Institute of Technology (INT). The study of the INT, a very traditional public institution controlled by the Ministry for S&T can be a good example of the efforts and experiments of public research institutions to adapt to a new management style. The INT is one of Brazil's most traditional research institutions. Created in 1921 by the Ministry of Agriculture, Industry and Commerce to investigate industrial processes for using fuels and ores, it was restructured in 1933 with the aim of studying technologies that would allow for better use of natural resources and national raw-materials, and also to train technicians. The vocation to develop technologies for industrial use was clearly indicated, although industrialization was incipient and the State still neither industrialist nor direct producer of basic input materials.

The INT's experience is quite similar to the situation analyzed by Bell (1993). He observes that the research institutions in developing countries have been set up generally on the initiative of the government. They have taken the form of institutes administered in a centralizing manner with the mission of lending support to the process of economic development. However, instead of structuring themselves to answer the companies'

Federal University of Rio de Janeiro
Getulio Vargas Foundation

Journal of Technology Transfer **24:** 149–157, 1999.

needs, they began to model themselves after the more advanced institutes overseas. This produced the erroneous idea that they could replace the companies' R&D departments developing technologies that would later on be transferred to the companies.

The first part of the paper discusses the elements of change that allow us to consider the emergence of a new paradigm for administering the research system in Brazil. The second part presents the changes introduced to the INT. The capacity to adopt new orientations seems to be a necessary condition for INT to remain within the Brazilian research system. Our hypothesis is that the changes now being made are meant to introduce new criteria for selecting projects and for evaluation and performance to create a dynamics of interaction with the economic agents. The use of new regulating instruments, redefinition of objectives and participation of different institutional agents reflect the emergence of a new paradigm of managing research systems.

2. The reorientation of Brazil's policy in support of industrial research

The Brazilian system of scientific and technological research was born in the early 70s at considerable remove from the productive system. Brazil's industrialization was based on incorporating foreign industrial technology brought in by multinational companies, built into equipment or bought in the form of bundled technology packages. This exclusive option for the use of foreign technology, along with an extremely protectionist industrial policy, seriously limited the national capacity to develop its own technological solutions and build partnerships between the research community and industry. With the rare exception of a few state-owned companies, no in-house R&D departments were created. Companies limited themselves to developing minor adaptive improvements and received no encouragement to invest in more innovative technological projects.

The universities were entrusted to train the human resources necessary for industrialization but were not asked by industry to provide solutions to technical problems and even less to develop new products or processes. It was the scientific community itself, along with the upper governmental administration, that determined the broad guidelines for Science and Technology policy. As a result, the public research institutions enjoyed considerable autonomy in defining their projects and mutual unfamiliarity between research institutions and industrial companies was the rule during the first phase of consolidation of the national research system, in the period between 1970 and 1985.

Moreover, from the 80s on, the resources of the National Fund for Scientific and Technological Development (FNDCT) stagnated or shrank. The FNDCT has been the main source of public financing in the past. In 1984, the new Program for Scientific and Technological Development (PADCT) was created, partly funded by the World Bank. Then, a growing share of the original FNDCT was allocated to the PADCT. The scientific community strongly disagreed with the change since it meant less autonomy for the individual researchers in defining their own projects. The PADCT was run on the basis of an intense competition between the research groups that applied for funding.

Creating the Ministry for Science and Technology (MCT) in 1985 represented a new landmark in S&T policy. For all its initial instability as an institution (Maculan, 1995), the MCT introduced new rules for research and professionalization of R&D management. It became a systematizing organ for the practices of the scientific community and endeavored to lend incentive for a new dynamic of gathering techno-scientific knowledge to respond to the demands of industrial competitiveness.

From 1989 on, the role of the State begins to be questioned, along with governmental action in the economy and the preservation of many public agencies or companies under direct control of the State. It was argued, in a rather simplistic manner, that investment in research was no longer needed, since the opening up of the economy would take care of providing Brazil's industries with more advanced technology. Strong criticism began to be leveled at the inability of R&D institutions to transfer the results of their research to the productive sector.

The opening of the economy after 1990 bore consequences on the formulation of Brazil's policy as to Science and Technology. An analysis of this policy reveals the swift multiplication of financing programs for research to be carried out jointly by public institutions and companies. In 1992 and 1993, two laws were passed aiming at encouraging companies to invest in research projects in cooperation with universities. In 1996, programs supporting innovation for small companies were defined to strengthen the regional dimension of the process of technological capacity-building in industry. Over 300 agreements between universities and companies were signed to apply these laws. Industry's share in Brazil's investments in R&D rose from 7% in the 80s to 32% in 1997.[3]

For the Ministry of Science and Technology, the concept of productivity and the possibility of applying the new knowledge that is generated are increasingly more common criteria to evaluate research activities. Parallel to this, access to financing through the system of public announcements, besides introducing competition among research groups, often entails the obligation to set up a partnership with companies. Such a mechanism leads to a more technological dimension being given to projects and to encouraging companies to take part in the financing of the project.

The new orientation set down by the Ministry impacts directly and immediately institutions that dedicate themselves to scientific and technological development. At present they are all involved in redefining their working principles, objectives, and especially their sources of funding. Adopting managerial control methods and performance indicators like those used by corporations aims at a profound restructuring of these public research institutions, introduces new parameters of quality and efficiency, and shows the need for acquiring new administrative competencies. Reformulation involves the sources and types of financing, setting up partnerships with national or foreign laboratories or companies, research objectives, methods of diffusion and valorization of results, and modalities of administration and management of human resources (see Table I).

Research institutions and universities are trying to establish a new pattern of relations with industry. They are slowly adopting criteria of performance of an economic nature and favoring research activities that are more geared to meet the requirements of the industry whose pattern of

Table I
The new orientation in research management

Research management	New orientation
Definition of projects	More rigorous prioritization of projects and more precise definition of objectives
Carrying out the research	Diversification of projects: commissioned research providing specialized technical services, consulting, cooperative research
Financing	Redefinition of the sources of funding and modalities of allocation
Inter-institutional relations	Setting up partnerships with companies, other research institutions and development agencies
Objectives	Developing technological solutions for industry

technological capacity-building and innovation calls for new formal mechanisms of absorption of knowledge. At present, administrative interface structures are created by the universities and research institutions for the purpose of speeding up direct or indirect commercialization of research results (Macedo and Maculan, 1997).

However, at the same time, part of the academic community has expressed its opposition or its lack of consensus with these institutional innovations. The changes are conflicting and not easy to introduce. Reducing the budget resources allocated by the federal government to research institutions has produced an atmosphere of uncertainty that has forced them to rethink how their activities should be organized and financed. These imposed changes have sowen fear amidst the research community because of the lack of a tradition of interchange between research institutions and industry. The community remains strongly attached to an idealistic vision of free non-oriented research.

Two recent examples illustrate the debate over these changes. Some years ago, a small group of professors and officers of the Technology Center took the initiative of creating a business incubator inside the campus of the Federal University of Rio de Janeiro (UFRJ). It took five years for the councils of the university to give their formal acceptation. It has been the same in the University of Brasilia and in many other institutions. The technology transfer units of the Engineering Institute of the UFRJ recently changed the overhead on consultancy and research contracts. The professors strongly reacted against this decision. As a result, an important discussion occurred among the professors and researchers and finally the decision had to be changed.

To sum up, the changes are having a deep effect on many dimensions of how to manage research institutions: access to financial resources, relations with companies, prioritization of products or results (commissioned services, publications, courses, patents), types of contracting, criteria for evaluating the performance of researchers, use of rewards and incentives, inter-institutional relations (with universities, regional development agencies, economic sectors), and criteria for allocating resources among groups belonging to the same institution.

This set of changes[4] in the organization of research activities, besides expressing a redefinition of the role of the State in developing research and innovation, changes the traditional relationship between the public and private sectors for the generation and use of new knowledge. The concepts of an economic nature that direct the policy supporting research express a gradual process of capitalization of knowledge.

The reorientation of the Brazilian policy does not constitute an isolated phenomenon, but rather fits into a broader context of changes brought to the S&T system concerning the need to incorporate scientific and technological knowledge with productive activities. According to Lundvall (1992), the most fundamental resource in the modern economy is knowledge. Competition between firms and competitivity between nations now are based on innovation capabilities and ask for more relationships between basic research and industry. Different sources of knowledge enter into the innovation process. Scientific results may be inputs to this process at different stages. That means that scientific activities and technical change are brought closer together and become interdependent activities and that research-industry relationships are more strategic.

3. The INT experience: the need for a new policy of management

Despite the fact that the INT possesses a strong institutional identity, it lacks the means for an effective strategy of regular interchange with industry. All through its history the INT has occupied a significant position amid research institutions. Its studies on corrosion or the use of alcohol as fuel are well known. It has been active in different fields of research, but in many cases this was done in fragmented fashion, without the necessary strategy of systematization to achieve decisive results and really transfer industrial technology.

The importance of the Institute can be evaluated by the fact that it lies at the origin of the creation of other public institutions in the Brazilian S&T network. In the 30s and 40s the INT drew up the first Brazilian Standard of Materials,[5] an initiative that led to the creation of the Brazilian Association of Technical Standards. In 1961 the Metrology Division of the INT saw the birth of the National Institute of Weights and Measures. Finally, in 1970 the National Institute of Industrial Property was founded as a result of INT efforts to demonstrate the importance of controlling trademarks and patents in the country's industrialization process. From 1974 onward the INT has developed significant research on sources of renewable energy for the chemical industry.

Throughout these seventy-five years of its existence, its situation within the federal administration has undergone many changes. It has been controlled successively by the Ministry of Agriculture, by the Ministry of Industry and Commerce, and finally by the Ministry of Science and Technology since 1986. This trajectory underscores the numerous oscillations of Brazil's policy for science and technology. Such institutional instability makes it hard to assess the true importance of research for the process of industrialization. During the 70s the Institute went through several crises, especially of a financial nature, which eventually led to a state of depletion.

When the INT was entrusted to the Ministry of Science and Technology in 1986, it began to guarantee its identity and institutional stability. The indefinitions and contradictions of Brazil's S&T policy stood as a direct menace to the INT, although up to 1990 the allocation of budget resources did show a favorable tendency. Federal financing plummeted until 1993 and then began to grow again, although the 1995 figure still stood lower than that of 1990.

Different changes in federal administration produced a chronic institutional instability that has had negative impacts on the possibility of contracting researchers and the availability of financial resources to invest in the modernization or expansion of laboratories. There were budget cuts and the federal administration was substantially knocked off balance by a governmental program that laid off civil servants. It became extremely difficult for the INT to maintain and renew its staff due to low wages and the large number of retirees among the more competent researchers. Lack of resources precluded modernizing the laboratories, while the administrative system was stiff, slow and bureaucratic. The excessive autonomy of the researchers in defining their own projects[6] intensified fragmentation of the lines of research and made it hard to accumulate knowledge and competencies.

Faced with the risk of complete disarray, the INT began a wide in-house debate to identify its positive and negative points and appraise its perspectives. In 1990, a first Strategic Plan was prepared with the recommendation of an important set of changes to the running of the Institute. It became clear that public R&D institutions have to prove that they have the capacity and competence to develop technological solutions that enhance the competitiveness of Brazilian industry. The INT defined its two main aims as: (a) raising own resources equal to 20% of the budget until 1992, at least 10% derived from R&D projects and sales of services, and (b) re-assessing lines of action in light of market opportunities.

Now, the chief objective is to generate funds to ensure the survival of the Institute and lessen its vulnerability to budget oscillations. The strategy of action traced must make it feasible to define measurable, realistic goals within the short-, medium- and long-term horizons, and allow for permanent evaluation of the Institute's performance as a center for research, development and technical assistance. Attending to market demands is clearly stated as a guideline. The INT begins to adopt a management policy based on administration systems directed towards results and seeks to raise funds from foreign sources in order to guarantee a minimum level of investments rather than remain completely dependent on governmental funding.

Major organizational changes are introduced. The INT is re-arranged internally around the areas where its highest competencies lie, namely, industrial chemistry, new materials and industrial engineering. Levels of hierarchy are done away with. The various areas are directly re-ordered around the directorate of the Institute so as to ensure greater alacrity in decision-making concerning allocation of resources.

New criteria for prioritizing projects are adopted. The concept of basic research was scrapped and technical assistance projects take over the predominant role. The projects that are deemed most important attend to a formal demand of some external client with transfer and effective use of technology. The privileged projects are: courses and training; technical analyses; audits; tests and experiments; and attending to small firms seeking technical solutions.

Priority areas are defined as follows: 1) development of new products and processes in the ambit of new materials and fine chemistry and prospective studies of new materials; 2) providing technological services that improve the management of production in small companies in terms of productivity and quality; 3) providing technological services and products in the areas of health and environment (development of equipment, quality certification, chemical control of pharmaceuticals); 4) projects to attend to governmental programs; and 5) services to provide technological information to companies.

The evaluation of the technological activities that are carried out is based on new criteria: 1) the results are really used by the companies; 2) projects foster the creation of micro-companies or the undertaking of lucrative and innovative projects; 3) projects allow new financial resources to be gathered. These criteria must ensure clear verification of the quality of the technological products developed and offer the possibility of developing an external image that has greater impact on the local business community and federal or state agencies for the development of research. On the other hand, raising financial resources has a direct repercussion on the material conditions of the work performed by researchers (wage complements, modernization of laboratories, and so on).

In order to make research results marketable, the INT defined a policy of training human resources and achieving competencies. The knowledge and competence of researchers and administrative personnel are updated. In 1995, 70% of the people who made up the administrative and technical area received some kind of training.

New routines and procedures are defined so as to guarantee the quality of the technological services that are offered. The laboratories have been modernized and eighteen of them have obtained a quality label from the National Institute of Metrology to carry out tests and analyses. Its main clients are large chemical and petrochemical companies. In this way, the INT has institutionalized a steady and autonomous source of income. The increase in the number of contracts with these companies may be credited to these investments.

The report on the results of management referring to the 1990/94 period (INT, 1995) presents some interesting data. There was growth of revenue from services, contracts and agreements compared to 1990, which reflects a greater capacity to valorize competencies. Accumulated revenue doubled, though it was still insufficient to ensure true financial autonomy for the Institute vis-à-vis resources of the federal government. There was a change in the INT's "clientele" with the reduction of orders from the government and state companies, these being gradually substituted by orders from private companies (see Table II). In 1994, more than 500 firms were attended by specialized technical services, compared with 286 in 1990.

Table II
Evolution of the source of demand
(% of the total number of projects)

	1993	1994	1995	1996	1997
Government	34,2	49,4	19,4	21,0	33,9
State companies	20,5	2,5	5,4	4,0	9,7
Private companies	45,2	48,1	75,2	75,0	56,5

Source: INT (1995), MCT (1998).

The INT not only consolidates its national level attending to the government programs but also strengthens interchange relations with other research institutions. These partnerships are indispensable for expanding, diversifying and renewing the base of accumulated knowledge. Cooperation agreements have been signed not only with the main Brazilian research universities but also with several R&D centers in Germany, United Kingdom, Japan and Portugal.

In addition to the interaction with academic institutions that concentrate much of the capacity of basic research and constitute the main generators of new knowledge, the INT endeavors to tighten ties with regional business associations. Possible market niches have been identified for supplying services and partnerships. One notes a reorientation of the activities in favor of small companies based on partnerships with the Brazilian Support Service for Small Companies (SEBRAE) which has opened an office within the INT as well as with the Federation of Industry of the State of Rio de Janeiro.

The revenue-generating technical areas diversified and the supply of technological services to the micro and small companies expanded. The INT has joined up with the technology information department set up by the Secretariat for Science and Technology of the State of Rio de Janeiro which channels the requests for technical assistance of small companies. At present there is a proposal to set up an incubator for technology-based firms. In 1997 three microenterprises started to exploit the technologies developped in the INT laboratories.

There has been a rise in the offer of adult education courses for technicians in some large firms with a remarkable increase in the number of participants. A course in technology management designed for engineers and administrators has recently been set up in tandem with the department of chemistry of the federal university.

The INT now has its own library specialized in chemistry and new materials. It has joined the technology data network of the Ministry for Science and Technology and is connected to several international data banks. This enables it to bring up to date its own data bases and commercialize them in the form fo CD-Roms.

Within this context of valorization and commercialization of knowledge, the INT is going to acquire the capacity to manage its projects, develop competence to negotiate with other institutions (large and small companies, industrial associations, other research institutions, regional development agencies), define valorization strategies and organize its material, technical and human resources. This transformation in the way of thinking and organizing research activities constitutes a change of managerial competence and skill. Assessment of the quality of research that is carried out, valorization strategy, selection of priority projects, criteria for allocating resources and training of human resources, all these elements become fundamental in the reorganization of the INT.

4. A new paradigm for management of the research system

INT's new management policy represents an important experience in the context of the changes in the administration of public research institutions. The INT is introducing organizational innovations to find new conditions for valorizing its competencies and accumulated knowledge. The option is for closer contact with the industrial sectors by adopting a new pattern of relations with companies and granting preference to attending small companies on a regional basis.

The INT appears to have defined objectively its space of action by definitively abandoning the proposal of generating innovations for industry. Instead, it is attempting to perform specialized technical tasks that enable the productive process to improve. These technologically less ambitious activities should allow the number of clients to increase and additional resources to be generated. In this way the INT eventually integrates better with a network of research institutions and performs work that helps companies to improve their technological base.

If we look at the intensity of these changes in the practices and organization of the INT research activities, we may draw the same conclusion as Etzkowitz (1989), i.e. that a veritable "academic revolution" is going on, with universities and research institutions transformed into actual agents of economic development in the absence of more pro-active industrial and technological policies. As a counterpart to this, conflicting interests grow around the definition of the macro-aims of research activities. The objective of commercializing knowledge leads research groups to become "quasi-firms" and change the way they organize their activities. Research institutions start administrating their projects in terms that resemble more and more the principles and rules of business management.

For Webster and Etzkowitz (1991), the productive system is evolving towards an economy characterized chiefly by permanent incorporation of new learning. This trend conditions each country's international competitiveness and obliges governments to take steps that encourage research institutions to make their results known among the industrial sector at large. But that is not enough; it is also crucial to draw up policies that stimulate corporations to invest in research, that is, in the production and assimilation of new knowledge. The participation of companies in academic research should also be promoted so that they can accompany technological progress and make innovations. The linear model of innovation, which moved from basic to applied research and to industrial development of products or processes must be replaced by a spiral dynamic: the search for solutions to the problems of industrial technology raises questions for basic research, while the solutions that are developed allow advances to be made in matters of research.

This is the scenario in which governments are reformulating their S&T policies to consolidate a national system of innovation by integrating agents from both public and private national and regional sectors. More intense interaction between research institutions and industry can be organized in several ways: 1) prioritizing research projects that offer greater potential for innovation; 2) diversifying the mechanisms for transferring knowledge to industry; and 3) linking public financing to company investments in R&D.

This new paradigm of regulating national research systems is based on defining technological change as the product of research carried out in conjunction with several interacting agents. This means that increasing financial resources are now allocated to more applied research projects and that new indicators of research outputs (training courses, cooperation with industry) are introduced in the evaluation process. The search for partnerships is now strongly incited as a condition for the continuity of public funding.

5. Conclusions

Starting in 1990, the INT set out to make deep changes in the way it was run. The changes implemented by the INT well illustrate the new concepts outlined above. The main reformulation concerns sources and volume of financing, ways of prioritizing projects, classification of research products and setting up inter-organizational relations. However, as the experience is recent and the INT is still adopting new measures, it is not yet possible to evaluate all the impacts of theses measures on the future performance of the Institute.

These changes around a strategy of more profitable action imply defining research projects with greater potential of being applied and attending to the needs of industry, with new criteria as to selection and prioritization of research, evaluation of the performance of the researchers and adopting contractual mechanisms in relations with companies.

These changes also demand that the institution acquire new managerial competencies and adopt a dynamic of accumulating knowledge. Peering arrangements with companies, defining a strategy of valorizing technological competencies on a contractual basis (with requirements as to the quality of attendance), more rigorous management of projects, adopting incentives for human resources as regards results, all this means that the institution is assimilating performance concepts of an economic nature that are more akin to those of corporations. The new pattern of transferring knowledge to industry seeks to attend to the needs of technological capacity-building of the small companies that have been neglected up to now.

Nonetheless, many challenges remain that relate to factors beyond the INT's sphere of decision, such as the continuity of public investments in research, alterations in the international economic scenario and the intensity of technological changes.

If changes are crucial for the future of the Brazilian innovation system and for the competitiveness of firms, some political elements seem to be lacking. The new political orientations are not clearly understood by the research institutions. On the other hand, the stimulus for firms to enter into joint activities with them seems to be low. Bridging mechanisms between science and technology is of crucial importance in a knowledge-based economy but it needs a two-way strategy. For both partners, the efforts to change should receive an evidently visible award and the way to change should be more clearly oriented.

Once largely limited to developing research projects on behalf of the federal government, without a clear strategy for industrial use, now public R&D institutions have to expand their skills and initiate regular economic relationships with industry. Research is now regarded as a tool for solving technical industrial problems and increasing competitiveness and economic growth. But the Brazilian government seems to have underestimated obstacles and resistance within research institutions as well as the companies' lack of interest.

Acknowledgement

The authors wish to thank Professor Henry Etzkowitz for his useful suggestions and comments but are obviously responsible for the limits and errors. This work benefited from financial support from the Brazilian CNPq and CAPES agencies.

Notes

1. The National Council for Scientific and Technological Development—CNPq—is a federal government agency that finances research projects and administrates research institutes.
2. The R&D centers of state companies (now undergoing privatization) and military research institutes are also subject to the same demands as to quality and efficiency (see Maculan and Mello, 1997).
3. According to a recent declaration of Vargas, the present S&T Minister.
4. The elements presented here are a synthesis of information gathered in documents of the MCT, the Journal da Sociedade Brasileira pelo Progresso da Ciência (Newspaper of the Brazilian Society for Scientific Progress), which expresses

the concerns of the scientific community, and interviews with people in charge of research institutes.

5. The specification of Portland Common Cement.

6. In 1987, 20% of the projects were defined individually by the researchers.

References

Bell, Martin, 1993, 'Integrating R&D with Industrial Production and Technical Change: Strengthening Linkages and Changing Structures', UN Economic and Social Council, Economic and Social Commission for Western Asia, Amman, Workshop on Integration of Science and Technology in the Development Planning and Management Process.

Etzkowitz, Henry, 1989, 'Entrepreneurial Science in the Academy: A Case of the Transformation of Norms', *Social Problems* **36** (1), 14–29.

INT, 1995, *Gestão 1990/94 Relatório de resultados*.

Lundvall, Bengt-Ake (ed.), 1992, *National systems of innovation. Towards a theory of innovation and interactive learning*, London: Pinter, pp. 1–19.

Macedo, Sebastiana and Anne-Marie Maculan, 1997, 'As Fundações dentro da relação Universidade-Empresa: A Criação da Fundação COPPETEC', VII Seminario Latinoamericano de gestión Tecnológica, ALTEC 97, La Habana, October 27–29.

Maculan, Anne-Marie, 1995, 'A Política Brasileira de Ciência e Tecnologia de 1970 a 1990. Balanço e Perspectivas da Pesquisa Científica e do Desenvolvimento Tecnológico', Novos Estudos Cebrap, 43, São Paulo: Brasileira de Ciências Ltda., pp.175–195.

Maculan, Anne-Marie, and Jose Manoel C. de Mello, 1997, 'Mudanças no Sistema de Pesquisa e Inovação das Indústrias de Infraestrutura: A Necessária Redefinição do Papel dos Centros de P&D das Empresas Públicas', Workshop Pronex-NEI, Universidade Federal do Rio de Janeiro, July 24–25.

Ministerio da Ciencia e da Tecnologia (MCT), 1995, *Despesa Realizada da União em C&T. 1980-1994*. Brasília: Sinopse.

Ministerio da Ciencia e da Tecnologia (MCT), 1996, *Relatório Estatístico. 1985 a 1995*, Brasília.

Ministerio da Ciencia e da Tecnologia (MCT), 1998, Brasilia: SECAV.

Webster, Andrew and Henry Etzkowitz, 1991, 'Academic-Industry Relations: The Second Academic Revolution: A Framework Paper for the Proposed Research Workshop on Academic-Industry Relations', Science Policy Support Group, Concept Paper 4.

[15]

Evaluating the Impact of Manufacturing Extension on Productivity Growth

Ronald S. Jarmin

Abstract

This article combines the richest data set and most rigorous econometric methodology used, to date, to evaluate the impact of manufacturing extension services on client productivity growth. Client plants are identified in the Longitudinal Research Database (LRD) and their labor productivity growth between 1987 and 1992 is compared to nonclients. Both simple OLS and two-stage models are used to estimate the effect of manufacturing extension on productivity growth. The first-stage probit model is informative about the client selection process. Results suggest that participation in manufacturing extension is associated with between 3.4 and 16 percent higher labor productivity growth between 1987 and 1992. The article demonstrates the usefulness of longitudinal microdata sets, such as the LRD, for program evaluation.

INTRODUCTION

Recent years have seen explosive growth in government-supported manufacturing extension programs that provide industrial modernization services to small and medium-sized manufacturers (SMEs).[1] The premise behind these programs is that the market has failed to provide SMEs with the optimal level of industrial modernization services, notwithstanding the critical roles these firms play in the overall economy [see Feller, 1997; National Research Council, 1993, for discussions on the justification for, and logic of, manufacturing extension programs]. As a result, SMEs have failed to adopt modern technologies and business practices that would help them improve productivity and enhance their competitiveness. In response to this, a nationwide network of manufacturing extension centers has been established to provide information and assistance to the nation's 370,000 small and medium-sized manufacturers.

Unfortunately, although there have been manufacturing extension centers in a handful of states for decades, virtually no credible and convincing empirical evidence

[1] The generally accepted definition of an SME is a plant with fewer than 500 employees. In my analysis, I restrict this definition to plants with between 20 and 499 employees.

Journal of Policy Analysis and Management, Vol. 18, No. 1, 99–119 (1999)

Published by John Wiley & Sons, Inc. CCC 0276-8739/98/01099-21

exists on the effectiveness of these programs. Policymakers, taxpayers, and researchers want to know if the tax dollars that support these programs yield the desired benefits. To address this need, this article provides much-needed empirical evidence on the effectiveness of these centers. The article combines the richest data set and most rigorous econometric methodology used, to date, to estimate the impact of manufacturing extension on productivity.

To do this, I match data from eight manufacturing extension centers in two states to plant-level data contained in the Census Bureau's Longitudinal Research Database (LRD). The matched client–LRD data are useful for evaluating manufacturing extension for two reasons. First, by matching client records to the LRD, it is possible to construct treatment and control groups. To perform a credible evaluation, it is critical to have a control group of nonclients against which to compare the performance of extension clients. Second, the LRD contains several variables useful for evaluation that are measured consistently across clients and nonclients, across different extension centers, and over time.

As a secondary goal, this article seeks to determine the applicability of micro panel data sets, such as the LRD, for program evaluation. The Government Performance and Results Act of 1993 requires all federal agencies to provide annual performance reports by the year 2000. Existing and continuously updated data sets like the LRD may provide a credible and cost-effective way to evaluate some programs. This article directly examines the efficacy of using the LRD to evaluate programs that affect business establishments.

The article demonstrates that a credible analysis can be performed with matched client–LRD data. The results suggest that the sample of manufacturing extension clients analyzed here enjoyed between 3.4 and 16 percent more growth in labor productivity between 1987 and 1992 than did nonclients. Importantly, these estimates are subject to fewer data and methodological caveats than those from previous studies of manufacturing extension. In addition, the two-stage model provides a rich description of the characteristics of plants that are likely to use manufacturing extension services. Although the results reported here are specific to the eight manufacturing extension centers analyzed, the data and methodology are applicable for evaluating a broad range of programs that seek to improve the performance of business establishments.

Description of Manufacturing Extension

It is well known that large manufacturing plants are more productive than small and medium-sized plants and that this gap is growing. An explanation that has been forwarded to explain this is that small plants do not employ up-to-date production techniques. Research on technology usage [see, for example, Dunne, 1994; Kelly and Brooks, 1991] has shown that small plants and firms are much less likely than their larger counterparts to adopt and employ modern manufacturing technologies. It is further argued that the reason SMEs fail to update their production processes is the result of a lack of unbiased information about the benefits of doing so [see National Research Council, 1993].

In an attempt to improve the performance of SMEs by providing unbiased information on modern production techniques, several states, and more recently the federal government, support the operation of manufacturing extension centers. Although a few states have operated centers for several decades, the real growth in manufacturing extension started in 1989 with the creation of the Manufacturing

Extension Partnership (MEP).[2] The MEP is administered by the National Institute of Standards and Technology (NIST) as part of its effort to improve the competitiveness of U.S. manufacturing industries.[3] Federal support for manufacturing extension activities has grown from $6.1 million in 1988 to $138.4 million in 1995 [U.S. General Accounting Office (GAO), 1995]. Federal support to individual centers must be at least matched by state and local sources. Thus, the total expenditure for manufacturing extension in the United States far exceeds these figures. Since NIST became involved, manufacturing extension has grown from a few state-supported centers to a large integrated network with centers in every state.

Manufacturing extension centers provide technical and business assistance to small and medium-sized manufacturers, much as agricultural extension agents do for farmers. This assistance often consists of providing "off-the-shelf" solutions to technical problems. Examples might include helping a plant install a CAD/CAM system or switching to newer, lower cost, higher performance materials. Manufacturing extension centers can also channel more recent innovations generated in government and university laboratories to SMEs that lack access to such information. Besides helping plants adopt modern manufacturing technologies, most centers also offer business, marketing, and other "softer" types of assistance.

There is considerable variation across extension centers in the types of services offered and the method of service delivery. Extension center staffs can either contract to do projects with clients or they can refer clients to private consulting firms, vendors, or other government agencies. Also, centers typically cater their service mix to the needs of the local economy. Even with the diversity that exists between centers, the idea is that extension services will help clients become more productive and compete more effectively in the international marketplace.

Issues in Evaluating Manufacturing Extension

To maximize the effectiveness of extension programs, it is crucial that program stakeholders (such as extension clients, extension centers, NIST, state and local governments, and Congress) have detailed information about past and current program performance and that a reliable evaluation framework be in place to analyze its future performance. However, different program stakeholders have different needs for, and concepts of, program evaluation. For example, center managers may want to evaluate the relative effectiveness of different service types, whereas federal policymakers are more interested in whether extension services enhance the competitiveness of the SMEs participating in the program. Addressing the needs of every program stakeholder in a single evaluation study is not practical. Therefore, I limit the focus of this article to evaluating the overall effectiveness of manufacturing extension in achieving its broad policy goals of improving the productivity and competitiveness of the nation's small and medium-sized manufacturers.

Methodologically, the soundest way to evaluate programs like manufacturing extension is to collect and analyze experimental data.[4] Namely, firms would be randomly assigned to treatment and control groups. Evaluation would then consist

[2] Although all eight of the extension enters used in the analysis are presently part of the MEP, this was true for only one of the eight during the period for which the data apply.

[3] Other parts of NIST's effort to improve competitiveness include the Advanced Technology Program (ATP) and the Malcolm Baldrige National Quality Award.

[4] See Heckman, Hotz, and Dabos [1987], LaLonde [1986], LaLonde and Maynard [1987], and Moffitt [1991] for discussions of program evaluation methodology.

of simple comparisons of the performance of treatment and control firms. Unfortunately, this has not been done, nor is it likely to be done, for manufacturing extension. Evaluation must, therefore, be carried out with nonexperimental data.

To understand the issues surrounding the evaluation of manufacturing extension with nonexperimental data, it is instructive to review the methods used in studies to assess the effectiveness of agricultural extension programs. This literature is more developed than that for manufacturing extension and, although significant differences exist between agricultural and manufacturing extension,[5] both programs have generically similar objectives (that is, improve farm/manufacturer performance through outreach and education), and share many of the same evaluation issues.[6]

Birkhaeuser, Evenson, and Feder [1991] review the literature on evaluating agricultural extension and report that most studies find that extension has positive and significant impacts on knowledge diffusion, technology adoption, productivity, and profits. However, they point out that the existing studies of agricultural extension are subject to several qualifications concerning data and methodology. Namely, these studies suffer from four major problems: (a) lack of a control group; (b) failure to control for the influence of nonextension services and secondary information flows; (c) failure to incorporate information about the characteristics of the services provided; and (d) selection bias.

Researchers assessing the effectiveness of manufacturing extension programs need to address these concerns to the greatest extent possible. Existing studies generally fail to do this, however. Shapira, Youtie, and Roessner [1996] review the literature on evaluating manufacturing extension and note that most studies involve analyzing simple metrics, such as activity levels or client satisfaction. More importantly, they find that most studies fail to compare the performance of client plants to a control group of nonclients. They argue that without appropriate control groups "it will not be possible to conduct long-term and reliable assessments of the performance and impact of industrial modernization programs and services" (p. 202).

To date, only limited work has been done to rigorously measure the impact of manufacturing extension programs. The studies by Luria and Wiarda [1996] and Oldsman and Heye [1997] employ the Industrial Technology Institute's Performance Benchmarking Service (PBS) Database to examine the impact of manufacturing extension programs in Michigan and New York, respectively. They exploit the PBS data to construct control groups of nonclient plants, and then they compare the performance of clients and nonclients along several dimensions. They find evidence of positive program impacts on sales, employment, and value-added growth but not on productivity growth.

However, it is not clear that the PBS database, which is primarily from plants in the Midwest, is an appropriate control group for clients from New York. Further, impact estimates in both studies may be subject to selection bias. Finally, although this data set contains an impressive array of metrics of interest for evaluation, its small sample size, limited time-series dimension, and narrow geographic and industry coverage limit its utility for evaluating the entire MEP system.

Advantages of This Study

The data and methodology I employ to evaluate the impact of manufacturing extension on productivity growth allow me to address all but one of the four concerns mentioned

[5] See Feller [1993] and Shapira [1990] for discussions about the differences between agricultural and manufacturing extension. See True [1969] for a history of agricultural extension in the United States.
[6] This discussion is based on Jarmin [Center for Economic Studies, 1995].

earlier. First, the LRD provides an excellent control group (problem 1). Namely, I use all nonclient plants in the two states in which the eight extension centers are located. The LRD contains complete industry and geographic coverage, unlike other data sets employed to evaluate manufacturing extension. Thus, the LRD can be used for nationwide evaluations and to compare the performance of geographically dispersed centers offering different types of services or service delivery mechanisms. More importantly, the LRD contains time-series data for all manufacturing plants. Time-series data are important to identify changes in client plant performance that may have resulted from participation in extension programs.

Second, a subset of manufacturing extension centers studied here included some information on the type and intensity of the services provided to each client (problem 3). Although I do not pursue this approach in the present article, this type of information allows evaluators to see if the effect of extension services varies by the resources devoted to them or by the type of service provided.

Third, I control for selection bias (problem 4) by estimating a Heckman-style two-stage model. Because plants self-select into the set of extension clients, standard estimates of program impact may be biased. Unfortunately, I do not have data on nonextension services (problem 2) that clients and nonclients may have received during the period in which extension services were provided.

The rest of this article is organized as follows: In the next section, I discuss the evaluation data set constructed by linking extension client records to the LRD. Then I outline the empirical models used to estimate the impact of extension services on client performance. Estimation results are discussed in the fourth section, and in the last section I offer my conclusions.

DATA

The ability of researchers to measure the impact of manufacturing extension has been hampered by the scarcity of appropriate data. Lacking experimental data, researchers ideally want measures of client performance (like productivity) before and after services are provided (that is, panel data). They also want similar measures for a nonclient control group. Finally, they need data on other plant characteristics that affect performance. With these data, one can estimate models that measure the impact of manufacturing extension, while controlling for both observed and unobserved characteristics of clients and nonclients.

Data Sources

The data used in this study come closer to the aforementioned ideal than any used to evaluate manufacturing extension to date. The data are from two primary sources: (a) confidential plant-level census data contained in the LRD and housed at the Center for Economic Studies (CES) of the U.S. Bureau of the Census, and (b) a small number of data items extracted from the administrative client records of the participating manufacturing extension centers. For this study, NIST/MEP asked eight manufacturing extension centers in two states if they would make client information available to CES on a confidential basis.

The client plants included in the analysis received extension services between 1987 and 1992.[7] This period is used because 1992 is the most recent year for which the

[7] Only one of these centers received support from NIST/MEP during this period, through all later became part of the MEP system.

LRD contains information on all manufacturing establishments in the United States. The LRD is made up of establishment data from the Censuses of Manufacturers and the Annual Surveys of Manufacturers (ASM). A complete census is conducted in years ending in 2 and 7. In the intervening years, the ASM provides data on all plants with more than 250 employees and on a probability sample of smaller plants. Only a small subset of the client plants is included in the ASM. Thus, 1992 is the last year that changes in performance can be analyzed for the most client plants.

A shortcoming of these data is that they apply to services provided by older extension centers and, therefore, may not be representative of the MEP system in its current form. In particular, the services provided by these centers during this period were focused to a greater degree on encouraging technology adoption than is common in the current MEP system. Currently, many manufacturing extension centers' energies are consumed helping clients write business plans and helping plants obtain various quality assurance standards, such as ISO 9000. One might expect that these services would have a smaller impact on productivity than those aimed at encouraging technology adoption. Unfortunately, detailed production data are not yet available for the smaller clients of centers created after 1992.

However, the advantages of the LRD for evaluating manufacturing extension are many. First, these are true panel data. That is, observations for individual plants at different points in time represent contemporaneously measured activity. Many evaluation studies generate data for two points in time by asking survey respondents to provide both contemporaneous and retrospective data [see, for example, Luria and Wiarda, 1996; Youtie and Shapira, 1997]. The quality of retrospective data is often questioned because of problems such as recall bias. Second, data in the LRD is measured consistently over time and across clients and nonclients. Finally, the LRD

Figure 1. Size distribution of extension clients and non-clients.

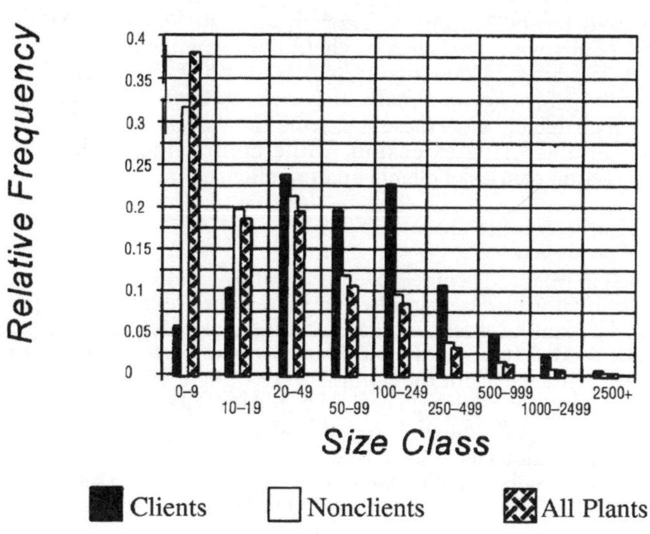

is the only data set available that can be used to generate national estimates of program impact that are comparable across different extension centers.

Data Matching

To carry out the analysis, records from the LRD and the client data must first be linked together. The U.S. General Accounting Office (GAO) requires a 70 percent match rate for studies such as this. The NIST/MEP adopted the GAO standard for the evaluation of the impact of manufacturing extension.

The LRD does not contain any variables, such as names and addresses, that can be used to match the client records. To link the client data to the LRD, I employed information contained in the Standard Statistical Establishment List (SSEL).[8] The LRD and SSEL share establishment identifiers so that once client records are matched to the SSEL, they can easily be linked to the LRD.

Linking the client and SSEL records is done by creating matching variables from one or more fields that are common between the two data sets. For example, a useful

Table 1. Summary statistics

	All plants		19<Employment <500	
Variable	Client mean	Nonclient mean	Client mean	Nonclient mean
Number of plants	1559	15,982	1196	7423
Age, 1992	15.97	16.04	15.86	15.84
Employment, 1992	170.21	71.70	121.58	92.94
Employment growth rate, 1987–1992	0.013	-0.088	0.004	-0.115
Sales, 1992	30,797,199	13,418,587	14,158,048	12,685,173
Sales growth rate, 1987–1992	0.052	-0.085	0.043	-0.088
Sales growth rate, 1982–1987	0.427	0.338	0.425	0.404
Annual wage, 1992	28,072	25,013	26,512	25,082
Production worker share, 1992	0.699	0.724	0.668	0.674
Value added per worker, 1987	53,042	50,853	52,004	54,665
Value added per worker, 1992	56,709	52,797	55,561	57,925
Labor productivity growth rate, 1982–1987	0.215	0.203	0.186	0.192
Labor productivity growth rate, 1987–1992	0.052	0.010	0.070	0.022
Percentage of plants in urban areas	83.3	81.5	84.1	80.4
Percentage of plants in urban areas with extension centers	63.1	44.2	62.6	41.7
Number of extension projects	3.82	—	3.54	—
Total project costs	63,787	—	66,254	—

Notes: Employment is the total number of employees from the Longitudinal Research Database (LRD). Sales is the total value of shipments from the LRD. Wages is payroll divided by employment from the LRD in 1992 dollars. Production worker share is the number of production workers divided by employment from the LRD. Labor productivity is measured as value added per worker from the LRD in 1987 dollars. The number of extension projects is the number of distinct project records per client from the extension client data. Total project costs is the total client investment as a result of its engagements with manufacturing extension. Real values for shipments and value added were obtained using the National Bureau of Economic Research's (NBER's) four-digit deflators [see Bartelsman and Gray, 1995].

[8] See Doms and Peck [1994] for a more detailed description of the SSEL.

matching variable consists of the concatenation of elements of the establishment's name and its zip code. The matching variable is then used to flag potential matches between the two data sets. These matches are then verified by hand.

To obtain a match rate greater than 70 percent, this procedure was repeated four times where a different matching variable was employed in each round. This resulted in 8516 of 11,343 client records from the eight extension centers being successfully linked to the SSEL. This yields a 75.1 percent match rate. However, this match rate is misleading because each client record refers to a project and individual clients often have multiple projects. There are 3972 clients[9] in the extension client data, 2807 of which were successfully matched to the SSEL.[10] Thus, the actual match rate is 70.7 percent, just over the 70 percent level required by the GAO.

Data Description

All of the client records used in this study included a measure of employment. The matched clients account for 78 percent of the total employment contained in the client records. The matched establishments also account for 20.7 percent of the total 1992 LRD employment in the two states where the client plans reside.

All variables used to estimate the impact of manufacturing extension on productivity growth are taken from the LRD. This ensures consistent measurement across clients and nonclients, across centers, and across time. Because a large portion of the clients are small plants with fewer than 250 employees, I use only census year data from the LRD. The model estimated later requires three observations per plant. Therefore, my analysis is restricted to plants that were in the LRD from 1982 to 1992 (that is, for each of the 1982, 1987, and 1992 Censuses of Manufacturers).

Because of this restriction, the number of client plants included in the analysis drops from 2482[11] to 1559, and the number of nonclient plants drops from 34,889 to 15,982. Figure 1 compares the 1987 size distribution of plants in the client and nonclient subsamples to the full population of plants in the two states where the eight extension centers operate. The figure shows that client plants tend to be larger than nonclient plants. Also, plants in the nonclient sample used in the subsequent analysis tend to be larger than plants in the entire establishment population. This is due to restricting the sample to plants that were in since, at least, 1982. This restriction omits new entrants that tend to be smaller than existing plants.

Table 1 provides some summary statistics for client and nonclient plants. The statistics for the all-plant sample in the first two columns show that, in addition to being larger, extension clients enjoyed more sales and labor productivity growth over the 1987 to 1992 period (the period in which clients received services) than did nonclients. However, clients also grew faster during the previous five-year period.

Table 1 also gives summary statistics for a subsample of plants that had between 20 and 499 employees in 1987. This group of plants comprises the bulk of the SME (fewer than 500 employees) population targeted by manufacturing extension programs. The SME subsample shown in Table 1 and used in my subsequent analysis

[9] A client is defined, in these data, as a unique name and address combination.

[10] Note that the definition of a client does not necessarily directly correspond to an establishment. For instance, it was often the case that more than one client was found to match a single establishment. These were often just different parts of the same plant. Also, there were cases where a client record matched to more than one plant. If the plants were all in the same zip code, I allowed the match.

[11] This is fewer than the 2087 clients mentioned earlier because of several clients being outside the two states to which analysis is restricted.

omits plants with fewer than 20 employees for two reasons. First, the client group has a very small proportion of plants with fewer than 20 workers relative to the nonclient group (see Figure 1). Second, data quality is lower for small plants in the LRD than it is for plants with more than 20 employees.[12]

The 20 to 499 employee SME subsample yields client (treatment) and nonclient (control) groups that have much more similar characteristics (age, number of employees, wages, and the like) than the all-plant sample. An interesting thing to note is that, in this subsample, client plants are less productive in level comparisons in both 1987 and 1992, and they experienced slightly lower productivity growth between 1982 and 1987.

METHODOLOGY

This section describes the three basic models used to estimate the impact of manufacturing extension services on productivity growth. Productivity growth is an excellent performance metric for this study for several reasons. First, all program stakeholders view productivity growth as an important goal of manufacturing extension. Second, in the case of productivity growth, there is no mismatch between policy goals and those of client plants. Further, productivity growth does not suffer from the "zero-sum" criticism as does other widely cited performance metrics, such as sales and employment growth.[13] Finally, the LRD is well suited for examining differences in productivity growth between clients and nonclients.

I specify a value-added production function and estimate its parameters by OLS, by a standard Heckman-style two-stage procedure where the second-stage OLS regressions are estimated on the client and nonclient subsamples separately, and by a two-stage procedure where the second-stage regression is estimated on the pooled sample. The first-stage probit equation is informative about the client selection process and is interesting on its own.

To estimate the impact of extension services on client productivity, I specify the following value-added production function:[14]

$$Y_{it} = Ae^{\delta Ext_{it}} K_{it}^{\beta} L_{it}^{\eta} e^{\varepsilon_{it}} \tag{1}$$

where Y_{it} is value added L_{it} is employment, K_{it} is the book value of the capital stock of plant i in period t, and ε_{it} is an error term. The extension variable Ext_{it} is a dummy variable that takes a value of 1 if plant i is a client in period t and 0 otherwise.

None of the client plants in the data set used for estimation received extension services before 1987, and all of them completed at least one project before the end of 1992. Thus, and given that analysis is constrained to Census of Manufacturers data,

[12] All plants with more than 20 employees and no plants with fewer than 5 workers are sent a form in census years. A number of factors are used to determine which plants with between 5 and 19 workers are sent forms. Plants that do not receive forms are called administrative records cases and much of their data is imputed.

[13] It may be the case that plants that increase sales and employment do so at the expense of other plants. Unless these losing plants are outside the policymaker's jurisdiction, the program may produce little or no net benefit to the policymaker's constituents. This simple transfer of activity from one group of plants (nonclients) to client plants produces a zero-sum outcome. Productivity growth, however, represents the ability to produce more wealth with the same resources. As such, it is not a zero-sum outcome.

[14] Because the goal of this study is simply to see whether manufacturing extension affects productivity, a more elaborate specification of the production technology is not required.

a natural strategy is to see if clients experienced more productivity growth between 1987 and 1992 than did nonclient plants. To do this, I take natural logarithms and rearrange equation (1) to get:

$$\Delta\log\left(\frac{Y}{L}\right)_i = \alpha + \delta Ext_i + \beta\Delta\log\left(\frac{K}{L}\right)_i + (\mu - 1)\Delta\log(L)_i + \varepsilon_i \tag{2}$$

where the Δs denote the change in the value of the variable between 1987 and 1992 (for example, $\Delta log(Y/L)_i = log(Y/L)_{i92} - log(Y/L)_{i87}$) and the parameter μ measures deviations from constant returns to scale. The dependent variable is the percent change in labor productivity between 1987, before clients received extension services, and 1992, after clients had completed at least one project. The parameter δ measures the percent difference in productivity growth between client and nonclient plants.

If all the differences between client and nonclient plants are controlled for, δ will be an unbiased measure of the impact of manufacturing extension. An important source of potential bias, however, is unobserved differences between clients and nonclients, for example in the abilities of plant managers. Taking differences mitigates the impact of unobserved plant fixed effects that may bias the estimate of the impact extension.

Another, perhaps more troubling, source of bias is the fact that plants were not randomly assigned to the client and nonclient groups. For example, it could be the case that plants with higher than average productivity growth self-select into the client group, so that overstates the impact of extension services.[15]

Although differencing (that is, estimating a fixed-effects model) can help reduce selection bias, a more general solution is given by Maddala [1983]. This involves estimating a two-stage model that accounts for the client selection process. The details

Table 2. OLS estimates: Productivity growth (absolute *t*-statistics in parentheses).

	All plants	19<*L* <500	All plants	19<*L*<500
Constant	0.009**	0.018***	0.012	0.015
	(1.995)	(2.677)	(1.193)	(0.941)
Ext	0.045***	0.043**	0.034**	0.038**
	(2.945)	(2.470)	(2.194)	(2.150)
Δlog(*K/L*)	0.148***	0.097***	0.148***	0.096***
	(29.164)	(14.007)	(29.057)	(13.962)
Δlog(*L*)	-0.196***	-0.133***	-0.198***	-0.138***
	(24.793)	(11.068)	(25.024)	(11.394)
Two-digit standard industrial classification (SIC) dummies	No	No	Yes	Yes
N	15,263	7782	15,263	7782
R^2	0.110	0.047	0.115	0.050

Notes: The model estimated is given in equation (2). The dependent variable is the growth rate in labor productivity for 1987 to 1992. *K/L* is the capital labor ratio. Capital is the book value of machinery and structures assets from the Longitudinal Research Database (LRD) deflated by two-digit Bureau of Economic Analysis (BEA) capital stock deflators.

**Statistically significant at the 5-percent level.

***Statistically significant at the 1-percent level.

[15] The econometric issues that arise as a result of the self-selection are well known in the literature. See Maddala [1983] for a large number of cities in the general applied econometrics literature, while Stromsdorfer [1987] and Moffitt [1991] provide reviews of the evaluation literature.

of this model are given in the Appendix. The first stage is a probit model of the client selection process. This is utilized to construct instruments that are used in second-stage regressions of the model given in equation (2) to correct for selection bias.

The variables used in the probit model include all the explanatory variables in the second-stage regressions plus several dummy control variables that were differenced out of equation (1) to obtain equation (2). These include two-digit industry, size class, state, and urban/rural dummies, and a dummy for whether the plant was owned by a multiplant firm.

To ensure identification of the two-stage model, I include a dummy for whether the plant is in a metropolitan statistical area (MSA) that contains a manufacturing extension center. It is very probable that being near a center would affect the likelihood of becoming a client. Statistics reported near the bottom of Table 1 support this notion. Client plants are only slightly more likely to be in urban areas than nonclients. However, they are much more likely than nonclient plants to be in a MSA where a manufacturing extension center is located.

Conversations with persons associated with some of the centers included in the analysis suggest that a significant portion of clients were plants that were experiencing growth (in sales) before becoming clients but were facing some productivity-related obstacle to further growth. Thus, I include measures of sales growth between 1982 and 1987, and the 1987 level of productivity in the probit model. Finally, I also include the 1987 capital-to-labor ratio to see if capital-intensive plants were more or less likely to become clients.

RESULTS

This section is broken into four subsections. First, I discuss OLS estimates of equation (2). Then, I present and discuss estimates of the first-stage probit model. Second-stage estimates are given for the separate subsample and pooled approaches in the last two subsections, respectively.

OLS Estimates of the Impact of Manufacturing Extension

Table 2 presents simple OLS estimates of the model in equation (2). The model is estimated for two sets of plants: all plants and the SME subsample where 1987 employment is between 20 and 499 (the bulk of the target population for extension services in this sample). I also estimate the regressions with (columns 3 and 4) and without (columns 1 and 2) two-digit industry dummies.[16] Even though the specification of equation (2) differences out time invariant variables, such as industry dummies, industry effects may be important.

The results indicate that value added per worker grew between 3.4 and 4.5 percent faster at client plants between 1987 and 1992 than it did at nonclient plants. Only two other studies have analyzed the impact of manufacturing extension on productivity growth. Luria and Wiarda [1996] find that Michigan Manufacturing Technology Center had no impact on productivity growth. Oldsman and Heye [1997] study the New York Manufacturing Extension Partnership and find that productivity growth is positively related to their measure of service intensity. However, because of the

[16] The estimates of the parameters of interest do not appreciably change when three- or four-digit industry controls are used.

Table 3. Probit estimates (absolute *t*-statistics in parentheses).

	All plants	19<*L* <500
Constant	-1.069***	0.386*
	(3.025)	(1.600)
Δlog(*K*/*L*)	0.28	0.017
	(1.399)	(0.720)
Δlog(*L*)	0.318***	0.357***
	(9.395)	7.939)
Urban	-0.162***	-0.082
	(2.167)	1.369)
Multiunit	-0.172***	-0.190***
	(3.758)	(3.785)
Age	-0.035	0.042
	(0.685)	(0.639)
Center	0.403***	0.348***
	(9.788)	(7.268)
Initial productivity level	-0.068**	-0.111***
	(2.473)	(3.223)
Previous sales growth	0.025***	0.017*
	(3.164)	(1.802)
Initial capital intensity level	0.031**	0.051**
	(2.239)	(2.486)
Two-digit standard industrial classification (SIC) dummies	Yes	Yes
Size dummies	Yes	Yes
N	14,941	7697
log*L*	-3808.25	-2749 45

Notes: The model estimated is given in equation (A.3) (see the Appendix for details). The dependent variable is *Ext*. *K*/*L* is the capital labor ratio. Capital is the book value of machinery and structures assets from the Longitudinal Research Database (LRD) deflated by two-digit Bureau of Economic Analysis (BEA) capital stock deflators. URBAN = 1 if inside a metropolitan statistical area (MSA). MULTI = 1 if owned by a multiunit firm. CENTER = 1 if located inside a MSA that contains an extension center.

*Statistically significant at the 10-percent level.

**Statistically significant at the 5-percent level.

***Statistically significant at the 1-percent level.

difference between their service intensity measure and the *Ext* variable used here, it is difficult to directly compare the estimates from the two studies.

The estimated impact of extension is slightly higher for the SME subsample and is slightly lower when industry controls are included. However, the positive estimated impact of manufacturing extension from OLS may be a result of higher productivity growth plants self-selecting into the group of client plants. To control for the effects of selection, I next estimate several variants of the two-stage model (given by equations (A.1) through (A.5) in the Appendix).

First-Stage Client Selection Probit Model

The first step is to estimate the first-stage client selection model (equation (A.3) in the Appendix) via probit maximum likelihood. The results, reported in Table 3, are interesting independent of the full two-stage model. Namely, they describe the characteristics of plants that exploited the services offered by manufacturing extension. This description is nearly as useful to policymakers and researchers interested in programs, such as manufacturing extension, as knowing about program effectiveness.

The estimates in Table 3 show that single-unit plants are more likely to become clients. This is consistent with Kelly and Brooks's [1991] results on external learning opportunities and technology adoption. Namely, multiunit plants with productivity concerns are likely to have access to additional information resources through their parent firms that are not available to single-unit plants. Single-unit plants, by definition, must go outside the firm for assistance. The results in Table 3 also suggest that plant age is not related to the likelihood of plants using manufacturing extension. This is similar to Dunne's [1994] finding that plant age was not related to technology usage.

The variable *CENTER* is a dummy for whether or not the plant is in an MSA with a manufacturing extension center. The variable is likely to be correlated with the likelihood of a plant being a client (close proximity to a manufacturing extension center would make a plant more likely to know extension services are available and to take advantage of them) but not correlated with productivity growth.[17] Thus, I use it as an instrument to identify the two-stage model. The results in Table 3 show that being in an MSA with an extension center is positively associated with client status.

For brevity, I do not report the coefficients for the size and industry dummies included in the first-stage probits. However, they do provide some interesting insights into the client selection process. The pattern of the size dummies indicates that small plants are much less likely to become clients than larger plants.[18] Thus, not only are smaller establishments less likely to employ modern technologies [as in Dunne, 1994], they also appear less likely to seek out information on new technologies than larger establishment. The estimates from the two-digit standard industrial classification

Table 4. Second-stage estimates: Clients only (asymptotic absolute *t*-statistics in parentheses).

	All plants	$19<L<500$	All plants	$19<L<500$
Constant	0.101**	0.0113*	0.084	0.106
	(1.965)	(1.783)	(1.464)	(1.460)
$\Delta\log(K/L)$	0.117***	0.073***	0.118***	0.073***
	(4.854)	(2.868)	(4.831)	(2.805)
$\Delta\log(L)$	-0.157***	-0.111**	-0.162***	-0.124**
	(3.910)	(2.191)	(4.000)	(2.405)
Mills ratio	0.032	0.038	0.052	0.082
	(0.924)	(0.827)	(1.335)	(1.494)
Two-digit standard industrial classification (SIC) dummies	No	No	Yes	Yes
N	1418	1105	1418	1105
R^2	0.062	0.025	0.070	0.036

Notes: The model estimated is given in equation (A.1) with the addition of the mIlls ratio obtained from the probit estimation of equation (A.3) (see the Appendix for details). The dependent variable is the growth rate in labor productivity for 1987 to 1992. *K/L* is the capital labor ratio. Capital is the book value of machinery and structures assets from the Longitudinal Research Database (LRD) deflated by two-digit Bureau of Economic Analysis (BEA) capital stock deflators.

*Statistically significant at the 10-percent level.

**Statistically significant at the 5-percent level.

***Statistically significant at the 1-percent level.

[17] Urban plants are generally larger and more productive than rural plants. However, the probit model controls for both size and the urban/rural status of plants. Thus, *CENTER* measures the impact of proximity to an extension center independent of these other effects.

[18] In the all-plant probit (column 1 of Table 3), plants in the three smallest size classes representing plants with less than 50 employees are significantly less likely to be clients than plants in the largest size class, representing plants with 2,500 or more employees.

Table 5. Second-stage estimates: Nonclients only (asymptotic absolute *t*-statistics in parentheses).

	All plants	19<*L* <500	All plants	19<*L*<500
Constant	-0.004	-0.002	0.013	0.007
	(0.580)	(0.193)	(1.156)	(0.585)
Δlog(*K*/*L*)	0.145***	0.094***	0.145***	0.097***
	(20.693)	(10.813)	(20.632)	(10.780)
Δlog(*L*)	-0.209***	-0.150***	-0.210***	-0.153***
	(20.517)	(9.085)	(20.435)	(9.010)
Mills ratio	0.088***	0.084**	0.030	0.057
	(2.884)	(4.083)	(0.917)	(1.271)
Two-digit standard industrial classification (SIC) dummies	No	No	Yes	Yes
N	13,523	6592	13,523	16592
R^2	0.114	0.051	0.118	0.054

Notes: The model estimated is given in equation (A.2) with the addition of the mIlls ratio obtained from the probit estimation of equation (A.3) (see the Appendix for details). The dependent variable is the growth rate in labor productivity for 1987 to 1992. *K/L* is the capital labor ratio. Capital is the book value of machinery and structures assets from the Longitudinal Research Database (LRD) deflated by two-digit Bureau of Economic Analysis (BEA) capital stock deflators.

**Statistically significant at the 5-percent level.

***Statistically significant at the 1-percent level.

Table 6. Second-stage estimates of extension impact (asymptotic absolute *t*-statistics in parentheses).

	All plants	19<*L* <500	All plants	19<*L*<500
$(\alpha_c - \alpha_n)$	0.105**	0.115*	0.071	0.099
	(2.020)	(1.791)	(1.214)	(1.345)
Two-digit standard industrial classification (SIC) dummies	No	No	Yes	Yes

Notes: Estimates of α_c and α_n are from Tables 4 and 5, respectively. $\text{Var}(\alpha_c - \alpha_n) = \text{Var}(\alpha_c) + \text{Var}(\alpha_n)$, where $cov(\alpha_c, \alpha_n) = 0$ since the estimates come from differenct samples.

*Statistically significant at the 10-percent level.

**Statistically significant at the 5-percent level.

(SIC) dummies suggest that plants in SIC major groups 30 and 33 through 38 are all more likely to become clients relative to other industries.[19]

The estimates of the impact of initial productivity levels and previous sales and growth are very interesting. These indicate that plants with high sales growth between 1982 and 1987 (the period prior to becoming a client) and lower productivity in 1987 are more likely to become clients. This pattern is consistent with reports from extension center personnel that high-growth, low-productivity plants made up a substantial portion of the extension clients in these states during the period under study. This is also consistent with the notion that the centers analyzed here focused more on services intended to promote technology adoption with the goal of improving productivity rather than on services, such as ISO 9000 certification, that are often intended to help plants keep or gain customers.

[19] Standard Industrial Classification (SIC) 30 is Rubber and Plastic Products, 33 is Primary Metals, 34 is Fabricated Metal Products, 35 is Nonelectrical Machinery, 36 is Electrical and Electronic Equipment, 37 is Transportation Equipment, and 38 is Instruments and Related Products.

Table 7. Pooled second-stage estimates: Endogenous dummy variable model (asymptotic absolute t-statistics in parentheses).

	All plants	19<L <500	All plants	19<L<500
Constant	-0.001	0.002	0.009	-0.001
	(0.168)	(0.176)	(0.869)	(0.073)
$\Delta\log(K/L)$	0.142***	0.094***	0.143***	0.094***
	(21.296)	(11.161)	(21.236)	(11.142)
$\Delta\log(L)$	-0.205***	-0.146***	-0.205***	-0.150***
	(20.783)	(9.268)	(20.722)	(9.334)
$\Phi(Z_i\hat{\gamma})$	0.160***	0.154***	0.082*	0.126**
	(2.866)	(2.940)	(1.843)	(2.144)
Two-digit standard industrial classification (SIC) dummies	No	No	Yes	Yes
N	14941	7967	14941	7967
R^2	0.109	0.047	0.113	0.050

Notes: The model estimated is given in equation (A.5) (see the Appendix for details). The coefficient on $\Phi(Z_i\hat{\gamma})$ measures the impact of extension. The dependent variable is the growth rate in labor productivity for 1987 to 1992. K/L is the capital labor ratio. Capital is the book value of machinery and structures assets from the Longitudinal Research Database (LRD) deflated by two-digit Bureau of Economic Analysis (BEA) capital stock deflators.

*Statistically significant at the 10-percent level.

**Statistically significant at the 5-percent level.

***Statistically significant at the 1-percent level.

Table 8. Pooled second-stage estimates: Partial and full selection models for the small and medium-sized manufacturers (SME) subsample (asymptotic absolute t-statistics in parentheses).

	Partial selection model		full model	
Constant	-0.024	-0.030	-0.020	-0.026
	(1.380)	(1.201)	(1.100)	(1.015)
$\Delta\log(K/L)$	0.094***	0.094***	0.127***	0.127***
	(10.951)	(10.934)	(9.981)	(9.991)
$\Delta\log(L)$	-0.148***	-0.153***	-0.126***	-0.130***
	(9.296)	(9.389)	(5.720)	(5.806)
$\Phi(Z_i\hat{\gamma})$	-0.156	-0.154	-0.129	-0.128
	(0.922)	(0.922)	(0.725)	(0.711)
$\Phi(Z_i\hat{\gamma})\Delta\log(K/L)$	—	—	-0.270***	-0.270***
			(3.867)	(3.895)
$\Phi(Z_i\hat{\gamma})\Delta\log(L)$	—	—	-0.229*	-0.234*
			(1.660)	(1.683)
$\phi(Z_i\hat{\gamma})$	0.358*	0.340*	0.323*	0.305
	(1.904)	(1.769)	(1.649)	(1.524)
Two-digit standard industrial classification (SIC) dummies	No	Yes	No	Yes
N	7697	7697	7697	7697
R^2	0.047	0.051	0.050	0.053

Notes: The model estimated is given in equation (A.5) (see the Appendix for details). The coefficient on $\Phi(Z_i\hat{\gamma})$ measures the impact of extension. The dependent variable is the growth rate in labor productivity for 1987 to 1992. K/L is the capital labor ratio. Capital is the book value of machinery and structures assets from the Longitudinal Research Database (LRD) deflated by two-digit Bureau of Economic Analysis (BEA) capital stock deflators.

*Statistically significant at the 10-percent level.

**Statistically significant at the 5-percent level.

***Statistically significant at the 1-percent level.

Overall, the first-stage probit estimates show that extension clients tend to be larger single-unit plants that are doing well from a sales perspective but have lagging productivity. These results hold for both the all-plant and SME subsamples. The characteristics of client plants, as described here, further suggest that productivity growth is an appropriate performance metric for evaluating manufacturing extension.

Second-Stage Regressions on the Client and Nonclient Subsamples

Tables 4 and 5 list estimates for second-stage regressions on the client and nonclient observations, respectively.[20] In this model, the estimate of the effect of extension is the difference between the constant terms of the client and nonclient regressions (that is, α_c-α_n). These are listed in Table 6.

The regressions in Tables 4 and 5 include estimates of the inverse Mill's ratio obtained from probit estimation of the first-stage client selection model as an instrument to control for selection bias. The estimated coefficients on these terms indicate that selection bias is a problem in two of the four models.[21] More importantly, controlling for selection yields considerably different estimates of the impact of extension on productivity growth.

The estimates of program impact (that is, α_c-α_n) listed in Table 6 are, in each case, higher than the OLS estimates in Table 2, where a simple extension dummy was employed. Again, we see that the impact of extension is slightly larger for the SME subsample. The interpretation of estimates from the regressions, in the last two columns of Table 6, that includes industry controls poses problems. The constant terms, in the regressions reported in Tables 4 and 5, are sensitive to the choice of the omitted two-digit industry.[22] Thus, different estimates of extension impact can be obtained depending on which industry is omitted.

Second-State Estimates on the Pooled Sample

The standard errors associated with the estimates of program impact in Table 6 are large. More efficient estimates can be obtained by estimating the second-stage regression on the pooled sample (for example, estimating the model given by equation (A.5) in the Appendix). The most basic pooled model is the endogenous dummy model where the extension dummy, Ext_i, from equation (2), is replaced with the estimate of the probability of plant i being a client (which is given by $\Phi(Z_i\gamma)$ in equation (A.5) in the Appendix) obtained from the first-stage probit model. These results are given in Table 7. The estimates of extension impact are both statistically and economically significant.

Table 8 presents estimates for the target SME subsample from the less restrictive partial and full selection models. In theory, parameter estimates from estimating the unrestricted model (that is, the full selection model) on the pooled sample should be equivalent to those obtained in separate second-stage regressions on the client and nonclient subsamples, as in Tables 4 and 5. In practice, however, this is not the case.

The problem is that $\Phi(Z_i\gamma)$, $\phi(Z_i\gamma)$, and the interaction terms are all highly collinear, by construction. As a result, the estimation procedure yields unreliable estimates. For example, it is implausible that the difference in the capital coefficients between

[20] The standard errors in the second-stage regressions are corrected as in Lee [1982].

[21] The estimates of σ_{nE} are significantly different from zero in the first two columns of Table 5.

[22] In the results reported here, the omitted group is SIC 35, the two-digit industry with the largest number of clients (and nonclients).

the client and nonclient subsamples (that is, $\hat{a}_c - \hat{a}_n$) is as large as −0.27. From Tables 4 and 5, this difference is only −0.027, which is a much more credible estimate. Therefore, the estimates in Tables 4 and 5 should be viewed as the more reliable estimates of the full unrestricted model. The most reliable estimates of extension impact are given in Table 6 (for the unrestricted model) and Table 7 (for the endogenous dummy variable model). However, the point estimates in Table 7, obtained using the pooled sample in the second-stage regression, are more efficient.

The estimates in Tables 2, 6, and 7 place the range of the impact of manufacturing extension on client productivity growth between 3.4 and 16 percent for the period 1987 to 1992.[23] This suggests that this sample of manufacturing extension client plants benefited significantly from the services provided by the eight extension centers in the two states analyzed here.

CONCLUSIONS

The primary goal of this article was to help meet the urgent need for credible empirical evidence about the effectiveness of manufacturing extension programs. These programs are intended to improve the productivity and enhance the competitiveness of the nation's SMEs. There has been tremendous and rapid growth in these programs recently, and taxpayers and policymakers want to know if the benefits from manufacturing extension are sufficient enough to merit continued public support.

To see if these services are effective, I compared the productivity growth of plants that participated in manufacturing extension to a control group of plants that did not. Eight extension centers in two states made administrative client records available for the analysis. These clients received extension services between 1987 and 1992. I matched the client records to the Census Bureau's Longitudinal Research Database (LRD). Once this was done, it was possible to identify client (treatment) and nonclient (control) plants in the LRD for each of the two states. The LRD contains performance metrics of interest to program stakeholders,[24] and several control variables that are measured consistently across clients and nonclients, across centers, and over time.

The OLS and two-stage estimation results presented earlier suggest that this sample of clients enjoyed between 3.4 and 16 percent more growth in labor productivity between 1987 and 1992 than did nonclients. The two-stage estimates control for selection bias, which may be a problem because plants self-select into the program. Interestingly, controlling for selection bias increases the estimated impact of extension.

The first-stage probit estimates provide a rich description of the characteristics of plants that are likely to be extension clients. These estimates indicate that larger single-unit plants that experienced high sales growth but had low productivity before being a client were more likely to become clients. Plants near extension centers were also more likely to become clients.

Although the results given here are subject to fewer data and methodological problems than previous studies of manufacturing extension, there are some caveats. First, the analysis covers only eight extension centers that were in operation before

[23] This range includes all but one of the point estimates in Table 6. Recall, however, that the estimate of 7.1 percent for the all-plant sample with two-digit SIC dummies is sensitive to the choice of the omitted SIC dummy.

[24] Although only one performance metric was employed in this article, work is underway to analyze the impact of extension on other measures, such as plant survival and employment growth.

1992. These centers may not be representative of the current MEP system. Further, the analysis looks only at the direct impact of extension services on a single measure of client performance, productivity. Additional data on secondary program benefits and program costs are needed to ascertain whether manufacturing extension provides positive net social benefits. Nevertheless, I believe that the results presented here represent the best available evidence on the impact of manufacturing extension on the productivity and competitiveness of SMEs.

Finally, a secondary goal of this article was to see if micro panel data sets like the LRD are useful for program evaluation. Assessing the usefulness of the LRD and similar data sets for program evaluation is important, given that the Government Performance and Results Act of 1993 will eventually require all federal agencies to submit annual performance reviews. This article clearly demonstrates that the LRD is an appropriate and effective tool for evaluating the impact of manufacturing extension programs. The data and methodology used here can be applied to evaluate virtually any program or policy (for example, economic development and export promotion programs, and changes in workplace safety and environmental regulations) that affects business establishments.

APPENDIX

The model, based on Maddala [1983], is given by:

$$\Delta\log\left(\frac{Y}{L}\right)_{ci} = \alpha_c + \beta_c\Delta\log\left(\frac{K}{L}\right)_{ci} + (\mu_c - 1)\Delta\log(L)_{ci} + u_{ci} \quad (A.1)$$

$$\Delta\log\left(\frac{Y}{L}\right)_{ni} = \alpha_n + \beta_n\Delta\log\left(\frac{K}{L}\right)_{ni} + (\mu_n - 1)\Delta\log(L)_{ni} + u_{ni} \quad (A.2)$$

$$Ext^*_i = Z_i\gamma + \varepsilon_i \qquad Ext_i = 1 \; if \; Ext^* > 0, \; Ext_i = 0 \; if \; Ext^* \le 0 \quad (A.3)$$

Subscript c denotes client observations and n denotes nonclient observations. We observe a client observation for plant i if $Ext_i = 1$ and a nonclient observation if $Ext_i = 0$. The variable Ext^*_i measures the propensity of plant i to become a client. However, we only observe the binary variable, Ext_i, which tells us whether a given plant is a client or not.

This model can be estimated using a Heckman two-stage procedure where the first stage involves estimating equation (A.3) via probit maximum likelihood. From this, I obtain estimates of the inverse Mill's ratio for each plant.[25] The Mill's ratio is then used as an additional instrument to correct for selection bias in second-stage OLS regressions of equation (A.1), the coefficients on the Mil's ratio terms estimate $\text{cov}(u_c\varepsilon) = \sigma_{c\varepsilon}$ and $\text{cov}(u_{nc}\varepsilon) = \sigma_{n\varepsilon}$ for the client and nonclient regressions, respectively. If these estimates are nonzero, then selection bias exists and the OLS estimates of the impact of extension services, σ, in equation (2), are biased.

[25] The inverse Mill's ratio is given by $-\phi(Z_i\gamma) / \Phi(Z_i\gamma)$ for client plants and by $\phi(Z_i\gamma) / (1-\Phi(Z_i\gamma))$ for nonclients where ϕ and Φ are the normal density and cumulative functions, respectively.

The impact of extension, in the two-stage model, is measured as $\alpha_c - \alpha_n$. However, rather than estimating second-stage regressions for clients and nonclients separately, it may be more convenient to estimate the second-stage regression on the pooled sample. Note that:

$$E\left(\Delta\log\left(\frac{Y}{L}\right)_i\right) = Prob(Ext_i^* > 0)E\left(\Delta\log\left(\frac{Y}{L}\right)_i \mid Ext_i^* > 0\right) \tag{A.4}$$

$$+Prob(Ext_i^* \le 0)E\left(\Delta\log\left(\frac{Y}{L}\right)_i \mid Ext_i^* \le 0\right)$$

which can also be written as:

$$\Delta\log\left(\frac{Y}{L}\right)_i = \alpha_n + \beta_n\Delta\log\left(\frac{K}{L}\right)_i + (\mu_n - 1)\Delta\log(L)_i + \tag{A.5}$$

$$\Phi(Z_i\gamma)\left[(\alpha_c - \alpha_n) + (\beta_c - \beta_n)\Delta\log\left(\frac{K}{L}\right)_i + (\mu_c - \mu_n)\Delta\log(L)_i\right] + (\sigma_{n\varepsilon} - \sigma_{c\varepsilon})\phi(Z_i\gamma) + \mu$$

Besides providing a direct measure of the main parameter of interest, $\alpha_c - \alpha_n$, equation (A.5) provides an easy way to impose cross-equation restrictions on equations (A.1) and (A.2). Estimating equation (A.5) on the pooled sample is algebraically equivalent to estimating equations (A.1) and (A.2) on the client and nonclient subsamples, respectively. Call the unrestricted case of equation (A.5) the full selection model. By imposing restrictions on equation (A.5), two additional models can be obtained. If, $(\beta_c = \beta_n)$, $(\mu_c = \mu_n)$, and $(\sigma_{n\varepsilon} = \sigma_{c\varepsilon})$, we get the endogenous dummy model where the extension dummy, Ext_i, in equation (2) is replaced with $\Phi(Z_i\gamma)$, the estimated probability of a plant being a client. The partial selection model is obtained from the endogenous dummy model by letting $(\sigma_{n\varepsilon} \ne \sigma_{c\varepsilon})$.

This research was funded by the National Institute of Standards and Technology's Manufacturing Extension Partnership. I thank Tim Dunne, Brad Jensen, Dan McMillen, Bob McGuckin, Ariel Pakes, John Redman, Ken Troske, and seminar participants at the Caed '96 in Helsinki, the University of Aarhus (Denmark), the Census Bureau, the 2nd Georgia Productivity Workshop, the 1996 Georgia Tech Manufacturing Modernization Evaluation Workshop, the Southern Economic Association (SEA) meetings in Washington, DC, and the Allied Social Science Association (ASSA) meetings in New Orleans for helpful comments and suggestions on earlier drafts of this article. Any findings, opinions, or conclusions expressed here are those of the author and do not necessarily reflect the views of the Census Bureau or the National Institute of Standards and Technology.

RONALD S. JARMIN is an Economist at the Center for Economic Studies, U.S. Bureau of the Census.

REFERENCES

Bartelsman, E.J., & Gray, W. (1995). The NBER manufacturing database. Washington, DC: Board of Governors of the Federal Reserve System (mimeo).

Birkhaeuser, D., Evenson R.E., & Feder, G. (1991). The economic impact of agricultural extension: A review. Economic Development and Cultural Change, 39, 607–650.

Center for Economic Studies, U.S. Bureau of the Census. (1995). Using matched client and census data to evaluate the performance of the manufacturing extension partnership. Washington, DC: R.S Jarmin.

Commission of Engineering and Technical Systems, Manufacturing Studies Board, National Research Council. (1993). Learning to change: Opportunities to improve the performance of smaller manufacturers. Washington, DC: National Academy Press.

Doms, M., & Peck, S. (1994). Examining the employment structure of firms in manufacturing. Washington, DC: Center for Economics, U.S. Bureau of the Census (mimeo).

Dunne, T. (1994). Plant age and technology use in U.S. manufacturing industries. Rand Journal of Economics, 25, 488–499.

Feller, I. (1993). What agricultural extension has to offer as a model for manufacturing modernization. Journal of Policy Analysis and Management, 12, 574–581.

Feller, I. (1997). Manufacturing technology centers as components of regional technology infrastructures. Regional Science and Urban Economics, 27, 181–197.

Heckman, J.J., Hotz, V.J., & Dabos, M. (1987). Do we need experimental data to evaluate the impact of manpower training on earnings. Evaluation Review, 11, 395–427.

Kelly, M.R., & Brooks, H. (1991). External learning opportunities and the diffusion of process innovations to small firms. Technological Forecasting and Social Change 39, 103–125.

LaLonde, R.J. (1986). Evaluating the econometric evaluations of training programs with experimental data. American Economic Review 76, 428–451.

Lee, L.-F. (1982). Some approaches to the correction of selectivity bias. Review of Economic Studies 49, 355–372.

Luria, D., & Wiarda, E. (1996) Performance benchmarking and measuring program impacts on customers: Lessons from the Midwest Manufacturing Technology Center. Research Policy 25, 233–246.

Maddala, G.S. (1983). Limited-dependent and qualitative variables in econometrics. New York: Cambridge University Press.

Moffitt, R. (1991). Program evaluation with nonexperimental data. Evaluation Review 15, 291–314.

Oldsman, E., & Heye, C.R. (1997). The impact of the New York manufacturing extension program: A quasi-experiment. In P. Shapira and J. Youtie (Eds.). Manufacturing modernization: Learning from evaluation practices and results. Atlanta, GA: School of Public Policy, Georgia Tech.

Shapira, P. (1990). Modernizing manufacturing: New policies to build industrial extension services. Washington, DC: Economic Policy Institute.

Shapira, P., Youtie, J., & Roessner, J.D. (1996). Current practices in the evaluation of U.S. industrial modernization programs. Research Policy 25, 185–214.

Stormsdorfer, E.W. (1987) An overview of recent findings and advances in evaluation methods. Evaluation Review 11, 387–394.

True, A.C. (1969). A history of agricultural extension work in the United States, 1785–1923. New York: Arno Press and The New York Times.

U.S. General Accounting Office (GAO). (1995). Manufacturing extension programs: Manufacturers' views of services. Briefing report to the Chairwoman, Subcommittee on Technology, Committee on Science, House of Representatives.

Youtie, J., & Shapira P. (1997). Tracking customer progress: A follow-up study of customers of the Georgia manufacturing extension alliance. Journal of Technology Transfer 22, 42–52.

[16]

ELSEVIER

Research Policy 30 (2001) 977–992

research policy

www.elsevier.nl/locate/econbase

US manufacturing extension partnerships: technology policy reinvented?

Philip Shapira*

School of Public Policy, Georgia Institute of Technology, Atlanta, GA 30332-0345, USA

Abstract

The US manufacturing extension partnership (MEP) is examined as an example of the new partnership paradigm in US technology policy. The MEP provides technology assistance services, particularly for small and medium-sized enterprises. Influenced by aims to reinvent government and reorient technology policy, the MEP seeks to be comprehensive, collaborative, and demand-driven. However, the MEP's partnered management style is constrained by political and industrial systems that continue to operate on traditional lines. After probing these tensions, the paper offers insights for the MEP's future development and for other technology and innovation policies that seek to emulate the MEP's partnership approach. © 2001 Elsevier Science B.V. All rights reserved.

Keywords: Manufacturing extension; Technology policy; Partnerships; United States

1. Introduction

In contemporary technology policy, public–private partnerships have emerged as a central organizing mechanism for promoting research and development and accelerating the diffusion of innovations. As defined by the OECD, such partnerships in the field of technology policy encompass "... any innovation-based relationship whereby public and private actors jointly contribute financial, research, human, and infrastructural resources, either directly or in kind." (Organisation for Economic Cooperation and Development, 1998, p. 8.) In the United States, as in other OECD member states, there are now numerous cooperative programs in the field of technology policy and technology transfer involving a wide range of public and private participants (Coburn and Berglund, 1995).

There is, of course, much variety in the structure and organization of technology partnerships (see, for example, the range of partnership models discussed in Etzkowitz and Leydesdorff, 1999). In some instances, private organizations take the lead in consortia that also involve universities and other public institutions; in other cases, public organizations lead technology partnerships that involve multiple public agencies, non-profits and private groups. After providing further context to the development of the partnership paradigm in the United States, this article draws on a partnership case of the second type: the US manufacturing extension partnership (MEP). The MEP is a collaborative initiative between federal and state governments that also involves non-profit organizations, academic institutions, industry groups, and a variety of public and private technology assistance and business service providers. The MEP aims to upgrade the performance and competitiveness of US industry, especially small and medium-sized manufacturing enterprises with fewer than 500 employees.

* Tel.: +1-404-894-7735; fax: +1-404-894-0535.
E-mail address: ps25@prism.gatech.edu (P. Shapira).

PII: S0048-7333(00)00168-2

On its face, the MEP's collaborative approach presents a model of "reinvented" technology policy that is comprehensive, locally-managed, and demand-driven — representing a contrast with the fragmented and "technology-push" methods of previous federal technology transfer programs. In many ways, the MEP program has been remarkably successful — a significant achievement in a policy arena with a mixed record of performance. However, as the paper discusses, many tensions are inherent in pioneering new technology policy management styles in the context of political, fiscal, industrial, and innovation systems that continue to operate on lines that are more traditional. These tensions are examined and insights are drawn both for the future development of the MEP and for other technology and innovation policies that seek to emulate the MEP's approach. Although the MEP represents a publicly-led model, the case illustrates trends and issues relevant to a broad range of public–private partnerships, particularly in highlighting the inter-organizational tensions that occur within partnership boundaries and the mission trade-offs that result.

2. New frameworks for partnerships in technology policy

Many writers have commented on the paradigm shifts that have occurred in US innovation and technology policies over the past five decades (for example, Tassey, 1992; Crow, 1994; Galli and Teubal, 1997). Following the end of World War II, there was an emphasis on building-up the capabilities of national labs, universities, and corporate research centers and promoting R&D for such government-oriented missions as defense, energy, space exploration, health, and agriculture. In the 1960s, a focus on technology transfer emerged, with the aim to promote greater civilian "spin-off" from mission-driven public R&D.

Now, a third shift is underway. Prompted by increased global economic competition and the end of the cold war, US technology policy is more attentive to explicit civilian commercialization goals. Perhaps even more significantly, new implementation concepts are being pursued which — eschewing traditional linear "pipeline" models of technology "push" and "pull" — are based on more complex and iterative perspectives on the technology development and diffusion process (Branscomb and Florida, 1998). In particular, new patterns of industry collaboration and commercialization are being promoted, through industry consortia, university–industry linkages, and public–private partnerships. Examples of the new policy approaches at the federal level in the United States include the development of the advanced technology partnership (ATP), the manufacturing extension partnership, the partnership for a new generation of vehicles, and the United States innovation partnership (Brody, 1996; Shapira et al., 1997).

Collaboration between laboratory researchers and potential industry users or between the public and private sectors to develop and diffuse technology is not, of course, a new idea. In the late 19th and early 20th centuries, state land grant universities throughout the United States were chartered to work with farmers through agricultural extension and with industry through engineering experiment stations (Combes, 1992). By the 1920s and 1930s, scientists in government laboratories and universities were already collaborating on joint research projects with industry — a pattern that was massively reinforced during World War II by the collaborative, large-scale government, university, and industry projects in such fields as atomic weapons, computing, and radar (Freeman and Soete, 1997). Ironically, as Freeman and Soete (1997, p. 300) go on to note, the very success of these "big science" projects buttressed the notion that public investment in basic R&D would result, linear fashion, in applications and innovations. This encouraged the build-up of basic R&D capabilities and the public sponsorship of mission-oriented research that predominated in the 1950s and 1960s.

The American strategy of "front-loading" policy attention and public resources to basic research and government missions, while leaving commercial innovation and technology diffusion to fare for themselves in the private sector, underwent increasingly critical scrutiny in the late 1970s and 1980s. During this period, Japan emerged as a fully-fledged industrial and technological power. Japan's achievement was seen, at least in part, as an outcome of public–private collaboration for the purposes of commercial innovation that was far tighter than then common in the United States. Japanese public policy had, in Morris-Suzuki's (1994) terms, promoted an effective "social network" of innovation, where the sharing of information and the collaborative matching of resources by the public sector, research institutions and private companies significantly accelerated not only technological progress but also commercial success. US industries, including those in knowledge-intensive high technology sectors, appeared relatively disadvantaged in terms of commercializing technology, despite America's evident strength in basic science and core research. This prompted US initiatives in the 1980s to encourage the cooperative transfer of research and development from federal laboratories to industry, establish university–industry research centers, and promote industry consortia in electronics, machine tools, and other industries (Office of Technology Assessment, 1990).

Despite these initial steps, and the considerable debate that accompanied them, it was not until the 1990s that a sustained effort to develop a new policy framework for public–private technology partnerships was put into place. On assuming office in 1993, the Clinton Administration defined new principles for the federal government's role in technology policy that included promoting a business environment to support private innovation and encouraging civilian technology development and commercialization (Executive Office of the President, 1993). The administration's implementation strategies included reallocating federal funds from military to civilian research, stimulating collaborative relationships between research performers and technology users, and measures to speed the deployment of new technologies to industry. These efforts to reorient federal technology policy through emphasizing public–private partnerships, civilian goals, and technology deployment have not been without criticism. For example, strenuous charges of unnecessary "corporate welfare" were levied at Clinton's new technology policies following the election of a Republican Congress in 1994. Moreover, it is evident that attention to establishing a new technology policy paradigm was greater in the administration's first term than in its second. Besides the difficulty of working with the Republican Congress, other factors at work included the strong performance of the American economy in the mid-to-late 1990s and perceived needs to re-boost defense spending, each in their own way "bumping-down" civilian technology policy's place on the broader policy agenda. Yet, these caveats notwithstanding, it does seem that a new era in US technology policy has taken root, and that attention to public–private collaboration and civilian technology goals will continue.

In understanding how this reorientation of policy came about, there is a natural tendency to focus on the changes that have occurred at the federal level. However, there is another important strand to the story, namely the role of states in establishing new concepts and frameworks for technology policy. During periods when the federal government has been reticent to take explicit civilian technology promotion measures, states have often done so under the guise of economic development. State universities and engineering experiment stations (and their successors) have long been engaged in collaborative, pragmatic relationships with local industries. States have sought to promote economic development by establishing regional technology clusters, among the most notable being North Carolina's Research Triangle. Most recently, US states have considerably increased their investments in technology policies, including university/non-profit research centers, joint industry–university research partnerships, direct financing grants, incubators, and other programs using research and technology for economic development (Coburn and Berglund, 1995; Berglund, 1998).

By the mid-1990s, states were spending about US$ 2.7 billion of state-derived funds (i.e. excluding federal and industry funds) on research and technology programs (State Science and Technology Institute, 1998). By comparison, the FY 1998 federal budget for the US National Science Foundation (NSF) was US$ 2.5 billion (for research support). This is, of course, not an entirely appropriate comparison: NSF has a far greater emphasis on fundamental science and engineering than do the states, although at least some of the state spending is used to match or leverage NSF awards. Nonetheless, it is symbolic that total US state research and technology spending now exceeds NSF's budget.

The relevance of the states is not simply that their budgets have increased, but also that they represent a significant, although at times under-emphasized, element in the reinvention of US technology policy. In America's decentralized federal system, the states are in their own right a countervailing source of policy intervention and experimentation. For example, from the late 1970s through to the mid-1980s, intense economic restructuring and growing international competition generated many debates in Washington about the need for new industrial and technological policies. But, for political and ideological reasons, there was little in the way of a comprehensive federal response. However, a series of states did put into place their own industrially-focused technology policies, including Pennsylvania's Ben Franklin Partnership and Ohio's Thomas Edison Program (Osborne, 1988). These state models influenced the federal technology policy measures in the landmark 1988 Omnibus Trade and Competitiveness Act (which, for the first time, designated federal agency responsibilities for *commercial* technology promotion and established the program of regional manufacturing technology transfer centers that evolved into the MEP) and also the subsequent technology policy initiatives of the Clinton administration. One of the essential elements of these state models was partnership between government, industry, and academia. Moreover, the states not only pioneered prototypes for current partnership models, but today also serve as stakeholders in many technology partnerships now being promoted by the federal government.

3. Rationality of technology partnerships

As just discussed, in the reorientation of US technology policy, a new emphasis is being placed on commercialization. Public–private partnerships have emerged as a principal organizational and implementing mechanism. It is appropriate to ask: why? In considering this question, it is important to distinguish between general arguments for public policy intervention in technology policy and specific rationalizations for the use of public–private partnerships as a mechanism for implementation.

In general, an active governmental role in support of science and technology is typically justified on grounds of market failure. For example, it is advanced that because of the difficulties of fully appropriating results, private firms are likely to under-invest in research and development, thereby leading to under-investment from the view of potential social returns (Tassey, 1992). Related justifications focus on the concepts of public good and strategic interest, where purely private efforts in such areas as medicine, environment, defense, or industrial competitiveness are judged unable to maximize social well-being or to meet political or community goals.

Although public sponsorship has long been seen as an appropriate response to private market shortfalls in supplying optimal or desired levels of basic research and development, it has often been assumed that private mechanisms coupled with open access to public research institutions and universities could adequately disseminate technological innovations. However, it is increasingly recognized that market failures and strategic interests also exist in the process of technology diffusion. Potential users face uncertainty, information and learning costs, and other externalities which may result in under-investment in available technologies; similarly, potential suppliers of information and assistance also face learning costs, may lack expertise, or face other organizational barriers in promoting the diffusion of potentially rewarding technologies. System-level factors, such as the lack of standardization, regulatory impediments, weaknesses in financial mechanisms, and poorly organized inter-firm relationships, may also constrain the pace of technology diffusion. In the context of rapid international flows of information and capital and increased global competition, it has also been argued that strategic national and regional efforts to maintain industrial competitiveness depend not only on innovativeness, per se, but more than ever on the diffusion, effective application, and further incremental improvement of known technologies (Office of Technology Assessment, 1990).

Where market failures or strategic interests in technology diffusion are established, a range of specific policy tools is available. Measures such as subsidies and tax breaks for individual firms, enhanced information provision, or the development of traditional functional technology transfer programs may be effective policy responses. Why then the current emphasis on partnerships? Although the justifications for public–private partnerships take on many variations, it is possible to identify three broad types of reasons.

3.1. Efficiency and scale (economic rationality)

Perhaps the most common justification for promoting public–private partnerships as an implementation strategy for technology policy is one of efficiency.

Usually, there is the desire is to optimize existing programmatic investments, by reducing overlaps and leveraging of scarce resources. It is also hoped that transaction and search costs are reduced. Given the historical context where individual programs and policies exist but do not necessarily coordinate well with one another, the efficiency motivation for partnerships is also a response to "government failure." The desire to form partnerships to improve the total efficiency of public technology policies is consistent with increased concerns by both public and private sector leaders about government performance enhancement. In addition to addressing government failure, partnerships can potentially be market correcting. If firms can be brought together in consortia, the risks involved in research and development can be shared, leading to a better balance between costs and returns. Typically, this argument is applied to generic, "pre-competitive" research. Yet, it also applies to the diffusion of technologies. For example, individual small firms in a locality may inadequately invest in training employees to use new technologies, perhaps because the cost of developing a new training program is too high for any single firm to justify. All firms, facing the same cost barriers, thus under-train and suffer from inadequate skills to use new technologies. However, if these firms are organized into a consortium, the training development costs can be spread over multiple firms, thereby lowering the unit costs to affordable levels for individual firms. In this case, with partnership, all firms benefit. Often, it takes an intermediary to stimulate the organization of firms into initial partnership.

3.2. Enhancement of scope and learning (social rationality)

A second reason for promoting public–private partnerships is to improve the scope and quality of action — to do things through collective processes that could not be done individually. Partnered initiatives often seek to offer combinations of services, for example by combining technology assistance, training, management, and marketing assistance in an integrated fashion to firms. This is a mix that may require expertise from multiple public and private service providers. Even more fundamentally, partnerships frequently aim to accelerate the learning and sharing of knowledge and to promote networks of information flow, technology diffusion, experimentation, and organizational innovation. Fountain (1998) highlights "shared resources, shared staff and expertise, group problem-solving, multiple sources of learning, collaborative development, and diffusion of innovation" among the benefits of inter-organizational partnerships and consortia. In this sense, partnerships aspire to serve as investments in social capital or "associational economies" to promote the processes of collective learning and innovation now seen as crucial in the formation of effective systems of innovation at national, regional, and sectoral levels (Cooke and Morgan, 1998).

3.3. Sustainability (political rationality)

In the "realpolitik" of public policy, it must also be admitted that partnerships have emerged as a political response to attacks on traditional concepts and instruments of government policy. The last decade or so in the United States has been one where any discretionary governmental activity has come under great skepticism and criticism. This is certainly true of technology policy. At the federal level, political momentum by conservatives to leave technology development and diffusion to the market was strong in the 1980s and remains powerful in the 1990s. These political forces frequently suggest that government cannot effectively guide technology policy, due to inefficiency, lack of skill, bureaucracy, or political favoritism. In response, technology policy proponents have advanced partnerships as a more effective, non-traditional approach that involves business input and matching funds, market incentives and fee for service, and an emphasis on performance. It is also true that in some cases, public agencies and institutions have promoted the formation of partnerships to serve their own organizational objectives. Partnerships can be seen as way of extending leverage and influence. In other instances, agencies seek partnerships for purposes of survival.

In theory, ideal public–private partnerships are ones that concurrently promote the triple goals of economic efficiency, social learning, and political sustainability. In practice, this ideal is hard to achieve. At best, most partnerships optimize between these three goals. If pushed too far in any particular direction, performance on one or both of the other goals suffers. Thus, if partnerships are too focused on efficiency and performance goals, then learning and innovation may be given short shrift. Yet, if too focused on social learning without efficiency outcomes, then the political system may become impatient at the lack of "visible" results and so curtail support. If political rationality results in partnerships that are mostly symbolic and which lack economic and social substance, this too is problematic, particularly in terms of economic and social opportunity costs and the credibility of public action. It might be noted, however, that political sustainability and symbolism cannot be ignored: some partnerships that have high political value can preserve themselves, even if less than fully effective in economic and social performance. Paradoxically, even if partnerships seem to demonstrate economic and social outcomes, their continuance can be uncertain if they lack appeal to political values (a problem that affected the ATP in the mid-1990s).

These contrasting rationalizations for partnerships, and the need for partnerships to navigate and negotiate between different objectives and needs, suggest that the partnership is a particularly complex category among the tools of technology policy intervention. The hyperbole of partnership promises cooperation, reciprocal benefits, and seemingly effortless leveraging of resources. The reality is not so simple: partnerships usually have costs as well as benefits and frequently have to spend time and energy to reconcile competing interests. Once new policy frameworks have been established to encourage partnerships, significant leadership and management resources then need to be invested to ensure net benefits and desired results. The kinds of issues, tensions, and management strategies that are involved in implementing partnerships in technology policy are illustrated in following sections, through the case of the US MEP.

4. The manufacturing extension partnership

The MEP is a network of technology assistance and service providers that aims to upgrade the performance and competitiveness of small and medium-sized manufacturers in the United States.[1] The MEP is a collaborative initiative between federal and state governments that also involves non-profit organizations, academic institutions, and industry groups. The National Institute of Standards and Technology of the US Department of Commerce is the MEP's federal sponsor.

The provision of technology and related business assistance to small and mid-size firms in the United States is not an entirely new activity. In the 1950s and 1960s, state industrial extension and technology assistance programs were established in several states, including Georgia, North Carolina, and Pennsylvania. These early programs diffused technical information and used professional engineers and other technical specialists to help local firms improve their use of technology (Shapira, 1990). Increased concerns about industrial competitiveness and regional economic development over the next two decades prompted the development of similar programs in other states. By 1990, manufacturing extension and technology transfer programs had been established in 28 states. However, of these only about one dozen states operated field service networks using industrial experienced staff able to work on-site with firms — a factor critical in being able to address specific manufacturing problems on the shop floor of small firms.

The federal government started its own direct support of industrial extension programs in the late 1980s and early 1990s. Under the auspices of the 1988 Omnibus Trade and Competitiveness Act, the National Institute of Standards and Technology set up a handful of manufacturing technology centers, working in an initial collaboration with selected states (Office of Technology Assessment, 1990; National Research Council, 1993). But the most dramatic expansion of federal sponsorship came with the Technology Reinvestment Program, first announced in 1993 and implemented in 1994 and 1995 (Advanced Research Projects Agency, 1993). Although this multi-billion dollar program was targeted towards the post-cold war conversion and restructuring of America's defense-industrial base, it did make significant funds available for the upgrading and deployment of technology in civilian industries, including small and mid-sized manufacturers. The National Institute of Standards and Technology was allocated a share of these resources to increase the number of manufacturing technology centers and to organize, through what was then the embryonic MEP, a

[1] Small and medium-sized manufacturers are defined in the US as those with fewer than 500 employees.

collaboratively-delivered set of industrial extension services to manufacturers throughout the country.

The Technology Reinvestment Program had a major impact on the structure and character of the MEP. Funding was awarded through a competitive review process and states and other applicants normally had to match one-half of proposed costs with their own or private funds. Additionally, applicants were guided to form partnerships of service providers. There was an explicit requirement that proposals for funding address a criterion entitled "Coordination and Elimination of Duplication." This criterion required the proposer to understand and link with related service providers in the service region, be consistent with existing state strategies, and not duplicate existing resources or services. Partnership proposals were judged in terms of the number, diversity, and skills of constituent service providers, geographic scope and coverage, cohesiveness, organization, and management structure (National Institute of Standards and Technology, 1994).

Boosted by the Technology Reinvestment Project and with subsequent support from the civilian budget through the US Department of Commerce, the MEP has now grown to a network of more than 70 centers in all 50 states. MEP centers are structured either as separate non-profit corporations or as part of other organizations, such as universities, stage agencies, technology centers, or economic development groups. A variety of partnership elements are fostered. The federal sponsor works with states and local centers in program management and development and in backing joint working groups, staff training, information and communications systems, and common tools. MEP's annual federal funding (about US$ 105 million for the fiscal year 2000, of which more than US$ 90 million goes to center support) is supplemented by more than US$ 130 million in state and private funds (latest estimates indicate that private sector revenues and fees contribute around US$ 70 million annually to MEP operations). In almost every case, individual centers operate with and through local networks of associated public and private service providers. Centers usually have governing or advisory boards that include local public and private-sector representatives.

Currently, about 30,000 firms are being assisted each year by the MEP through assessments, technical assistance projects, information workshops, and other services. Two-thirds of assisted companies have fewer than 100 employees. MEP centers reach and serve these firms directly through over 400 local offices and through a partnership network of more than 3100 affiliated public and private organizations that deliver or support the delivery of services to small and medium-sized manufacturers.

The most common partnership relationships are with economic development organizations and universities. About 95% of centers have relationships with these types of organizations.[2] The next most common type of organizational relationship, for two-third percent of the centers, is with community or vocational colleges and technical institutes. Almost sixty percent of centers have relationships with industry associations and small business development centers, and about one-half with private consulting companies. To a lesser extent, partner relationships are also reported with federal laboratories, larger companies, utilities and training organizations (Table 1). Although there are issues of data comparability, the number of MEP partners in 1997 was more than three times greater than the 750 affiliated organizations reported by 40 centers at the end of 1995. It suggests that MEP center affiliations with partnership organizations have grown.[3]

MEP organizational partnerships are diverse. Many MEP centers have arrangements where other service providers act as program affiliates to perform particular operating functions, such as marketing to prospective customers, or provide specialized services, for example in helping manufacturers with environmental compliance. MEP centers have also established collaborative initiatives with industry associations, large manufacturers, technology centers, and other groups through which information, training, networking, technology diffusion, or other special projects are

[2] Calculated form of reports to NIST by 68 MEP centers (June 1997), with removal of duplicative information. There are variations in how different centers define and report their affiliates. Some centers do not report information about organizations that staff informally used to provide assistance to manufacturers. In addition, data from seven mostly newer MEP centers is not included in this analysis.

[3] Further details about the characteristics and organizational forms of MEP partnerships are found in Shapira and Youtie (1998a). The balance of this section of the article draws upon this study.

Table 1
US manufacturing extension partnership: affiliated organizations[a]

Type of organization	Percent of centers reporting affiliation
Economic development organization	97
University or 4-year college	95
Community or vocational college	66
Industry association	59
Small Business Development Center	59
Other non-profit business assistance organization	57
Consulting company	48
Federal laboratory	38
Other government agency	38
Other extension service (cooperative, industrial)	31
Large company	31
Electric power or other utility	31
Training organization	29
Other for-profit organization	26
Vendor (of equipment or software)	10

[a] Source: analysis of manufacturing extension partnership center reports to the National Institute of Standards and Technology, June 1997. Based on reports from 68 centers.

targeted to small and medium-sized manufacturers in a particular locality, industry, or supply chain.

Perhaps most frequently, MEP centers use other service providers on a subcontract or referral basis. About one-quarter of MEP's technical assistance projects involve outside service providers.[4] In such cases, center staff typically conduct an assessment of a customer's needs, propose a project, and then recommend qualified outside service providers or consultants to assist in implementation. Centers tend to use other service providers in fields both outside of and within traditional MEP core competencies. Human resource projects, where most MEP centers do not have in-depth expertise, are most likely to involve outside service providers. However, the second most common area for third party projects — process improvement — is a central MEP core competency. Here, the involvement of outside partners to provide services presumably leverages the number of projects within their field of expertise that center staff can manage. Other common areas for third party projects include business systems and management, market development, and quality.

While partnerships between MEP centers and other organizations are often informal, increasingly the tendency is for these relationships to be structured in writing, through memoranda of understanding, performance agreements or binding contracts. Formal agreements are universal where money changes hands. But there is no single contractual model for the whole system; each center has considerable flexibility in organizing partnership arrangements within allowable legal, auditing and sponsor criteria. MEP centers may entirely underwrite the cost of activities or specialized services by partners, although this mode of partnership is becoming less prevalent as MEP centers face greater pressure to generate fee revenues. In other cases, MEP centers and partners share costs (at times with in-kind as well as cash contributions) or collectively obtain resources for a special project from NIST, the state or another sponsor. With the aim of generating revenues, some centers seek management fees from outside service providers who implement referred projects with MEP customers. In other instances, vendors, corporations, or large private consultants may donate cash, equipment, in-kind or pro-bono services in liaison with MEP centers.

In several senses, the MEP is an organizational hybrid. It is not a purely private partnership, as its leading players are public and non-profit entities. Yet, private service providers and, of course, private manufacturers are integrally involved. Second, the MEP is not a purely national effort, in contrast to the advanced technology partnership — a federal program that promotes national consortia of private companies, laboratories, and other institutions to develop and commercialize leading-edge technologies; nor is the MEP a purely local effort, as is the case with state technology initiatives like Pennsylvania's Ben Franklin partnership or the Georgia Research Alliance — programs that are concerned only with technology development in individual states, without regard to national impact.[5] Rather, the MEP embodies a partnership between different levels of government — with the relationships between these levels undergoing change over time. Finally, the MEP is not a purely technological initiative, in the "hard" meaning of this

[4] Analysis of 8443 technical assistance projects of 8 h or more with companies completed by 59 MEP centers in 1996 shows that outside service providers were involved in 24% of projects.

[5] For comparison with state technology initiatives and partnerships, see the program descriptions available through the State Science and Technology Institute at http://www.sti.org/.

term. Rather, considerable weight is given to issues of management, training, marketing, and strategy, as well as to technology per se.

Yet, these variations notwithstanding, the MEP certainly incorporates most of the design principles articulated in government reform proposals and advanced by advocates of new public–private technology policy partnerships (Osborne and Gaebler, 1993). First, the program seeks a cooperative relationship between the public and private sectors. The private sector is involved not just as a recipient, but also as a service partner and an advisor. Second, the program is decentralized and flexible, with individual centers able to develop strategies and program services which are appropriate to state and local conditions. Third, the MEP seeks not to duplicate existing resources. Rather than provide services directly from the federal level, MEP awards are designed to get existing service providers, whether they be consulting firms, non-profit organizations, academic institutions, public agencies or trade associations, to cooperate and coordinate in their efforts to assist local manufacturers.

5. Partnership outcomes and tensions

Many advantages of scale, scope, and shared learning are claimed from the MEP's efforts to promote partnership and service coordination to aid the industrial modernization of small and mid-sized manufacturers (US Department of Commerce, 1999a). These include reduced duplication, access to special skills, greater flexibility, the leveraging of public and private dollars, and improved service to clients. To what extent are these professed public and private benefits actually achieved, and at what cost?

Such questions were probed in a study of MEP service coordination conducted in two phases between 1995 and 1998. The first phase of the study examined the initial development, operation, and effects of service coordination in the MEP system (Shapira et al., 1996a). On-site cases were conducted of six MEP centers and their partnership relationships. The second phase of the study examined subsequent changes in how MEP centers coordinated services with partner organizations (Shapira and Youtie, 1998a). The design of the second phase involved revisiting four of the original case study centers (about 2 years had elapsed between the first and second rounds of case studies, allowing changes in partnership arrangements to be observed). Two new case study MEP centers were added. In both phases of the study, structured interviews were conducted with MEP staff, partner organizations, small business customers, and state sponsors. These case studies were augmented by reviews of program documents from each center and its affiliates, an analysis of partnership information from MEP's national records, interviews with national program staff, and input (in Phase I) from an expert advisory panel. [6] In summary, this research found these benefits:

1. Enhanced service partnering and coordination stimulated by the MEP has made available a wider range of expertise to firms and, in many instances, a more systematic approach to providing assistance.
2. Involving other partners has allowed MEP centers to increase flexibility and has particularly helped newly established centers to ramp-up their services fairly quickly by building on and expanding existing resources. MEP centers have been able to draw upon other well-established organizations, such as economic development organizations, to conduct marketing and outreach campaigns.
3. Resource sharing has occurred. Facilities at community colleges have been used for business training programs and for demonstrating new technologies. Experts at universities and federal labs have been involved in helping firms to resolve specific technical problems.
4. New working relationships have been forged with private consultants through whom MEP centers have been able to broker a range of

[6] The research on service coordination in the MEP system was directed by Philip Shapira and conducted with Jan Youtie, Gordon Kingsley, and Marc Cummins. The eight MEP centers and networks that were the focus of case studies were: the Chicago Manufacturing Center, the Florida Manufacturing Center, the Georgia Manufacturing Extension Partnership, the Great Lakes Manufacturing Technology Center (Cleveland, OH), the Industry Network Corporation (a multi-state franchise serving five states including New Mexico, Arizona, and Nevada), the Manufacturing Extension Partnership of Southwest Pennsylvania (Pittsburgh region), the Minnesota Manufacturing Technology Center, and the Oklahoma Alliance for Manufacturing Excellence. The cited project reports (1996, 1998a) contain full details of the research methodology, cases, and findings and are available through the Worldwide Web site of the Georgia Tech Policy Project on Industrial Modernization at http://www.cherry.gatech.edu/mod.

business-oriented services to small and mid-sized firms. Centers have also used partnerships to develop new service offerings. In affiliation with third party organizations, MEP centers have won grants to develop new tools, training, and group service programs and to extend services in critical fields, including environmentally conscious manufacturing and human resources.

5. Small and medium-sized customer firms indicated that it was preferable to deal with one organization able to offer a range of needed business services from public and private sources — as opposed to numerous single-function government programs or private vendors promoting only their own products.

So far, so good: it does seem that the emphasis on partnership has influenced how services are organized and delivered to small and mid-sized manufacturers through the MEP system. Moreover, most companies report that they take action on MEP project recommendations: one state study shows that that over two-thirds of customers act on the recommendations provided by program staff (Youtie and Shapira, 1997), while in a national MEP post-project survey almost three-quarters of customers report that interactions with MEP centers cause them to take actions they would not have done otherwise (Voytek, 2000). Companies further report that these actions lead to improvements in manufacturing processes and skills (Shapira and Youtie, 1998b). Several studies have found that MEP-prompted actions lead to cost savings, additional sales, higher profit margins, or improved productivity (see, for example, the evaluations conducted by the US General Accounting Office, 1995, and Jarmin, 1999). However, other studies report more mixed results. For instance, Luria (1997) found that Michigan MEP customers improved to a greater extent than non-customers in sales growth, employment growth, and adoption of certain process improvements and technologies. But, center customer growth in wage rates, profitability, and labor productivity were not significantly different from that of non-customers.

Overall, the weight of the evaluation evidence (particularly of studies that include non-customer controls) is that MEP projects typical have positive but mostly modest effects on companies (Shapira, 2000). To an extent, this conclusion is a reflection of the genuine difficulties that companies have in reporting quantitative effects from project interactions, particularly in terms of dollar impacts. But it may also reflect the fact that many manufacturing extension projects are of small scale and, by themselves, cannot be expected to have substantial effects. A majority of the interactions that the MEP has with companies involve less than 8 h of total staff time (a threshold below which projects are often deemed too small to evaluate). Conversely, those cases where MEP interactions have had significant strategic impacts on companies frequently involve in-depth, long-term engagements (which may extend over several years), and are often associated with internal management changes (Youtie, 1997; US Department of Commerce, 2000). Several analysts have recommended that the MEP needs to give greater attention to strategic offerings that will assist firms become more distinctive and specialized in their marketplaces, rather than offer standardized, often short-term, services that simply emphasize efficiency (Luria, 1997; Oldsman and Heye, 1998). Indeed, MEP management acknowledges that while their program has established a national network, is reaching significant numbers of firms, and has many individual success stories, it has yet to achieve hoped-for levels of center performance and customer improvement across the whole MEP system. "We have done a lot, but only touched the tip of the iceberg," notes the director of the program.[7]

Why has the MEP's achievement in rapidly realizing scale in its service delivery system been paralleled by mostly modest performance outcomes for customers? It can reasonably be argued that such outcomes come with the territory: the pragmatic adoption of improved technologies and methods by existing manufacturing firms will typically lead to incremental, rather than dramatic, results. At the same time, it is also evident that the performance outcomes experienced to date reflect, at least in part, inherent tensions in the program's partnership model. While at first glance, partnership appears to offer a "free ride" that can leverage public investment, further analysis shows that, in fact, partnership tends to be accompanied by tensions that reduce desired leveraging effects. Such tensions derive from conflicts and tradeoffs between

[7] Plenary remarks by Kevin Carr, Director, NIST Manufacturing Extension Partnership, 2000 Manufacturing Extension Partnership National Conference, 8 May 2000, Orlando, FL.

the three underlying rationalizations for partnership formation noted earlier. For example, although MEP centers have formed partnerships and, in so doing, have pursued the economic rationalization of leveraging resources, this process has both direct (i.e. MEP) and indirect (i.e. non-MEP) costs. Field interviews conducted as part of the service coordination study found that MEP programs actively engaged in service coordination incurred significant transaction costs, including the expense of identifying and qualifying outside providers, information exchange, contracting, consulting, and monitoring. Furthermore, in most instances, the other program resources "leveraged" by MEP centers were not "free" in that they had to be paid for by other public or private sources. Additionally, the inter-organizational tensions associated with partnership promotion efforts required the expenditure of significant amounts of "political" capital, for example in resolving concerns about clients being "stolen" or about one program working in another's territory. In short, some collective resources that (in theory) might have been available to re-allocate to more strategic assistance to firms were (in practice) expended internally in processes of organizational adjustment or were not truly available to be leveraged for different purposes.

Similarly, aspirations for partnerships to improve overall service quality were not so easily achieved. The case study interviews found that while the differential characteristics of program partners added new capabilities to the system, efforts to promote tighter service coordination also revealed limitations among those partners (Shapira et al., 1996a; Shapira and Youtie, 1998a). These affected how various partners performed in delivering modernization services to manufacturers. For example, economic development organizations often offered general referral services but typically could not deliver technological or longer-term project assistance to firms. Federal laboratories and university researchers possessed particular technical capabilities in specific fields, but drawing on these capabilities was hampered by asynchronous time horizons and administrative barriers within these large institutions. Small business development centers provided needed business planning capabilities, but their lack of manufacturing background posed problems in face-to-face dealings with manufacturers (see also the evaluation of MEP-Small Business Development Center partnerships in Yin et al., 1998). Private-sector consultants did have an orientation towards manufacturing needs. However, their expense rates and operational styles were often geared to large-manufacturer budgets. Finally, the involvement of community colleges promised additional institutional resources for local manufacturing extension partnerships, but these sometimes proved ephemeral as college administrators (under continual funding pressure) focus on their own priorities, rather than the MEP's.

To an extent, some of these problems could have been avoided with better selection of partners. However, the political rationality of partnerships is such that lead organizers have to work with the spectrum of organizations they find within the jurisdictions defined as eligible for funding. Moreover, it is not always apparent what the limitations of partners are until real work is underway. Of course, many of these operational issues are addressable — and can be seen as teething problems that settle down as partnerships evolve over time. Indeed, the case studies found that MEP partnerships went through successive stages, during which not only was there changes in the balance of benefits and costs but also much learning about the how partnerships could be most effectively structured and managed to accommodate developments in customer needs, technology, and policy. MEP partnerships were first formed under conditions of increased federal and matching funds, with guidance to demonstrate a high level of coordination and service partnership. Under these conditions, MEP centers entered into a wide-ranging set of service partnerships, as our analysis of MEP-affiliated organizations illustrated. However, as MEP centers subsequently operationalized their partnerships, they have better understood the strengths and weaknesses of particular affiliates. This has led to changes in arrangements. In many cases, relationships have been scaled-down or ended. In other situations, links have continued but important adjustments have been made. To take one example, to address the high cost of some private consultants, one center has negotiated reduced rate structures which take into account the fact that the center bears the marketing costs and that center referrals often generate opportunities for follow-on work (Shapira and Youtie, 1998a).

As they have gained more experience with partners and partnerships, we now see many MEP centers undertaking a substantial restructuring of their service coordination arrangements. The outcome has been to focus on tighter links with a smaller set of the most capable partners — with the ability to adjust partner arrangements, as customer business and technological needs change. Links with more marginal partners are being reduced, usually — although not always — in an amicable manner that maintains communication and allows collaboration on an as-needed basis. This trend has been accelerated by the reduction in the federal contribution to MEP center costs — from about one-half of each center's budget in the program's early years to a planned steady-state level of one-third. As federal resources become tighter, centers have to reduce their own costs as well as generate additional revenues from customers and other sponsors. Partnerships with other service providers continue to be important. But centers have recognized that they must more exactly specify what each service provider is expected to contribute to the partnership, how partnership performance will be monitored, and under what conditions partnerships will be renewed or, if necessary, terminated.

The challenges faced by MEP affiliates in achieving more substantial performance outcomes are compounded by difficulties in designing assessment and evaluation systems that can match the differential objectives pursued by sponsors, service providers, and customers. Among the sponsors, state and federal aims in co-funding partnerships do not necessarily coincide. A principal federal interest in promoting manufacturing extension is the improvement of industrial competitiveness and productivity for the nation as a whole. States, on the other hand, are particularly concerned about what happens within their own boundaries. They seek new jobs, although in reality enhanced productivity may reduce employment in some instances, and have few qualms about supporting "their" businesses, even at the cost of jobs in another state (for example, by assisting low-wage small firms, some extension efforts may help the shift of high-wage jobs from manufacturing regions in other states — a process that may be zero-sum or worse from a national perspective, but which may be judged worthwhile from a state standpoint). Similarly, from the viewpoint of individual partner organizations and assisted firms, each represents different stakeholders and interests that may or may not consider MEP evaluative metrics as important benchmarks of performance or outcomes. Thus, community colleges may seek to boost trainee enrollments in their institutions (an own goal), rather than upgrade the internal training processes of firms (which relates to the MEP partnership goal of overall firm upgrading). This example illustrates a sense in which partners in the MEP network can also be competitive in terms of customer procurement and maximizing credit for their own programs.

Diversity in objectives is found, of course, in many policies and programs whether operated through partnerships or not. However, it is apparent that as partnerships grow, so does the multiplicity of stakeholders and differential interests, thereby adding to the complexity of what and whose objectives evaluation should measure. Moreover, it is also apparent that some organizations were added to particular MEP partnerships to bolster local political support, head off potential "turf battles" and augment outreach efforts, rather than provide distinct service delivery capabilities. Evaluative results may show that these partners contribute little on standard measures, but underestimating the role of these partners could impair a federal sponsor's ability to maintain ongoing in-state political support for inter-governmental partnership initiatives such as MEP.

A further tension arises between efforts to standardize performance metrics for MEP affiliates to promote efficiency, effectiveness, and public accountability (a desired goal for reinvented government), and the complex, non-standard, and hard-to-quantify character of customized services offered to promote innovation, networking and learning (elements of the social rationality of the partnership). Standardized evaluation metrics for manufacturing extension providers have stressed business outcomes, such as increased sales or reduced costs, and economic development impacts, particularly effects on jobs (Oldsman, 1996; Shapira et al., 1996b). Illustrative of this trend are the four performance measures emphasized by the MEP to meet the US Government Performance and Results Act: increased sales, labor and materials savings, capital investment, and inventory savings (US Department of Commerce, 1999b). However, differences in the specialized services that various partners offer present challenges to the use of such narrowly defined efficiency measures across the whole partnership. Diverse organizations are involved in MEP programs, including universities, colleges, economic development organizations, technology centers, federal labs, utilities, and different kinds of private sector consultants. Some partners offer services that were more likely to have quantifiable effects than others. Some are more likely to produce short-term gains, whereas others promise long-run effects on an industry or regional economy. For example, partners who provide human resource and training services or help firms develop supplier linkages and inter-industry networks frequently have less tangible effects on a firm's bottom-line than organizations, which show firms how to make fast savings in energy consumption or fix a particular machine problem. The risk here is that partners whose activities have deeper long-term impacts may be unfavorably compared to service providers whose results are more readily measurable now.

In a sense, there is a tension here that goes beyond differences in partner roles and to the basic expectations ascribed to partnerships themselves. On the one hand, partnerships are promoted as innovative mechanisms, able to address structural, long-running problems and needs better than previously fragmented approaches and services. On the other hand, partnerships are also expected to efficiently and quickly produce visible results. There are undeniably possibilities for partnerships to simultaneously make progress on both of these fronts, but realistically, as we found in our cases, tensions may be apparent between what can be measured now and what is of long-term value.

A final set of tensions is found in the intergovernmental relationship between the federal government, the states, and the MEP centers. The federal-state relationship is simultaneously a collaborative effort to share resources to promote modernization objectives and a contractual arrangement between the federal government and the states for the delivery of certain services and outcomes. While the former role requires openness and information sharing, the latter tends to encourage the withholding of "bad news" by centers and requires conformity to established bureaucratic norms and regulations. On the one hand, the MEP aims for learning, flexibility, and experimentation — leading to interactions with companies that have strategic outcomes. On the other hand, the MEP is also under pressure to achieve wholesale coverage, serve large numbers of firms, and quickly produce "visible results." The former reflects the social learning rationality in the MEP's partnership design. The latter reflects the economic and political rationalizations of achieving scale and institutional survival.

There is much discussion of this tension within the MEP system (see, for example, Oldsman and Luria, 1998), with many proposals for change. However, it has been rather hard to make progress on these proposals. In earlier phases of development (roughly the period 1988 through 1994), the MEP was arguably more open to the innovation and discursive dialogue (and also the risk of failure) that characterizes the social learning pole. After a heightened concern with political sustainability (around 1995 to 1996, with the election of a Republican-led Congress), the MEP is now under pressure to emphasize economic efficiency and measurable performance. In this reality, the drive for visible results tends to get priority, leading to many MEP projects that are short-term, of limited duration, but not always of strategic import. These projects seem to produce outcomes that are adequate to secure ongoing federal political support and funding. However, as the MEP's federal investment declines to one-third, it remains to be seen whether a focus on short-term economic rationality will generate the increased state and private revenues necessary to sustain a national system. The former may require more emphasis on political sustainability, as adjustments are made to MEP partnerships to garner greater state and local political support and the funding this brings. The latter may be best served by initiating interactions and projects with companies that are more customized, long-term, and strategic, so as to generate significantly more fee revenue. This would change the service mix, shifting from large numbers of small projects to somewhat fewer larger projects, potentially excluding many of the smallest manufacturers and also regional or industry network initiatives from which it is hard to generate income. In short, it is probable that in the future the MEP will continue to grapple with how it emphasizes and meets contrasting goals and objectives, within the three-axis framework of efficiency and scale, sustainability, and scope and learning.

6. Conclusions

The experience of the MEP exemplifies how partnership initiatives negotiate, are stymied by, or otherwise manage the tensions and challenges posed by new technology policies that emphasize inter-organizational and inter-sectoral collaboration. In part, the story here is of conflict between traditional and reinvented approaches to government; in part, it is also about the tensions inevitably inherent within the new models of technology partnership. The net product is to make partnerships complex mechanisms to manage well, with varied and usually not straightforward outcomes. As proponents hoped, the MEP case studies did find real benefits associated with partnered service coordination. These included avoiding the duplication of services, tapping specialized skills, spreading development costs of new tools, broader marketing to new industrial customers, improving access to particular industries and areas, flexibility in staffing and the delivery of services, improving service quality, enhancing visibility in the locality, and strengthening state and local support. At the same time, while partnership and service coordination had significant advantages, the cases also identified costs and tensions. These included increased transaction costs (including the expense of identifying service providers, information sharing, contract management, and monitoring projects), difficulties in maintaining quality across partner organizations, delays in timely service delivery, and inter-organizational tensions through unresolved conflicts over client and service territories.

With increased attention being paid to promoting public–private partnership in industrial modernization and other areas of technology policy, there should be a careful assessment of it's benefits and costs. In the MEP examples, it was found that new partnership arrangements resulted in significant advantages, but this should not lead policymakers and program managers to overlook the reality that there are expenses and tensions associated with greater coordination. Investments of resources, time, people, technology, and political capital are needed to make public-partnerships work well.

As J.M. Keynes observed more than half a century ago, "the difficulty lies, not in the new ideas, but in escaping from the old ones." Much of the high-level debate about partnerships in technology policy assumes away the real difficulties of implementing such approaches, particularly where partnerships are laid on top of existing systems and organizational arrangements. Yet, at the same time, the "grounded" case study literature on organizational partnerships — in management as well as public administration — is replete with as many stories of failure as success. Interestingly, the management literature emphasizes that the most successful business partnerships and alliances are those where there is a strategic commitment to learning (see, for example, Burgelman et al., 1996, p. 593–595). In the world of public–private partnerships, learning is also expressed as a major goal. As the MEP cases indicated, there is learning in these partnership over time, and improvements occur. However, it is apparent that other goals related to economic rationalization and political expectations constrain the commitment to learning too, leading to less than the strategic outcomes originally anticipated (no doubt, similar processes are at work in the business world too).

In this context, the extent, the identification and diffusion of "best practices" in partnership operation is a very critical element — since this process has the potential to shorten aggregate learning cycles, by transferring successful experiences, methods, and tools between partnerships and locations. Promoting the transfer of best practice is invariably not straightforward and sometimes not possible because of the particular tensions, constraints, and circumstances found in individual partnerships. Nonetheless, best practice transfer is one of the important practical steps that can be taken — with the program design lesson that mechanisms and funds for such activities should be built in when new partnership initiatives are designed. Indeed, it has been possible to identify a set of best practices that appear to influence positive outcomes from MEP partnerships (Shapira and Youtie, 1998a; Westra, 1997; Yin et al., 1998).

Recognition should also be paid to the tradeoffs that are inherent as partnership stakeholders multiply. For example, it has been suggested that the ATP should be more closely tied to states and to regional goals (Hill, 1998). The payoff is seen as a higher local profile for the ATP that might then be translated into greater support in Washington, as well as the potential to garner state and local funding matches. Yet, as the MEP case illustrates, this would also lead to a changes and conflicts in ATP goals, as state desires for local economic development rose in prominence against national objectives for industry-wide technological commercialization.

Overall, it is to be recognized that US technology policy has progressed in substantive ways compared with a decade or so ago. The MEP surely illustrates this. Whereas the US once lagged most other industrialized economies in the scale and scope of its industrial modernization efforts, today there is a nationwide system in place that is comparable (and in several areas more innovative) than found elsewhere. The convergence of a new emphasis on civilian technology diffusion with the reorientation of governmental implementation towards decentralized public–private partnerships has paved the way for the establishment of the MEP. Such public–private partnerships by themselves do not "waive away" longstanding and broader tensions in the relationships between the public and the private, between technological innovation and diffusion, between competing organizations, and between different elements of a federal political system. But if implemented with realistic program design and effective management, they do offer a renewed framework to elucidate and address these issues, and to explore opportunities for improvement.

Acknowledgements

Special acknowledgments are due to Jan Youtie for collaborative contributions with the author to several of the underlying research projects cited in this paper. An earlier version of the paper was presented at the Conference on Civilian Technology Policy in the United States and Europe, Atlanta, 9–10 April 1999.

References

Advanced Research Projects Agency, 1993. FY93 Technology Reinvestment Project, Program Solicitation. US Department of Defense, Arlington, VA.

Berglund, D., 1998. State Funding for Cooperative Technology Programs. Batelle Memorial Institute, Columbus, OH.

Branscomb, L., Florida, R., 1998. Challenges to technology policy. In: Branscomb, L., Keller, J. (Eds.), Investing in Innovation: Creating a Research and Innovation Policy that Works. MIT Press, Cambridge, MA.

Brody, R., 1996. Effective Partnering: A Report to Congress on Federal Technology Partnerships. Office of Technology Policy, US Department of Commerce, Washington, DC.

Burgelman R., Maidique M., Wheelwright, S., 1996. Strategic Management of Technology and Innovation. Irwin, Chicago, IL.

Coburn, C., Berglund, D., 1995. Partnerships: A Compendium of State and Federal Cooperative Technology Programs. Batelle Memorial Institute, Columbus, OH.

Combes, R., 1992. The Origins of Industrial Extension: A Historical Case Study. Master's Thesis, School of Public Policy, Georgia Institute of Technology, Atlanta, GA.

Cooke, P., Morgan, K., 1998. The Associational Economy: Firms, Regions, and Innovation. Oxford University Press, Oxford.

Crow, M., 1994. Science and technology policy in the US: trading in the 1950 model, Science and Public Policy, August.

Etzkowitz, H., Leydesdorff, L., 1999. Symposium: the triple helix of university–industry–government relations. The Journal of Technology Transfer 24 (2/3), 111–254.

Executive Office of the President, 1993. Technology for America's Economic Growth: A New Direction to Build Economic Strength. The White House, Washington, DC, February.

Fountain, J., 1998. Social capital. In: Branscomb, L., Keller, J. (Eds.), Investing in Innovation: Creating a Research and Innovation Policy that Works. MIT Press, Cambridge, MA.

Freeman, C., Soete, L., 1997. The Economics of Industrial Innovation, 3rd Edition. MIT Press, Cambridge, MA.

Galli, R., Teubal, M., 1997. Paradigmatic shifts in national innovation systems. In: Edquist, C. (Ed.), Systems of Innovation: Technologies, Institutions, and Organizations. Pinter, London.

Hill, C., 1998. The Advanced Technology Program. In: Branscomb, L., Keller, J. (Eds.), Investing in Innovation: Creating a Research and Innovation Policy that Works. MIT Press, Cambridge, MA.

Jarmin, R., 1999. Evaluating the impact of manufacturing extension on productivity growth. Journal of Policy Analysis and Management 18 (1), 99–119.

Luria, D., 1997. Toward Lean or Rich? What Performance Benchmarking Tells us About SME Performance, and Some Implications for Extension Center Services and Mission. In Shapira, P., Youtie, J. (Eds.), Proceedings of Third Workshop on the Evaluation of Industrial Modernization Programs on Manufacturing Modernization: Learning from Evaluation Practices and Results. Georgia Institute of Technology, Atlanta, GA.

Morris-Suzuki, T., 1994. The Technological Transformation of Japan. Cambridge University Press, Cambridge, UK.

National Institute of Standards and Technology, 1994. Technology Reinvestment Project Deployment Activity Areas: Lessons Learned. In: Materials presented at workshops in Oakland, CA and Atlanta, GA, 28 March, 1994. Manufacturing Extension Partnership, Gaithersburg, MD.

National Research Council, 1993. Learning to Change: Opportunities to Improve the Performance of Smaller Manufacturers. Commission on Engineering and Technical Systems, Manufacturing Studies Board. National Academy Press, Washington, DC.

Office of Technology Assessment, 1990. Making Things Better: Competing in Manufacturing. USGPO, Washington, DC.

Oldsman, E., 1996. Does manufacturing extension matter? An evaluation of the industrial technology extension service in New York. Research Policy 25 (2), 215–232.

Oldsman, E., Heye, C., 1998. Waste not, want not — a look at the impact of manufacturing extension centers. Journal of Technology Transfer 23 (1), 37–42.

Oldsman, E., Luria, D., 1998. Integrating strategy and evaluation in the manufacturing extension partnership. In: Shapira, P., Youtie, J. (Eds.), Manufacturing Modernization: Implications of Evaluation Results for Program Improvement and Policy Development. School of Public Policy, Georgia Institute of Technology, Atlanta, GA.

Organisation for Economic Cooperation and Development, 1998. Public/private partnerships in science and technology: an overview, STI Review, p. 23.

Osborne, D., 1988. Laboratories of Democracy: A New Breed of Governor Creates Models for National Growth. Harvard Business School Press, Boston, MA.

Osborne, D., Gaebler, T., 1993. Reinventing Government: How the Entrepreneurial Spirit is Transforming the Public Sector. Plume, New York.

Shapira, P., 1990. Modern times: learning from new state initiatives in industrial extension and technology transfer. Economic Development Quarterly 4 (3), 186–202.

Shapira, P., 2000. Evaluating manufacturing extension services in the United States: experiences and insights. In: Paper presented at the US–European Workshop on Learning from Science and Technology Policy Evaluation Bad Herrenalb, Germany, 11–14 September 2000.

Shapira, P., Youtie, J., 1998a. Manufacturing Partnerships: Coordinating Industrial Modernization Services. Phase II Final Report. Georgia Institute of Technology, School of Public Policy and Economic Development Institute, Atlanta, GA.

Shapira, P., Youtie, J., 1998b. Evaluation industrial modernization: methods, results, and insights from the Georgia manufacturing extension alliance. The Journal of Technology Transfer 23 (1), 17–28.

Shapira, P., Youtie, J., Kingsley, G., Cummings, M., 1996a. Coordinating Industrial Modernization Services: Impacts and Insights from the US Manufacturing Extension Partnership. Georgia Institute of Technology, School of Public Policy and Economic Development Institute, Atlanta, GA.

Shapira, P., Youtie, J., Roessner, D., 1996b. Current practices in the evaluation of US industrial modernization programs. Research Policy 25 (2), 185–214.

Shapira, P., Kingsley, G., Youtie, J., 1997. Manufacturing partnerships: evaluation in the context of government reform. Evaluation and Program Planning 2 (1), 103–112.

State Science and Technology Institute, 1998. Survey of State Research and Development Expenditures: Fiscal Year 1995. State Science and Technology Institute, Westerville, OH.

Tassey, G., 1992. Technology Infrastructure and Competitive Position. Kluwer, Norwell, MA.

US Department of Commerce, 1999a. Solutions for Manufacturers: Manufacturing Extension Partnership. National Institute of Standards and Technology, US Department of Commerce, Gaithersburg, MD.

US Department of Commerce, 1999b. Annual Performance Plan, Fiscal Year 2000. Technology Administration, US Department of Commerce, Washington, DC.

US Department of Commerce, 2000. More Transformed Firms: Case Studies. Manufacturing Extension Partnership. National Institute of Standards and Technology, US Department of Commerce, Gaithersburg, MD.

US General Accounting Office, 1995. Manufacturing Extension Programs: Manufacturers' Views of Services, GAO/GGD-95-21-216BR. GAO, Washington, DC.

Voytek, K., 2000. Where are we and where are we headed? In: Presentation by NIST MEP Evaluation Director at 2000 Manufacturing Extension Partnership National Conference, 9 May 2000, Orlando, FL.

Westra, R., 1997. Partnership Improvement Project: Advanced Technology Centers and Manufacturing Extension Centers. Center for Occupational Research and Development, Waco, TX.

Yin, R., Merchlinksy S., Adams-Kennedy, K., 1998. Evaluation of MEP-SBDC Partnerships. Report prepared by Cosmos Corporation for the Manufacturing Extension Partnership, National Institute of Standards and Technology, Gaithersburg, MD.

Youtie, J., 1997. Toward a cross-case analysis of outcomes of exemplary engagements by MEP centers. In: Shapira, P., Youtie, J. (Eds.), Proceedings of Third Workshop on the Evaluation of Industrial Modernization Programs on Manufacturing Modernization: Learning from Evaluation Practices and Results. Georgia Institute of Technology, Atlanta, GA.

Youtie, J., Shapira, P., 1997. Tracking customer progress: a follow-up study of customers of the Georgia manufacturing extension alliance. Journal of Technology Transfer 22 (2), 43–52.

[17]

The Effectiveness of Government Promotion of Advanced Manufacturing Technologies (AMT): An Economic Analysis Based on Swiss Micro Data

Spyros Arvanitis
Heinz Hollenstein
Stephan Lenz

ABSTRACT. This paper describes the approach applied and the results of an evaluation of the effects of the Swiss government programme to promote the diffusion of Advanced Manufacturing Technologies (AMT) from 1990 to 1996. The method used is based on specifying and econometrically estimating with firm data simultaneously an *adoption equation* which, besides the main explanatory variables as proposed by the theory of technology diffusion, includes a policy variable, and a *policy equation* with a set of firm characteristics and an adoption variable as regressors. The results are consistent with a positive impact of promotion on adoption of AMT, particularly a more intensive adoption of AMT for firms which did not use AMT when the programme started.

1. Introduction

The importance of innovation and technology policy has considerably grown for most OECD countries in the last years. Such a policy covers a wide spectrum of activities, ranging from the direct support of basic reseach to more indirect measures aimed at improving the capability of firms to innovate and to use new technologies. Recently, new developments in technology policy with an increased emphasis on technology diffusion and adoption, organizational change and innovative behaviour have raised new methodological challenges for the evaluation of these policies (see e.g. OECD, 1997a; Georghiou and Roessner, 2000). Accordingly, there is an increased interest in OECD countries in the issue of evaluation of government programmes and policies. Evaluation issues are central to improving the effectiveness of policy and formulating "best policy practices" (see OECD, 1997b, 1998).

The main feature of Swiss technology policy is the low weight it places on direct measures for fostering innovation in the economy. It is primarily oriented towards creating a favourable environment for the introduction of new products and production techniques, whether such innovations rely on firm-internal research and development or on the adoption of novelties generated by other firms or institutions. This framework-oriented policy is supplemented by a number of specific measures to stimulate rapid diffusion of selected basic technologies which are considered to be relevant for a broad spectrum of industrial activities. A typical example of the last approach is a programme to promote the use of "Advanced Manufacturing Technologies" (AMT) launched in 1990 and ended in 1996. It offered the firms information and training services as well as subsidies for consultancy and development projects; the latter were in most cases based on joint ventures between firms or between firms and research institutions embedded in regional networks. The concept of this type of policy measures is to strengthen the firms' ability to undertake the techno-organizational adjustments necessary for successfully (and rapidly) adopting the new technology which, in the case of AMT, poses a challenge for many (smaller) firms of most industries.

This paper presents the approach applied and the results of an evaluation of the effects of this programme which is described in somewhat more detail in section 4. The method used is based on

Final version accepted on August 15, 2001

Swiss Federal Institute of Technology Zurich
Weinbergstrasse 35
ETH Zentrum
CH-8092 Zurich
E-mail: arvanitis@kof.gess.ethz.ch

Small Business Economics **19**: 321–340, 2002.

specifying and econometrically estimating with firm data simultaneously an *adoption equation* which, besides the main explanatory variables as proposed by the theory of technology diffusion, includes policy variables which discriminate government supported from non-supported firms, and a *policy equation* which contains a vector of factors influencing the selection procedure of supported firms (or, the other way around, the decision of firms to participate to the government programmes); the policy equation takes also account of firms' prior experience in the supported technology, thus including the dependent variable of the adoption equation as an additional regressor. The analysis is based on data of 463 firms (96 supported and 367 non-supported firms). Our results suggest that promotion did lead to a more intensive adoption of AMT especially for firms which did not use AMT when the programme started.

In section 2 we characterize our approach and point to some differences to other methods of evaluation. In section 3 we introduce the (theoretical) model of technology adoption underlying our investigation. In section 4 we describe the data base and some features of the diffusion process of AMT and the firms' use of the policy measures to be evaluated. In sections 5 the specification of the empirical model of adoption and econometric estimates are presented. The empirical confirmation of the adoption model is a necessary condition for a successful application of our model-based evaluation approach. Section 6 contains the specification of the policy equation and the corresponding econometric estimates. Section 7 is devoted to the policy evaluation in the narrow sense. We describe and discuss the results of the simultaneous probit estimation of the adoption and policy equation, each of them also including the dependent variable of the other one. We conclude by an assessment of the proposed procedure and some recommendations for evaluating specific policy measures.

2. Evaluation concept and econometric implementation

The majority of evaluations of public support for technology diffusion in Switzerland is primarily directed to the efficiency of such programmes in a rather narrow sense. The topics covered by such evaluations are typically the following: Is the target group well informed on the support measures available? Do firms to which policy is targeted take part to a satisfactory degree? Are management and procedures of a programme efficient? What are the motives for participating? Are barriers of diffusion as perceived by firms sufficiently addressed by the policy measures? etc.[1] This type of evaluation, though useful, does not yield an assessment of the economic effects of diffusion-oriented measures, because it concentrates on "programme immanent" performance measures and does not take account of firms not participating (no control group analysis).[2] Recently, an evaluation methodology based on econometric models with firm-level data has been proposed and partly applied on specific problems such as public support to business R&D, manufacturing technology centres, adoption of new technlogy, etc. (see e.g. Capron and Van Pottelberghe de la Potterie, 1997; Jarmin, 1998; Geyer et al., 2000). Our evaluation concept is based on this type of methodology.

The present investigation is focussed on the results of policy intervention. More specifically, it is asked whether the primary goal of public support, i.e. a more rapid and broader diffusion of AMT compared to firms standing aside (i.e. the control group), is actually attained. However, we do not investigate the impact of the government progamme on firms' performance per se, which would imply a different analysis framework as the one used in this paper.

The envisaged type of evaluation requires an econometric analysis based on the theory of technology diffusion using micro-level firm data for a sample which contains supported firms as well as not-supported ones. More specifically, we estimate equations of technology adoption based on general factors determining the use of new production techniques as well as policy variables to identify the marginal effect of policy intervention.

Assumming that the goal variable (e.g. adoption of a certain technology) of a firm i in period t denoted $Y_{i,t}$ is determined by a vector of theory-based variables $X_{i,t}$ (e.g. diffusion theory) the evaluation procedure may be presented formally by

$$Y_{i,t} = \alpha_0 + \beta_1 X_{i,t} + \beta_2 P_i + \textit{control variables} + e_{i,t} \tag{1}$$

where P_i is a vector of policy instruments with takes some non-zero value for programme participants receiving policy support and zero for non-participants, control variables for the sectors or industries the firms belong to and $e_{i,t}$ is a stochastic term.

In equations of type (1) the impact of policy promotion is measured directly and can be interpreted straightforwardly based on the signs of β_2. If a policy measure aims at increasing Y, a positive (and statistically significant) parameter indicates an advantage of the supported firms over the non-supported ones with respect to the goal variable, a negative one shows a disadvantage.[3] In both cases we can derive clear indications with regard to the impact of the policy measures: a positive coefficient means that policy is effective in the direction targeted by policy makers; in the case of a negative sign the influence of policy runs opposite to the policy goals bringing promoted firms away from target. Consequently, statistically insignificant coefficients hint to an ineffectiveness of the policy instruments.

If enough randomness with respect to the allocation of policy support to firms and projects exist, data for the supported firms as well as for similar non-supported firms would provide the evaluator with a basis for causal, econometric analysis, because under these circumstances the statistical preconditions for this type of analysis (hypotheses on type of the distribution functions of residuals, etc.) would be fulfilled. However, given the many factors involved in the process of political decision-making that determines the allocation of policy support, random allocation seems a rather unrealistic assumption; in this case the evaluation results may be biased due to self-selection (see e.g. Maddala, 1983). Any evaluation procedure will have to take this problem explicitly into account.

In case that only *cross-section data* are available, the more widely-used selection-correction method introduced by Heckman (1979) can be applied, which is a *two-equation approach* built up on an equation of type (1) above and a selection equation. Based on assumptions on the joint distribution of the residuals of these two equations, the coefficient β_2 in (1) can be estimated by a simultaneous or a sequential approach. In this study we choose a *two-equation framework* by adding to the basic evaluation equation (1) a second equation explaining the policy variable P_i.

A particular advantage of this approach is that important features of the political economy of firm participation to government programmes can be explicitly taken into consideration in the evaluation procedure. However, there is no standard theory explaining whether or not a firm obtains support, because the selection procedure is mostly a specific one, depending not only on the priorities of government agencies with respect to some categories of firms to be especially supported (e.g. small firms, firms belonging to hightech sectors, etc.) but, among other things, also on the earlier experience of firms with other support programmes as well as with the technologies which are promoted by a certain programme (see section 6).

In this study two equations were estimated simultaneously, one for technology adoption A (goal variable) and a second one for government support P (policy equation):

$$A_i = \alpha_0 + \beta_1 X_i + \beta_{2,i} P_i + \textit{control variables} + e_{i,t} \qquad (2a)$$

$$P_i = \gamma_0 + \gamma_1 Z_i + A_i + u_i \qquad (2b)$$

where Z_i is a vector of factors influencing the selection procedure of supported firms or, the other way around, the decision of firms to participate to the government programmes. Further, it is reasonable to assume that firms with prior experience in the technology supported by policy measures could be more eager to participate to the technology programme than firms without such experience; thus, the second equation also contains the goal variable as regressor.

For this special version of a simultaneous probit model the estimation method was based on a "mean- and covariance-structure model" (Browne and Arminger, 1995) and was implemented in the software programme MECOSA 3 (Arminger, 1995). A two-step procedure (first, estimation of the coefficients of the reduced form of the original simultaneous equation system, then estimation of the structural parameters based on the covariance matrix of the reduced form coefficients) was applied to estimate the simultaneous probit model according to the algorithm implemented in the above-mentioned software.

3. A model of technology adoption

The objective of this section is to formulate an equation explaining the decision to adopt some element(s) of AMT based on a set of mainly firm-specific factors determining the profitability of new technology. Our theoretical approach is based on the general framework proposed by Karshenas and Stoneman (1995). According to these authors the main (neoclassical) models of diffusion may be categorized in four groups (epidemic, rank, stock and order model) which can be fused to a "higher order model" containing the specific features of the differrent approaches.[4] Within this general conceptual framework our approach belongs rather to the category of "rank models" emphasizing the heterogeneity of firms as determinant of inter-firm diffusion patterns. In this view it is assumed that potential users of a new technology differ from each other in important dimensions so that, at a certain point in time, some firms obtain a greater return from new technology than others do. The larger the net advantages resulting from the technology adoption, the stronger the tendency to adopt this technology earlier and more intensively.

We distinguish several groups of factors which potentially influence (positively or negatively) a firm's profitability from adopting new technology and therefore the decision to introduce it at a certain time. A first one includes a set of anticipated benefits of new technology such as savings of inputs, general cost reductions, higher flexibility, improvement of product quality, etc. (see e.g. Clark, 1987 or Milgrom and Roberts, 1990 for theoretical treatments of this aspect). The new technique may save labour or some specific labour skills; it may reduce capital needs, for example, through increased utilization of equipment, reduction of inventories or space requirements, etc. It may also lead to higher product quality or better conditions for product development. Moreover, it may increase the flexibility of the production process allowing the exploitation of economies of scope. For this group of variables we expect a positive influence on the adoption decision (i.e. early and intensive use of new technology is favoured).

A second category of variables refers to anticipated barriers to the adoption of new technology (see e.g. Cainarca et al., 1990). We identify four main types of such hindrances: financial (e.g. general liquidity constraints, cost of new technology and manpower training) and human capital restrictions (e.g. lack of engineers and technicians); informational and know-how barriers (e.g. uncertainty with respect to the performance of new technology or the utilization of new production capacity); organizational and managerial barriers (e.g. resistance to new technology within the firm; insufficient attention of the management) and sunk costs barriers; the latter refer to the high substitution costs firms have to incur in order to introduce the new technology, for example, in case of insufficient compatibility of new technology with existing equipment, existing organization or existing product portfolio (see e.g. Link and Kapur, 1994). We expect such barriers to be negatively related to adoption.

A third category of explanatory variables is related to the (product) market conditions under which the firms are operating, particularly the competitive pressures they are exposed to. Mostly, market concentration, a structural variable, is taken to reflect competitive pressures. In the game-theoretic literature the impact of market structure upon the schedule of adoption dates is shown to depend critically on the difference of profit rates preceding and following adoption (see e.g. Reinganum, 1981). This dependence being quite complicated, most studies do come to theoretically ambiguous results with respect to the effects of market concentration on adoption (see Reinganum, 1989 for a review of this literature). Thus, whether positive effects in the tradition of Schumpeter are stronger than negative "free competition effects" has to be resolved at the empirical level. Therefore, we make no prediction concerning the sign of the concentration effect. Another line of thought argues that it is the elasticity of demand faced by a firm in its specific market that induces innovative or imitative activity (see Kamien and Schwartz, 1970 for the original argument). In those markets where competition pressure is greater, demand elasticities can be expected to be higher because of the existence of close substitutes, thus driving firms to innovative activity or rapid new technology adoption (see e.g. Majumdar and Venkataraman, 1993). In accordance to this line of reasoning, we have proxied in an earlier study on the determinants of process innovation competitive pressures through the intensity of price and non-price competition at the product market and postulated a positive relationship to innovative activity (see Arvanitis and Hollenstein, 1994); we apply the same argument for new technology adoption.

A further group of variables serves to characterize a firm's products and production technique. The main idea is that product type and existing production technique of a firm play an important role in determining extent and limits of the use of AMT (see e.g. Taymaz, 1991; Evers et al., 1990). On the product side, it is expected that the benefits from applying AMT in producing differentiated goods would be rather small as compared to standardized ones; therefore, a negative impact of (the degree of) product differentiation is expected. With respect to production technology a positive relationship between the "length of production run" and adoption is postulated, because AMT can fully unfold its potential primarily in plants with large-batch or mass production ("scale effects"). The development of more sophisticated AMT during the last decade, however, favoured production flexibility, thus enabling firms to adapt quickly and efficiently to external or internal changes. This "flexibility effect" works in the opposite direction to the (traditional) "scale effect" thus weakening the overall influence of variables characterizing the type of product and process technology on AMT adoption.

The firm's ability to absorb knowledge from external sources and exploit it for its own innovative activities is a major determinant of innovation performance in general and of technology adoption in particular (see Cohen and Levinthal, 1989 or, specifically for the case of AMT adoption, Baldwin and Rafiquzzaman, 1998). We consider two important aspects of a firm's absorptive capacity: firstly, the firm's (overall) ability to assess technological opportunities in (or around) its fields of activity which depends primarily of the endowment with human and knowledge capital, and, secondly, the embedding of the firm in knowledge networks facilitating access to information and resources relevant for technology adoption. Both elements of absorptive capacity should be positively related to (early and intensive) use of AMT.

Finally, firm size, an explanatory variable used in most studies of adoption behaviour (see Karshenas and Stoneman, 1995), is also included in the present study; it is expected to be positively related to adoption.

4. Database and some characteristics of adoption and promotion of AMT

The analysis is based on firm data collected in the course of the Swiss Innovation Survey 1996 as a supplement to the standard questionnaire, thus allowing the combination of AMT-specific information with basic data on innovation and technology use. The available variables are to a large extent qualitative in nature, i.e. categorical or ordinal measures (mostly on a five-point Likert scale). As far as AMT is concerned the questionnaire yields data on the time profile of the introduction of nineteen AMT-elements, the linking of these technologies, the assessment of a whole series of objectives pursued by introducing AMT as well as the significance of factors impeding its application, the impact of AMT on competitiveness, employment requirements and organizational structure and, finally, information on government promotion of AMT.[5]

The survey has been addressed to manufacturing firms based on a sample stratified by industry (17 2-digit industries) and firm size (3 industry-specific size classes) with full coverage of the upper size class in each industry. Additionally, the questionnaire was sent out to firms not belonging to this panel which had participated to the programme of AMT promotion. The response rate has been about 34% with a somewhat higher percentage for non-panel firms. 80% of the respondents used at least one AMT element in 1996 with a median AMT-intensity of seven technologies and 20% have got promoted by one or more element(s) of the government support programme. The final data set used in the econometric estimations contains 463 firms (all of them already using AMT in 1996 or planning to use it up to 1999), fairly representative for the manufacturing industries and firm size classes in the original sample (see Table A.1 in appendix I).

Owing to limited resources we could not perform a non-response analysis, so we cannot exclude that some kind of selectivity bias as to the adoption behaviour of the responding firms may exist in our data. However, we minimized the risk of being confronted with this kind of bias by building our empirical analysis not on the (presumably biased) information on adopting AMT in general, but on the specific use of some AMT elements for firms already being an adopter of AMT (see section 5).

Table I contains some information on the adoption rates of 19 elements of AMT in Swiss manufacturing since 1990 (including planned application of these technologies till 1999 as assessed in fall 1996 when the data was collected). The adoption rate (percentage of firms using a certain technology) in 1999 and the diffusion velocity (percentage increase of the share of firms using a certain technology element in the period 1990–1999) varies quite strongly among the technology elements listed in Table I. For example, the diffusion of CNC machines being already an "old" technology (the first use in our sample dates back to 1968) was quite high already in 1990 and changed "only" by about 35% between 1990 and 1999. At the other extreme, "new" technologies such as "simulation, rapid prototyping", FMS or inter-company computer networks (ICCN) were used only by a small fraction of firms in 1990 but this share quadrupled until 1999 in the first two cases and increased by factor 14 in the case of ICCN.

In the late eighties it was a widespread view among industrialists and policy makers in Switzerland that the adoption of AMT was slower in Swiss industry than in many other countries

TABLE I
Adoption of Advanced Manufacturing Technologies (AMT) (Percentage of manufacturing firms; 1999: planned adoption)

Technology element	1990	1993	1996	1999
Design				
Computer-aided design and/or engineering (CAD/CAE)	37.3	50.6	57.0	60.8
Computer-aided design/manufacturing (CAD/CAM)	23.3	34.4	42.5	47.8
Simulation, rapid prototyping	2.6	5.0	7.7	11.4
Planning				
Digital firm data recording	32.2	40.1	48.7	64.2
Computer-aided (manufacturing) planning (CAP)	32.9	40.8	50.7	62.9
Fabrication				
Computer numerically controlled machines (CNC/DNC)	40.6	48.9	52.8	55.0
Materials working lasers	5.7	8.8	10.7	13.1
Pick-and-place robots	14.2	19.9	23.2	26.7
Complex robots	5.3	8.1	11.8	13.4
Flexible manufacturing cells (FMC)	6.1	10.1	13.8	20.0
Flexible manufacturing systems (FMS)	4.2	8.6	13.1	18.8
Handling				
Automated storage and retrieval systems (AS/RS)	17.8	25.0	32.2	41.5
Transport systems (AGVS)	6.4	10.8	13.2	16.5
Quality control				
Quality control (CAQ) on material	14.5	22.4	28.7	41.2
Quality control (CAQ) on final products	14.2	21.9	30.0	42.8
Communication				
Local area network (LAN) for technical data	16.2	25.9	36.2	47.1
Local area network (LAN) for factory use	13.6	22.2	31.8	44.1
Production planning systems (PPS)	31.8	41.9	54.8	67.3
Inter-company computer networks (ICCN)	2.6	6.6	15.8	37.3

(especially as far as SMEs were concerned). The experts preparing the promotion programme identified serious knowledge deficits with respect to the potential of AMT, the use of these technologies in the narrow sense as well as its interaction with the organization of production (see CIMEX, 1988). Therefore, it is not surprising that deficiencies with respect to (highly) qualified manpower was seen as the crucial bottleneck hampering the diffusion of AMT. In addition, the experts were convinced that part of the problem was the weak interaction between polytechnics and firms, particularly SMEs. Based on this appraisal following guidelines were formulated for the promotion programme: first, the programme had to take into account the interrelatedness of technical, organizational and human (knowledge) aspects of the adoption of AMT; second, promotion should take place at the regional level to make use of local spillover potentials[6] and lead in the long run to an increase of networking among firms and among firms and research/education institutions; thirdly, the programme should only help to overcome bottlenecks in the early phase of adoption, therefore it was to avoid that expectations of permanent subsidizing would arise.

In order to operationalize its concept the government agency in charge initiated seven regional AMT centres which started to work in 1990. Usually a (regional) polytechnical institute (or a network of such institutes) was the nucleus of such a centre to which different types of actors (leading companies, experts, consulting firms, etc.) were connected. The programme consisted of three types of measures which were either directly delivered or arranged by the AMT centres: first, information and training ranging from "one day information" up to "two years full-time training courses"; second, consulting aiming either at supporting the realization of a specific AMT project or – more often – at preparing the introduction of AMT in a firm by means of an analysis of its specific needs and potentials for this technology which led to the formulation of an "adoption plan" integrating technical, organizational and training aspects. Third, the programme subsidized also development projects in the field of AMT typically based on joint-ventures of firms with polytechnical institutes (where the firms had to bear at least 50% of the project costs). The programme ended in 1996 with the majority of the AMT centres being integrated in (groups of) polytechnics.

Table II contains some information on the extent of government support of AMT in the period 1990–1996 by industry and firm size.

TABLE II
Government promotion of AMT

	All firms	Government-supported firms		Type of government support		
				T	C	R&D
	N	N	%	(% of supported firms)		
Total manufacturing	463	96	20.7	37.5	43.1	60.2
Groups of industries						
Metalworking	99	23	23.2	43.5	62.5	43.5
Machinery/vehicles	101	28	27.7	25.8	34.5	76.7
Electr.machinery/electronics	95	25	26.3	32.1	25.0	71.4
Other industries	168	20	11.9	54.6	57.1	40.9
Firm size (number of employees)						
Less than 50	83	22	26.5	18.5	53.9	63.0
50–99	87	13	14.9	50.0	61.5	66.7
100–199	127	20	15.8	55.0	57.9	45.0
200–499	100	20	20.0	55.0	31.6	31.6
500 and more	66	21	31.8	24.0	20.0	88.0

Note: T: training; C: consulting; R&D: R&D projects.

About 20% of the firms in the final data set have been supported by some element of the AMT promotion programme. Subsidizing of R&D projects connected with the introduction and/or extension of AMT has been the type of support most frequently applied for (60% of the firms); training and consulting services were claimed by about two fifths of the firms respectively. There are considerable differences among industries and firm size classes with respect to promotion frequency and mode of promotion. Not surprisingly, mechanical engineering/vehicles and electrical engineering/electronics having a very high potential for AMT use benefited most from government support. For these industries promotion was concentrated primarily in R&D projects, whereas in metalworking and other industries the support was focussed on consulting and, somewhat lesser, on training schemes. Very small (less than 50 employees) and large firms (more than 500 employees) have received AMT support more frequently than medium-sized firms. There is also a close relationship between mode of promotion and firm size: small and large firms received more-than-average support in R&D, medium-sized ones in training; firms with up to 200 employees were stronger promoted through the offering of consulting services.

5. Specification and econometric estimation of the adoption model

The model

Throughout the paper we use an ordinal measure of the change of the AMT intensity between 1990 and 1996 (DAMTINT) as adoption variable. Thus, we implicitly presume that government promotion became effective already during the first year. The AMT intensity is measured as the number of technology elements (out of a list of 19 such elements in Table I) in use in a firm in a certain year. We constructed a three-level ordinal variable which contains the following categories: no change or change by 1 AMT element (29.9% of the firms which had adopted at least 1 AMT element till 1996); change by 2 to 4 elements (45.2% of the adopting firms); change by more than 4 elements (24.9% of the adopting firms). A sensitivity analysis was conducted with a two-level and a four-level ordinal variable; both of them yielded virtually the same results with respect to the explanatory variables as the three-level variable. We chose the three-level variable on grounds of a better statistical fitting. By defining the adoption variable as a measure of intensity change we avoid to some extent identification problems which arise, first, because we do not know which specific technology is supported in every single case of promotion and, secondly, we do not dispose of data for the explanatory variables which are differentiated by technology and time.

In concordance with the theoretical discussion in section 3 we distinguish several groups of explanatory variables (besides one adoption-specific control variable and four industry dummies; see Table III). A first group of determining factors refers to objectives of and motives pursued by the firms for the adoption of AMT which we interpret as proxies for anticipated revenue increases due to the use of new technology.[7] The six metric variables listed in Table III under the heading "objectives/motives" are the factor scores resulting from a principal component factor analysis of 26 single objectives of AMT included in the questionnaire (see Table A.2 in appendix I). Two of these factors are related primarily to expected product improvements (QUAL, DEV) covering a wide spectrum of possible gains from AMT use such as higher product quality and variety, introduction of "intelligent" products, higher flexibility at the market as well as improved conditions for the development of new products. Another two variables seem to be connected more closely to expected changes in the production technique: various sources of productivity gains related to the introduction of AMT as mentioned in technical literature are reflected by the variables COST (labour saving; capital saving, i.e. reduction of inventories and space requirements, higher utilization of equipment, shorter production time, etc.) and FLEX (higher flexibility of work organization, flexibility within the firm in general, etc.). A fifth factor (BEST) refers to anticipated revenue potentials as a result of securing technological competitiveness (technology lead, keeping to "best practice", etc.). For these five variables covering several aspects of potential benefits arising from applying AMT we expect to find a positive influence on the adoption variables. The

TABLE III
Specification of the adoption model

Variable	Description	Sign
Dependent variable		
DAMTINT	Change of the AMT intensity (i.e. number of AMT elements used) in the period 1990–1996	
Independent variables		
1. *Objectives of/motives for the adoption of AMT*		
(Scores of a principal component factor analysis of 26 objectives of AMT; six factors)		
FINCOMP	Favourable financial conditions; competitive pressure	?
COST	Cost reduction	+
FLEX	Higher flexibility	+
DEV	Improving product development	+
QUAL	Better product quality	+
BEST	Securing technological lead/"best practice"	+
2. *Impediments to the adoption of AMT*		
(Scores of a principal component factor analysis of 26 barriers to AMT; six factors)		
TECH	High technological costs/uncertainties	–
KNOWPERS	Lack of knowledge/lack of adequately qualified personnel	–
RESIST	Resistance to new technology within the firm	–
INVCOST	High investment costs	–
UTILIZ	Uncertainty with respect to capacity utilization	–
COMPAT	Compatibility problems (e.g. with installed machinery, etc)	–
3. *Market conditions*		
IPC	Intensity of price competition in the product market (five-point Likert scale)	+
INPC	Intensity of non-price competition in the product market (five-point Likert scale)	+
CONC16–50	Three dummy variables for market concentration based on the	?
CONC11–15	number of principal competitors in the world (product) market	?
CONC1–10	(16 to 50, 11 to 15, 1 to 10 competitors; firms with more than 50 competitors as reference group)	?
4. *Type of production technique/products*		
Product characteristics (dummy variables with 'standardized products' as reference group)		
PDMARKET	Product differentiation	–
PDUSER	Products according to user specifications	–
Process characteristics (dummy variables with 'single-piece production' as reference group)		
SBATCH	Small-batch production	+
LBATCH	Medium-batch/large-batch production	+
CONTFLOW	Continous flow/mass production	+
5. *Absorptive capacity*		
HUMCAP	Percentage share of highly qualified employees	+
COOP	Cooperation in R&D activities (dummy variable)	+
6. *Firm size*		
L, L^2	Number of employees and its square	+/?
7. *Control variables*		
INT90	Intensity of AMT in 1990 (starting year of the programme)	?
Industry Dummies	Metalworking, machinery, electrical machinery/electronics, chemicals/plastics (with "other industries" as reference group)	?

sixth factor (FINCOMP) has a somewhat different character and cannot be linked directly to revenue increases. It contains elements of a "defensive" adoption strategy (making use of favourable conditions for financing AMT introduction; competition pressure manifested through declining market share; etc.). It is not obvious whether these motives lead to early or late technology adoption; thus, the sign of the coefficient of this variable is a priori undetermined.

A second group of variables is related to factors impeding AMT adoption. Again six metric variables were constructed by a principal component factor analysis of 26 impediments to adoption (measured on a five-point Likert scale), which can be interpreted as proxies for several types of anticipated adjustment costs (related to the introduction of AMT) probably leading to late (less intensive) adoption (see Table A.2 in appendix II). Thus, we expect a negative sign for these variables covering all categories of obstacles identified in section 3: INVCOST and KNOWPERS reflect costs/financial barriers and risks and impediments due to the lack of qualified personnel respectively, TECH and UTILIZ mostly informational and know-how barriers, RESIST organizational and managerial weaknesses, and COMPAT stands for obstacles related to sunk costs (several types of incompatibility).

A third category of explanatory variables represents conditions on the firm-specific product markets: intensity of price and non-price competition (IPC and INPC; measured on a five-point Likert scale); market concentration (CONC; three dummies for firms confronted with markets with varying numbers of competitors). We expect a positive sign for the two variables measuring the intensity of competition, whereas the influence of the market concentration can be positive or negative.

Next we hypothezise that the degree of product differentiation and the "length of production run" exert an influence on adoption, which should be negative in the first case and positive in the latter. Product differentiation is measured by two dummies (0,1) representing the production of customer-specific (PDUSER) or otherwise differentiated products (PDMARKET) with standardized products as reference group. The "length of production run" is proxied by three dummies (0,1) which stand for small-batch (SBATCH) and large-batch (LBATCH) as well as mass production or continous flow production (CONTFLOW) with single-piece production as referende group. We already mentioned that the flexibility potential of more recent vintages of AMT could weaken the overall impact of these two groups of variables.

The firm's ability to absorb knowledge from external sources, which we expect to be positively related to early/intensive adoption, is measured by two variables: HUMCAP, the share of employees with qualifications at the tertiary level, should capture the overall ability of the firm to assess technological opportunities of AMT, whereas COOP, a binary measure of R&D cooperation (yes/no), stands for the advantages in adopting AMT by being integrated in knowledge-related networks of innovative firms and research institutions.

Firm size (SIZE), which should be positively correlated with AMT adoption, is specified as a polynom with respect to the number of employees (linear L and quadratic term L^2; see Table III).

Finally, we included the AMT intensity in 1990, the start year of the promotion programme (INT90) and some industry dumies as control variables.

Econometric estimate

Table IV contains the ordered probit estimates of the adoption equation (DAMTINT as dependent variable) based on 463 observations (total sample) for firms having adopted at least one AMT element. The coefficients of an estimate of the full technology adoption model as specified above are listed in column 1. Column 2 contains only those coefficients which were statistically significant at the 10%-level (restricted model). A circumspection of the results in column 1 and 2 shows that the estimates are robust.[8]

The main contribution to the explanation of adoption behaviour (in addition to the control variables) comes from three groups of variables: objectives of adoption of AMT reflecting anticipated revenue increases related to the application of AMT, capacity to absorb external knowledge and firm size.

Four out of six variables representing the influence of adoption objectives yield the expected

TABLE IV
Ordered probit estimation of the adoption model; total sample

Explanatory variables	DAMTINT (1)	(2)
Objectives		
FINCOMP	–0.15** (0.06)	–0.14** (0.06)
COST	0.11* (0.06)	0.12** (0.05)
FLEX	0.22** (0.06)	0.23** (0.06)
DEV	0.06 (0.06)	
QUAL	0.09* (0.06)	0.09* (0.05)
BEST	0.20** (0.06)	0.18** (0.06)
Impediments		
TECH	0.03 (0.06)	
KNOWPERS	0.04 (0.06)	
RESIST	0.00 (0.06)	
INVCOST	–0.11* (0.06)	–0.13** (0.06)
UTILIZ	0.05 (0.06)	
COMPAT	–0.05 (0.05)	
Market conditions		
IPC	0.09 (0.06)	
INPC	–0.06 (0.06)	
CONC16–50	0.10 (0.17)	
CONC11–15	0.00 (0.15)	
CONC01–10	0.11 (0.14)	
Type of production		
PDMARKET	0.07 (0.12)	
PDUSER	–0.02 (0.14)	
SBATCH	0.05 (0.12)	
LBATCH	0.32** (0.12)	0.33** (0.11)
CONTFLOW	–0.04 (0.15)	
Absorptive capacity		
HUMCAP	0.00 (0.01)	
COOP	0.27** (0.12)	0.26* (0.12)
Firm size		
L	0.34** (0.15)	0.32** (0.14)
L^2	–0.02* (0.01)	–0.02 (0.14)
Control variables		
INT90	–0.15** (0.02)	–0.15** (0.02)
Metalworking	0.04 (0.16)	0.05 (0.16)
Machinery/vehicles	0.00 (0.18)	0.10 (0.16)
Electr. machinery/electronics	0.24 (0.19)	0.36** (0.17)
Chemicals/plastics	0.42* (0.23)	0.48** (0.22)
N	463	463
McFadden R^2	0.114	0.106
LR statistic (χ^2)	113	104
% concordance	72.1	71.3
Equal slope test (χ^2)	31	19

Note: Equation (1) contains all model variables as specified in section 5 (see Table III); equation (2) contains only the variables with statistically significant coefficients at the 10%-level (restricted model). Standard errors are included in brackets (**, * indicate statistical significance at the 5%-level and 10%-level resp.). Intercepts have been throughout omitted.

positive and statistical significant coefficients, one of them (FINCOMP) a negative one. The variables FLEX ("higher flexibility") and BEST ("remaining on top of techological improvements") seem to have the largest weight among these variables. Among the anticipated benefits those related to factor-saving or flexibility-enhancing improvements of production processes (COST, FLEX) are more important than those referring to expected gains on the product side (DEV, QUAL). We could not find any significant effect for the variables reflecting adoption impediments except for INVCOST ("high investment costs of new technology") which yielded a statistically significant negative coefficient.

In accordance with earlier results for the innovation behaviour of Swiss manufacturing firms the variables reflecting market conditions show no significant influence on adoption behaviour (see Arvanitis and Hollenstein, 1994); in the case of the variable for market concentration (CONC) this result probably reflects the countervailing effects mentioned in section 3 above.

The modes of existing production technology and existing types of products do not seem to play a major role for explaining the change of the intensity of AMT use; only firms with medium-/large-batch production (variable LBATCH) adopt, as expected, AMT more intensively than other firms. There are two contrary effects (scale economies vs. flexibilty possibilities) affecting the overall influence of these variables which seemingly counterbalance each other.

The variable COOP (dummy for R&D cooperation) used here as a proxy for the ability of a firm to absorb new technological knowledge exerts a significantly positive influence on the change of the intensity of AMT absorption; for the variable HUMCAP we find also a positive but not statistically significant effect.

As far as firm size is concerned the linear term is positive and statistically significant in both model versions. There is some evidence that adoption intensity is increasing less than proportionately with firm size but it is rather weak (negative coefficient of L^2 in equation (1)).

Finally, we obtained a negative coefficient for the control variable INT90. Moreover, we found positive and statistically significant coefficients of the dummy variable for chemicals/plastics (also for electrical machinery/electronics in the restricted model in equation (2)).

In sum, the anticipated new revenue potentials are seemingly much more relevant for the firm's decision to adopt (or intensify the use) of AMT than the costs which are associated with the introduction and adjustment of these new technologies to a firm's specific needs. A high capacity to efficiently absorb and apply new knowledge supports an earlier and/or increasing intensity adoption of AMT. Financing the investment for the new technology is found to be a problem especially for small firms causing a postponing of the adoption of AMT. Large firms seem to have a general advantage over small ones.

6. Specification and econometric estimation of the policy equation

To model the impact of government AMT promotion (policy effect) we constructed a variable based on a firms' assessment of the overall stimulus of government support on a five-point Likert scale (computed as a mean of the reported stimulus in each of the three types of activities – training, consulting, R&D projects – reported here; see appendix II for the exact wording of this question in our questionnaire). These stimulus measures were subsequently transformed to a binary variable (CIMTHM; value 0 for stimulus of 1 or 2, value 1 for stimulus of 3 to 5 ("high stimulus") on the original five-point scale).[9] This binary variable has been used throughout in this study.

A policy equation was specified on grounds of "ad hoc" plausibility arguments. Specifically, several firm-specific factors were taken into consideration. First, firm size was inserted as an independent variable in the equation. Small firms may have a larger incentive to claim such support than large ones because of limited financial resources; moreover, the promotion of small firms has been an explicit goal of the Swiss AMT programme.[10] Thus, we expect a negative correlation of the policy variable with firm size. Second, industry dummies were used as control variables. Third, we considered two institutional characteristics of firms which may be related to the probability of being supported by a government programme: status as affiliate or parent-house and status as affiliate of a foreign enterprise. Although it is not a priori clear in which direction these variables could influence the policy variable, it seems reasonable to control for these institutional characteristics. Fourth, firms confronted with serious problems of financing innovation projects would be more inclined to claim government support than other firms. We constructed a proxy for "limited financial resources for developing/adopting new technology" by computing the mean of the assessments of the responding firms with respect to five impediments of innovative activity related to financial problems (measured on a five-point Likert scale).[11] We expect a positive sign for this variable. Finally, we included a dummy variable measuring the previous experience of a firm with other government technology promotion programmes. We expect a positive sign also for this variable because of the greater awareness of such firms of the benefits of technology promotion and the bettter knowledge of the administrative procedures in applying for policy support.

Probit estimates of the policy equation are found in Table V. Accordings to these results there is only a weak negative relation between firm size and the probability of being a government-supported firm; the coefficient of L is not statistically significant. Important factors which correlate positively with the probability of goverment support are the lack of financial resources and, partly, previous experience with government support in other programmes and the status of being an affiliate of a foreign enterprise. The effect of industry dummies is negligible (except for the metalworking industry).

TABLE V
Probit estimation of the policy equation; total sample

Explanatory variables	CIMTHM (stimulus)	
Firm size		
L	–0.15	(0.20)
L^2	0.03	(0.02)
Groups of industries		
Metalworking	0.49**	(0.23)
Machinery/vehicles	0.25	(0.23)
Electrical machinery/Electronics	0.16	(0.24)
Chemicals/plastics	–0.32	(0.40)
Ownership status		
Affiliate company	–0.09	(0.17)
Firm in foreign ownership	0.37*	(0.22)
Financing difficulties	0.26*	(0.15)
Previous experience with government support	0.29*	(0.17)
N	463	
McFadden R^2	0.060	
LR statistic	21	
% concordance	66.9	

Note: **, * indicate statistical significance at the 5%-level and 10%-level resp.; intercepts have been throughout omitted.

7. Econometric estimation of the simultaneous model and policy evaluation

We estimated a simultaneous probit model with an adoption and a policy equation, each of them containing as additional explanatory variable the dependent variable of the other one (see section 2). We conducted the simultaneous estimation only with the independent variables which in the single equations in Tables IV and V resp. yielded statistically significant coefficients at the 10%-level (restricted models in column 2 of both tables). Columns 1 and 2 of Table VI contain the estimates of the simultaneous probit model based on all available observations ("total sample"; N = 463). Columns 3 and 4 show the estimates of the model using only a sample of those firms which at the launching of the promotion programme in 1990 did not yet apply AMT in production ("reduced sample"; N = 330). Testing for the policy effect with the reduced sample is thus more restrictive than testing it with the total sample.

The basic pattern of the model estimates for both equations is the same as in the single equations (adoption and policy equation in Tables IV and V resp.) indicating a certain robustness of the underlying relationships. The coefficient of DAMTINT (adoption variable) in the policy equation is positive but for both samples not statistically significant at any relevant test level (columns 2 and 4). We could thus not find any evidence that the firms supported by the policy

TABLE VI
Simultaneous probit estimation of the adoption and the policy equation

Explanatory variables	DAMTINT	CIMTHM	DAMTINT	CIMTHM
	Total sample		Reduced sample[1]	
	(1)	(2)	(3)	(4)
FINCOMP	–0.14** (0.07)	–	–0.14* (0.08)	–
COST	0.11** (0.05)	–	0.11 (0.07)	–
FLEX	0.23** (0.06)	–	0.27** (0.07)	–
QUAL	0.10** (0.05)	–	0.13** (0.06)	–
BEST	0.19** (0.06)	–	0.20** (0.07)	–
INVCOST	–0.12** (0.06)	–	–0.06 (0.07)	–
LBATCH	0.33** (0.11)	–	0.26** (0.13)	–
COOP	0.25** (0.12)	–	0.22 (0.14)	–
L	0.33** (0.16)	–0.03 (0.27)	1.53** (0.63)	–1.68 (1.30)
L^2	–0.02 (0.02)	0.01 (0.04)	–0.42 (0.54)	0.08 (0.17)
INT90	–0.15** (0.02)	–	–0.20** (0.03)	–
Firm in foreign ownership	–	0.46** (0.23)	–	0.28 (0.34)
Financing problems	–	0.13 (0.16)	–	0.17 (0.20)
Previous experience with governmenr support	–	0.41** (0.17)	–	0.51** (0.22)
CIMTHM (policy variable)	0.08 (0.12)	–	0.28* (0.16)	–
DAMTINT (adoption variable)	–	0.07 (0.13)	–	–0.09 (0.15)
N	463		330	
$Q_T(\Theta: \aleph^2 = 21.1; df = 14)$	24.5		15.8	
ρ	0.10 (0.09)		0.21** (0.12)	
R	0.250	0.087	0.239	0.087

Note: [1] Reduced sample: firms which at the start of the promotion programme in 1990 did not yet apply AMT. Standard errors are included in brackets (**, * indicate statistical significance at the 5%-level and 10%-level resp.). Intercepts and industry dummies have been throughout omitted.

measures were those who were able to intensify the use of AMT stronger than non-supported ones in the period 1990–1996. For policy evaluation the relevant estimation result is related to the effect of the variable CIMTHM in the adoption equation (columns 1 and 3). For the total sample no significant influence of the policy variable on adoption intensity could be found. In the estimates based on the reduced sample (only firms which adopted AMT for the first time after 1990) the coefficient of CIMTHM (policy variable) in the adoption equation ist positive and statistically significant (10%-level). This result for the more restrictive case amounts to a clear hint that participation to the government programme may have contributed to a more intensive adoption of new technology in the case of supported firms relative to non-supported ones. Therefore, government promotion has been to some extent effective, especially for firms using AMT for the first time, which was one of the goals of the programme.

The policy effect is significant also in estimates not shown here which were based on the subsample of small firms with less than 200 employees. On the contrary, we could not find a significant policy effect for the subsample of large firms (200 and more employees). In futher estimations for stimulus variables differentiated by type of support (training, consulting, R&D projects) not presented here we found similar results for information/training and consulting. In the case of R&D-support we obtained a statistically significant positive coefficient of the policy variables only for the subsample of firms with less than 50 employees (see Arvanitis et al., 1998, ch. 6).

8. Conclusions

This paper applies a procedure of analyzing the impact of public promotion of AMT on the intensity of diffusion of such technologies based on the simultaneous estimation of an adoption equation specified within the framework of standard diffusion theory and a policy equation capturing important aspects of the selection procedure of supported firms. The firm data used in this study were collected especially for this evaluation.

According to our results the most important factors influencing the adoption behaviour of a firm are, besides industry effects: positively, a series of anticipated new revenue potentials (e.g through cost reduction, higher flexibility, etc), the high capacity to efficiently absorbing and applying new technology and firm size; negatively, investment costs which are associated with the introduction and adaption of these new technologies.

What about the effectiveness of the Swiss government programme to promote AMT? The evidence suggests that the policy goal to foster the diffusion of AMT was attained in the case of firms adopting AMT for the first time or characterized by a low intensity of AMT use (and for small firms with less than 200 employees with some overlapping between these two categories).

The main advantage of our approach (and of microeconometric approaches in general) relative to alternative methods is, on the one hand, the explicit formulation, based on economic theory, of causal relationships between the goal variables of a policy support programme (in this case: the change of the intensity of technology adoption) and the factors, including policy instruments, influencing these goal variables. On the other hand this approach takes explicitly account of factors related to the political economy of the selection and participation respectively of firms with respect to the support programme in question.

The most serious shortcomings of our analysis (and of microeconometric methods in policy evaluation in general) are related to data limitations. Mostly, the evaluators do not dispose of time series of data from "before to after" the policy support programme or only data on a few variables not allowing an adequate modelling of the underlying relationships. Further problems may arise if adequate economic theory is lacking in the fields covered by an evaluation.

Keeping the definition of the goal variable rather *narrow* (in this case we used the narrow *intermediate* goal of public support of a more rapid and broader diffusion of AMT instead of the rather broad *final* goal of the programme of enhancing the performance of the participating firms, e.g. in terms of productivity, etc.) may be helpful in tackling such problems. Taking broader goals into consideration would often involve a two- or three-stage modelling and/or the explicit consideration of externalities.

In spite of the shortcomings of the micro-econometric evaluation approaches sketched above there are still many advantages on their side which force us to consider them a core element of policy evaluation. Improvements of the database would significantly increase its reliability because most of the weaknesses of the approach lie at the empirical level. Therefore, it is crucial that the preconditions for an evaluation should be secured from the very beginning of policy implementation, i.e. at the preparatory stage of a promotion programme. This means, among other things, that the institutions responsible for policy implementation should be obliged to collect the necessary data and have the authority to enforce the participating firms to deliver the required information.

Acknowledgements

The research reported in this paper was supported by the Swiss Federal Ministry of Economic Affairs. The authors have much benefited from comments and suggestions of the participants of the international conference on "The Economic Evaluation of Technological Change", organized by the U.S. Department of Commerce in Georgetown University Center, Washington D.C., 15–16 June 1998.

Appendix I

TABLE A1
Description of the data set

	N	%
Total manufacturing	463	100
Industry		
Food, beverage	19	4.1
Textiles	14	3.0
Clothing, leather	5	1.1
Wood	10	2.2
Paper	15	3.2
Printing	13	2.8
Chemicals	14	3.0
Pubber, plastics	24	5.2
Glass, stone, clay	16	3.5
Metals	99	21.4
Machinery	84	18.1
Electrical machinery	27	5.8
Electronics, instruments	83	17.9
Vehicles	17	3.7
Other manufacturing	23	5.0
Firm size (number of employees)		
5–49	83	17.9
50–99	87	18.8
100–199	127	27.4
200–499	100	21.6
500 and more	66	14.3

TABLE A2
Factor analysis with the objectives of/motives for the adoption of AMT

Objectives/motives	Factor loadings (at least 0.40)					
	(1)	(2)	(3)	(4)	(5)	(6)
Availability of financial funds	0.80					
Existing equipment already depreciated	0.67					
Favourable macroeconomic prospects	0.67					
Competitors introduce AMT	0.57					
Declining market share	0.53					
Reduction of production time		0.66				
Need of reducing costs in general		0.62				
Labour saving		0.58				
Higher utilization of equipment		0.55				
Securing delivery in time		0.45				
Need to improve the organization of production		0.40				
Higher quality of product development			0.73			
Reduction of time needed to develop new products			0.69			
Introduction of "intelligent products"			0.49		0.49	
Improving information for cost calculations			0.48			
Reduction in space requirements			0.43			
Fundamentally new production concept				0.72		
Higher flexibility of work organisation				0.69		
Higher flexibility within the firm in general				0.43	0.41	
Reduction of inventories				0.42		
Higher product variety					0.79	
Higher product quality				0.43	0.54	
Higher flexibility at the market					0.53	
Adjusting to "best practice"						0.78
Becoming familiar with a new technology						0.62
Securing technological lead						0.41
Statistics						
Number of observations						501
Kaiser's measure of sampling adequacy (MSA)						0.860
Variance accounted for by the six components						0.546
Root mean square off-diagonal residuals (RMSE)						0.060
Variance accounted for by each factor	6.70	2.02	1.67	1.40	1.29	1.12
Final communality estimate						14.2

Characterization of the six factors based on the factor pattern

(1) Favourable financial conditions; high competitive pressure (FINCOMP)
(2) Cost reduction (COST)
(3) Improving product development (DEV)
(4) Higher flexibility (FLEX)
(5) Higher product quality (QUAL)
(6) Securing technical lead/"best practice"(BEST)

TABLE A3
Factor analysis with the impediments to the adoption of AMT

Impediments	Factor loadings (at least 0.40)					
	(1)	(2)	(3)	(4)	(5)	(6)
Lack of engineers/technicians	0.72					
Lack of IT specialists	0.67			0.41		
Lack of other qualified personnel	0.66			0.42		
Intra-muros training too expensive	0.64					
Lack of relevant know-how in general	0.61					
Insufficient support by suppliers of AMT	0.58		0.41			
Investment volume too large		0.82				
Pay-back period too long		0.74				
Liquidity constraints		0.63				
Technology/software too expensive		0.61				
Trend price of technology falling		0.42	0.41			
Technology not yet developed far enough			0.71			
Performance of AMT uncertain			0.65			
Software development too expensive			0.61			
Information costs/problems			0.56			
Difficulties with software or interfaces			0.49			
Resistance to new technology within the firm				0.75		
Organizational problems				0.67		
Insufficient attention of the management				0.67		
Concept for the adoption of AMT not well-defined						
Installed equipment (rather) new					0.74	
Utilization of new production capacity uncertain					0.68	
Extra-muros training too expensive	0.42				0.45	
Insufficient compatibility with existing machinery						0.84
Insufficient compatibility with product portfolio						0.76
Insufficient compatibility with organization						0.65
Statistics						
Number of observations						495
Kaiser's measure of sampling adequacy (MSA)						0.899
Variance accounted for by the six components						0.596
Root mean square off-diagonal residuals (RMSE)						0.053
Variance accounted for by each factor	8.26	2.18	1.73	1.47	1.04	1.00
Final communality estimate						15.7

Characterization of the six factors based on the factor pattern
(1) Lack of knowledge/lack of qualified manpower (KNOWPERS)
(2) High investment costs (INVCOST)
(3) High technological costs and uncertainties (TECH)
(4) Intra-firm resistance to new technology (RESIST)
(5) Capacity utilization uncertain (UTILIZ)
(6) Compatibility problems (COMPAT)

Appendix II

Wording of the Policy Question:
(question 6 of the questionnaire)

Have you got *financial or other support by a government agency* for introducing or extending the use of Advanced Manufacturing Technologies (AMT) in your enterprise?

Yes O No O

If yes, how strong has been the *impulse you got from the support* for the introduction or extension of AMT in your firm?

Type of support				Stimulus: very weak 1	2	3	4	very strong 5
– Information/training	no O	yes O	→	O	O	O	O	O
– Consulting	no O	yes O	→	O	O	O	O	O
– R&D projects	no O	yes O	→	O	O	O	O	O

(For a full version of the questionnaire see Arvanitis et al. 1998 or www.kof.ethz.ch)

Notes

[1] This is by far not only for Switzerland the case. Typical examples for Switzerland are Freiburghaus et al. (1990) who evaluated the promotion of joint research of universities and firms, and Balthasar et al. (1997) who assessed the effectiveness of the Swiss participation in technology programmes of the EU; for Austria see, for example, Polt et al. (1994) who assessed the public support measures for the introduction of AMT. A recent report by OECD (1995) with respect to the diffusion of information technologies in SMEs, drawing on a number of country studies, is also primarily based on this type of evaluation, which does not sufficiently take account of economic performance measures or changes of technology adoption (an exception, to some extent, is the country study "Canada"). See OECD (1997b) for an assessment of the state of the art in technology evaluation and Shapira et al. (1996) for a survey of the evaluation praxis in the United States.

[2] The same type of critique was put forward by Stoneman and van Diederen (1994) for the United Kingdom: "The DTI in the United Kingdom, for example, legitimates its diffusion policy with reference to market failure but evaluates its diffusion programmes predominantely in terms of the efficiency of their management, the accuracy of targeting, the appropriateness of their tool mix and the appreciation of the recipients of information" (p. 928).

[3] The argument runs the other way around if a policy measure aims at decreasing Y.

[4] For a recent review of the literature on the theory of technology diffusion in general see Sarkar (1998).

[5] The questionnaire is available in German, French and Italian and is available on request or can be downloaded from www.kof.ethz.ch.

[6] The importance of such regional effects is confirmed by a recent econometric study of the adoption of CNC machines and microprocessors in the U.K. (Baptista, 2000).

[7] Such an interpretation can be justified on ground of some evidence on the degree of the attainment of the pursued firm objectives related to AMT adoption (measured on a five-point Likert scale): about 48% of the firms having adopted AMT reported a (very) high degree of attainment of their objectives; some 40% of them reported a middle degree of attainment and only 12% of the AMT adoptors answered that their degree of goal attainment had been (very) low (see Arvanitis et al., 1998, ch. 4).

[8] See Arvanitis and Hollenstein (2001) for a full account of the results of the econometric estimation of the adoption model used in this paper.

[9] Overall 26% of promoted firms report a "high" policy stimulus; the corresponding figures for training, consulting and R&D are 28.9%, 22.7% and 15.0% respectively.

[10] One could also argue the other way round stating that large firms have better chances to get government subsidies than small ones because they can spend more in lobbying and have possibly greater experience in this field. However, we use a specific variable to cover "promotion experience", thus we expect that with respect to firm size the "small firm effect" will dominate.

[11] These impediments are: high innovation costs, long payback period for innovation projects, lack of internal and external financial resources and high tax burden.

References

Arminger. G., J. Wittenberg and A. Schepers, 1996, *Mean- and Covariance-Structure Analysis (MECOSA), User Guide*, Friedrichsdorf.

Arvanitis, S., L. Donzé, H. Hollenstein and S. Lenz, 1998, *Wirksamkeit der diffusionsorientierten Technologieförderung des Bundes. Eine Analyse anhand von Firmendaten*, Berne: Bundesamt für Wirtschaft und Arbeit.

Arvanitis, S. and H. Hollenstein, 1994, 'Demand and Supply Factors in Explaining the Innovative Activity of Swiss Manufacturing Firms. An Analysis Based on Input-,

Output- and Market-oriented Innovation Indicators', *Economics of Innovation and New Technology* **3**(1), 15–30.

Arvanitis, S. and H. Hollenstein, 2001, 'The Determinants of the Adoption of Advanced Manufacturing Technology. An Empirical Analysis Based on Firm-level Data for Swiss Manufacturing', *Economics of Innovation and New Technology* (forthcoming).

Baldwin, J. R. and M. Rafiquzzaman, 1998, 'The Determinants of the Adoption Lag for Advanced Manufacturing Technologies', *Working Paper No. 117, Micro-Economic Analysis Division*, Ottawa: Statistics Canada.

Balthasar, A., G. Reger, C. Bättig and S. Bührer, 1997, *Evaluation der Schweizer Beteiligung an den FTE-Rahmenprogrammen der Europäischen Union*, Zurich: vdf Hochsulverlag AG.

Baptista, R., 2000, 'Do Innovations Diffuse Faster Within Geographical Clusters?' *International Journal of Industrial Organization* **18**(3), 515–535.

Browne, M. W. and G. Arminger, 1995, 'Specification and Estimation of Mean- and Covariance-Structure Models' in G. Arminger, C. C. Clogg and M. E. Sobel (eds.), *Handbook of Statistical Modeling for the Social and Behavioural Sciences*, New York: Plenum Press, pp. 185–249.

Cainarca, G. C., M.G. Colombo and S. Mariotti, 1990, 'Firm Size and the Adoption of Flexible Automation', *Small Business Economics* **2**(2), 129–140.

Capron, H. and B. van Pottelberghe de la Potterie, 1997, 'Public Support to Business R&D: A Survey and Some New Quantitative Evidence', in OECD (ed.), *Policy Evaluation in Innovation and Technology. Towards Best Practices*, Paris: OECD, pp. 35–47.

CIMEX, 1998, *Wirtschaft, Wissenschaft und Staat auf der Suche nach neuen Lösungsansätzen zur Förderung der computerintegrierten Fertigung (CIM)*, Bericht der CIM-Expertengruppe, Berne: Bundesamt für Konjunkturfragen.

Clark, K., 1987, 'Investment in New Technology and Competitive Advantage', in D. J. Teece (ed.), *The Competitive Challenge*, New York: Harper & Row, pp. 59–81.

Cohen, W. M. and D. Levinthal, 1989, 'Innovation and Learning: The Two Faces of R&D', *Economic Journal* **99**(4), 569–596.

Ewers, H.-J., C. Becker and M. Fritsch, 1990, *Wirkungen des Einsatzes computergestützter Technologien in Industriebetrieben*, Berlin: Walter de Gruyter.

Freiburghaus, D., W. Zimmermann and A. Balthasar, 1990, *Evaluation der Förderung praxisorientierter Forschung (KWF)*, Berne: Bundesamt für Konjunkturfragen.

Geyer, A., C. Rammer, W. Pointner, W. Polt, H. Hollenstein, L. Donzé and S. Arvanitis, 2000, *Evaluierung des ITF-Programms FlexCIM*, Seibersdorf: Österreichisches Forschungszentrum Seibersdorf.

Georghiou, L. and D. Roessner, 2000, 'Evaluating Technology Programmes: Tools and Methods', *Research Policy* **29**(4–5), 657–678.

Heckman, J., 1979, 'Sample Selection Bias as a Specification Error', *Econometrica* **47**(1), 153–161.

Jarmin, R. S., 1998, 'Manufacturing Extension and Productivity Dynamics', *CES Working Paper No. 98-8*, Washington D.C.: U.S. Bureau of the Census.

Kamien, M and N. Schwartz, 1970, 'Market Structure, Elasticity of Demand and Incentive to Invent', *Journal of Law and Economics* **13**(1), 241.–252.

Karshenas, M. and P. Stoneman, 1995, 'Technological Diffusion', in P. Stoneman (ed.), *Handbook of the Economics of Innovation and Technological Change*, Oxford: Blackwell, pp 265–297.

Link, A. N. and P. Kapur, 1994, 'A Note on the Diffusion of Flexible Manufacturing Systems Technology', *Economics Letters* **46**, 357–362.

Maddala, G. S., 1983, *Limited Dependent Variables and Qualitative Variables in Econometrics*, Cambridge: Cambridge University Press.

Majumdar, S. K. and S. Venkataraman, 1993, 'New Technology Adoption in U.S. Telecommunications: The Role of Competitive Pressures and Firm-level Inducements', *Research Policy* **22**(5–6), 521–536.

Milgrom, P. and J. Roberts, 1990, 'The Economics of Modern Manufacturing: Technology, Strategy, and Organization', *American Economic Review* **80**(3), 511–528.

OECD, 1995, *Information Technology (IT) Diffusion Policies for Small and Medium-Sized Enterprises (SMEs)*, Paris: OECD.

OECD, 1997a, *Industrial Competitiveness in the Knowledge-Based Economy. The New Role of Governments*, Paris: OECD.

OECD, 1997b, *Policy Evaluation in Innovation and Technology*, Paris: OECD.

OECD, 1998, *Technology, Productivity and Job Creation. Best Policy Practices*, Paris: OECD.

Polt, W., E. Buchinger, L. Jörg, A. Kopcsa, H. Leo, L. Mustonen, F. Ohler and S. Patsios, 1994, *Evaluierung des ITF-Förderschwerpunktes 'Flexible Computerintegrierte Fertigung' (FlexCIM)*, Seibersdorf: Österreichisches Forschungszentrum Seibersdorf.

Reinganum, J. F., 1981, 'Market Structure and the Diffusion of New Technology', *Bell Journal of Economics* **12**(2), 618–624.

Reinganum, J. F., 1989, 'The Timing of Innovation: Research, Development and Diffusion', in R. Schmalensee and R. Willig (eds.), *Handbook of Industrial Organization*, Vol. 1, New York: Elsevier Science Publishers, pp. 849–908.

Sarkar, J., 1998, 'Technological Diffusion: Alternative Theories and Historical Evidence', *Journal of Economic Surveys* **12**(2), 131–176.

Shapira, P., J. Youtie and J. D. Roessner, 1996, 'Current Practices in the Evaluation of U.S. Industrial Modernization Programmes', *Research Policy* **25**(2), 185–214.

Stoneman, P. and P. van Diederen, 1994, 'Technology Diffusion and Public Policy', *The Economic Journal* **104**(2), 918–930.

Taymaz, E., 1991, 'Flexible Automation in the U.S. Engineering Industries, *International Journal of Industrial Organization* **9**(5), 557–572.

[18]

Transmission of Risk-Averse Behavior in Small Firms

Dennis Patrick Leyden
Albert N. Link

ABSTRACT. Small-sized firms are typically more entrepreneurial and engage in more innovation and risk-taking behavior. For that reason they are considered the engines of future economic growth. One policy for stimulating such activity is to provide government contracts for small firms. However, such contracts as typically written result in increased risk-averse behavior on the part of small firms out of a desire by government officials to shift the risk to the firms. This, in turn results in a reduced level of innovative and entrepreneurial activity. To eliminate the disincentive to engage in innovation and entrepreneurial activity, government needs to bear the risk associated with such contracts. One possible solution, given the natural risk aversion of elected officials, is to engage in a portfolio approach to small firm contracts by which the government can diversify away some of the risk.

1. Introduction

Small-sized firms are typically more entrepreneurial and engage in more innovation and risk taking than their larger counterparts. For that reason they are considered the engines of future economic growth and, as a result, are often the target of various governmental policies designed to encourage their activity. Paradoxically, however, the practice of elected officials targeting small firms with governmental contracts may stimulate small-firm activity at the expense of the entrepreneurial activity that makes them so attractive.

This paper presents a positive analysis of the effect of governmental contracts on small-firm behavior. Because of characteristics unique to small firms, we find that such contracts will result in a reduction in the willingness of small firms to engage in risk taking, and with that reduction in risk taking is a general reduction in entrepreneurial and innovative activity.

This result is important because it suggests that public policies that seek to stimulate entrepreneurial and innovative activity by targeting small firms for governmental contracts will at best achieve a Pyrrhic victory that increases small firm activity but actually reduces the level of entrepreneurial and innovative activity. Assuming that the stimulation of entrepreneurial and innovative activity is a desirable policy objective, the problem is not whether government should target small firms at all, but rather how governmental contracts could be designed to insulate the small firm from direct influence by elected governmental officials and yet provide sufficient levels of oversight.

Our results are also important from a private-sector perspective. We find that small firms are particularly susceptible to reduced levels of entrepreneurial activity and innovation if a significant part of their revenue comes from government contracts. As a result, small firms interested in avoiding such reductions need to examine closely the structure of their government contracts if they wish to avoid the problems often attributed to public-sector enterprises.

This paper is outlined as follows. Section II characterizes the special nature of small firms that differentiates them from their larger counterparts. It also explains the relationship between a firm's revenue function and risk-averse behavior. Section IIII then provides an analysis of the relationship between influence by elected governmental officials and risk-averse behavior on the part of small firms. Finally, Section IV contains a brief summary and some observations about the implications for public policy.

Final version accepted on November 27, 2002

Department of Economics
University of North Carolina at Greensboro
P.O. Box 26165
Greensboro, NC 27402-6165
U.S.A.
E-mail: leyden@uncg.edu
al_link@uncg.edu

Small Business Economics **23**: 255–259, 2004.

II. A model of small firm behavior

It is well known that small firms are typically exposed to greater competitive pressures and uncertainty in the production process than their larger counterparts.

Consider, then, a small firm producing in a perfectly competitive, output market in which its production process is uncertain.[1] Because the revenues for this firm will be a positive, linear function of output Q, we can represent revenues by the function:

$$R = R(Q) \tag{1}$$

where:

$$\frac{\partial R}{\partial Q} > 0 \qquad \frac{\partial^2 R}{\partial Q^2} = 0. \tag{2}$$

Because the firm encounters uncertainty in the production of the good, let the level of output Q be an uncertain function of the firm's effort E:

$$Q = Q(E, \theta) \tag{3}$$

where θ is an index variable representing the presence of the production risk and assumed to be a random draw from a set of possible states-of-nature with known distribution function $f(\theta)$. Let θ take on non-negative values with greater values of θ implying increasingly lower values of Q:

$$\frac{\partial Q}{\partial \theta} < 0 \qquad \frac{\partial^2 Q}{\partial \theta^2} \leq 0. \tag{4}$$

For a given level of effort, define the maximum output level to be the output level associated with θ equal to zero. Finally, assume that Q is a positive, strictly concave function of E and that increases in θ reduce the marginal productivity of E:

$$\frac{\partial^2 Q}{\partial \theta \partial E} < 0 \tag{5}$$

The cost of producing Q is assumed to increase at an increasing rate with the total level of effort expended:

$$C = C(E) \tag{6}$$

such that:

$$\frac{dC}{dE} > 0 \quad \text{and} \quad \frac{d^2C}{dE^2} = 0. \tag{7}$$

Hence, the small firm's problem is to maximize expected profits:

$$\max_E \int (R(Q(E, \theta)) - C(E))f(\theta)d\theta. \tag{8}$$

The solution to the small firm's problem is an unique, optimal level of effort characterized by the first-order condition that expected marginal revenue should equal marginal cost:

$$\int \frac{dR}{dQ}\frac{\partial Q}{\partial E} f(\theta)d\theta = \frac{dC}{dE}. \tag{9}$$

Whether or not the firm engages in risk-averse behavior will depend on whether such behavior increases the expected profits of the firm. Within the context of our model, the firm's decision centers on the level of output to produce. We define risk-averse behavior to be a costly act that reduces the firm's exposure to future uncertainty in order to better achieve its objective. Within the structure of our model, such behavior will allow the small firm to produce the expected quantity $Q^e = \int Q(E, \theta)f(\theta)d\theta$ with reduced uncertainty, perhaps to the point of certainty.

In practice, this reduction in uncertainty can be achieved through a variety of methods. Besides the familiar purchasing of insurance, the firm can engage in greater monitoring of the production process, hiring outside agents to manage or perform the production process, and so forth. All such actions increase costs either directly through increased payments to others or indirectly by slowing the production process and thus increasing the costs borne by the firm.

In order for risk-averse behavior to be rational, it must be that the gains associated with reducing exposure to risk exceed the increased cost associated with reducing that exposure. However, for the small firm in a competitive market, there will be no gain associated with engaging in risk-averse behavior because such behavior is costly and because there is no associated gain in expected revenues. To see that, consider the case illustrated in Figure 1 of a small firm faced with an uncer-

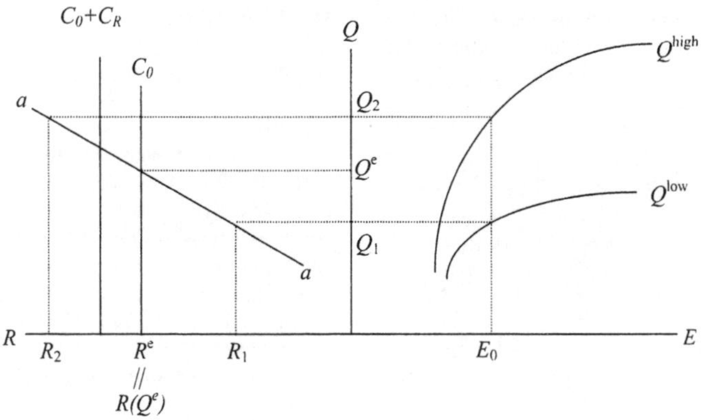

Figure 1.

tain production function that takes the form $Q^{low}(E)$ with probability p and the form $Q^{high}(E)$ with probability $(1 - p)$.

Suppose further that the firm has incurred the (certain) cost $C_0 = C(E_0)$ based on its efforts E_0, and is now awaiting either a low output, defined as $Q_1 = Q^{low}(E_0)$ (with probability p), or a high output, defined as $Q_2 = Q^{high}(E_0)$ (with probability $(1 - p)$). In such a situation, expected output will be $Q^e = pQ_1 + (1 - p)Q_2$ and expected revenues will be $R^e = pR(Q_1) + (1 - p)R(Q_2)$. Because the revenue function $R(Q)$ is linear in output (recall equation (1)), the relationship between output and revenue can be represented by the straight line aa in Figure 1, and an inspection of that figure reveals that the expected revenues associated with producing output with uncertainty, R^e, will be equal to the revenues $R(Q^e)$ associated with producing the expected output Q^e with certainty.

Because risk-averse activity increases total costs (represented by C_R in Figure 1), the firm will find that engaging in risk averse activity will increase its costs from C_0 to $C_0 + C_R$, leave revenues unchanged at Re, and thereby reduce profits. As a result, there is no advantage to the firm engaging in risk-averse behavior.

III. The effect of government contracts

Now suppose that the small firm receives its revenues not from selling in a competitive market but from contracting with elected officials for the production of some good or service. Because of the difficulty observing the entire production process, elected officials typically monitor a small subset of the attributes of the good or service (Lindsay (1976)). To make this more concrete, assume that the contract requires that the firm produces some quantity of a good or service by an agreed date. If the firm meets the target quantity and date, the firm will receive its contracted payment, sometimes with some small additional reward if the firm finishes early or if the output is greater than originally specified. However, if the firm fails to meet its target, there are often large penalties that significantly reduce the total revenue received from the government. Such contracts are designed to assure that the target is hit and are based on a general risk-aversion on the part of elected government officials concerned with their chances of reelection.[2] Thus, the relationship between the firm's output and its total revenue will no longer be linear as noted in equation (1) but will instead be more or less concave. Assume, therefore, that the revenue function associated with a government contract is:

$$R = \hat{R}(Q) \tag{10}$$

such that:

$$\frac{\partial \hat{R}}{\partial Q} > 0 \qquad \frac{\partial^2 \hat{R}}{\partial Q^2} < 0. \tag{11}$$

The degree of concavity can of course vary, but in general it will increase with the importance that the government puts on the output and the degree to which the elected officials are risk averse.

Under such circumstances, there is a potential advantage to the firm engaging in risk-averse behavior because such behavior allows the firm to produce Q^e with greater certainty and thereby increase its expected revenues. To see that, consider once again the simple production process in which output takes the form $Q^{low}(E)$ with probability p and the form $Q^{high}(E)$ with probability $(1 - p)$. Assuming that the revenue function can be represented by the curve *bb* in Figure 2, the concavity of bb results in expected revenues $R = pR(Q_1)+(1 - p)R(Q_2)$ being less than the revenues associated with producing expected output $Q^e = pQ_1 + (1 - p)Q_2$ with certainty.

For the firm, this means that there is a revenue gain associated with engaging in risk-averse behavior. As a result, if the cost of risk-averse behavior is less than this gain in revenues associated with such behavior, the firm will find such behavior profitable and will therefore choose to act in a risk-averse manner. In Figure 2, the gain to the firm associated with eliminating production uncertainty completely would be an increase in revenue from R^e to $R(Q^e)$. Because the increase in cost associated with risk-averse behavior in Figure 2 is assumed to be some C_R, which is less than the revenue increase, the firm will find it to its advantage to engage in risk-averse behavior and increase its profits by $R(Q^e) - R^e - C_R$.

IV. Conclusion

If the stimulation of entrepreneurial and innovative activity in small firms is a desirable policy objective, the above analysis suggests that the typical contract with elected officials may not work. The fundamental problem is that the elected officials seek to reduce the risk associated with the production process through the design of incentive contracts that result in a concave revenue function.

Moreover, while one might imagine that having a government bureaucracy issue the contract would be better, that is not, in fact the case. As Leyden and Link (1993) show, bureaucracies are also affected by the risk aversion of elected officials. As a result, when they contract with a private-sector firm for the production of some good or service, they will generally insist on the firm taking actions to indemnify the bureaucracy from production risks. Such contracts with bureaucracies typically result in either a fixed-fee contract with full insurance or a cost-plus type contract and thus do not generate a concave revenue function for the firm. However, the effect is the same because such contracts increases the costs of the firm and induce it to take risk-averse actions either directly in the production process or indirectly through the purchase of some form of insurance such as performance bonds.

How, then, can elected officials stimulate small firm activity without creating the disincentives to engage in innovative and entrepreneurial activity?

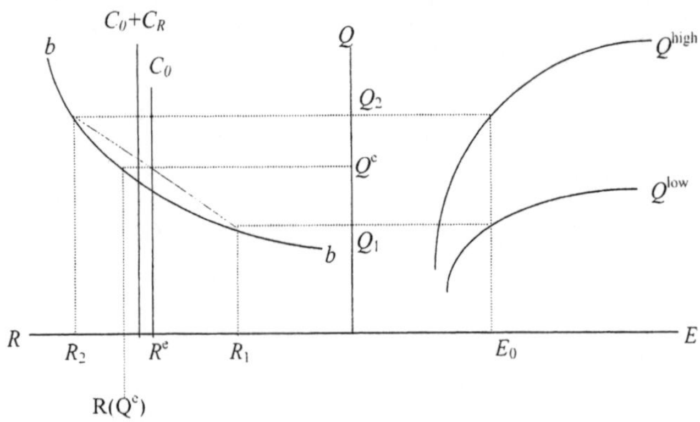

Figure 2.

The key would seem to be to offer small firms contracts similar to what they would get in competitive markets with, in particular, linear revenue functions. This, in turn suggests that elected officials must either be willing to bear the risk associated with such contracts or to reduce that risk through other means. One possibility is to reduce the risk by contracting with a number of small firms so that, on average, performance on contracts is closer to what the elected officials desire. Because this "portfolio" approach requires a fairly large volume of contracts, such an approach is more likely to be feasible at the state and especially federal level than at the local government level.

Notes

[1] Bozeman and Kingsley (1998, p. 10) define risk as "the exposure to the chance of loss from one's actions or decisions." Within the context of our model, such exposure centers on the production process that the firm engages in.

[2] Governmental contracts may also be designed by bureaucrats. However, because bureaucrats utltimately depend on elected officials for their budgets and compensation, they too have an incentive to act in a risk-averse manner. See Bendor, Taylor, and Van Gaalen (1985) for a more detailed examination of the penalties and rewards that elected government officials design, and see Leyden and Link (1993) for a more detailed explanation of how bureaucrats transfer risk to private-sector firms in the contracts they design.

References

Baldwin, J. N., 1987, 'Public versus Private: Not that Different, Not that Consequential', *Public Personnel Management* **16**, 181–193.

Bellante, D. and A. N. Link, 1981, 'Are Public Sector Workers More Risk Averse than Private Sector Workers?', *Industrial and Labor Relations Review* **34**, 408–412.

Bendor, J., S. Taylor and R. Van Gaalen, 1985, 'Bureaucratic Expertise versus Legislative Authority: A Model of Deception and Monitoring in Budgeting', *American Political Science Review* **79**, 1041–1060.

Blank, R. M., 1985, 'An Analysis of Workers' Choice between Employment in the Public and Private Sectors', *Industrial and Labor Relations Review* **38**, 211–224.

Borcherding, T. E., W. W. Pommerehne and F. Schneider, 1982, 'Comparing the Efficiency of Private and Public Production: The Evidence from Five Countries', *Zeitschrift fur Nationalokonomie Supplement* **2**, 127–156.

Bozeman, B. and G. Kingsley, 1998, 'Risk Culture in Public and Private Organizations', *Public Administration Review* **58**, 109–118.

Gibbons, R., 1998, 'Incentives in Organizations', Working Paper 6695, Cambridge: National Bureau of Economic Research.

Katz, E. and A. Ziderman, 1986, 'Toward a Theory of Incremental Wage Scales', *Applied Economics* **18**, 1047–1050.

Khursheed, A. F. and T. E. Borcherding, 1998, 'Organizing Government Supply: The Role of Bureaucracy', in F. Thompson and M. T. Green (eds.), *Handbook of Public Finance*, New York: Marcel Dekker, pp. 43–91.

Leyden, D. P. and A. N. Link, 1993, 'Privatization, Bureaucracy, and Risk Aversion', *Public Choice* **76**, 199–213.

Lindsay, C. M., 1976, 'A Theory of Government Enterprise', *Journal of Political Economy* **84**(5), 1061–1077.

MacCrimmon, K. R. and D. A. Wehrung, 1986, *Taking Risk: The Management of Uncertainty*, New York: Free Press.

Mueller, D. C., 1989, *Public Choice II*, Cambridge: Cambridge University Press, p. 257.

Niskanen, W. A., 1971, *Bureaucracy and Representative Government*, Chicago: Aldine-Atherton.

Tucker, I. B., 1988, 'Entrepreneurs and Public-Sector Employees: The Role of Achievement Motivation and Risk in Occupational Choice', *Journal of Economic Education* **19**, 259–268.

Wilson, J. Q., 1989, *Bureaucracy: What Government Agencies Do and Why They Do It*, New York: Basic Books, pp. 69–70, 132, 191–192.

Part IV
Knowledge Flows from Universities and Laboratories

[19]

GROWTH PATTERN OF ACADEMIC ENTREPRENEURIAL FIRMS[1]

JÉRÔME DOUTRIAUX
UNIVERSITY OF OTTAWA

EXECUTIVE SUMMARY

Academic entrepreneurship, the creation of new business ventures by university professors, technicians, or students, is increasingly being promoted by university-based innovation centers and university business offices. It is seen as an efficient university–industry technology transfer mechanism, and, in some cases may contribute to university revenue. Whereas most entrepreneurs leave the university at time of start-up, others keep their academic postings as full-time or part-time professors. "Part time" entrepreneurship may be interesting from a university point of view, because (i) it keeps in the laboratory a creative individual, (ii) it may provide through part-time academic positions for a more efficient use of university resources, and (iii) it encourages more contacts between faculty, students, and the business world. However, manufacturing firms led by part-time entrepreneurs do not seem to be as aggressive and growth-oriented as "independent" firms. The university professor interested in the successful transposition of an idea, an invention, to the commercial sector, may therefore have to do it through licensing, or resign from the university to devote all his time and energy to the development, manufacturing, and commercialization process. And the university interested in investing in a new "academic firm" created to exploit commercially an invention made in its own laboratories should beware of keeping the academic entrepreneur on staff, or, if he stays on staff, of involving him with the management of the new company.

These are the main conclusions of a study of the evolution from 1980–1981 to 1984–1985 of 38 young firms originally created by entrepreneurs while at various universities in several Canadian provinces. The results, supported by the observation of a small control group of 8 Canadian high-technology companies, bring out a dichotomy: for firms offering technical services, the continued

---

Address correspondence to Jérôme Doutriaux, Faculty of Administration, University of Ottawa, Ottawa, Ontario, Canada K1N 6N5.

[1]This research project was partially supported by a grant from the Technological Innovation Program, Technology Branch, Department of Regional Industrial Expansion, Ottawa, Canada. Significant contributions to this project were made by Colleen Bigelow and Branko Peterman. Their help was greatly appreciated.

Journal of Business Venturing **2**, 285–297 (1987) 0883-9026/87/$3.50

academic posting of the firm's creator does not appear to overly affect its operations and growth; however, manufacturing firms completely independent from the university tend to grow faster. Other results support the negative correlation that seems to exist between the growth rate of sales and closeness to academic life, as measured, for example, by the use of university laboratories and equipment. Whether they are still closely connected to a university or not, the entrepreneurs in our sample do note many potential positive benefits from continuing contacts with the academic world. It is possible that the negative correlation between growth and university links observed in the sample results more from the individual characteristics of the entrepreneurs than from a negative influence by the university. As venture capitalists have long noticed, one of the main determinants of growth and success is the entrepreneur himself. The entrepreneurs staying close to the university and using university services may be less aggressive and less risk-oriented than the others.

INTRODUCTION

One of the most striking characteristics of the current mutation of the economic structure of western economies is the emphasis put on entrepreneurship, on the creation and development of small and medium-sized businesses. This is not really surprising because, in the USA in particular, these businesses account for a significant share of the 25 million jobs created since between 1968 and 1983, whereas larger corporations, such as the Fortune 500, have reduced employment in the recent past.

Small businesses, and particularly newly established small businesses, form a very volatile group with a high rate of exits and entries. Typically, only about a third of new firms are still alive four years after founding (Dun and Bradstreet 1967). Fortunately, the number of new start-ups, particularly in times of economic slowdown, generally more than makes up for these disappearances. The countercyclical behavior of small businesses has an especially good effect on job creation and economic growth.

According to David Birch of MIT, there is a strong correlation between the existence of a university population specifically trained in R&D, and the creation of jobs through small and medium-sized business (Birch 1984). This is probably due to the 10% or 15% of new ventures with unlimited upside potential, based on innovations that are commercially accepted. They have a higher-than-average growth rate and a lower-than-average disappearance rate (Doutriaux 1984). These high-technology or "knowledge" enterprises are generally founded by experienced professionals with advanced university education. Universities therefore have a real but indirect effect on the creation and growth rate of innovative small business ventures, in addition to their general effect on the level of knowledge and on the technical and business training provided for larger businesses or governments.

Our knowledge of the direct effect of universities on the creation of new business ventures through academic entrepreneurship is rather limited at this time. Indeed, the number of start-ups coming directly from the university is small: 10% in a sample of 60 Swedish companies (Utterback, Reitberger, and Martin 1982), and only 3.3% in a sample of 243 Silicon Valley start-ups (Cooper 1972). It is, however, appropriate to investigate whether the special links that these firms have with the academic community have a positive impact on their evolution and whether they should receive special attention from venture capitalists, governments regulators, and university administrators. A recent study has shown a positive correlation between quality of academic research output and the number of patents applied for by the researchers (McQueen and Wallmark 1984): should these researchers be encouraged to participate in the industrialization of their patents while staying at the university, or should they license them out?

This paper is based primarily on the observation of the evolution of 38 academic entrepreneurial firms between 1980–1981 and 1984–1985 and on the evaluation of the effect of their links with a university. A detailed analysis of the first years of existence of forty-eight Canadian firms including the present sample, was done in 1981 (Doutriaux and Peterman 1982). It showed that academic start-ups are quite similar, on the average, to individual start-ups, and that, in spite of their university connections, the entrepreneurs encounter similar problems and difficulties, dividing their time between their academic duties and their business venture, and being less concerned about the financial success and even survival of their company than would entrepreneurs with no other source of income. This may be due to the fact that, in 1981, very few Canadian universities had formal rules on the commercial exploitation of university research and on joint ventures with their own staff.

A small sample of eight young Canadian high-technology firms created by university-based entrepreneurs is used as control group. The analysis of the growth behavior of these companies confirms the growth rate differential between firms still linked with a university and those that are now fully independent.

In a first section of the paper, past studies of academic entrepreneurship are reviewed. The methodology and research findings of our two samples are then presented. The conclusion outlines the main implications of this study and calls attention to areas in urgent need of further research.

PAST RESEARCH ON ACADEMIC ENTREPRENEURSHIP

The number of studies dealing specifically with academic entrepreneurship—the start-up of a new venture by a member of the faculty or of the technical staff of a university—is rather limited. Some studies deal with entrepreneurship as a by-product; others, prepared by innovation centers or by the members of technical universities, describe the venture creation process in institutions directly supportive of the spin-off concept; and only a small number of studies deal with general start-ups by university faculty in institutions that are not actively supporting such activities.

A recent study at Linköping University (Olofsson and Wahlbin 1984) dealing with 12 technology-based young firms found that nine were using results of original research by the founders. Seven of the 12 firms are led by entrepreneurs who are still members of the staff of the university. Not surprisingly, the companies with the highest growth are those that have received external financing and that are led by entrepreneurs who have left the university.

Three other studies, at Texas A&M University (Rodenberger and McCray 1981), and in Canadian (Doutriaux and Peterman 1982) and American (Vesper and Crosswhite 1983) universities, include consulting, technical services, and manufacturing start-ups. Whereas technical start-ups by Texas A&M University were generally not related to research interests or technological development coming from research at the university, many Canadian start-ups were the direct outcome of university research. Logically, science and engineering schools lead most of the time to technical service or technical manufacturing firms, whereas business schools lead to activities with low technical content. Full professors seem to start more firms than other professoral ranks, but smaller ones. And technical firms, particularly technical manufacturing firms, have higher potential for growth than consulting or service firms (Table 1). As shown in Table 1, manufacturing–academic–entrepreneurial firms using technologies developed partially at the university (Doutriaux and Peterman's study) seem to grow at a faster rate than similar firms using technologies not related to university work (Texas A&M study). Their rate of growth is similar to the rate obtained by Tyebjee and Bruno (1982) with a sample of California high-technology firms.

TABLE 1 Growth of Academic Start-ups and of High-Technology Firms

	Texas A&M		Doutriaux and Peterman		Tyebjee and Bruno California high-tech firms
	All firms	Manufacturing only	All firms	Manufacturing only	
Number of firms	32	15	37	22	197
Average age (years)	10(1980)	12(1980)	6.3(1981)	4.8(1981)	2.3(1981)
Age range (years)	1 to over 20	1 to over 20	1–16	2–12	1–3
Median sales ($1000)	250	500	190	400	235
Average sales ($1000)	3000	2100	1253	2160	1000
Percent of firms above $1 million sales	30%	40%	27%	45%	19%

These results are consistent with what one would have expected: higher growth rate in technical manufacturing than in technical consulting, and higher growth in "high-technology" manufacturing than in general manufacturing. Does that necessarily mean that academic entrepreneurs have a comparative advantage over other entrepreneurs and that venture capitalists should favor them over other types of entrepreneurs? The empirical evidence provided by this study seems to suggest that it is not necessarily so and that, if he is growth-oriented, the entrepreneur should leave his university and its protected environment.

EVOLUTION OF ACADEMIC ENTREPRENEURIAL FIRMS, 1981–1984

The Sample

To test the effect of continuing university links on the development of Canadian technically based academic–entrepreneurial firms, a new survey of the firms studied by Doutriaux and Peterman in 1980–1981 was organized. All these firms had been created by entrepreneurs while teaching, doing research, or studying at a number of universities in various parts of Canada. Their names and addresses were obtained from University administrators and heads of departments. The sample of 38 firms retained for this second study was subdivided into several groups, each representing one type of activity and the current contractual links of the owner–entrepreneur with a university:

- Firms offering technical services to the community or doing R&D activities, having their owner still on university payroll (TSU, Technical Service, University), or completely independent (TSI);
- Firms involved in some type of manufacturing activities, with the entrepreneur still on university payroll (MU) or not (MI).

Eleven firms included in the 1980–1981 study and involved in management consulting or pure retailing activities were eliminated from the present study. The technical service companies that did participate in this new survey (Table 2) are active in testing, research, development, or related activities in chemical engineering, biology, electrical engineering, computers. The manufacturing firms in the sample operate in fields as diverse as computer equipment, industrial ceramics, electrical systems, and machinery. The manufacturing firms led by professor–owners (MU) are relatively small operations with little dynamism.

TABLE 2 Size of the Academic Entrepreneurial Firms Surveyed, 1984—1985

			Sales ($1000)			Employment (persons)[a]		
Type of firm	Number of firms with data	Average age (years)	Average	Average without largest firms	First and third quartile	Average	Average without largest firms	First and third quartile
TSI	6	8.8	494	173	45–500	9	4	1–13
TSU	9[b]	9.3	1070	107	20–250	20	3.2	1.5–7
MI	8	11.3	2000	1145	560–1900	41	30	16–60
MU	3	6	88	33	30–70	3	2	2–4

[a]For TSU firms, the entrepreneur-professor being counted as one.
[b]Including a large firm involved in a merger in 1983.

Two thirds of the entrepreneurs have Ph.D. degrees in technical fields; all the others have at least a Master's degree in science or engineering, except in the MI group, which includes two B.Sc. degrees and two B.Sc.–MBA combinations. A little over half of the entrepreneurs had some nonacademic work experiences prior to start-up.

Survival Rate

Only about 18% of the 38 firms considered in the analysis have disappeared between 1981 and 1984–1985: two firms (one TSU, one MI) were actually closed, two were merged or acquired (including a large TSU firm), and three firms could not be found (Table 3). This is not very different from the 14% disappearance rate found by Shapiro and Giglierano (1982)[2] in Columbus, Ohio, for firms operating in similar sectors and having the same average age. Globally, our 21 independent firms (TSI, MI) had a disappearance rate of 14%, a little lower than our "university" firms (23%). Individually, however, the TSI and MU groups are at 30% and 25%, respectively: these differences are difficult to explain and are probably due to the small size of the sample.

Level of Success and Type of Firm

The level of success of the surviving firms is not uniform: some have experienced strong growth in sales and output per employee between 1981 and 1984–1985; others are facing difficulties and have reduced the level of activity. As shown in Table 4, the owner–professors

TABLE 3 Rate of Survival of Academic Entrepreneurial Firms, 1980–1981/1984–1985

		Number of companies				
			Disappeared			
Type of firm	Total number of companies (1980)	Surviving (well and not so well)	Merged	Closed	Lost	Percentage disappeared
TSI	9	9	—	—	—	0%
TSU	13	9	1	1	2	30%
MI	12	9	1	1	1	25%
MU	4	4	—	—	—	0%
Total	38	31	2	2	3	18.4

[2]Shapiro and Giglierano report a 12% disappearance rate for business consulting firms 5–9 years old, and a 16% disappearance rate for machine shops and industrial testing laboratories 7–11 years old.

TABLE 4 Perceived Performance and Real Growth of Sales (Number of Firms in Each Case)

		Perceived performance: Successful and well		Perceived performance: Poor, unsatisfactory		Total
Real growth	Positive	TSI: 4	MI: 5	TSI: —	MI: 1	Independent: 10
rate of sales		TSU: 5	MU: 2	TSU: 1	MU: 1	University: 9
1980–1981/1984–1985	Negative	TSI: 2	MI: 2	TSI: —	MI: —	Independent: 4
		TSU: 1	MU: —	TSU: 1	MU: 1	University: 3
	Total:	Independent: 13		Independent: 1		
		University: 8		University: 4		

TABLE 5 Aggregate Growth of Sales, Employment, Productivity, 1980–1981/1984–1985

All the Firms Having Complete Reports [in Parentheses, Results for All But the Largest Firms (1984–1985)]

Type of firm	Number of firms	Mean level of sales ($1000) 1980–1981	Mean level of sales ($1000) 1984–1985	Mean level of employment (number of persons) 1980–1981	Mean level of employment (number of persons) 1984–1985	Average annual growth rate[a,b]: Total sales (%)	Total employment (%)	Productivity (%)
TSI	6(5)	190(68)	494(173)	8(3)	9(4)	24(23)	3(7)	21(16)
TSU	9(8)	423(38)	1070(107)	11(3.2)	20(3.2)	23(26)	15(0)	8(26)
MI	9(8)	1308(721)	1800(1030)	30(21.2)	37(26.6)	7.9(8.9)	5.1(5.6)	2.7(3.3)
MU	3(2)	12(10)	88(33)	3.(2.6)	3(2)	54(26)	4(− 10)	51(36)

All the Firms, Arranged by Size

1984–1985 sales (dollars)	Number and type of firms	Average age in 1985 (years)	Mean level of sales ($1000) 1980–1981	Mean level of sales ($1000) 1984–1985	Mean level of employment (number of persons) 1980–1981	Mean level of employment (number of persons) 1984–1985	Average annual growth rate[a]: Total sales (%)	Total employment (%)	Productivity (%)
Over 1 million	5 MI	12	2250	3670	50.7	63.6	12.2	5.6	6.6
	1 TSI								
	1 TSU								
200000 to 1 million	2 MI	8	102.5	382.5	7.2	15.2	32.9	18.6	14.3
	1 MU								
	1 TSI								
	2 TSU								
Less than 200000	1 MI	9	31.6	78.0	2.6	2.2	22.5	−4.5	27
	2 MU								
	4 TSI								
	5 TSU								

[a]Average rate of total sales; because of the mix of small and medium-sized firms in the sample, the average of the rate of growth of individual firms would be very different

[b]Productivity growth of 17%, 30%, and 3.2%, respectively, for TSI, TSU, and MI firms if the largest and the smallest (1984–1985) firms are removed from each category.

had mixed feelings about the success of their firms, whereas the independents expressed more satisfaction. That difference in perception cannot be explained by the real growth rate of sales (Table 5), the volume of sales, or the growth rate of sales per employee over the period, the two groups being quite similar. Rather, it may reflect the more positive thinking of the independents, ready to react with confidence, even if things are not as good as they could be. As shown in table 4, four independents (two TSI, two MI) who felt that their operations were doing well actually had a reduction in real sales over the 1981–1984 period; on the other hand, two professors who felt that their firms were doing poorly actually had good increases in sales level. Growth of sales is clearly a very partial measure of success.

It is difficult to evaluate the relationship between the type of firm (MI, MU, TSI, TSU) and its growth rate between 1980–1981 and 1984–1985. On the average, the growth rate of employment for all the firms (7.8%) is very close to the rate of 7% reported by Dunkelberg and Cooper (1982) for a large sample of firms of similar age. As shown in Table 5, our MI firms tend to be larger and have the lowest average annual growth rate for sales, employment, and productivity, and the MU firms are the smallest and have the highest growth rates over the period. As this study includes only survivor, these results were expected, growth being generally negatively correlated with size.

Globally, however, if one looks at the overall relationships between sales, employment, and age, differences between types of firms can be observed. Figures 1, 2, and 3 show the

FIGURE 1 Effect of the age of a company on the evolution of its sales level.

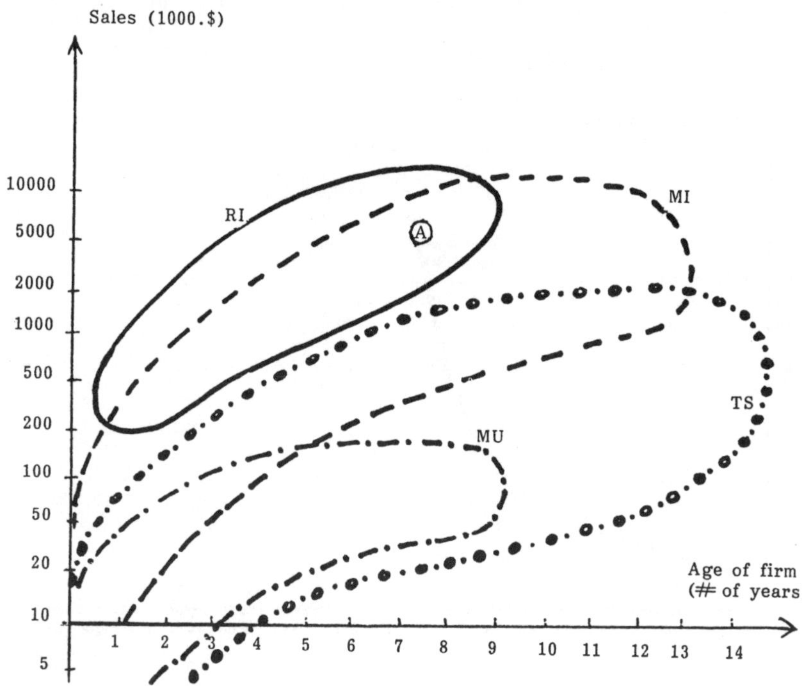

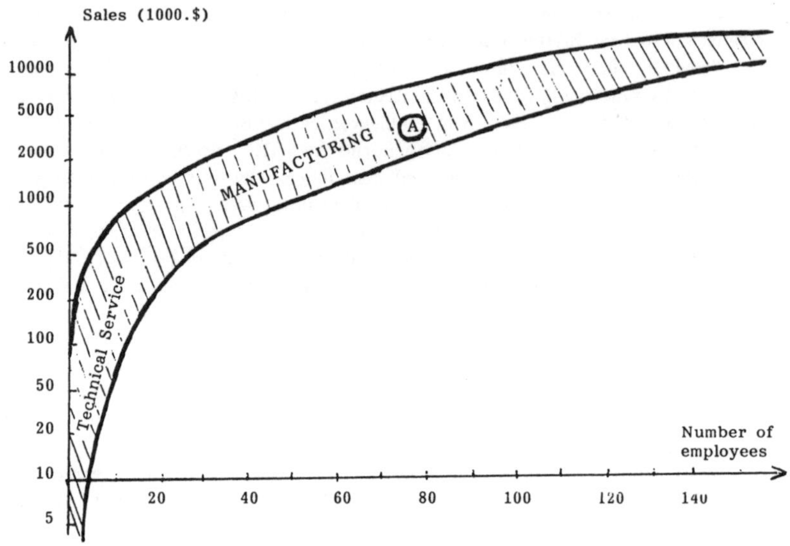

FIGURE 2 Sales and level of employment.

FIGURE 3 Employment and age of the firms.

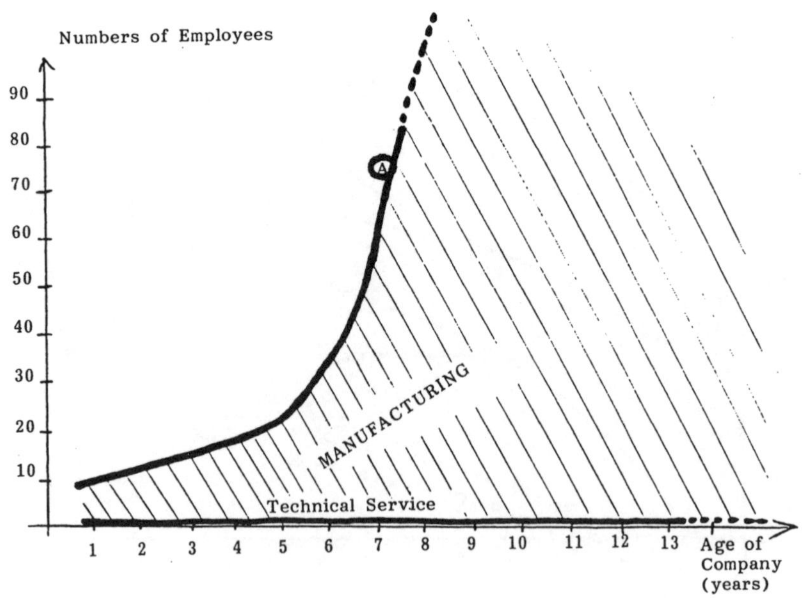

cluster of all the observations in the sample. The data do not permit distinguishing between technical service firms created by an entrepreneur still on university payroll (TSU) or having left the university (TSI). In both cases, we have mostly a mix of very small and small companies employing less than ten people and, if we exclude the two largest firms, having reached an average level of sales of $140,000.00 after 9 years of existence. As shown on Figure 2, these firms have a very high marginal rate of productivity per employee, sales growing very quickly with the number of persons employed. Two firms stand out because of their size: one TSI with over 30 employees is an applied research company with industrial and government clients that occasionally does some manufacturing to prepare a prototype or develop its instrumentation; the other one, a TSU with now close to 200 employees, is involved in engineering consulting. (This latter firms was not included in the TS cluster of Figure 1; its 1980–1981 sales and employees levels appear as point"A" on Figures 1,2, and 3.) In terms of characteristics, these two TS firms are closer to the manufacturing firms included in the sample than to the other TS firms.

Whereas manufacturing firms led by an entrepreneur still on university payroll seem to stay small and not to differ much in terms of sales and level of employment from companies offering technical services, the other manufacturing firms in the sample tend to become much larger (clusters MI and MU, Figure 1). These firms have, on the average, a lower marginal rate of productivity per employee than TS firms (Figure 2), the probable effect of the lower average skill level of their manpower, which includes proportionately more production workers and fewer professionals than their technical service competitors.

Links with a University

As noted earlier, 13 "technical services" and 6 "manufacturers" are still on the payroll of Canadian universities as full-time (11 cases) and part-time (8 cases) full and associate professors. Seven of the entrepreneurs who started their business as students or postdoctoral fellows are still associated with their university as professors (three cases) or as adjunct professors or occasional lecturers. As shown in Table 6, TS firms in particular take good advantage of their links with a university: as a source of "cheap" qualified labor in particular, and as a source of space and technical equipment. It should be noted that among those renting office space, are two successful MI; one of them is the only firm in the sample in which the university is actually represented as a shareholder.

A relatively large number of small firms seem happy to be able to take advantage of university laboratories and equipment, for a fee or for free. Such facilities are generally too expensive for smaller firms to have themselves, so access to university laboratories and equipment is particularly helpful to the small firms. Manufacturing companies taking ad-

TABLE 6 Entrepreneurs–University Links (Number of Firms with Such Links)

Type of firm	Number of respondants	Use of space		Laboratory/equipment		Hire students	Offer services
		Free	Rented	Free	Rented		
TSI	7	0	0	2	1	6	2
TSU	10	1	1	0	4	3	1
MI	8	0	2	1	2	4	0
MU	4	0	0	1	2	1	2
Total (%)	29	3%	10%	14%	31%	48%	17%

TABLE 7 Growth of Sales and Use of University Laboratories and Equipment (For Free or for a Fee)

Type of firm	Companies NOT using university laboratories: Number	Median sales 1984–1985 ($1000)	1980–1984 Median average annual growth rate	Companies using university laboratories: Number	Median sales 1984–1985 ($1000)	1980–1984 Median average annual growth rate
ONLY Companies Created before 1978 (in Parentheses, Averages)						
TSI	2	1100 (2750)	24% (26)	1	180 (185)	14% (60)
TSU	2			2		
MI	4	1900 (3250)	8% (64)	3	520 (815)	8% (10)
Total	8	1900 (3000)	17% (45)	6	220 (500)	13% (35)
ALL Companies Having Provided the Data						
TSI		174	24%	3	85	13%
TSU				3		
MI		1900	8%	3	160	16%
MU				3		
Total		500	23%	12	110	13%

vantage of university laboratories and equipment tend to be smaller than the others, but have a comparable rate of growth (Table 7). This is, however, not the case for TS firms using such university services, which while smaller than the others, also have a lower median growth rate: would the easy availability for such services be counterproductive, keeping alive in some cases firms that are not really viable, or reducing the initiative of their owners?

A little over 50% of the first product or service offered by the 25 firms that provided this information (Table 8) was developed at the university. This is an interesting measure of the efficiency of the university–industry technology transfer. MU firms are characterized, as expected, by a very high level of university-based product development. It is, however, difficult to explain the large difference between TSI and TSU firms: perhaps interpretation of the wording of the question "developed in university laboratory" was difficult, or perhaps TSU entrepreneurs wished to avoid (or not reveal) conflicts of interest. More interesting and significant is the weak inverse relationship observed between university-based product de-

TABLE 8 Percentage of First Product Developed in University Laboratory

	Type of Firm: TSI	TSU	MI	MU
Number of answers	6	8	8	3
Average % (s.d.)	73% (41)	26% (40)	43% (44)	97% (6)

	Size of firms (sales, $1000): less than 200	200 to 1000	over 1000
Type of firms	1MI,2MU,4TSI,5TSU	2MI,1MU,1TSI,2TSU	5MI,1TSI,1TSU
Average % (SD)	70% (41)	42% (43)	27% (43)
Average % (SD), MI and MU only	100% (0)	46% (40)	38% (47)

velopment and size, for all the firms and also for the subgroup of manufacturing firms. As the firms in most of our subgroups have similar average ages, size can be used as a proxy for average growth rate: there seems to exist a weak negative correlation between the growth rate of a firm and the percentage of first product developed in university laboratories. Another study would be necessary to investigate the reasons behind that seemingly negative effect of extensive university-based product development: type of product and newness of the technology, type of market, personality of the entrepreneur, reduction of the risk and of the urgency felt by the product developer. . . .

At least four of the founders still on university payroll have been involved with incubator centers affiliated with a university. Views on the usefulness of such centers vary; for example, one successful technical service firm notes that the center was very helpful, while another judges that, even though the university was supportive, it got little from the center. Only one of these four companies (two manufacturing, two technical services) is doing well.

The entrepreneurs listed a number of advantages resulting from their special relationships with the university. The largest number of respondents (16 out of 31) noted the research field: exchange of ideas, discussion of problems, scientific expertise, new ideas. This aspect of the university–industry interface, noted mostly by the technical service firms but also by a few manufacturing firms, is an illustration of the transfer of knowledge and of the need for good communication required by the innovation process. The other advantages of close links with a university include access to larger instrumentation and access to cheap qualified manpower with graduate and undergraduate students.

GROWTH OF HIGH-TECHNOLOGY FIRMS CREATED BY UNIVERSITY-BASED ENTREPRENEURS

The Sample

Detailed information was obtained in 1985 on the start-up phase and early growth of 70 young Canadian high-technology entrepreneurial firms. These firms were part of the group of 76 entrepreneurial firms founded between 1965 and 1980 and currently involved in some degree of manufacturing found among the 163 respondents to a short survey of 508 canadian high-technology companies.

Only eight of these 70 companies were created by individuals while they were at a university as professors, postdoctoral fellows, or doctoral students, a proportion that is very close to the 10% of a sample of Swedish companies (Utterback et al. 1982) and higher than the 3.3% similar firms found in a large sample of Silicon Valley start-ups (Cooper 1972). Our analysis is based on the investigation of these eight companies.

Analysis

The eight companies are between eight and 16 years old (9.8 years on the average). Two were created by Ph.D. candidates, all the others by professors of engineering and science. In five cases, the entrepreneurs left the university at time of start-up or very soon after and experienced a moderate rate of growth (after 10 years, median sales of 2.6 million dollars, average sales of 11 million dollars, median employment of 75 persons, average employment of 190 persons). A sixth company with one founder still at university experienced rates of growth of sales and of employment similar to the average of the previous group. This company was created on a near-equal-ownership basis by four partners, of whom three were university based: the ownership pattern has barely changed over the years, but the partner who kept his academic position has very little to do with the management and operations of the company. The two other companies in the sample experienced the slowest growth and development of the group. One was created by two university professors who kept their university postings until at least 1985. One has now resigned to take full-time care of his company. The other company, fully owned by a university, was created by three professors and researchers who are no longer connected with the firm. Even if these two firms have now reached or passed the half million dollar sales level, they have been growing at a much slower pace than most others in the sample.

This sample of eight canadian high-technology firms is not large enough to ascertain external validity, the generalization of the "slow-growth syndrome" to all the firms linked to a university. It does, however, increase the credibility of the results obtained in the study described in the first part of this paper and leads one to seriously ponder the cause-and-effect relationships in Canada between some universities and the technical firms linked to academia through partial or full ownership, or academic postings.

CONCLUSIONS

The major points illustrated by this study of two samples of Canadian academic entrepreneurial firms are 1) the difference in growth behavior and dynamics between technical service firms and manufacturing firms, and 2) the seemingly negative effect on the growth and development of manufacturing firms of continuing contracts with the university. Manufacturing firms created by a member of the academic community while at a university to exploit some results of his research seem to behave similarly to most high-technology firms if the entrepreneur leaves his (her) academic posting. However, this does not seem to hold for "part-time" entrepreneurs who stay on university payroll. These conclusions result from the analysis of two relatively small samples and may be due more to the personal characteristics of the entrepreneurs than to any real impact of the academic community on young firms. They do, however, warrant more extensive research because they have important implications for the academic scientific community.

Whereas academic entrepreneurship is certainly an efficient university–industry technology transfer mechanism, these results pose crucial questions to university administrators and to venture capitalists. For the owner–professor, remaining on university payroll does reduce risk and provide easy access to equipment, laboratories, and student labor. It may result in the creation of small businesses by people who, otherwise, would not be entrepreneurs. For the universities, the part-time professor/part-time businessman route may provide for an efficient use of academic resources and increase contacts with the business world for the benefit of students and of applied research. For the venture capitalist or for the university wanting to maximize the revenue accruing from an invention or an innovation, the data indicate, however, that licensing out the invention may be the wisest course, or that, if a company is created to exploit it, that company should be as independent as possible from the university.

REFERENCES

Birch, D. October 1984. Communication at the *International Symposium on the Evolution of Small and Medium Sized Businesses in an Ever-changing World*. Québec, Canada: Université du Québec à Trois-Rivières.

Cooper, A. C. 1972. Incubator organizations and technical entrepreneurship. In Cooper and Komives, eds., *Technical Entrepreneurship, A Symposium*. Milwaukee: Center for Venture Management, pp. 108–125.

Doutriaux, J. April 1984. Evolution of the characteristics of (high tech) Entrepreneurial firms, *Proceedings of the 1984 Frontiers of Entrepreneurship Research Conference*. Babson College and Georgia Institute of Technology, p. 369.

Doutriaux, J., and Peterman, B. 1982. Technology transfer and academic entrepreneurship at Canadian universities, *Technological Innovations Studies Program. Ottawa, Canada: Technology Branch, Department of Industry Trade and Commerce.*

Dun and Bradstreet 1967. New York, p. 3. *Patterns for Success in Managing a Business.*

Dunkelberg, W.C., and Cooper, A.C. 1982. Patterns of small business growth, *Academy of Management Proceedings* 1982:411.

McQueen, D.H., and Wallmark, J.T. April 1984. Innovation output and academic performance, *Proceedings of the 1984 Frontiers of Entrepreneurship Research Conference*. Babson College and Georgia Institute of Technology, p. 175.

Olofsson, C., and Wahlbin, C. April 1984. Technology-based new ventures from technical universities: A Swedish case, *Proceedings of the 1984 Frontiers of Entrepreneurship Research Conference*. Babson College and Georgia Institute of Technology, p. 192.

Rodenberger, C.A., and McCray, J. April 1981. Start-ups from a large university in small town, *Proceedings of the 1981 Frontiers of Entrepreneurship Research Conference*. Wellesley: Babson College, p. 84.

Shapiro, A., and Giglierano, J. April 1982. Exits and entries: A study in Yellow Pages journalism, *Proceedings of the 1982 Frontiers of Entrepreneurship Research Confererence*, Babson College, p. 113.

Tyebjee, T.T., and Bruno, A.V. April 1982. A comparative analysis of California start-ups from 1978 to 1980, *Proceedings of the 1982 Frontiers of Entrepreneurship Research Conference*. Wellesley, MA: Babson College, p. 163.

Utterback, J., Reitberger, G., and Martin, A. April 1982. Technology and industrial innovation in Sweden, *Proceedings of the 1982 Frontiers of Entrepreneurship Research Conference*. Wellesley, MA: Babson College, p. 177.

Vesper, K. H., and Crosswhite, J. April 1983. Faculty start-ups from schools without entrepreneurship courses, *Proceedings of the 1983 Frontiers of Entrepreneurship Research Conference*. Wellesley, MA: Babson College, p. 72.

[20]

Growing the Linköping Technopole—A Longitudinal Study of Triple Helix Development in Sweden

Magnus Klofsten
Dylan Jones-Evans
Carina Schärberg

ABSTRACT. In the context of successful examples of regional development in Sweden, the Linköping region is often named. Whilst maintaining a historical reputation for culture and learning, Linköping has developed, during the last thirty years, as a major centre within Sweden for technological growth, especially in the creation and development of new technology-based firms. This high technology environment has been further enhanced by the presence of a growing international university, a number of significant multinational high-technology companies and the location of several public sector research establishments. Previous research on regional development suggests that this milieu could be characterised as a 'technopole,' namely an arena where these organisations co-operate with each other in a formal or informal way to develop the technological capability of the region. This paper sets out to examine and describe the different 'actors' that have contributed to the recent development of Linköping as one of the fastest growing technological regions in Europe. The linkages between the actors are examined in detail. The results of the study indicate that one of the most important actors behind the positive development is the university. In addition, the development that has taken place over the last thirty years can be likened to a spiral where success begets success to foster a positive entrepreneurial climate.

1. Introduction

Rapid technological advance within a region can result in increased competitive advantage in local, national and international markets, which can subsequently lead to an expansion in industrial output, employment and prosperity (Cooke, 1996; Maillat, 1995; Saxenian, 1994; Smilor *et al.*, 1990). However, the industrial competitiveness of a region and success in accelerating growth and increasing employment depends on the capacity of various regional 'actors' to innovate in response to changing external conditions, especially the continuing rapid pace of technological development. Consequently, various approaches have been developed to initiate and sustain the development of technological innovation at a regional level, ranging from designed systems such as regional innovation strategies (Landebaso, 1996) to more specific organisational developments, such as Business Innovation Centres. However, one approach that has seen increased interest during the last few years is the concept of a regional industrial complex, commonly known as technopoles (Gibson and Smilor, 1988). There are varying definitions for this phenomenon, but they can be broadly described as planned centres for the promotion of high-technology industry whose main aim is to generate the basic materials of the informational economy (Castells and Hall, 1994). As a result, many cities and regions can become critical agents of economic development.

As Cooke and Morgan (1993) have emphasised, there are a number of key elements of such a regional network. These include a thick layering of public and private industrial support institutions, high-grade labour market intelligence and associated vocational training, rapid diffusion of technology transfer, a high degree of inter-firm networking and, above all, receptive firms well-disposed towards innovation. In particular, one of the crucial indicators of successful development through the presence of a regional technology network, such as a technopole, is the type of 'actor' involved within the network. These can include research institutions and universities (Segal, 1986), small technology-based firms (Jones-

Centre for Innovation and Entrepreneurship
Linkoping University
Linkoping, 581 81 Sweden

Journal of Technology Transfer **24**: 125–138, 1999.

Evans and Westhead, 1996), intermediary organisations (Klofsten and Jones-Evans, 1996) and science parks (Hassink, 1996). However, the presence of these actors alone, working in isolation, will not lead to the success of the region, and it is important to understand the different linkages that exist between these organisations in developing the innovative potential of a region.

This paper will discuss and present an analysis of the development of one specific 'technopole,' namely that of the city of Linköping in Sweden. It will concentrate on examining the role of, and links between, the different economic actors that have made the city one of the success stories of Swedish industry during the last thirty years. In particular, it will highlight the role of the university as a direct catalyst in this development. This paper will also describe the sequential development of the Linköping technopole and the changing roles played by different actors over time. These actors form the 'triple-helix' within the Linköping region, namely industry, the university and government (in this case, the public sector at a local level). Therefore, this paper will demonstrate, in a non-quantitative way, the formal and informal network relationships that exist between the different groups of actors. It will also attempt to determine whether there have been any changes over time in the relationship between actors, the development of network itself, and the main driving force behind the development of the technopole.

2. The development of the Linköping region

The city of Linköping, and its evolution to become one of the most successful European examples of technology-based regional development, is one of Sweden's best-kept secrets. Whilst the Cambridge phenomenon in the UK and Sophia-Antipolis in France have been highly publicised as examples of successful technopoles developments within Europe (Castells and Hall, 1994; Segal, 1986; Laffitte, 1985), very little is known, outside of the region, of the growth of this former administrative centre and agricultural market town.

The industrial process in Linköping can be divided into a number of steps and dates back to the beginning of the century. The first step follows the decline of the furniture and textiles sectors in Sweden, which had earlier played a considerable role in the development of local industry. In the early 1910s, the electricity industry was established in Sweden, and with the completion of a regional railway network, Linköping became an important centre for industries associated with mechanisation and electrification. In addition, the metal industry became dominant in Linköping during the 1930s, and there was a heavy growth with the development of manufacturing in railway wagons, railbuses, buses, thermo-technology and foundry. The second historical step occurred during the 1940s, when Saab developed the Swedish aircraft industry. This was initially focused on military aircraft but later diversified into civilian aircraft. Saab was soon to become the major employer in the city, and its prime position in the development process of Linköping was further underlined with its role as the catalyst for the decision to establish a university in the late 1960s, which led to the third step in development of the city.

With the establishment and subsequent growth of the university to encompass three faculties (arts and sciences, health and sciences and technical), further high technology organisations were attracted to the Linköping region. These included international firms such Fujitsu, Bull and Ericsson. Meanwhile, an increasing number of small spin-off firms was created in the vicinity surrounding the university, which were characterised by products or services based on advanced technology or technological expertise, such as image processing, fibre composites, electronics, network security, sensors and vacuum technology. These firms, which operate in a diversity of areas and employ more than 13,000 people have a combined turnover of more than 11.6 billion Skr (US$1.7 billion), and include large firms such as Saab's Aircraft Division and Ericsson (Klofsten, 1996). In addition, the Swedish government began to support the relocation of its research institutes during the 1970s. As a result, the Swedish National Road and Transport Research Institute and the Swedish Defence Research Establishment were moved to Linköping, mainly to take advantage of the links with both the university and the knowledge-intensive industrial environment. During the 1980s, a regional initiative sup-

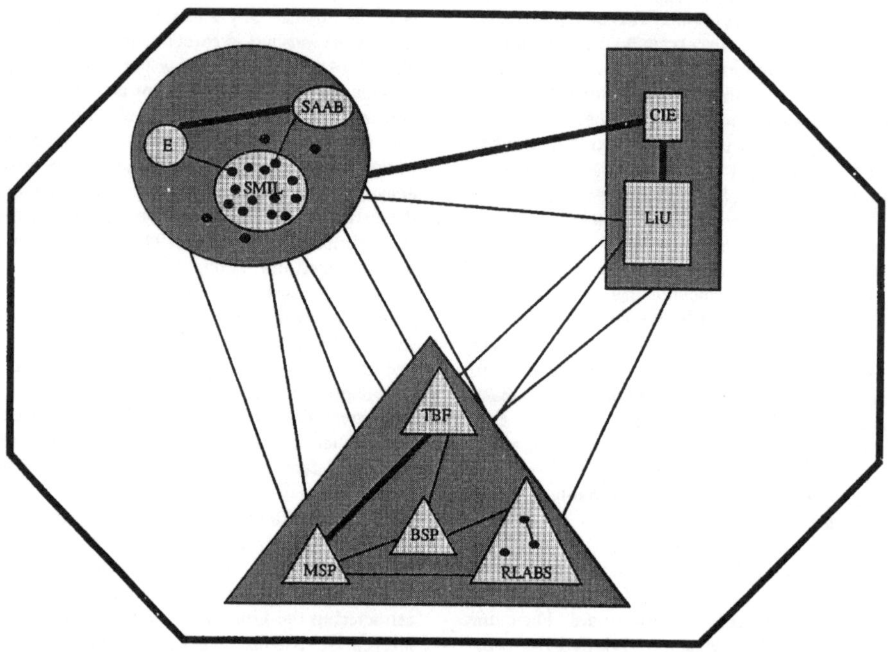

Index

BSP: Berzelius Science Park
CIE: Centre for Innovation and Entrepreneurship
E: Ericsson
LiU: Linköping University
MSP: Mjärdevi Science Park
R-Labs: Government Research Labs
SAAB: Saab Aircraft Division
SMIL: Business Development in Linköping
TBF: Technology Bridge Foundation

University organisation
Public sector bodies
Private industrial firms
Formal links
Informal links

Figure 1. The Linköping technopole.

ported by both private and public sector actors led to the creation of both the Mjärdevi and Berzelius Science Parks.

Whilst the development of the Linköping technopole has been fairly recent, it has more in common with the older types of technopole, built on the basis of innovative milieu, such as Silicon Valley in southern California (Saxenian, 1994) or Route 128, Boston (Scott, 1986), than the other types such as science cities, technology parks, or technopole programmes (Castells and Hall, 1994). As Lundgren (1995) notes, industrial milieu or networks can be defined as consisting of actors linked together by their performance of complementary or competitive industrial activities, employing or consuming economic sources to process other resources. The remainder of the paper will discuss the main actors involved in the process of development of the technopole in Linköping.

3. The technopole—Actors and linkages in Linköping

Private industrial firms

Ericsson. Ericsson's organisational structure changed at the beginning of 1997 and the company's central organisation now consist of three main business areas. Two of these are represented in Linköping, namely Business Area Mobile Systems (which consists of cellular systems, mobile data communications/personal pager systems, and private radio networks) and Mobile Phones and Terminals, which is a new division aimed at the end-user market, mainly mobile telephones.

Within the technopole, Ericsson has formal industrial links with Saab in the company Ericsson-Saab Avionics, with each of the firms owning 50% of this joint venture. This has been established to concentrate knowledge in some of the priority areas of the Swedish defence industry. Although there are no developed links with SMIL as an organisation, informal links exist in the shape of vendor-supplier relationships. Services are bought from SMIL members—SoftLab and Frontec—as well as larger Linköping firms such as SAAB Combitech and Cap Gemini. Ericsson is also represented on the board of Mjärdevi Science Park, and whilst there is some collaboration regarding real estate development on the site of MSP, there is no co-operation in business operations. Ericsson has no links with Berzelius Science Park (because of the medical focus of the park's activities), any of the government research laboratories, or the Technology Bridge Foundation.

Apart from the collaboration with Saab, Ericsson's closest links with other regional actors seem to be with the university. This link is not only through recruitment of graduates but also through representation on the boards of the Institute of Technology, the Faculty Board for Programme in Applied Physics and Electrical Engineering and Programme for Computer Science and Engineering. In particular, there is close R&D co-operation between Ericsson and the departments of Computer and Information Science and Electrical Engineering, which is set to continue and strengthen during the next few years.

Saab Aircraft Division. Saab Aircraft is principally an aircraft and defence company, the aircraft side focusing on the Gripen and Viggen military programs and the 340 and 2000 commercial airliners programs. In addition to military aircraft, the company's defence products include guided weapons, electronics, optronics and training equipment. Saab is also active in developing a number of commercial high-technology niches for space, industry, traffic systems, data communications and telecommunications. The company has approximately 8,000 employees and sales in 1995 totalled almost US$ 1,120 million.

As with Ericsson, there are informal links with the other industrial firms in the region, including SMIL members. However, unlike Ericsson, Saab has a long-standing relationship with a number of the government research laboratories, especially the National Defence Research Establishment), which is becoming increasingly formal with the increase in the need for co-operation. However, there are no direct links with the other public sector bodies, including the Technology Bridge Foundation or the science parks. There is considerable contact with the university departments such as the Departments of Mechanical Engineering, Computer and Information Science, Sys-

Table I
Actors of the technopole

Actor	Founded	Employees	Activity
The University			
Linköping University (LiU)	1969	17000 students 2600 staff	Today there are three faculties which conduct multi-disciplinary research and development activities: Institute of Technology, Faculty of Arts and Science, Faculty of Health Sciences
Centre for Innovation and Entrepreneurship (CIE)	1994	4	Creates a fruitful integration and stimulation of activities for technology-based firms, and teaching and research of technology-based entrepreneurship within the university
Private firms			
Ericsson	1981	2300	Manufacturing of mobile telephones and provider of cellular systems and data communication
SAAB	1937	8000	Principally an aircraft defence company focusing on the Gripen and Viggen military programs and the Saab 340 and Saab 2000 commercial programs.
Business Development in Linkoping (SMIL)	1984	—	An association of business leaders which provide support for technology-based companies in the region
The Public Sector			
Mjärdevi Science Park (MSP)	1984	2.5	Represents a broad spectrum of high-tech expertise and specialised skills including incubator services and provision of seed capital
Berzelius Science Park (BSP)	1997	1	Entrepreneurial support in development of firms of medicine and/or medicine technology
Technology Bridge Foundation (TBF)	1994	2.5	Goal is to build bridges between industrial organisations and universities, and to initiate and finance projects for further product development
National Laboratory of Forensic Science (SKL)	1975	160	Central laboratory for forensics in police investigations
The Swedish Institute of Production Engineering Research (IVF)	1982	10	Non-profit organisation which works in the field of applied research and technological development of manufacturing companies
Swedish Geotechnical Institute (SGI)	1975	80	Geotechnical centre providing Swedish and and international clients with expert knowledge
The National Defence Research Establishment (FOA)	1978	300	Conducts research for the total defence and support of disarmament and international security
The Swedish National Road and Transport Research Institute (VTI)	1975	220	Commissioned research for the Swedish Transport and Communication Research Board and the Swedish National Road Administration
Institute of Forensic Genetics (IFG)	1977	30	Analysis of blood to be used mostly in instance such as Paternity cases
Defence Material Administration, Testing Directorate (FMV)	1933	300	Testing of military aircraft

tem Technology and Mathematics. Activities include teaching, R&D industrial postgraduates and adjunct professors. Saab also have representatives on the faculty board for the Institute of Technology, which has resulted in the creation of the Saab-Linköping Institute of Technology Council, which brings together managers from Saab and professors from the Institute of Technology to discuss ways of developing education and R&D.

SMIL—Foundation for Business Development. SMIL was founded in 1984 by a group of business leaders and people from Linköping University. SMIL is actually a continuing development from a previous spin-off firm club founded by Industrial Liaison Office (ILO). Its origin stems from the realisation by the university and the member firms that the club needed more structure and organisation to be an effective support organisation. At this time (the first half of the 1980s), there were frequent spin-off activities occurring in the Linköping business community, with approximately 40 firms being established. The majority of these originated from the University, but others came from various high technology industrial firms in the region, such as the Saab Aircraft Division, Ericsson Radio, and Swedish Defence Research Establishment. Indeed, the first members were recruited from these organisations. The initial activities of SMIL consisted of breakfast meetings where lecturers were invited to speak on subjects of current interest to the founders of the small technology-based firms. Over time, these activities have continued to grow and develop into a number of different events. At the start there were 20 members which has evolved into a network of approximately 150 firms in the group which are all located in the vicinity of Linköping. These firms are offered a number of well-defined activities.

The supporting network in the Linköping area is highly conducive to the development of innovation within, and the growth of new technology-based firms. During the last 10 years, a network has evolved which incorporates a close interaction between the small technology-based firms—represented by SMIL—and Linköping University—represented by CIE. Therefore, the main link is with CIE, which operates through formal co-operation on programmes and other activities, with CIE contributing structural knowledge, credibility, personnel and financial resources. The link to the TBF is new and informal, and the natural contact point for most SMIL members is through project financing. The link with MSP has been built up during a longer period of time and is focused mainly on the incubator activities provided by the park. Indeed, the majority of the SMIL firms are now located in the science park, which implies a more natural contact and strengthens the link between the two organisations. Whilst some of the research laboratories belong to the SMIL network, they are not active and see their membership as a way of gathering information on what is happening in emerging high-technology small businesses in the region.

Public sector bodies

The Science Parks

Mjärdevi Science Park (MSP). In the early 1980s, a number of industrial firms, who were enthusiastic about working with the increased technological capabilities at Linköping University, expressed an interest in developing some sort of infrastructural instrument for implementing this relationship. At the same time, the local authority in the city of Linköping—the Municipality of Linköping—had begun to offer incubator facilities and basic services for new companies spinning out from Linköping University. In this environment of closer links between academia, industry and the local authority, it became quickly apparent that a further development was necessary if these different initiatives were to grow successfully. Thus the impetus for the development of a science park became a major priority in developing the industrial base of the city. The close co-operation between the different partners, headed by the Municipality of Linköping, resulted in unanimous decision to build a science park in Linköping, and in less than a year, Mjärdevi Science Park had become a reality. The Municipality of Linköping currently fully owns Mjärdevi Science Park AB, a company created to co-ordinate all achievements for the Mjärdevi-area. In addition, the Municipality also functions as the main developer at the Science Park.

Originally started in 1984 with six companies employing 150 people, MSP now has 110 companies situated on-site employing over 2500 people. Whilst the original aim of the Mjärdevi Science Park was to represent a broad spectrum of high-technology expertise and skills, the technological base of the region, coupled with the particular scientific expertise of the university, has led to a cluster of information technology firms in the region. These concentrate on the development of aspects as diverse as electronic image processing, telecommunication systems and software development. During the last couple of years, over 1000 jobs have been created at the park, with the largest growth in employment coming from large firms such as Ericsson Radio Systems (800 jobs), whose facility produces mobile phones and manufactures radio base stations for mobile phone systems and personal pagers. Other multinational corporations sited at MSP include ICL, Digital Equipment and Bull. Since it was established, the park has also positioned itself as an incubator for spin-off companies, with more than half (55%) of these having less then five employees. In addition, smaller niche firms with their origins in Linköping University make up a third of the companies on MSP. Indeed, the Municipality has encouraged the development of small technology-based firms through a number of initiatives, including the provision of business incubators on the science parks and through the availability of seed and venture capital.

Mjärdevi Science Park has only one formal strong link, namely with the Technology Bridge Foundation, although this has only become prominent during the last couple of years. However, this is not surprising, as the TBF is situated in near proximity to the Science Park, and they have a close working relationship. Indeed, one senior employee shares his time between the two organisations. The Science Park also has informal links with most of the other major actors in the technopole. Its linkages with the other industrial development in Linköping—Berzelius Science Park—is mainly characterised by informal contacts and co-operation, although the managing director of MSP provided advice and support with the development of the park. Similarly, there is contact with the various research laboratories situated in the Linköping region, in particular the Industrial Microelectronics Centre. The main industrial contact of MSP is with Ericsson, which is not surprising as they are situated on the park, and two of their senior managers sit on the board of the science park. With regard to the university, the main contacts are with CIE who are involved in the provision of business support programs and incubator facilities, although these are not formal. There are also a number of other contacts, especially with regard to the IKP (Department of Mechanical Engineering).

Berzelius Science Park (BSP). This Park was established with the goal of creating more companies in medicine and medical technology within the region by supporting entrepreneurs in business establishment and helping them to grow via the development of incubators. It was therefore developed to bring researchers in the medical field into contact with innovators and entrepreneurs. It has a number of ongoing projects in partnership with the University Hospital, the National Co-operation of Swedish Pharmacies, Mjärdevi Science Park (MSP) and the Technology Bridge Foundation (TBF). There are close informal links between the two science parks in the city of Linköping. Whilst there are linkages with smaller firms in the region (mainly through SMIL, the technology-based firms' network), there are no contacts with the two major industrial firms—Saab and Ericsson—mainly because they are not involved in the development of any applications in the medical industry. Similarly, BSP has no links with any of the government research laboratories. Although not as close as in the case of Mjärdevi Science Park, Berzelius Science Park has links with the Technology Bridge Foundation. These are expected to become more formalised as companies increasingly demand support with financing. BSP also has close links with the university hospital and medical faculty, as well as IMT (Department of Biomedical Engineering) and IKP (Department of Mechanical Engineering) at the Institute of Technology. As the park develops further, then these relationships will strengthen. Whilst it is aware of the activities of CIE, it has no links with this organisation.

Government Research Laboratories. There are a number of government-run research laboratories in the Linköping region.

National Supercomputer Centre (NSC). The NSC, which provides supercomputer resources to academic users in Sweden, has been financed, since 1987, by Saab and the Swedish Meteorological and Hydrological Institute (SMHI), which receives its weather forecasts from this institution. The co-operation with Saab consists of maintaining the company's information technology capability and those of the users of Saab's technology. There is very little contact with local high-technology firms, with most of the clients of the laboratories being national firms such as Volvo and SKF. There is also no contact with Ericsson, and interaction with other laboratories tends to be on an informal personal level between staff, rather than any co-ordinated links between the organisations. There is also very little contact with the university, and none with the Technology Bridge Foundation and the two science parks.

Institute of Forensic Genetics (IFG). As with the NSC, the Institute of Forensic Genetics has only minimal co-operation with other research laboratories, most notably the National Laboratory of Forensic Science. While IFG has very little contact with MSP, they have close co-operation with Berzelius Science Park as they are both involved in the development of medical technology. However, there is almost no medical industry in Linköping and as a result, there are no links with Saab, Ericsson or the Technology Bridge Foundation. The closest links seem to be with the university, with senior members of the Institute teaching within the Faculty of Health Sciences and chairing one of the research schools.

National Laboratory of Forensic Science (SKL). No real formal links between SKL and other research laboratories, although they have worked with organisations such as the National Defence Research Establishment. There is very little contact with Saab and Ericsson, and none with the science parks. Again, the strongest links are with the university, and SKL support a number of postgraduates in the department of Mechanical Engineering, and the department of Physics and Measurement Technology.

The Swedish Institute of Production Engineering Research (IVF). IVF is a non-profit organisation that works in the field of applied research and technological development. Its objective is to enhance the efficiency and competitiveness of manufacturing companies by developing and disseminating knowledge of new technology. It has wider links than some of the other laboratories (although no strong links with similar organisations). It has collaborated with the Department of Mechanical Engineering, as well as the Department of Computer and Information Science. It has also worked closely with the ILO on a number of occasions during the last few years. Whilst IVF is a member of SMIL, this is for information gathering only, and it has only a minimal interest in both science parks and the Technology Bridge Foundation.

Swedish Geotechnical Institute (SGI). The SGI is the official Swedish agency for geotechnical research and is active in research and safety matters. The Institute is under the administration of the Ministry of Transport and communications. It has close links with two other research laboratories namely FOA (The National Defence Research Establishment) and VTI (Swedish National Road and Transport Research Institute). The collaboration with the latter organisation has increased considerably during the last few years due to the complementary competences of both laboratories in the environmental area. Again, there is close relationship between this laboratory and the university, in particular the Department of Mechanical Engineering, TEMA Institute Department of Water and Environmental Studies, Quality Technology and Management, and Department of Physics and Measurement Technology. There is no contact with the other actors in Linköping.

The National Defence Research Establishment (FOA). The aim of the FOA is to undertake research for the total defence and support for disarmament and international security, and a major client is the Department of Defence. There is only a low level of co-operation with industry, mainly on the level of information exchange and minor contract work—there is some co-operation with SMIL members, although this is generally non-R&D activity. There is also an agreement with both Saab and Ericsson on project co-operation, but only on a minimal scale. Again, the university is seen as the major collaborative partner, with exchange of staff between the universities, and co-operation in over ten projects. Again, there are no contacts with actors such as the Technology Bridge Foundation, Mjärdevi and Berzelius Science Parks or, in this case, the other government laboratories.

The Swedish Road and Transport Research Institute (VTI). VTI conducts commissioned research for the Swedish Transport and Communications Research Board and the Swedish National Road Administration. It was relocated to Linköping in order to increase co-operation with the university sector. It currently has links to over ten departments in different ways, including lectures on courses, study visits, projects work, and tuition, as well as the sharing of staff between the university and the laboratory. One of the strongest formal links is with the Swedish Geotechnical Institute, in which administration and research are shared in the geotechnical area. The links with the industrial firms are not high, and it is a member of SMIL in order to receive information, although it buys some services from SMIL members. There are even fewer contacts with either Ericsson or Saab. Again, there is almost no contact with Mjärdevi and Berzelius Science parks or TBF.

Therefore, in general, there is no clear relationship between the size or age of the research laboratories, and the frequency or level of formalisation of their linkages with other actors. Certainly, there appears to be an "inner circle" of the more active research laboratories in the technopole i.e. IVF and SGI. What is also evident is the low intra-activity between the research institutes themselves, with very little co-operation between the different laboratories. For the research laboratories, the university is the actor with which they have the strongest relationship and their links with private industrial firms and the science parks are few. The latter could be related to the fact that the research labs are specialist bodies and have no incentive to create any links if there is no shared interest in their particular area. Therefore, the research laboratories have developed links with a limited number of actors in which they have a specific interest.

Technology Bridge Foundation. The Swedish Government has initiated seven technology bridge foundations, which are connected to the universities within the country. The goal of these foundations is to build bridges between industrial organisations and universities, and initiate and finance projects that lead to product development, predominantly through supporting industrially relevant R&D, encouraging SMEs to use the university's facilities and knowledge, and supporting co-operation between small and large firms. TBF has a formal link with MSP, mainly through shared personnel and co-operation in the development of companies on the Science Park. Indeed, the roles of the two partners are seen as complementary, with the Foundation focusing on delivery of financial packages for firms, and the science park managing the other various resources required by start-up and growing high technology small firms. TBF also has a similar, if not as prominent, relationship with BSP in the development of spin-off firms.

The industrial ties of TBF with the larger firms of Saab and Ericsson, as the name of the organisation suggests, is mainly as the link for SMEs looking to develop links with these organisations. However, with regard to SMIL, the co-operation is mainly through Mjärdevi Science Park although with regard to the public sector bodies, there are no current linkages between the government research laboratories and TBF, although this is expected to change in the future. There are also links with both CIE and Linköping University. The Technology Bridge Foundation is a minor node in the technopole as it is a relatively young actor focusing on the financial side. Therefore, this actor acquires its strength through its uniqueness in being the only regional organisation that focuses on the financing of innovation.

University organisations

Linköping University. Since its creation at the beginning of the 1970s, Linköping University has grown to become an institution that currently teaches over 17,000 students, employs over 2,600 people, and has an annual income (in 1996) of 1548 MSEK (US$194M). More importantly, the institution has been active in various multi-disciplinary research and development activities, particularly within the new technologies. Many of these have led to spin-off firms initiated by researchers and students that have emerged directly from departments at the university. There are three faculties at the University, from which a number of small technology firms have been initiated.

The Faculty of Health and Sciences. This faculty was initiated in 1986 as a result of a unique co-operation between the University of Linköping and Ostergötland County Council, and includes the medical program and several health-care education programs. Postgraduate research and scientific training are conducted within forty health-care disciplines, and the integrated nature of this work contributes to the development of close-knit teams. This helps participants to gain a wider perspective of health-care development and management. The research within the faculty is also inter-disciplinary, with close ties between biomedical and clinical researchers.

The Faculty of Arts and Sciences. The main aim of this faculty is to provide education in economics, social and behavioural sciences, and the liberal arts, and contains the School of Education, which provides seven teacher-training programs in basic and continuing education for teachers from the pre-school to adult level. The faculty also includes an interdisciplinary graduate school, the Institute of Tema Research, which specialises in technology and social change, health and society, communication, water and environmental studies and child studies. The Institute therefore integrates the main themes from various disciplines to examine diverse and complex modern societal problems.

The Institute of Technology. This is the largest faculty, and annually awards around 500 Masters of Science degrees in engineering and natural science programmes. In addition, a rapidly growing number of specialised graduates earn Certificates of Engineering after two years of study. Postgraduate research and scientific training are conducted in forty fields at eight large-scale departments. It has gained an international reputation for research and development expertise in a number of technological areas, including bioengineering, electronics, materials science, information technologies and industrial engineering. In particular, there is a deep appreciation of the need for inter-disciplinary research both within the Institute and in connection with other academic units, and research projects between the natural and computer sciences was one of the Institute's early innovations. The Institute's various departments are involved in a number of collaborative projects with local industrial partners.

There are some difficulties in grasping the importance and strength of the links between Linköping University and the other actors, as the university consists of three faculties as well as a large number of departments and centres. Each of these has their own contacts with the external world, mainly as a result of internal tradition, policy and management. At an individual level, many researchers or teachers have many contacts, although there are several differences in the characteristics of these links. Some researchers are involved in spin-off firm activities and consulting whereas others consider external activities as an interference with their academic research. A few departments have a strong tradition of creating spin-off firms and thus actively working on their links with the other actors in the technopole. These include the Department of Computer and Information Science, Department of Physics and Measurement Technology, Department of Mechanical Engineering and the Department of Management and Economics (Klofsten, 1996). Historically, since its founding, the university has increasingly developed its role as the anchor of the technopole. However, during the last five years, this has increased due to the stronger emphasis by the Swedish government on the importance of commissioned R&D. Originally, this role of initiator belonged to Saab, but today, in many respects, Linköping University is the centre of the technopole.

Centre for Innovation and Entrepreneurship. The Centre for Innovation and Entrepreneurship (CIE) is an autonomous unit at Linköping University that is conducting activities intended to stimulate the growth and development of technology-based firms. Over the years, a number of different activities have been developed by CIE, with the main emphasis on examining the different problems that occur at different times during a firm's development. With tailored activities, firms gain access to resources relevant to their particular stage of development. These activities are planned and conducted in close co-operation with SMIL whose membership is composed of small technology-based firms in the Linköping area. The main purpose of CIE is to create a fruitful integration between stimulation activities for technology-based firms, and the teaching and research of technology-based entrepreneurship within the university. These can be further divided into specific initiatives for SMIL members, including programmes for start-up firms, programmes for more established firms, and network activities. The success of these activities is as dependent on the contribution of the SMIL firms themselves to the group, as it is on the efforts of CIE in leading the activities. Each of the activities has a co-ordinator who, together with resource persons engaged for the project (who can be both academics or entrepreneurs), is responsible for the implementation of the activity.

Therefore, the main (and formal) link for CIE is with SMIL, with both organisations jointly organising business support programmes for technology-based firms. In this linkage, CIE handles the financing of the activities and personnel resources and is linked to SMIL as the managing director of the centre is also secretary of the board of SMIL. Furthermore, CIE is the direct link into the university for the SMIL-members. There is also an informal link to MSP through the co-operation of the incubator facilities. TBF is financing some of the business support programmes organised by CIE.

4. Linkages within the Linköping Technopole

Formal links and the technopole

In the Linköping technopole, the age of the actors seems to have little effect on the creation of formal links within the network, as demonstrated by the co-operation in the Saab-Ericsson Avionics joint venture. Indeed, this formal cooperation was enhanced by a mutual interest in a specific area. This is also supported by the formal co-operation between two public research laboratories.

The size of actors, in terms of the number of employees seems to affect the creation of formal links, with the larger actors tending to only have formal links with themselves. It also seems as if the degree of mutuality affects the probability to create formal links, with formal links existing within all three sectors in the technopole. This formalisation is either built on the fact that they have complementary services or mutual dependence. In a formal link, the link is built up because of several reasons, as compared to the informal one, where the incentives are fewer. In terms of incentives for creating a formal link, the most developed link is that between MSP and TBF, which has been created in order to meet similar economic, judicial, technical and social objectives.

Informal links and the technopole

It could be seen that informal links develop incrementally over a period of time and tend to be driven by the actors' internal interest. This was particularly noticeable within the public support organisations. Although younger, they seem have developed more informal links than the research laboratories. The public support organisations' activities and the need to develop an informal network within their target group could explain this. The lack of links from the private industrial sector towards, for example, the government research laboratories, could be explained by the fact that the private firms tend to find their client within the same industrial sector, whilst the research institutes are internally oriented. Furthermore, the public support organisations have self-interest in developing a network and they thus form the uniting glue in the technopole.

Linköping University has the largest number of activities of any of the actors in the regions and has developed the most informal links. As the university has grown, and the strength of certain actors in the private industrial sphere has decreased, its role as initiator or the motor of the technopole has overtaken those of specific industrial firms. Finally, it seems as if the activity performed by the actor creates the foundation of the actor's specific role in the technopole. Whilst the university is the central node in the technopole, the glue tends to be primarily the governmental support organisations as part of the public sector bodies. The private industrial firms have a dual role, primarily as discussion partner for the two other types of actors, but also as major regional recipient of graduates from the university.

5. Discussion

One of the primary questions posed by Gibson and Smilor (1988) is to what degree can the emergence of a technopole be speeded up by a local research university that has a favourable policy towards collaboration with private high-technology companies in conducting research and technology transfer? More importantly, does a technopole happen more rapidly when a policy favours that a research university assists high technology firms, than when this collaborative relationship is not planned or directed?

Unlike the case of the Cambridge phenomenon (Segal, 1988), it can be argued that the university has, not only through agencies such as CIE and SMIL, but also through research departments' strong links with other actors, played the major part in the development of a technopole in the Linköping region. In addition, the public sector, through the creation of the Mjärdevi Science Park and other resources, has contributed significantly to the provision of an infrastructure for the growth of the technopole. This is again very different to Cambridge, where neither central nor local government played any direct role in the emergence of the phenomenon. Thus, the interaction between the university and other actors existing within a regional network becomes of critical importance in technology-based regional development. The environment and external linkages identified by previous researchers which are considered to be of fundamental importance for high technologically-based regional development are indeed apparent in the Linköping region.

A number of reasons have been put forward as to why Linköping University has become the main driving force behind the technological success of the region. As a relatively young organisation, the university has encouraged a variety of different methods of undertaking academic research, especially in relation to industrial needs. This is reflected in the multi-disciplinary nature of the research being undertaken at the university, which has greater commercial application. For example, a project team within the department of physics and measurement technology is developing, in association with the computer assistance in engineering team from the department of computing and information science, a project designing programming environments to enable sensor-equipped industrial robots to make intelligent decisions. This leads to opportunities for a variety of spin-off activities, including the establishment of new technological ventures—as various studies have demonstrated, the computer services sector is dominated by small firms, and is often the easiest type of small technology based firm to establish.

The primary research interest of the university is within computer and information science, and it has established an international reputation for its activities in areas such as expert systems, artificial intelligence, and programming environments. While this expertise on its own would not be sufficient to guarantee commercial success, the very nature of this work being carried out within the Institute of Technology is geared towards the development of specific computer-based solutions to industrial and business problems. As a result, the university has established and encouraged a tradition of technology transfer by its staff. For example, one of the first firms to spin out from the university was OFO Electronics, which was established in 1977 by academics from the department of management and economics to design and manufacture professional sound and light equipment for orchestras and institutions. Other early successful firms include Sectra Secure Transmissions AB, which emerged from the electrical engineering department of the university in 1978. Through developing various services in the area of digital communications, the company now employs 110 people and has a current turnover of 90 MSEK (US$11.5M).

In addition to the attraction of large multi-national firms to the region, over 400 small technology-based spin-offs have been established in the Linköping region, with approximately 110 of these emerging directly as a result of academic research activities (Klofsten, 1996). The number of employees varies considerably within the firms, with the largest academic-based firm employing 2000 people, whilst a few others have also passed the 100 employee mark. However, the majority has between 1–5 employees and tends to offer consultant services on a limited basis. In total, the firms provide a labour market contribution of nearly 6000 employees, with the three largest firms together employing over 3000 people. The success of these firms, and the retention of a close relationship with the university (e.g. various professors are sitting on the boards of these firms and many of the firms' employees are former Ph.D students from different departments), has legitimised the process of academic entrepreneurship within Linköping University.

6. Conclusion

After having presented and discussed the role of, and links between, the different actors in the growth and development of the Linköping technopole, it appears that there are a number of different actors needed for creating a technopole. These actors, however, did not form a critical mass of knowledge until the foundation of the university. This new institution then acted as the driving force at the centre of the technopole as it has one of the absolute necessary resources for a technopole development—an extensive knowledge-base to be exploited by other actors.

In particular, it is worth noting the seemingly temporal dimension of the university's regional role, evolving over time into the primary catalyst for regional development. Rome was not built in a day, and neither is a technopole. In the development process, a number of steps can be identified. In this case the evolution before Saab, that is before 1940, can be characterised by an increased importance of SMEs and the rise and fall of geographically centralised companies. The Saab era in the technopole can be characterised as more of the driving force from 1940–1970, when the influence of the Saab group reached its peak. Thus, it seems as if the importance of the different actors in the technopole has been changing over the years both in numbers and in power of influence. Primarily, the number of actors within the technopole has increased continually, with only Saab and a number of unorganised SMEs were active in Linköping until the 1970s. Then the university started growing and the research labs were localised to Linköping. There was a clear need for co-operation between the SMEs and the SMIL network was formed. Since the middle of the 1980s, it can be said that the last cornerstones have been put in place in the technopole, namely the financing and support services for technology-based firms in the shape of SMIL, CIE, MSP, BSP and the TBF. They formed the glue in the technopole. This indicates a kind of interaction between the actors, having their rise and fall as time goes by. Saab had its peak as the primary actor and was followed by the university as the main catalyst in the technopole. The interplay between the actors is increasing continually which, according to the evidence presented here, is also something all the actors wish to develop further as part of their strategic development.

However, as the Cambridge Phenomenon has proved, there is probably no exact formula for developing a successful technopole. Whilst lessons can be learnt from the Linköping technopole, it is up to each nation, region and city to work out an appropriate strategy for technological development with as much vision and imagination as it can conjure up (Castells and Hall, 1994).

References

Castells, M. and P. Hall, 1994, *Technopoles of the World—The Making of 21st Century Industrial Complexes,* London: Routledge.

Cooke, P., 1996, 'Building a Twenty-First Century Economy in Emilia-Romagna', *European Planning Studies* **4**(1), 53–62.

Cooke, P. and K. Morgan, 1993, 'The Network Paradigm: New Departures in Corporate and Regional Development', *Environment and Planning D: Society and Space* **11**, 543–564.

Hassink, R., 1996, 'Technology Transfer Agencies and Regional Economic Development', *European Planning Studies* **4**(2), 167–184.

Johannessen, J., J. Dolva and B. Olsen, 1997, 'Organizing Innovation: Integrating Knowledge Systems', *European Planning Systems* **5**(3), 331–349.

Jones-Evans, D. and M. Klofsten, 1997, *Technology Innovation and Enterprise—European Experience*, London: Macmillan.

Jones-Evans, D. and P. Westhead, 1996, 'High Technology Small Firm Sector in the United Kingdom', *International Journal of Entrepreneurial Behaviour and Research* **2**(1), 15–35.

Klofsten, M., 1996, *Ideas that Really Mean Business. An Overview of High-Tech Firms in the County of Östergötland, Sweden*, 6th edition, Linköping: Centre for Innovation and Entrepreneurship, Linköping University.

Klofsten, M. and D. Jones-Evans, 1996, 'Stimulation of Technology-Based Small Firms—A Case Study Of University-Industry Co-Operation', *Technovation* **16**(4), 187–193.

Klofsten, M. and C. Schärberg, 1997, 'Small Business Support: Attitudes from the Organisers', RENT XI Research in Entrepreneurship and Small Business 11th workshop, Mannheim, Germany, Nov. 27–28.

Laffitte, P., 1985, 'Sophia Antipolis and Its Impact on the Côte d'Azur', in J. M. Gibb (ed.), *Science Parks and Innovation Centres: Their Economic and Social Impact*, New York: Elsevier.

Landebaso, M., 1996, 'Towards a New Regional Policy for the promotion of Innovation', Paper presented at the National Innovation Day Conference, Dublin, Ireland, Nov. 4.

Lundgren, A., 1995, *Technological Innovation and Network Evolution*, London: Routledge.

Maillat, D., 1995, 'Territorial Dynamic, Innovative Milieus and Regional Policy', *Entrepreneurship and Regional Development* **7**, 157–165.

Saxenian, A., 1994, *Regional Advantage—Culture and Competition in Silicon Valley and Route 128*, London: Harvard University Press.

Scott, A. G., 1986, 'High Technology Industry and Territorial Development: The Rise of the Orange County Complex, 1955–1984', *Urban Geography* **7**(1), 3–45.

Segal, N., 1986, 'Universities and Technological Entrepreneurship in Britain—Some Applications from the Cambridge Phenomenon', *Technovation* **4**, 189–204.

Smilor, R. W., G. Kozmetsky and D. V. Gibson 1988, *Creating the Technopolis—Linking Technology, Commercialization and Economic Development*, Cambridge, MA: Ballinger Publishing Company.

Tödtling, F., 1994, 'Regional Networks of High-Technology Firms—The Case of the Greater Boston Region', *Technovation* **14**(5), 323–343.

[21]

Firm Size, University Based Research, and the Returns to R&D

Albert N. Link
John Rees

ABSTRACT. This paper compares university-based research relationships between small and large firms as an explanation for the difference in innovative activity across firm sizes. We test the hypothesis that there are diseconomies of scale in producing innovations in large firms due to the inherent bureaucratization process which inhibits both innovative activity as well as the speed with which new inventions move through the corporate system towards the market. By utilizing university-based research relationships, small firms are able to avoid bureaucratic inefficiencies.

I. Introduction

Over the past decade scholars have studied the role of small-sized firms in the innovation process. A number of important conclusions have come forth as a result of these inquiries.[1] First, small firms are more innovative (in terms of the number of product innovations) relative to their size than large firms. Second, product innovations coming from small firms appear to be more significant than those coming from large firms. Surprisingly, no studies to date have sought to explain, or even speculate, why small firms have this innovation-related advantage.[2]

This paper compares university-based research relationships between small and large firms in an effort to identify one factor that might explain this noted difference in innovativeness. Our hypothesis is that innovation-based diseconomies of scale exist in large firms owning to the fact that bureaucratization in the innovation decision making process inhibits not only inventiveness but also slows the pace at which new inventions move through the corporate system toward market. Small firms who utilize university-based research relationships and are as a result more efficient in their internal R&D, partially avoid such problems.

This paper is outlined as follows. In Section II, the data that form the basis for our empirical investigations are described. In Section III, we provide an overview of firms' involvement in university-based research relationships. The empirical analysis in Section IV demonstrates that small firms are able to leverage their internal R&D activity through their research relationships with universities to a greater extent than large firms and, thus, enjoy a higher return on their research investments.

II. Description of the data

A. *The sample of firms*

In 1986/87 we assembled a data set related to firms' involvement with university-based research programs. Based on preliminary interviews with directors from both university and state research centers, several broad industry groups were identified to be the major 'users' of such external research relationships. These industry groups included computing equipment, machine tools, and aircraft and components. From these broad industry groups, a population of 1046 firms was identified from the 1986 DUNS file of the Dun and Bradstreet Corporation. After an initial mail survey to vice presidents of production/engineering, and follow-up telephone resurveys, complete information (defined below) was obtained on 209 firms.[3]

When surveyed, these firms were asked to

Final version accepted on August 22, 1989

Department of Economics
Department of Geography
University of North Carolina at Greensboro
Greensboro, NC 27412, U.S.A.

Small Business Economics **2**: 25—31, 1990.

classify themselves into one industry category based on their primary line of business. From their classification, these firms could be placed into five major SIC industry groups within the U.S. manufacturing sector: metalworking machinery (SIC 354), office and computing machinery (SIC 357), electronic components and accessories (SIC 367), aircraft and parts (SIC 372), and engineering and scientific instruments (SIC 381). The distribution of firms across these five industry groups is shown in Table I. Table II presents the distribution of these sample firms by size category. Along with the number of firms in each size category, the average number of employees per firms is also reported in that table.

TABLE I
Distribution of sample firms by industry

Industry	No. Firms
SIC 354	15
SIC 357	69
SIC 367	82
SIC 372	19
SIC 381	24
	209

TABLE II
Distribution of sample firms by size category

No. Employees	No. Firms	Avg. No. Employees
< 100	40	31
100 to 249	83	118
250 to 499	19	328
500 to 999	17	653
1000 to 9999	22	2,930
> 10,000	28	76,556

B. *Innovation-related characteristics of the firms in the sample*

Although there are many ways to characterize the innovativeness of a firm, one dimension relates to self-financed R&D activity. The sample firms were classified as R&D-active or not based on two separate criteria: R&D expenditures and R&D personnel. *A priori*, there was no reason to believe that these two indices would be perfectly correlated. For example, a firm that relies heavily on contracted research may not have an R&D budget proportional to its R&D staff. Likewise, especially in smaller firms, the R&D budget may be so insignificant both in absolute and relative terms that the category 'R&D personnel' is not meaningful. Or, the accounting system may not be refined sufficiently to separate R&D expenditures from other investments even when personnel are classified as related to R&D. Nevertheless, 93 percent of the firms in this sample expended funds on R&D in 1986 and 88 percent of all firms had at least one individual classified under the heading of R&D personnel. Table III shows the percentage of sample firms involved in R&D using each criterion. With the exception of firms with more than 10,000 employees, there is a marked similarity between the percentage of firms with an R&D budget and those with classified R&D personnel. In this largest category, 93 percent of the firms reported an R&D budget but only 68 percent reported having R&D personnel. Perhaps, and the data do not permit an investigation of this point, these largest firms rely most heavily on contracted research which is paid internally and conducted externally. For the entire sample of firms, the correlation coefficient between total R&D expenditures and total R&D personnel is 0.65 (significant at the 0.01 level or better).

TABLE III
Percentage of sample firms involved in R&D by size category

No. Employees	$ Percentage	Emp. Percentage
< 100	88%	83%
100 to 249	93	90
250 to 499	95	95
500 to 999	100	100
1000 to 9999	100	95
> 10,000	93	68

Table IV presents the percentage of sales devoted to R&D activity by size category for all firms in the sample. It appears that small firms devote a greater percentage of their sales to R&D than do large firms. While the percentage differences do not seem to be significant between the middle size categories, they are distinct between the categories of firms with less than 100 em-

TABLE IV
Percentage of sales devoted to R&D by size category

No. Employees	Percentage
<100	13.3%
100 to 249	10.4
250 to 499	12.2
500 to 999	12.3
1000 to 9999	10.5
>10,000	5.0

ployees and with more than 10,000 employees. For all firms in the sample, the average percent of sales allocated toward R&D is 10.6 percent.

A similar pattern across size categories for all firms in the sample is shown in Table V. There, the percentage of total personnel involved in R&D decreases from 16.1 percent for firms with fewer than 100 employees to 7.9 percent for firms with more than 10,000 employees. The variation between the middle categories is again not striking.

TABLE V
Percentage of total personnel involved in R&D by size category

No. Employees	Percentage
<100	16.1%
100 to 249	12.1
250 to 499	15.1
500 to 999	11.4
1000 to 9999	11.9
>10,000	7.9

III. Overview of firms' university-based research relationships

Sixty-nine percent of the sample firms were involved with at least one university-based research program in 1986. As shown in Table VI, the degree of university involvement appears to increase with firm size. Whereas just over 50 percent of the smallest firms (less than 250 employees) were active in at least one research relationship with a university in 1986, about 90 percent of the firms with more than 1,000 employees were so involved.

TABLE VI
Involvement with university-based research programs by size category

No. Employees	% Firms
<100	59%
100 to 249	51
250 to 499	74
500 to 999	94
1000 to 9999	86
>10,000	100

Information on three specific categories of involvement with a university-based research program was collected: faculty used as technical consultants (Consultants), contracted research projects (Contracts), and graduate students used as research assistants (Research Assts.). If a sample firm participated in at least one of these dimensions, then it was classified in Table VI as involved in a university-based research program. The percentage of firms active in each of these three types of activities is shown in Table VII by size category. In general, firms in the larger size categories make greater use of university faculty as technical consultants; however, in all three cases the percentage of firms with more than 10,000 employees who are involved in any dimension is greater than for any of the other size categories.

The existing literature on industry-university research relationships suggests that these types of relationships are fostered by firms for two major reasons: it is a mechanism to reduce research costs and a method to identify potential productive employees. To investigate this issue further, each firm was asked to indicate which of the following

TABLE VII
Involvement with university-based research programs by type of activity and by size category

No. Employees	Consultants	Contracts	Research Assts.
<100	41%	15%	47%
100 to 249	44	22	38
250 to 499	67	44	72
500 to 999	76	29	59
1000 to 9999	77	54	64
>10,000	96	96	82

were incentives (expected results from the relationship) to their participating in a university-based research relationship: 'problem solving in production processes' (Pbl. Sol.), 'product development' (Prd. Dev.) , 'use of university computing facilities' (Compt.), 'use of other university facilities' (Facil.), and 'gaining access to students as future employees' (Emplmt.). The percentage of firms noting each of these as incentives is shown in Table VIII by size category.

With the exception of firms in the smallest two size categories, fewer than 100 employees and 100 to 249 employees, the potential for solving production process problems appears to be an important reason for firms to forge research relationships with universities. The importance of this potential as an incentive for such collaboration does not vary much by size category beyond firms with 250 or more employees. Over 60 percent of the firms in the sample view product development as an important incentive for engaging in a research relationship with a university. The use of university relationships as a vehicle to gain access to computing facilities appears to be primarily a small-firm (more than 500 employees) phenomenon. It may be the case that large firms have the in-house computer capabilities to conduct the requisite research operations. Access to other university facilities as an incentive for engaging in a university-based research relationship is important to some firms, but it does not seem to be systematically related to the size of these firms. In accordance with anecdotal information, gaining access to students as future employees is a significant incentive for firms of all sizes to pursue university-based research relationships.

The last column in Table VIII reports firms' responses to a question regarding the importance of 'federal tax incentives as a motivation for engaging in collaborative research' with a university. While responses vary over size categories, only in the largest size category, firms with more than 10,000 employees, did more than 50 percent of the firms respond affirmatively.

Three response categories were used to determine firms' overall success with their university research relationships. The lion's share of the firms were satisfied with their collaborative research experience, as reported in Table IX.[4]

This overview of the primary data suggests several preliminary patterns of firm behavior. One, firms in all size categories were engaged in university research relationships to use, in general, faculty as technical consultants; and firms in the larger size categories do this to a greater degree than firms in the smaller size categories (Table VII). This collaboration tends to be oriented primarily toward product development and secondarily toward problem solving in areas related to production (Table VIII). Two, in addition to research expertise, firms in all size categories viewed access to students as future employees as a significant incentive for engaging in a university-based research relationship (Table VIII).

IV. The empirical analysis

This section presents the results of two empirical investigations into aspects of firms' participation in university-based research activity. In Part A, the trend noted in Table VI that firms in the larger size categories were more active in university-based

TABLE VIII
Incentives to engage in university research relationships by size category

No. Employees	Pbl. Sol.	Prd. Dev.	Compt.	Facil.	Emplmt.	Tax
< 100	3%	61%	5%	23%	55%	24%
100 to 249	19	69	18	19	65	16
250 to 499	57	71	29	57	79	40
500 to 999	37	63	6	19	69	29
1000 to 9999	42	47	5	16	84	11
> 10,000	63	77	13	62	93	54

TABLE IX
Firm's overall success in university research relationships by size category

No. Employees	"Very Satisfied"	"Somewhat Satisfied"	"Not Satisfied"
< 100	29%	64%	7%
100 to 249	38	62	0
250 to 499	71	29	0
500 to 999	44	56	0
1000 to 9999	46	54	0
> 10,000	25	75	0

research relationships than firms in the smaller size categories is investigated statistically. The specific question considered is: Is the probability of involvement with a university-based research program related to firm size? In Part B, the impact of external research relationships on the rate of return to firms' internal R&D is examined. Does the rate of return to R&D vary by firm size? Does the rate of return to R&D vary according to research participation with a university?

A. *The propensity to engage in external research relationships*

An inspection of the descriptive information in Table VI suggests that the propensity to engage in a university-based research relationships is related to firm size. The percentage of firms active with a university-based research program increases with category size.

To test formally for the influence of size on the propensity to engage in an external research relationship, a probit model was estimated. The independent variables in this model were firm size (SIZE) measured in terms of firm sales ($millions), industry concentration (CR),[5] and a binary variable equalling 1 if the firm was involved in basic research and 0 if it was not (BASRES).

The probit results, with asymptotic t-statistics in parentheses, are:

$$F^{-1}(P) = \underset{(0.31)}{0.114} + \underset{(2.94)}{0.016}\,\text{SIZE} - \underset{(-0.44)}{0.004}\,\text{CR} + \underset{(0.63)}{0.19}\,\text{BASRES}$$

$$-2 \times \log \text{ of the likelihood function} = 185.92$$

where $P = F(\alpha + \beta_0 \text{ SIZE} + \beta_1 \text{ CR} + \beta_2 \text{ BASRES}) = F(z)$ for F being the cumulative probability function. These regression results complement the pattern of activity shown in Table VI. The probability of participating in a university-based research program does increase with firm size.[6] The estimated coefficient on SIZE is significant at the 0.01 level or better. Industry concentration and involvement in a basic research program have no explanatory power in this specification.

B. *Firm size, university-based research relationships, and the returns to R&D*

A framework frequently used by researchers in economics for estimating the returns to R&D reduces to the following regression model:[7]

$$\text{TFPG} = \alpha + \beta\,(\text{RD}/Q) + \varepsilon$$

where TFPG represents total factor productivity growth, (RD/Q) is the ratio of R&D spending to firm sales, and β is the estimated rate of return to R&D which could be interpreted as an index of R&D efficiency.

To estimate this model, data were needed for the calculation of total factor productivity over a defined period. Sufficient data for these calculations were not available for all firms in the sample. TFPG over the period 1982 to 1987 could be calculated for only 158 R&D-active firms of the 209 firms in the sample. The 51 firms deleted from the analysis were mostly small firms with fewer than 100 employees. Overall, the rate of return to the 158 firms in this subsample was 26.1 percent. This result is reported in Table X.

Several versions of the basic model were estimated. First, to test for differences between the rate of return to R&D in large versus small firms, a second regressor was included in the above equation. Its equalled a binary variable interacted with the (RD/Q) term where the binary variable was given the value 1 for firms with less than 500 employees and 0 otherwise. The estimated least-squares coefficient on this term was not statistically different from zero, implying that there was no statistical difference between the returns in the two size groups.[8] This result is also reported in Table X.

Second, a similar specification was estimated to account for possible differences in the rate of

TABLE X
Estimated rates of return to R&D expenditures

Category	Estimated Rate of Return
Subsample of 158 firms	26.1%
Large firms	26.0%
Small firms	26.1%
Firms involved in university research	34.5%
Firms not involved in university research	13.2%
Large firms	
Involved in university research	29.7%
Not involved in university research	14.1%
Small firms	
Involved in university research	44.0%
Not involved in university research	13.9%

return to firms engaged in and not engaged in a university-based research relationships. For this, the binary variable equalled 1 if the firm was so engaged and 0 otherwise. As reported in Table X, the estimated returns to R&D in firms involved in university-based research relationships are more than twice those of firms that are not — 34.5 percent versus 13.2 percent.[9]

Finally, segmenting both by size and by university involvment by including two regressors in the original specification (one with a size dummy and the other with an university-based research dummy), the returns to R&D in small, university-based research-active firms was found to be quite large. As reported in Table X, the estimated rate of return to R&D in this group of firms was 44.0 percent compared to (1) 29.7 percent in large university-based research-active firms, (2) 14.1 percent in large non-university-based research-active firms, (3) 13.9 percent in small non-university-based research-active firms. Small firms appear to be able to transfer knowledge gained from their unviersity research association most effectively, compared to large firms, to increase the returns to their internal R&D activities.

V. Conclusions

While the results presented in this paper by no means explain fully why small firms have an innovation-related advantage over large firms, they do point out one interesting difference between an aspect of large and small firm research behavior. Although large firms are more active in university-based research *per se*, small firms appear to be able to utilize their university-based associations to leverage their internal R&D to a greater degree than large firms.

The analysis presented here did not take into account many of the other important factors associated with R&D efficiency, and so the results presented in Table X should be interpreted with caution. Still, the findings are noteworthy enough to encourage other investigators to investigate in more detail the ways in which firms internalize external technical information.

Acknowledgement

This paper was prepared for the Conference on Innovation and Technological Change: An International Comparison, organized by the Wissenschaftszentrum Berlin für Sozialforschung, Berlin, August 10—11, 1989. Funding for this research came from the U.S. Small Business Administration under contract SBA-89-989.

Notes

[1] Much of this research is summarized in U.S. Small Business Administration (1986) and in Link and Bozeman (1987).

[2] Relatedly, Acs and Audretsch (1987a, 1987b, 1988) show that the market environment most conducive for innovation is similar for both large and small firms. Also, they show that industry structure influences large firms' ability to innovate relative to small firms' ability to innovate, other things remaining equal.

[3] Whenever possible reported survey information (e.g., sales data) was verified against published data (e.g., Form 10-K data) to insure response reliability. When explainable differences occurred (e.g., a survey respondent reporting sales in \$ millions rather than \$ thousands) the primary data were corrected.

[4] There is not sufficient variation between the three response categories to conduct a more detailed investigation of inter-firm differences in success with university-based research.

[5] $0 < CR < 100$. These data came from Weiss and Pascoe (1986).

[6] A non-linear size variable was included in separate regressions, but the associated coefficient was not significant. As well, two other independent variables were considered. An index of foreign competition was included in a separate regression. This variable, based on data from the International Trade Commission and the Bureau of Industrial Economics, was constructed as the ratio of industry (four-digit) imports

divided by the value of industry shipments plus imports less exports (Link and Bauer, 1989). It exhibited no statistical influence on the estimated probability. And, a vector of three-digit SIC industry dummies was included in the various versions of the model. As a group, these dummies were not significantly different from zero and thus were deleted. Similar results were obtained when SIZE was measured in terms of employees.

[7] This model is fully explained in Griliches (1979). See Link (1987) for a review of the empirical literature. Using cross-sectional firm data from the U.S. manufacturing sector, the estimated rate of return to internal R&D is in the neighborhood of 20 percent.

[8] Link's (1980) analysis of the rate of return to R&D among firms in the chemicals industry found that the return increased with firm size to a modest threshold level, and then remained constant.

[9] This finding may not be inconsistent with the findings of others that the returns to basic research are greater than for other categories of R&D spending. Generally, research conducted at universities is toward the basic end of the R&D spectrum; however, the underlying data (see Tables VII and VIII) do not allow us to separate clearly this form of research relationship. This finding also corresponds favorably with that of Link and Bauer (1989). They report, based on a sample of 92 manufacturing firms, the rate of return to R&D in firms involved in cooperative research programs with other firms is nearly three time that of firms not so involved — 37.7 percent versus 12.9 percent.

References

Acs, Z. J. and D. B. Audretsch, 1987a, 'Innovation in Large and Small Firms', *Economics Letters* **23**(1), 109—112.

Acs, Z. J. and D. B. 1987b, 'Innovation, Market Structure, and Firm Size', *Review of Economics and Statistics* **69**(4), 567—574.

Acs, Z. J. and D. B. Audretsch, 1988, 'Innovation in Large and Small Firms: An Empirical Analysis', *American Economic Review* **78**(4), 678—690.

Griliches, Z., 1979, 'Issues in Assessing the Contribution of Research and Development to Productivity Growth', *Bell Journal of Economics* **10**, 92—116.

Link, A. N., 1980, 'Firm Size and Efficient Entrepreneurial Activity: A Reformulation of the Schumpeter Hypothesis', *Journal of Political Economy* **88**, 771—782.

Link, A. N., 1987, *Technological Change and Productivity Growth*, London: Harwood Academic Publishers.

Link, A. N. and L. L. Bauer, 1989, *Cooperative Research in U.S. Manufacturing*, Lexington, Mass.: Lexington Books.

Link, A. N. and B. Bozeman, 1987, 'Firm Size and Innovative Activity: A Further Examination,' final report to the Office of Advocacy, U.S. Small Business Administration.

U.S. Small Business Administration, Office of Advocacy, 1986, *Innovation in Small Firms*, Washington, D.C.

Weiss, L. W. and C. A. Pascoe, 1986, *Adjusted Concentration Ratios in Manufacturing, 1972 and 1977*, Washington, D.C.

[22]

Links and Impacts: The Influence of Public Research on Industrial R&D

Wesley M. Cohen • Richard R. Nelson • John P. Walsh
Department of Social and Decision Sciences, Carnegie Mellon University, Pittsburgh, Pennsylvania 15213
School of International and Public Affairs, Columbia University, New York, New York 10027
Department of Sociology, University of Illinois at Chicago, Chicago, Illinois 60607
wc02@andrew.cmu.edu • rrn2@columbia.edu • jwalsh@uic.edu

In this paper, we use data from the Carnegie Mellon Survey on industrial R&D to evaluate for the U.S. manufacturing sector the influence of "public" (i.e., university and government R&D lab) research on industrial R&D, the role that public research plays in industrial R&D, and the pathways through which that effect is exercised. We find that public research is critical to industrial R&D in a small number of industries and importantly affects industrial R&D across much of the manufacturing sector. Contrary to the notion that university research largely generates new ideas for industrial R&D projects, the survey responses demonstrate that public research both suggests new R&D projects and contributes to the completion of existing projects in roughly equal measure overall. The results also indicate that the key channels through which university research impacts industrial R&D include published papers and reports, public conferences and meetings, informal information exchange, and consulting. We also find that, after controlling for industry, the influence of public research on industrial R&D is disproportionately greater for larger firms as well as start-ups.
(*R&D; Innovation; Universities; Spillovers; Start-ups*)

1. Introduction

This paper reports findings from the Carnegie Mellon Survey on Industrial R&D on the contributions of university and government research labs—what we will call public research—to industrial innovation. By advancing our understanding of the contribution of public research to industrial R&D, we hope to deepen our understanding of the determinants of technological change broadly, and speak to assumptions that have guided policy discussions over the past two decades concerning the economic impact of public research.

Understanding the impact of public research on industrial R&D is central to understanding the innovation process itself. The so-called "linear model" of innovation, reflected most notably in Vannevar Bush's (1945) *Science—The Endless Frontier*, conceived of industrial innovation as proceeding from basic to applied research and then to development and commercialization. In this traditional view, public research—particularly university research—proceeds upstream and independently of technological development, which, however, draws from the pool of research results. A richer characterization of the innovation process has been developed over the past two decades by scholars such as Gibbons and Johnston (1975), Kline and Rosenberg (1986), Nelson (1990) and von Hippel (1988), among others. This conception is of a more interactive relationship where public research sometimes leads the development of new technologies, and sometimes focuses on problems posed by prior developments or buyer feedback. In this view, industrial innovation emerges from a complicated process where fundamental research need not play an initiating role, nor, at times, any role. In this study, we both build on and probe this perspec-

0025-1909/02/4801/0001$5.00
1526-5501 electronic ISSN

tive by considering the contribution of public research to technical advance in the context of the different sources and types of knowledge that may stimulate and inform industrial R&D.

In the post–World War II period, the presumption that basic research proceeds independently of, but often drives, technological developments strongly colored policymakers' commitments to public research generally (even though public research also encompasses a good deal of applied research and development). Over the past two decades, motivated by fiscal constraints and stiffening international economic competition, numerous policymakers have, however, effectively eschewed the linear model as they have encouraged universities and government labs to embrace the cause of technology commercialization. Reflecting the sentiment that public research is too distant from industry in the majority of industries (with the notable exception of biomedical research), policymakers have called on universities and government R&D labs to make their science and engineering more relevant to industry's needs. The National Science Foundation has embraced this mission with the creation of the Science and Technology Centers, the Engineering Research Centers and other programs that tie government support of university research to industrial participation. In order to stimulate regional economic development, numerous state governments have followed suit. Federal policymakers have even changed the nation's intellectual property policy to increase incentives to commercialize public research (e.g., the Bayh-Dole Amendment).

In this paper, we use data from the 1994 Carnegie Mellon Survey (CMS) of Industrial R&D to characterize the extent and nature of the contribution of public research to industrial R&D. We first assess how public research stands relative to other sources of information affecting industrial R&D.[1] In this context, we report on how public research tends to be used in industrial R&D labs. Second, we consider the overall importance of public research, as well as that of specific fields of basic and applied research and engineering. Third, we consider the importance of the different pathways through which public research may impact industrial R&D, including publications, informal interactions, consulting, and the hiring of university graduates. Finally, we consider what roles different kinds of firms (e.g., large versus small and start-ups versus established firms) play in bridging public research and industrial R&D.

In brief, our data suggest that public research has a substantial impact on industrial R&D in a few industries, particularly pharmaceuticals, and is generally important across a broad segment of the manufacturing sector. Like previous studies, we find that the contribution of public research to industrial R&D varies across industries. The uses of public research are also observed to vary. Consistent with a more interactive conception of the innovation process, our data show that public research is used at least as frequently to address existing problems and needs as to suggest new research efforts. The most important channels for accessing public research appear to be the public and personal channels (such as publications, conferences, and informal interactions), rather than, say, licenses or cooperative ventures. Finally, we find that large firms are more likely to use public research than small firms, with the exception that start-up firms also make particular use of public research, especially in pharmaceuticals.

2. Background

Two decades ago, scholars concerned with the study of technological advance had a limited understanding of the impact on technical advance of the kind of research that was conducted at universities and government labs. The discussion at that time tended to involve an attempt to sort out "demand pull" from "opportunity push" as factors influencing technological advance, where "opportunity push" was typically intended to capture the impact of new scientific and technological knowledge. Research that was "upstream" relative to commercial application was seen as key to initiating new inventive efforts. Since that time, most scholars have come to understand that this view—the linear model—does not adequately characterize the role of fundamental understandings emerging from science or engineering.

[1] We are using the word "information" here to refer to the understandings, facts, and theories that influence decision making and problem solving in industrial R&D.

Research by Gibbons and Johnston (1975), Klevorick et al. (1995), Mansfield (1991), Nelson (1990), Rosenberg (1992), Mowery and Rosenberg (1979), Rosenberg and Nelson (1994), Feller (1990), Hounshell (1996), Narin et al. (1997), and others have advanced our understanding of the role of public—particularly university—research in technological advance beyond the old debate about demand pull versus opportunity push. From this research, we learned that, while upstream research can spawn new research projects, downstream technical advances can also instigate and shape upstream research. Moreover, innovative efforts are often guided by perceptions of need and demand, although more fundamental understandings will often then shape how and with what success those efforts proceed.

As scholars were rethinking the linear model, policymakers' presumption of the social value of public research was weakening. In this context, empirical economists began assessing the contribution of academic research to technical advance. Little consensus has emerged, however, on the extent or nature of that contribution. The "Yale survey"—which laid the groundwork for the current study—found that firms in the majority of industries reported that the direct contribution of research conducted by universities and government R&D labs to their R&D activities was slight compared to sources within the "industrial chain" (i.e., firms in the industry itself, suppliers, and buyers). Science in general was also widely reported to be relatively more important. Nelson (1986) and Klevorick et al. (1995) interpreted their findings to suggest that recent university research (i.e., conducted in the prior 10–15 years) had little direct effect on industrial R&D outside of a few technologies such as drugs, other areas of medicine, sophisticated organic chemical products, and some areas of electronics. They argued that the impact of university research was nonetheless substantial, though indirect; that university research and science used in most industries' inventive efforts tended to be relatively mature and typically exercised influence through the more applied sciences and engineering fields or via the training of industry's scientists and engineers.

Using other methods and data sources, Mansfield (1991), Jaffe (1989), Adams (1990), and Narin et al. (1997) added other elements to the evolving picture. Mansfield's survey research study of 76 firms spanning information processing, electrical equipment and instruments, drug, metal, and oil firms suggested that 10 percent of new products and processes would have been delayed a year or more in the absence of academic research conducted within the prior 15 years. While relatively small, this impact was large in absolute terms, conceivably denying industry $24 billion in sales in 1985. Estimating the elasticity of corporate patenting with respect to university research to be as high as 0.6, Jaffe (1989) judged the influence of university R&D to be important overall and strongest in the drug industry. On the basis of a regression analysis of bibliometric data, Adams similarly finds the effects of academic R&D on manufacturing productivity to be important, though, in the case of academic research in the basic sciences, to take approximately 20 years to become manifest. More recently, using firms' patent citations to the scientific and engineering literature, Narin et al. (1997) concluded that the linkages between industrial R&D and current public research (conducted in either academia or government labs in the prior 10 years) grew dramatically between the late-1980s and early-to-mid-1990s.

Thus, Klevorick et al. (1995) find the direct impact of recent university research in most industries to be small when assessed relative to other sources of information or scientific knowledge generally. Not necessarily inconsistent with this evaluation, others find the impact to be substantial when assessed as elasticities or in absolute terms.

Further complicating the picture is the deepening of ties between public research institutions—particularly universities—and industry over the last two decades that may signify a recent change in the impact of public research. For example, U.S. patents granted to universities increased from 589 in 1985 to 3,151 in 1998 (National Science Board 2000). During this same period (and especially during the 1980s), industry funding of university research increased from $630 million to $1.896 billion (NSB 2000). Moreover, ties between industry and universities have generally deepened, as reflected in the 60% growth in university-industry R&D centers in the 1980s (Cohen et al. 1998) and the more than eightfold increase in university technology transfer offices between 1980 and 1995 (Association of University Technology Managers 1999).

Notwithstanding the different frames of reference and perspectives on the contribution of public research to technical advance, some understandings are reasonably solid. For example, there is broad consensus that while the linear model may capture key aspects of the innovation process in some settings, its applicability is limited. Previous studies also raise several questions. To the extent that academic research impacts industrial R&D, what are the channels of that impact? Also, how do industrial R&D labs actually use public research? What is the effect of firm size on the exploitation of public research, and do start-ups provide an important bridge between public research and industrial R&D? Using the CMS data, we now try to address these and related questions. At the outset, we note that our examination of the extent and nature of the contribution of public research to industrial R&D is incomplete. The innovation process is complex, often subtle and varies across industries, technologies, and over time in ways that one survey cannot hope to reflect fully.

3. Data

The data come from a survey of R&D managers administered in 1994. The population sampled are all the R&D units located in the U.S. conducting R&D in manufacturing industries as part of a manufacturing firm. The sample was randomly drawn from the eligible labs listed in Bowker's Directory of American Research and Technology (1994) or belonging to firms listed in Standard and Poor's COMPUSTAT, stratified by three-digit SIC industry.[2] We sampled 3,240 labs, and received 1,478 responses, yielding an unadjusted response rate of 46% and an adjusted response rate of 54%.[3] Our survey data are combined with published data on firm sales and employees from COMPUSTAT, Dun and Bradstreet, Moody's, Ward's, and similar sources.

The survey provides several measures of the influence of public research on industrial R&D that allow us to examine different dimensions of that influence. As noted above, much of the literature relevant to this paper focuses on the impact of university research on industrial R&D. In contrast, we ask respondents to consider the impact of research produced by universities *and* government research institutes and labs. Citations on U.S. patents to the scientific and engineering literature provided by the National Science Board (1998) suggest, however, that of public research institutions, universities dominate in their impact on industrial R&D, suggesting, in turn, that our findings characterize predominantly the impact of university research.[4]

For the analysis in this paper, we restricted our sample to firms whose focus industry was in the manufacturing sector and were not foreign owned, yielding a sample of 1,267 cases. This sample includes firms ranging from less than 25 to over 100,000 employees, with annual sales ranging from less than $1 million to over $60 billion. The median firm has 2,263 employees and annual sales of $412 million. The average firm has 20,263 employees and sales of $4.2 billion. The business units (defined as a firm's activity in a specific industry) range from fewer than 10 employees to over 50,000, with annual sales from zero to over $30 billion. The median business unit has 400 employees and $100 million in sales. The average business unit has 6,095 employees and sales of $2.1 billion. The average R&D intensity (R&D dollars divided by total sales) for the firms is 3.23%. Presented in Table 1, the size distribution of our firms

[2] We also oversampled Fortune 500 firms.

[3] The results of a nonrespondent survey showed that 28% of nonrespondents were ineligible for the survey because they either did no manufacturing or did no R&D. Excluding these from our denominator, as well as respondents who should not have been sampled, yields an adjusted response rate of 54% of eligible respondents.

[4] Citations on U.S. patents to the scientific and engineering literature provided by the National Science Board (1998) distinguish between the influence on industrial R&D of university research versus that conducted by government. As of 1994, patent citations to publications originating from academia exceed those originating from the federal government by 5.7 times. This multiple varies, however, across fields, from 4.9 in the combined fields of biomedical research, clinical medicine, and biology, to 8.4 in engineering and technology, suggesting that the impact of university research may be almost an order of magnitude greater than that of government research in industries outside the biomedical fields.

Table 1 Size Distribution of Business Units and Firms (By Employment)

(A) Business Units	Frequency	Percent	Cumulative Frequency	Cumulative Percent
Less than 100	262	24.2	262	24.2
100–500	302	27.8	564	52.0
500–1,000	114	10.5	678	62.5
1,000–5,000	229	21.1	907	83.6
5,000–10,000	71	6.5	978	90.1
10,000+	107	9.9	1,085	100.0
Frequency missing = 276				
(B) Firms				
Less than 100	215	17.7	215	17.7
100–500	212	17.4	427	35.1
500–1,000	64	5.3	491	40.3
1,000–5,000	229	18.8	720	59.1
5,000–10,000	111	9.1	831	68.2
10,000+	387	31.8	1,218	100.0
Frequency missing = 49				

and business units, expressed in numbers of employees, shows that 35% of our sample are "small" firms with less than 500 employees (and almost 18% have fewer than 100 employees). Over half of our business units have fewer than 500 employees.

In our survey, we asked R&D unit managers to answer questions with reference to the "focus industry" of their R&D unit, defined as the principal industry for which the unit was conducting its R&D.[5] For the purpose of presentation in the tables, our observations are grouped into 34 ISIC groups which are aggregated at the two- or three-digit level and span the manufacturing sector. For some analyses, we disaggregate our data more finely into 64 industries, defined roughly at the three-digit SIC level.

4. Sources of Knowledge Used in Industrial R&D

We begin our analysis by examining the importance of a broad range of information sources, of which the R&D conducted in university and government R&D labs is one. We asked respondents to indicate whether information from a source either suggested new R&D projects or contributed to the completion of existing projects over the prior three years. For each of these two functions, we listed a broad range of possible knowledge sources, including university and government R&D labs (i.e., "public research"), competitors, customers, suppliers, consultants/contract R&D firms, joint or cooperative ventures, and the firm's own manufacturing operations.

The "linear model" suggests that upstream research—and hence the institutions of public research that conduct much of it—should play an initiating role as a key source of the ideas leading to industrial innovation. Our results suggest, however, that a preponderance of industrial R&D projects are initiated in response to information from buyers (consistent with von Hippel 1988) or from the firm's own manufacturing operations. In virtually all industries, our respondents listed buyers and the firm's own manufacturing operations as the predominant sources suggesting new projects, with 90% listing customers and 74% listing the firm's manufacturing operations. Also, as shown in Figure 1, knowledge from public research labs does not play the central role in suggesting new projects; 32% of the respondents list public research as such a source, and in the majority of industries, public research plays only a modest role in suggesting new projects. In a few industries, however, public research was rated as a relatively important stimulus to the initiation of new R&D projects. With 58% of respondents reporting public research as a source for new project ideas, the pharmaceutical industry is the most heavily influenced by public research by this measure. For suggesting new projects, public research is also relatively important in the petroleum, steel, machine tool, semiconductor, and aerospace industries, each with 50% or more of the respondents reporting public research as a source of new project ideas.

Figure 2 shows that the sources of information contributing to R&D project completion is less concentrated than those suggesting R&D projects. No one source so dominates as did buyers in the prior case. Nonetheless, and consistent with the notion that downstream needs and feedback often guide

[5] By asking respondents to answer questions in terms of a single, identified industry, we reduce the measurement error that comes from having firms' responses reflect conditions in some ambiguous combination of the several industries in which they may be active.

Figure 1 **Information Sources Suggesting New Projects**

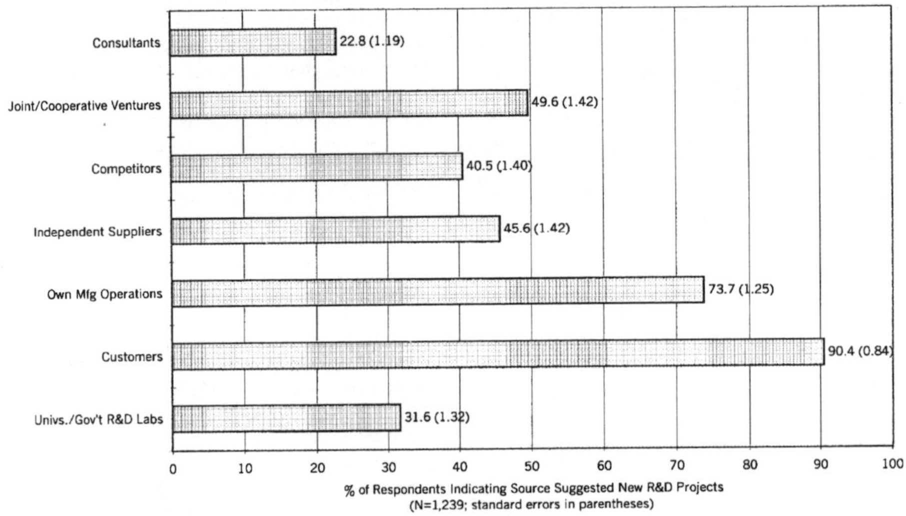

Figure 2 **Information Sources Contributing to Project Completion**

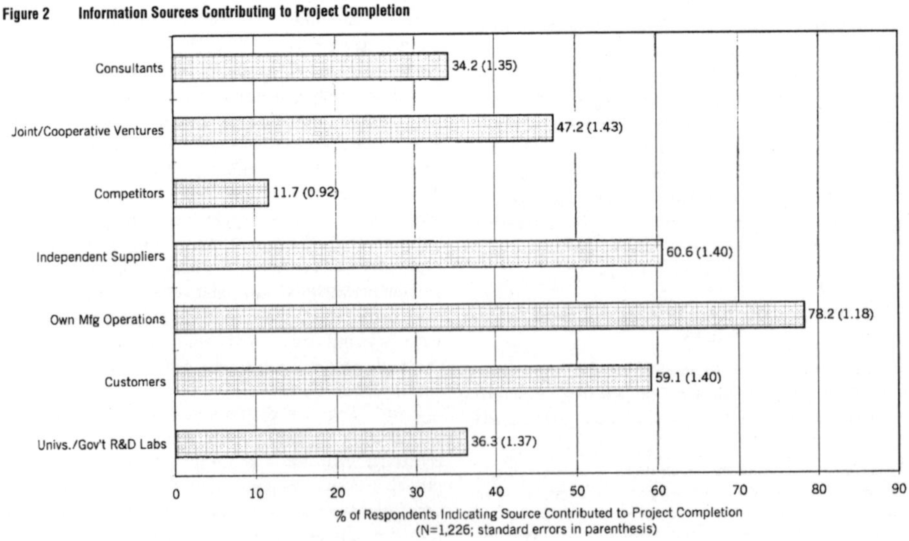

industrial R&D (cf. Kline and Rosenberg 1986), one source—namely the firm's own manufacturing operations—is more important than the others. About 78% of the respondents report that knowledge from manufacturing operations contributes to R&D project completion. The key role of manufacturing operations as a knowledge source is unsurprising given its centrality to process innovation and the importance of "manufacturability" for product innovation. The impacts of buyers and suppliers are now comparable, with close to 60% of respondents reporting use of each.

As noted above, the "linear model" of innovation would suggest that upstream research should principally suggest new project ideas. We observe, however, public research playing a slightly more important role overall as a knowledge source for R&D project completion rather than for project initiation, with 36% of respondents reporting public research as a source of knowledge for this former function versus 32% for the latter. This suggests that, at least as often as not, public research provides the means to achieve some technological goal, which itself emerges, if not from the firm's own R&D, then typically either from buyers' or the firms' own manufacturing operations. The industries where public research is relatively important in helping firms to execute projects (i.e., with scores exceeding 45%) include food, paper, drugs, glass, search/navigation equipment, car/truck, and aerospace. Of these industries, drugs, car/truck, and aerospace score the highest, all with scores above 55%.

We observed above that the pharmaceutical industry stands out in the degree to which public research both suggests new R&D projects and contributes to R&D project completion. The pharmaceutical industry is also distinctive with regard to its sources of knowledge. For example, in response to a follow-up question about which of these sources was the *most* important, the drug industry was the only one where a majority of respondents did not list customers as their most important source of new project ideas.[6] Only 31% of drug industry respondents listed customers as the most important source—the least of any industry (in contrast to 66% for the whole sample, $p < 0.0001$, using Wilcoxon rank sum test), though 20% listed public research as the most important (in contrast to 3% for the whole sample, $p < 0.0001$). Regarding the sources contributing to the completion of R&D projects, where firms' own manufacturing operations tend to be most important (chosen by 32% of the sample), only 19% of drug industry respondents reported their manufacturing operations to be the most important knowledge source ($p < 0.05$), scoring in the lowest quartile (along with three other chemical industries) of the 34 industries that span the manufacturing sector.[7]

We can only conjecture why customers and the firms' own manufacturing operations are less important as sources of knowledge for R&D in the drug industry as compared to others. Perhaps drug firms do not have to go to customers to determine what kinds of drugs will be most profitable, but can rely upon broad professional awareness of the relative incidence of different diseases and medical conditions to inform their expectations. With regard to the subordinate role of manufacturing operations, manufacturing processes have historically been rather routinized in the drug industry, at least until recently. Our data also suggest that if the linear model applies to any industry, drugs would be the best candidate, although even here feedback from customers (i.e., patients and physicians) commonly stimulates upstream research.

Figures 1 and 2 show that public research plays a less important role as a knowledge source than a number of other sources. It is useful, however, to divide our consideration of knowledge sources between those that lie directly in what may be called the firm's vertical chain of production and sale, including suppliers, buyers, and the firm's own manufacturing operations, versus those that do not. While the importance of public research never compares favorably to those sources of knowledge residing in the vertical chain (consistent with Klevorick et al.

[6] For the food industry, 49.45% chose customers.

[7] At the same time, 19% of our drug industry respondents claimed that public research, along with consultants/contract R&D, were the most important sources contributing to project completion. In contrast, the average for the total sample for public research is 5% ($p < 0.0001$) and 7% for consultants/contract R&D ($p < 0.0001$).

1995), its importance is comparable to knowledge sources that lie outside of it, namely competitors, consultants/contract R&D firms, and joint or cooperative ventures. In the aggregate results, joint/cooperative ventures (with other firms or universities) score consistently higher ($p < 0.0001$, using paired Wilcoxon rank sign test) than public research—around 50% for both suggesting project ideas and contributing to project completion. Public research outscores, however, consultants/contract R&D as a source of knowledge for both suggesting new R&D projects ($p <$ 0.0001) and contributing to project completion (n.s.).[8] Although rivals constitute a more important source for project ideas than public research institutions (41% versus 32%, $p < 0.0001$), public research institutions are markedly more important than rivals as a source of knowledge contributing to project completion—36% for public research versus 12% for competitors ($p < 0.0001$), suggesting that the impact of public research on firms' R&D is at least comparable to that of rivals' R&D. To provide perspective on the importance of the contribution of public research to industrial R&D, consider that empirical studies in economics suggest that "R&D spillovers" from competitors (roughly comparable to our notion of information flows from rivals) contribute significantly to technical advance and productivity growth within industries (Bernstein and Nadiri 1989, Griliches 1992).

5. What and How Much Does Industry Get from Public Research?

In our survey, we probed how public research contributes to technical advance in industry. The question is whether industry benefits from the creation of prototypes of, or idea-designs for, new products and processes, or rather from the less tangible and more intermediate input of disembodied knowledge. Scholars of technical advance, notably Price (1984) and Rosenberg (1992), have also highlighted another form of contribution of public research to technical advance in industry—the development of new instruments and techniques which find use in industrial R&D. Accordingly, we asked respondents to report the percentage of their R&D projects which, over the prior three years, "made use" of the following outputs from public research: (1) research findings; (2) prototypes; and (3) new instruments and techniques.

For each category of output, Table 2 presents the industry averages weighted by respondent lab R&D spending.[9] Since the response scale is defined in terms of broad categories (below 10%, 10%–40%, 41%–60%, 61%–90%, over 90%), we use category midpoints to compute the percentages except in the lowest category, "below 10%," to which we assign a value of 0%.

Responses to the survey suggest that the contribution of public research to industrial R&D is principally via research findings, and this contribution is far greater than that of "prototypes." The weighted average for the percentage of R&D projects using prototypes generated by public research is only 8.3%. Prototypes generated by public research falls in the 20%–35% range in only three industries (glass, TV/radio, and motors/generators), and in the 10%–20% range in another three (drugs, machine tools, and aerospace). In contrast, research findings were considered useful in 29.3% of our respondents' R&D projects. As Table 2 shows, public research findings are judged to be useful in over a quarter of respondents' R&D projects in the petroleum, pharmaceuticals, chemicals (nec), aerospace, communication, search/navigation equipment, TV/radio, and semiconductor industries. Our survey also finds the instruments and techniques developed by university and government labs to be useful to 22.2% of industrial R&D projects. While not as pervasive in their impacts as research findings, 20% or more of R&D projects make use of such instruments and techniques across a range of industries, including drugs, petroleum,

[8] Given that some joint or cooperative ventures are between firms and public research institutions, we believe that the contribution of joint and cooperative ventures to industrial R&D partially reflects that of public research. This conjecture is supported by significant ($p < 0.01$), within-industry correlations between the contribution of university/government labs and that of joint/cooperative ventures both to suggesting new R&D projects ($r = 0.18$) and to R&D project completion ($r = 0.20$).

[9] Weighting by R&D spending permits more accurate estimation of the overall impact of public research on industrial R&D, which requires, for example, giving the response of a $100 million R&D unit more weight than one that spends only a million.

Table 2 Use of Public Research Outputs in Industrial R&D

		Weighted Percentage of R&D Projects Using Output		
Industry	*N*	Research Findings	Prototypes	Instruments and Techniques
			Mean	
1500: Food	80	16.0	1.1	15.0
1700: Textiles	16	14.2	3.0	3.0
2100: Paper	24	24.7	3.3	26.6
2200: Printing/publishing	7	2.8	0.2	3.9
2320: Petroleum	13	34.1	0.1	27.5
2400: Chemicals, nec	57	28.9	5.8	28.0
2411: Basic chemicals	33	22.6	0.6	17.5
2413: Plastic resins	25	16.2	0.0	7.9
2423: Drugs	54	41.4	12.3	35.4
2429: Miscellaneous chemicals	25	19.3	9.9	9.4
2500: Rubber/plastic	29	21.1	8.1	12.4
2600: Mineral products	15	21.8	0.6	21.8
2610: Glass	6	24.7	23.7	48.1
2695: Concrete, cement, lime	6	7.7	0.0	11.6
2700: Metal, nec	7	15.4	8.1	14.2
2710: Steel	8	9.2	7.6	7.7
2800: Metal products	35	3.6	1.4	3.3
2910: General purpose machinery, nec	59	18.7	2.9	4.3
2920: Special purpose machinery, nec	57	22.1	0.7	1.1
2922: Machine tools	8	21.8	15.2	7.6
3010: Computers	23	23.5	3.2	22.6
3100: Electrical equipment	20	1.0	0.2	0.3
3110: Motor/generator	21	15.3	32.3	13.7
3210: Electronic components	22	3.2	1.5	3.2
3211: Semiconductors and related equipment	23	36.4	6.8	18.6
3220: Comm equipment	30	42.9	2.8	14.4
3230: TV/radio	7	45.3	23.9	46.8
3311: Medical equipment	66	17.7	2.4	6.3
3312: Precision instruments	32	12.9	6.3	13.3
3314: Search/navigational equipment	33	31.2	8.3	17.2
3410: Car/truck	8	21.4	7.3	14.8
3430: Auto parts	26	15.7	3.9	8.0
3530: Aerospace	42	28.6	13.4	24.9
3600: Other manufacturing	76	63.2	25.2	44.4
All	993	29.3	8.3	22.2

chemicals (nec), aerospace, glass, TV/radio, and computers.

The response scale used here—namely the percentage of R&D projects using public research output—allows us to go beyond purely subjective or relative notions of importance by affording the opportunity to quantify—however crudely—the impact of public research in terms of the magnitude of industrial R&D affected. In the 1991–1993 period, the U.S. manufacturing sector performed an average of $88 billion (in constant 1992 dollars) of R&D per year. To the extent that our sample reflects the range of R&D performing firms and industries in the manufacturing sector, and assuming that the percentage of projects using public research is equivalent to the percentage of R&D effort expended, our result that 29.3% of industrial

R&D projects used public research findings implies that an annual average of about $26 billion worth of industrial R&D used public research findings in the 1990–1993 period. Similarly, our finding that 22.2% of industrial R&D projects used instruments and techniques generated by public research implies that an annual average of almost $20 billion worth of industrial R&D used such instruments and techniques.[10] While we should interpret these estimates cautiously, the results suggest that the overall impact of public research on industrial R&D is substantial in absolute terms.[11]

6. What Fields of Public Research Contribute to Industrial R&D?

Building on a question posed in the earlier Yale Survey (Klevorick et al. 1995), we asked firms to evaluate, by field, the importance to their R&D (on a four-point Likert scale) of the contribution of public research conducted over the prior 10 years for each of ten fields, including: biology, chemistry, physics, computer science, material science, medical and health science, chemical engineering, electrical engineering, mechanical engineering, and mathematics.

Our first approach to using these responses to evaluate the importance of public research is to determine whether a preponderance of an industry's respondents (i.e., >50%) report public research to be at least moderately important (scoring at least three on a four-point scale). Table 3 presents, by industry, the percentage of respondents scoring each of the fields at least "moderately important." Consistent with the earlier Yale survey (Klevorick et al. 1995), we find that, of the basic sciences, only chemistry has a reasonably broad impact on industrial R&D, with over 50% of industry respondents reporting it to be at least moderately useful in the food, petroleum, metals, and several chemical industries, including drugs. To the extent that the other three examined fields of basic science, namely physics, mathematics, and biology, are relevant, they tend to impact a specific industry. Biology is particularly important for pharmaceuticals, with 64% of respondents reporting it to be at least moderately important, and physics is particularly relevant to semiconductors, with 62% reporting it to be at least moderately important. Following Nelson (1986) and Klevorick et al. (1995), we note that the relatively low scores for the basic sciences other than chemistry do not mean that research in these fields is unimportant for technical advance, but that their effect may be expressed through the applied sciences and engineering fields that they inform.

As may be expected, more respondents consider research in the engineering fields to contribute importantly to their R&D than research in the basic sciences, except for chemistry. Table 3 shows that the impact of public research in chemical engineering is most apparent in the petroleum and selected chemical industries. The impact of public research in mechanical engineering is most evident in general purpose machinery, glass and somewhat in the transportation-related group of industries. Electrical engineering is considered to be particularly important across the range of industries concerned with electronics, including computers, semiconductors, communications equipment, and instruments, as well as autoparts and glass.

Our findings on the impact of public research in the applied sciences suggest that the influence of medical and health science is strong in the drug and medical equipment industries. The impact of the other two applied sciences considered, namely materials and computer science, is much more pervasive—indeed more pervasive than any basic science or engineering field. More than half of respondents rate computer science research as at least moderately important in the glass, printing/publishing, communications equipment, search/navigation equipment, aerospace and, of course, computer industries. The field with

[10] Since projects using research findings and those using instruments and techniques derived from public research may overlap, we cannot aggregate the two types of impact.

[11] To the extent that we oversampled large firms (among the R&D performing population) these estimates may be biased upward (see results on firm size below). However, since this is an R&D, and therefore size, weighted average, the inclusion of more small firms would likely have only modest effects on the averages. Also, while our sample represents the manufacturing sector broadly (since we stratified our sample by industry), this estimate is sensitive to the distribution of firms across industries since industries vary in their use of public research.

Table 3 Importance of Public Research by Academic Discipline

Industry	N	Bio	Chem	Phys	CompSc	MatSc	Med	ChemE	EE	MechE	Math
		Percentage of Respondents Indicating Research "Moderately" or "Very" Important									
1500: Food	94	38.3	51.1	6.4	40.4	21.3	29.8	34.0	9.6	14.9	8.5
1700: Textiles	23	8.7	43.5	13.0	39.1	43.5	4.4	34.8	8.7	8.7	13.0
2100: Paper	31	22.6	45.2	22.6	35.5	51.6	16.1	48.4	12.9	16.1	12.9
2200: Printing/publishing	12	0.0	50.0	8.3	50.0	25.0	0.0	41.7	25.0	33.3	8.3
2320: Petroleum	18	11.1	55.6	22.2	33.3	44.4	5.6	44.4	11.1	38.9	22.2
2400: Chemicals, nec	75	13.3	52.0	8.0	24.0	22.7	17.3	34.7	1.3	5.3	5.3
2411: Basic chemicals	42	14.3	47.6	7.1	23.8	23.8	16.7	40.5	2.4	4.8	2.4
2413: Plastic resins	30	13.3	56.7	13.3	30.0	50.0	6.7	46.7	3.3	3.3	6.7
2423: Drugs	70	64.3	74.3	7.1	30.0	26.5	75.7	22.9	5.7	5.7	4.3
2429: Miscellaneous chemicals	32	12.5	62.5	9.4	31.3	46.9	12.5	37.5	3.1	12.5	9.4
2500: Rubber/plastic	35	2.9	45.7	28.6	31.4	60.0	5.7	48.6	11.4	22.9	11.4
2600: Mineral products	19	10.5	26.3	21.1	31.6	68.4	5.6	31.6	15.8	21.1	15.8
2610: Glass	6	0.0	33.3	33.3	83.3	83.3	16.7	33.3	66.7	66.7	16.7
2695: Concrete, cement, lime	10	30.0	30.0	30.0	30.0	60.0	10.0	30.0	10.0	30.0	10.0
2700: Metal, nec	9	11.1	55.6	0.0	22.2	77.8	0.0	44.4	0.0	22.2	0.0
2710: Steel	10	10.0	0.0	20.0	20.0	40.0	0.0	20.0	30.0	30.0	10.0
2800: Metal products	51	2.0	15.7	14.0	27.5	45.1	5.9	23.5	18.0	28.0	7.8
2910: General purpose machinery, nec	79	1.3	13.9	10.1	29.1	53.2	5.1	21.5	26.6	59.5	10.3
2920: Special purpose machinery, nec	74	10.8	23.0	25.7	35.1	38.4	5.4	20.3	31.1	36.5	14.9
2922: Machine tools	11	0.0	0.0	0.0	36.4	36.4	0.0	0.0	27.3	36.4	0.0
3010: Computers	28	3.6	21.4	35.7	67.9	50.0	0.0	14.3	64.3	53.6	21.4
3100: Electrical equipment	23	0.0	13.0	8.7	8.7	21.7	8.7	8.7	17.4	21.7	8.7
3110: Motor/generator	24	0.0	4.2	12.5	29.2	41.7	0.0	4.2	58.3	33.3	8.3
3210: Electronic components	28	3.6	25.0	28.6	32.1	53.6	7.1	10.7	63.0	50.0	28.6
3211: Semiconductors and related equipment	26	11.5	46.2	61.5	46.2	76.9	11.5	30.8	65.4	42.3	26.9
3220: Comm equipment	37	2.7	8.1	29.7	54.1	27.0	2.7	5.4	70.3	37.8	24.3
3230: TV/radio	9	0.0	11.1	33.3	44.4	55.6	11.1	22.2	66.7	33.3	22.2
3311: Medical equipment	76	35.5	34.2	21.1	30.3	47.4	76.3	18.4	29.0	29.0	15.8
3312: Precision instruments	38	15.8	18.4	21.1	39.5	31.6	15.8	5.3	52.6	39.5	23.7
3314: Search/navigational equipment	41	2.4	12.2	34.2	53.7	41.5	4.9	12.5	68.3	43.9	36.6
3410: Car/truck	9	11.1	22.2	33.3	44.4	55.6	11.1	22.2	33.3	44.4	22.2
3430: Auto parts	34	2.9	14.7	23.5	41.2	54.6	2.9	20.6	50.0	58.8	23.5
3530: Aerospace	51	2.0	31.4	31.4	54.9	68.6	3.9	27.5	45.1	54.9	28.0
3600: Other manufacturing	97	4.1	25.8	17.5	38.1	45.4	13.4	18.6	28.9	39.2	13.4
All	1,252	14.5	33.7	18.6	35.9	42.6	17.8	25.2	27.4	30.2	14.0

the most pervasive direct impact on industrial R&D is materials science. Half or more of industry respondents scored materials science as at least "moderately important" to their R&D activities in 15 of our 33 manufacturing industries, spanning the chemicals, metals, electronics, machinery, and transportation equipment industries. If any discipline can be awarded the title of a "general purpose" research field for the manufacturing sector, it is materials science.

When we identify industries in which a majority of respondents report some research field to be at least moderately important, we observe a pervasive impact of public research on industrial R&D. In 26 of the 34 industries, half or more of the respondents reported at least one public research field to be at least moderately important, and half or more of the respondents in 14 of the industries reported public research from at least two fields to be at least moderately important. If we compute the simple average of scores across respondents by field, a somewhat different picture emerges. One or more fields score an average of 3.0 or higher in only eight of the 34 industries, including drugs, semiconductors, medical equipment, communications equipment, aerospace, TV/radio, search/navigation equipment, and glass.

Another way of summarizing these data is to ask whether there is any field of public research that is especially important for a firm's R&D. To evaluate the importance of these most important fields by industry, we compute the average of the maximum field scores by respondent (notwithstanding the field) across all respondents for each industry. On average, respondents rate the most important field of public research to be "moderately important," as reflected in an average score of 3.04. Of the 34 industries, 21 report an average of this maximum score to equal at least 3.0. A handful of industries—drugs, glass, metals, computers, semiconductors, and medical equipment—show scores of 3.30 and above.

Thus, the majority of our results on field-specific impacts suggest that public research conducted over the prior 10 years is critical to a handful of industries and is pervasively useful across the manufacturing sector.

Of all our survey questions regarding the influence of public research, the question discussed in this section most closely resembles one posed in the Yale survey (Klevorick et al. 1995) and offers the best opportunity for comparison. There are, however, some important differences between the two surveys' samples, questions, and response scales. First, the CMS inquired about the importance of the research of universities and government labs and institutes, while the Yale survey inquired about university research alone. Second, we requested respondents to evaluate, by field, the importance of public research to *their own R&D activities* while the Yale survey asked respondents to report, by field, the relevance of university research to *technical progress for the respondent's industry as a whole*. Third, while both surveys use Likert scales, ours is four point while the Yale survey's was seven. Fourth, the fields covered in the CMS are a subset of those covered in the Yale survey; our survey did not include geology, agricultural science, applied math/operations research, and metallurgy. Finally, the populations surveyed differed. The Yale survey covered only larger, public firms; the CMS provided substantial representation from small and private firms as well. While we can do little about the questions or response scales (other than rescaling), we can make the two samples more comparable by restricting ours to only the larger firms (with sales over $500 million). For the comparison, we also conform our industry definitions to the Yale industry definitions and restrict our attention only to those 33 of the 130 four-digit SIC level industries surveyed by the Yale team that have at least four observations in each sample.

To consider whether the two surveys yield different results, we make several comparisons. First, we compute the average of the maximum field scores by respondent (notwithstanding the field) for each industry. For 21 of the 33 "comparison" industries in the CMS, this average maximum score equals at least three of a maximum of four (reflecting "moderate importance"). In comparison, for 23 of the 33 industries in theYale survey, this average maximum score equals at least 5.25 (the rescaled equivalent to 3.0 in the CMS). When we compute simply the average score by field for each industry, 12 of the 33 industries in the CMS have at least one field attaining an average score of 3.0 or above, and 13 industries in the

Yale survey have at least one field attaining an average score of 5.25. When we compare the number of industries in which half or more of the respondents reported at least one public research field to be at least moderately important, we again observe very similar results across the two samples.[12] Overall, the earlier Yale survey's findings of the field-specific influence of university research on industrial research appear to be comparable to those of the Carnegie Mellon Survey.

Can we conclude from this comparison that the extent of the contribution of public research to industrial R&D has changed little between 1983 when the Yale survey was administered and 1994 when the CMS was? Although our comparison provides little support for the position that the influence of public research has become more pervasive, we are reluctant to conclude that little has changed. Our comparison is necessarily based on the subset of industries where comparison was possible. Also, in light of the differences between the questions and response scales, any assessments of intertemporal changes in influence within industries are limited.[13] In any case, most of the results considered in this section suggest that, as of the mid-1990s, the contribution of public research to industrial R&D is widespread.

7. Dimensions and Incidence of Influence of Public Research

In the prior three sections, we examined the influence of public research along different dimensions. We examined: (1) how public research affects industrial R&D in the sense of whether it suggests new R&D projects or contributes to the completion of existing ones; (2) the influence of three different types of output of public research, including research findings, instruments, and techniques and prototypes; and (3) the general importance of the impact on industrial R&D by research field, scored along a semantic scale. In this section, we consider how these three dimensions of influence relate to one another, and what those relationships suggest about the influence of public research more generally. To address these questions, we compute simple correlations across the different dimensions of influence, measured at the industry level for 64 industries defined at the three- or four-digit ISIC level.

Across the three dimensions of influence, at the industry level, we typically observe significant correlations in the 0.25 to 0.45 range.[14] One telling exception to this pattern, is that, between each of the three types of output from public research (research findings, instruments and techniques, prototypes) and the frequency with which public research contributes to the completion of existing projects, the correlations are low (around 0.10) and insignificant.[15] Given that public research importantly affects industrial R&D by contributing to R&D project completion, we interpret these weak relationships to suggest that public research may contribute to project completion in ways or forms that are not reflected in the three output types considered, namely research findings, prototypes, or instruments and techniques. These may include, for example, the kind of knowledge that is conveyed via interpersonal interactions like consulting or informal relationships.[16] If true, this

[12] In 26 of the 33 comparison industries in the CMS, half or more of the respondents reported at least one public research field to be at least moderately important (scoring three out of a four-point scale). In the Yale survey, half or more of the respondents reported a score of five or six (on its seven point scale) in, respectively, 28 and 18 industries. Since the CMS number falls in the upper end of the range between the two Yale frequencies, the numbers appear to be similar.

[13] To the extent that there is an expressed difference from the earlier Yale survey in the evaluation of the importance of public research based on these field-specific scores, it is difficult to determine whether the authors of the earlier study may have understated the scope of the impact of public research or whether, between the dates of the two studies, the importance of public research increased (as Narin et al. 1997, for example, believe).

[14] The "importance score" (out of a one- to four-point Likert scale) for that field of public research which is most important to an industrial lab's R&D is most highly correlated with the other dimensions of influence (in the 0.30 to 0.45 range).

[15] In contrast, the correlations between the percentage of R&D projects affected by each of the three examined output types and the frequency with which public research suggests new projects are significant, in the range of 0.21 to 0.28.

[16] Consistent with this conjecture, we observe strong industry-level correlations between the degree to which public research contributes to project completion and the importance of the channels of

implies that our estimated percentages of industrial R&D projects affected by the research findings or instruments and techniques originating from public research understate the overall impact of public research.

Given that our findings show that public research can do different things in different industries, it should not be surprising that it is not just the R&D of "high tech" industries that benefits from public research. For example, among the six industries with 50% or more respondents reporting public research as a source of new project ideas, we find steel, machine tools, and petroleum—three industries considered to be "mature." Similarly, among the seven industries with 45% or more of the respondents reporting public research contributing to the completion of existing projects, we find three other mature industries, namely food, paper, and car/truck. To probe more systematically the relationship between our various measures of the influence of public research and industry innovative activity, we correlate our measures of influence with industry R&D intensity, measured as R&D weighted averages of business unit R&D expenditures to sales. As expected, the strength of the correlations vary considerably across the different dimensions of influence of public research, with the strongest relationships between industry R&D intensity and the percentage of industrial R&D projects making use of research findings ($r = 0.31$) or new instruments and techniques ($r = 0.29$) originating from public research. We observe the weakest relationship between industry R&D intensity and the degree to which public research contributes to the completion of existing projects ($r = 0.05$), which is not surprising if one considers that public research may partly substitute for a firm's own R&D.[17] If public research substitutes for industrial R&D and possibly obviates the need to undertake some R&D projects, then estimates of the percentage of R&D projects affected by public research outputs will only provide a partial, intermediate sense of the impacts of public research on technical advance. This provides yet another reason to believe that our estimates in §6 of the impact of public research outputs on the share of industrial R&D projects may underestimate the overall effect of public research on technical advance and productivity growth.

information flow between industrial R&D labs and public research (discussed below) of consulting and informal interaction (0.54 and 0.46, respectively). Also consistent with this conjecture, the channel of consulting is only weakly associated with the use of any of the three output types (all three correlations are 0.12 or less and n.s.). The pathway of informal interaction is, however, significantly associated with the use of research findings ($r = 0.28$).

[17] If it only substituted for R&D, we would expect a negative correlation.

8. Pathways of Knowledge Flow

The Carnegie Mellon Survey evaluated how useful information moves from universities and other public research institutions to industrial R&D facilities. We asked our respondents to report on a four-point Likert scale the importance to a recently completed "major" R&D project of each of 10 possible sources (or channels) of information on the R&D activities of universities or government R&D labs or institutes. The information sources considered include patents, informal information exchange, publications and reports, public meetings and conferences, recently hired graduates, licenses, joint or cooperative ventures, contract research, consulting, and temporary personnel exchanges. Table 4 presents the percentage of respondents indicating that a given channel is at least "moderately important." As shown, publications/reports are the dominant channel, with 41% of respondents rating them as at least moderately important. Informal information exchange, public meetings or conferences, and consulting follow in importance, with aggregate scores for each of these channels in the range of 31% to 36%.[18] After that point, there is a sharp dropoff in the scoring to another set of channels, namely recently hired graduates, joint and cooperative ventures and patents, with aggregate scores in the range of 17% to 21% of respondents indicating these channels to be at least moderately important. Licenses and personnel exchange are the least important channels, with scores of less than 10%.

[18] Thursby and Thursby (2001) similarly find that industry licensing executives identified publications and presentations, and especially personal contacts, as their most important sources for university technologies.

Table 4 Importance to Industrial R&D of Information Sources on Public Research

Industry	N	Patents	Pubs./ Reps.	Meetings or Conferences	Informal Interaction	Recent Hires	Licenses	Coop./ JV's	Contract Research	Consulting	Personnel Exchange
			Percentage of Respondents Indicating Research "Moderately" or "Very" Important								
1500: Food	93	9.7	51.6	37.6	44.1	21.5	10.8	22.6	30.1	46.2	7.5
1700: Textiles	23	13.0	26.1	26.1	21.7	21.7	0.0	13.0	8.7	13.0	0.0
2100: Paper	31	9.7	45.2	35.5	32.3	9.7	0.0	19.4	35.5	22.6	3.2
2200: Printing/publishing	12	16.7	33.3	25.0	16.7	8.3	8.3	0.0	16.7	25.0	0.0
2320: Petroleum	18	0.0	38.9	50.0	27.8	11.1	11.1	11.1	22.2	44.4	0.0
2400: Chemicals, nec	73	24.7	35.6	28.8	20.6	16.4	8.2	15.1	20.8	24.7	9.6
2411: Basic chemicals	41	17.1	36.6	26.8	39.0	17.1	2.4	14.6	17.1	34.2	2.4
2413: Plastic resins	28	14.3	35.7	32.1	21.4	21.4	0.0	3.6	10.7	14.3	0.0
2423: Drugs	68	50.0	73.5	64.7	58.8	30.9	33.8	41.2	52.9	58.8	8.8
2429: Miscellaneous chemicals	32	25.0	34.4	25.0	31.3	21.9	3.1	3.1	12.5	25.0	0.0
2500: Rubber/plastic	35	5.7	17.1	14.3	11.4	14.3	2.9	11.4	8.6	22.9	0.0
2600: Mineral products	19	5.3	26.3	21.1	21.1	31.6	5.3	10.5	10.5	26.3	10.5
2610: Glass	6	33.3	50.0	50.0	50.0	50.0	16.7	50.0	33.3	33.3	0.0
2695: Concrete, cement, lime	10	30.0	50.0	30.0	20.0	30.0	30.0	10.0	10.0	10.0	10.0
2700: Metal, nec	8	25.0	62.5	62.5	87.5	25.0	0.0	25.0	37.5	50.0	12.5
2710: Steel	11	18.2	36.4	54.6	45.5	18.2	18.2	36.4	54.6	36.4	18.2
2800: Metal products	51	19.6	25.5	13.7	25.5	17.7	7.8	13.7	9.8	21.6	3.9
2910: General purpose machinery, nec	76	15.8	30.7	26.3	30.3	14.5	7.9	10.5	13.2	31.6	1.3
2920: Special purpose machinery, nec	72	19.4	33.3	33.3	27.8	16.7	11.1	16.7	15.3	30.6	2.8
2922: Machine tools	11	9.1	36.4	45.5	45.5	18.2	0.0	9.1	18.2	36.4	0.0
3010: Computers	29	13.8	41.4	37.9	34.5	34.5	3.5	6.9	6.9	24.1	3.5
3100: Electrical equipment	23	8.7	30.4	21.7	21.7	0.0	0.0	8.7	13.0	8.7	0.0
3110: Motor/generator	25	4.0	40.0	36.0	44.0	12.0	0.0	20.0	12.0	28.0	4.0
3210: Electronic components	27	18.5	37.0	33.3	33.3	29.6	11.1	14.8	11.1	30.8	3.7
3211: Semiconductors and related equipment	25	20.0	60.0	48.0	54.2	36.0	12.0	20.0	20.0	40.0	4.0
3220: Comm equipment	37	5.4	48.7	32.4	32.4	27.0	8.1	8.1	16.2	29.7	18.9
3230: TV/radio	9	22.2	66.7	33.3	33.3	33.3	11.1	33.3	22.2	22.2	11.1
3311: Medical equipment	74	27.0	40.5	36.5	47.3	18.9	17.6	23.0	23.0	44.6	6.8
3312: Precision instruments	39	23.1	46.2	41.0	41.0	10.3	12.8	18.0	7.7	33.3	5.1
3314: Search/navigational equipment	40	7.5	52.5	50.0	50.0	20.0	12.5	27.5	32.5	42.5	12.5
3410: Car/truck	9	33.3	33.3	11.1	33.3	11.1	11.1	22.2	33.3	22.2	11.1
3430: Auto parts	32	9.4	43.8	31.3	25.0	18.8	9.4	21.9	18.8	21.9	9.4
3530: Aerospace	49	14.3	57.1	51.0	55.1	20.4	8.2	40.8	36.7	40.8	6.1
3600: Other manufacturing	93	12.9	33.3	34.4	31.2	17.2	5.4	9.7	16.1	19.4	7.5
All	1,229	17.5	41.2	35.1	35.6	19.6	9.5	17.9	20.9	31.8	5.8

These aggregate scores show that the most important channels of information flow between public research institutions and industrial R&D labs are the channels of open science, notably publications and public meetings and conferences. Moreover, these channels, as well as the next most important channels of informal information exchange and consulting, are relatively decentralized in the sense that they do not typically reflect formal institutional links. With the exception of consulting, these most important channels are also not mediated through any sort of market exchange.

The relatively high score for consulting underscores the importance of this little-studied vehicle through which public research impacts industrial R&D. Both more and less R&D intensive industries consider consulting to be important. Industries where over 40% of respondents report consulting to be at least moderately important include food, petroleum, drugs, metals, semiconductors, aerospace, and medical and other equipment.

Surprisingly, the hiring of recent graduates is of small importance overall. This result requires some elaboration. First, we are inquiring about the impact on industrial R&D of public research, not the about the broader economic impacts of university training. Also, if one looks at the industry-level results, one observes recent graduates playing a somewhat important role relative to other channels in selected industries, notably in the electronic component, communications equipment and computer industries, although publications are still more important in these industries.

One assumption underlying current policy is that cooperative or joint ventures between universities and industry will encourage the use of academic research by industry.[19] Since, as noted above, our data reflect predominantly the impact of university research, they suggest that such cooperative ventures have typically not contributed as importantly to industrial R&D as other channels, at least not in any direct fashion. Exceptions to this overall pattern, where a third or more of our respondents report joint and cooperative ventures to be at least moderately important, include drugs (where all channels tend to be more important), glass, steel, TV/radio, and aerospace. In some industries such as steel (where another channel, contract R&D, is an even more important channel than joint or cooperative ventures), such cooperative relationships may have become substitutes for industry R&D as upstream and corporate R&D have been cut back.

Intellectual property policy affecting public research has also changed over the past two decades. The Bayh-Dole Act permits universities, government research labs and other nonprofit institutions to obtain patent rights to the output of federally sponsored research. The impetus behind Bayh-Dole and related legislation was the assumption that there was a stock of underexploited, valuable knowledge residing in universities and other research institutions receiving federal funding, and that patents would "incentivize" the private sector to undertake the downstream R&D and related investment necessary for commercialization (Mazzoleni and Nelson 1998).

Although our results cannot speak to the effects of patents and licensing on industry's incentives to use and commercialize public research, they suggest that licenses and patents are subordinate means of conveying the content of that research to industry. Our data indicate that, even though the impact of public research is reasonably pervasive, patents and licenses appear to be useful mechanisms of technology transfer in only a few industries. In most industries, patents and licenses are not nearly as important as other channels for conveying public research to industry, including publications, conferences, informal information exchange, or consulting. Even in some high tech industries reporting substantial public research impacts, such as communications equipment and aerospace, patents and licenses achieved scores that were at best average. The main industry where patents and licenses appear to be important for conveying public research is the pharmaceutical

[19] Since the 1970s, for example, we have witnessed the growth of the NSF's Science and Technology Centers and Engineering Research Centers that tie government support for university research to industry participation. Stimulated by the example of such programs and by the desire to tap into industry and tied government support, university-industry R&D centers have grown rapidly. Of the 1,056 estimated university-industry R&D centers existing as of 1990, almost 60% were established in the prior decade and by 1990 they were the beneficiaries of almost 25% of all government support of university research (Cohen et al. 1998).

Table 5 Industry-Level Factor Analysis of Channels-of-Information Flow from Public Research to Industrial R&D

	Factor Loadings	
Channel	Factor 1	Factor 2
Publications/reports	**0.85**	0.04
Informal interaction	**0.84**	0.09
Public meetings or conferences	**0.84**	0.16
Contract research	**0.79**	0.23
Consulting	**0.70**	0.23
Joint or cooperative ventures	**0.63**	0.29
Patents	0.34	0.20
Personnel exchange	0.00	**0.70**
Licenses	0.21	**0.69**
Recently hired graduates	0.25	**0.51**
Eigenvalue	3.87	1.49

Table 6 Correlations Between Importance of Information Channels and Role of Public Research, by Industry ($N = 64$)

	Role of Public Research	
Information Channels	Suggests New R&D Projects	Contributes to Project Completion
Patents	0.08	0.08
Publications/reports	0.23	0.49**
Public meetings or conferences	0.37*	0.46**
Informal interaction	0.25**	0.46**
Recent hires	0.55**	0.22
Licensed technology	0.20	0.28*
Cooperative/joint ventures	0.15	0.35**
Contract research	0.46**	0.56**
Consulting	0.45**	0.54**
Personnel exchange	−0.00	−0.05

*Significant at the 0.05 confidence level.
**Significant at the 0.01 confidence level.

industry—with 50% and 34% of respondents reporting patents and licenses, respectively, to be at least moderately important as channels. Even here, however, informal channels and the channels of open science were still more important.

Thus, even if one accepts the premise of the Bayh-Dole Amendment that privatization of public research motivates industry to use it—which some scholars (e.g., Mowery et al. 2000) do not—we should not assume that patents and licenses provide the grist for the mills of industrial R&D. Rather, it is typically the public expressions of public research (i.e., publications, meetings, and conferences), or even informal interactions and consulting that centrally convey the content of that research to industry.

One question regarding the channels through which public research impacts industrial R&D is how the use of different channels, or at least the respondents' evaluations of their importance, relate to one another. To address this question, we conducted a factor analysis at the industry-level. Presented in Table 5, the factor analysis shows the industry average scores for the channels loading on two factors. The channels that load on the first factor include publications, conferences and meetings, informal interaction, consulting, contract research, and joint or cooperative ventures. Licenses, the hiring of recent graduates, and personnel exchange load on a second factor. Patents do not load on either factor. The loadings on the first factor suggest that person-to-person interactions tend to be used with and perhaps complement more public channels such as publications or meetings. Indeed, prior work in the sociology of science suggests that direct exchanges often complement the published literature as a means of conveying information (Walsh and Bayma 1996). In the case of industrial R&D, such interactions may stem, for example, from an engineer seeing an article and then contacting the author about how the research might be used. Or, communications with faculty might point to published research of use to the firm's R&D. We suspect that both processes are common and help explain the high correlations among these channels.[20]

In addition to examining the relationships across the channels of influence of public research, our survey data permit examination of the relationship between the importance of the different channels

[20] Gibbons and Johnston (1975) similarly found that the technical literature and "personal contact" were most beneficial to technological innovation when used together. Faulkner and Senker (1995) confirm this finding in their study of public sector research and industrial innovation in biotechnology, engineering ceramics, and parallel computing. Jensen and Thursby (2001) similarly report that successful licensing of university technologies to industry typically requires, not only the license, but additional input by the university-based inventor due to the embyronic state of development of the licensed technology.

and how public research affects industrial R&D—i.e., whether it suggests new R&D projects or contributes to the completion of existing ones. Table 6 presents the correlation coefficients between the industry average scores for the importance of each channel to a firm's R&D and the frequency, by industry, with which public research either suggests a new R&D project or contributes to project completion. We observe that the importance of the channels of consulting, contract research, public conferences, and meetings are significantly and comparably correlated (i.e., within 0.10 of one another) with both of these effects of public research. We find, however, much stronger correlations (in the range of 0.46 to 0.49) between the importance scores of publications, informal information exchange, and joint/cooperative R&D projects with the contribution of public research to project completion than to the suggestion of new projects (in the range of 0.15 to 0.25). This finding suggests that industrial R&D personnel seek out academics, search the literature, or form cooperative ventures with public research institutions more commonly to address particular needs or problems than to generate new project ideas.[21]

9. Firm Characteristics and the Influence of Public Research

Having considered the aggregate and industry-level impacts of public research on industrial R&D, we now consider within-industry, cross-firm differences in those impacts. Following Link and Rees (1990) and Acs et al. (1994), we focus on whether there is a relationship between the degree to which a firm might exploit public research and its size. On the basis of an analysis of geographically defined, state-level industry R&D expenditures and university research expenditures, Acs et al. (1994) infer that small firms' R&D spending responds disproportionately more to university research, and conclude that small firms have a "comparative advantage at exploiting spillovers from university laboratories." For a sample of firms in machine tools, computing, and aircraft and components, Link and Rees (1990) find that large firms are more likely to have ties to university research. However, their analysis of rates of return to R&D shows the returns to R&D of small firms' involved in university research to be higher than that of large firms, which they interpret to mean that small firms are able to transfer knowledge gained from their university research associations more effectively than larger firms.[22]

The CMS data offer a range of direct—though self-reported—measures of the knowledge inflow from university and government R&D labs. As described above, our data include percentages of firms' R&D projects that make use of public research, as well as whether or not knowledge from university or government labs either suggests new R&D projects or contributes to the completion of existing projects in the prior three years. We use an ordered logistic regression model to analyze within-industry variation in the percentages of R&D projects that use public research since the response scale is expressed as range intervals (i.e., below 10%, 10%–40%, 41%–60%, 61%–90%, over 90%). We employ a logistic regression to analyze factors that might affect whether public research suggests new R&D projects or contributes to project completion because each of these two latter variables is binary (i.e., yes/no). To focus on the effect of the within-industry variation of selected firm characteristics, we include 33 industry dummies to control for industry-fixed effects.

The key independent variable we consider is firm size, measured as log of firm sales. We believe that

[21] The only pathway with an importance score more highly correlated with the contribution of public research to suggesting new projects is "recently hired graduates with advanced degrees."

[22] Both of these papers were trying to consider why smaller firms' R&D productivity is greater than that of larger firms by appealing to the possibility that small firms rely more heavily on extramural research, and particularly university research, in lieu of internal R&D. Cohen and Klepper (1996) show, however, that the lower average R&D productivity of larger firms can easily be explained by the fact that larger firms simply undertake more R&D projects at the margin because, in most industries, they earn a higher return per R&D dollar due to their spreading of the fixed costs of R&D over greater levels of output. Thus, the differences in the average R&D productivity of larger versus smaller firms do not necessarily reflect any differences in R&D efficiency related to firm size, nor need they reflect greater exploitation of extramural knowledge flows on the part of smaller firms.

our sample is well suited to testing the impact of firm size since it includes firms from across the entire range of the firm size distribution, with over a third of our sample firms having fewer than 500 employees, which is the conventional cutoff to reflect small firms. We also wish to consider whether public research affects technical advance particularly by influencing the R&D of start-up firms since start-ups are sometimes viewed as a key vehicle for transferring university research into commercial products.[23] Thus, we also include on the right-hand side a dummy variable, START-UP, signifying whether a firm is a start-up, defined as a firm that is no more than five years old, with fewer than 500 employees in 1993, and typically active in only one industry. Our analysis of the impact of public research on start-up firms is, however, limited. Because our sample was not designed to test the role of university research in start-up companies, we did not oversample start-up firms and so have relatively few observations—22, which are concentrated in only three industries; five observations are in pharmaceuticals, four in computers, and four in medical equipment.

Presented in Table 7, the logistic regression results show a highly significant, positive effect of the log of firm size across all three dependent variables, suggesting that larger firms are more likely to use public research. One might argue that since larger firms tend to conduct more R&D and be involved in more R&D projects, one might expect that public research is more likely to suggest new R&D projects or contribute to the completion of existing ones for larger firms just on a random basis. Perhaps more informative, then, is the column (1) result that larger firms use public research in a larger percentage of their R&D projects.[24] In addition to the size effect, we find a positive and significant coefficient estimate on START-UP in columns (1) and (3). START-UP is only significant at the 0.10 confidence level in column (2), suggesting that while start-up firms use public research in a greater percentage of their R&D projects than other firms and are more likely to use public research for completing existing R&D projects, they do not clearly rely more than other firms on public research as a source of new project ideas.[25]

Table 7 Within-Industry Determinants of the Influence of Public Research (Logit Regressions)

	1	2	3
Dependent Variable	% of R&D Projects using Public Research	Suggesting New R&D Projects	Contributing to R&D Project Completion
Intercept 1	−6.64**	−3.36**	−4.06**
	(0.54)	(0.56)	(0.55)
Intercept 2	−5.32**		
	(0.51)		
Intercept 3	−4.36**		
	(0.50)		
Intercept 4	−2.60**		
	(0.48)		
Log firm size	0.26**	0.23**	0.34**
	(0.05)	(0.06)	(0.06)
START-UP	1.23**	0.84ᵃ	1.52**
	(0.43)	(0.51)	(0.51)
Number of observations	1,193	1,186	1,176

Notes. (1) Standard errors in parenthesis.
(2) Industry dummy coefficient estimates are omitted.
(3) ᵃ: Significant at 10%; *: Significant at 5%; **: Significant at 1%.

[23] The Association of University Technology Managers 1999 Licensing Survey reports that 62% of university licenses were to small firms (12% were to companies started as a result of the license), and that university research led to the founding of 344 companies (AUTM 1999).

[24] We also find that larger firms are more likely to score higher on MAXSCI—the importance of public research from the most relevant field. The MAXSCI regression results are available upon request.

[25] We checked to see if our results were robust to the inclusion of R&D intensity (measured as the ratio of business unit R&D employment over total business unit employment), the Ph.D./M.D. intensity of the lab, and the effort dedicated by the R&D lab to the monitoring or gathering of extramural scientific and technical information. All the results are robust, except for the already weak effect of "START-UP" in column 2, which becomes insignificant. These results are not featured due to the likely endogeneity of all three variables, and controlling for that endogeneity is beyond the scope of the present paper.

We also ran all specifications with MAXSCI as a dependent variable. These results suggest that start-ups are not significantly more likely to rate any field public research as more important. To consider the possibility that start-ups disproportionately focus on basic research, we created a new dependent variable, MAXBASIC, which is the maximum score among the basic research fields of biology, chemistry, physics, and mathematics. We replicated the Table 7 regressions for this variable to see if start-ups score higher in terms

To probe the role of start-ups in more detail, we ran a series of comparisons between start-up firms and the rest of our sample for the three industries where we have at least four start-up firms per our definition: drugs, computers, and medical equipment. We use a nonparametric (Wilcoxon rank sum) test for differences in group means between start-ups and the rest of the firms in each industry.[26]

For the pharmaceutical industry, we find that start-ups are more closely tied to public research than established firms. For example, while about half of the non-start-up drug firms report public research suggesting new projects or contributing to existing projects, 100% of start-up firms said yes to both questions ($p < 0.10$ for each). Comparing the percent of projects that use the public research outputs of research findings, prototypes, or instruments and techniques, we again find that start-up firms score higher than more established drug companies. On average, 72% of start-up firm projects made use of research findings, while only 36% of established firm projects did ($p < 0.05$). For prototypes, the averages are 27% for start-ups and 13% for non-start-ups (ns). For instruments/techniques, start-ups are again higher (23% v. 20%) though the gap is small.[27]

For our other two industries, medical equipment and computers, the results on the use of public research by start-ups versus more established firms are mixed. For example, in medical equipment, established firms score higher on public research suggesting projects (36% v. 25%) and start-ups score higher on public research helping to complete projects (50% v. 42%), though neither difference is statistically significant. Medical equipment start-ups report a significantly greater percent of projects making use of research findings (50% v. 21%, $p < 0.05$), prototypes (33% v. 10%, $p < 0.05$), and instruments/techniques (33% v. 15%, n.s.).[28]

For computers, evidence of a disproportionate use of public research by start-ups is weaker. For suggesting projects, there is no difference (25% of start-ups v. 23% of non-start-ups). For completing projects, 50% of start-ups say, "yes", versus 19% of non-start-ups ($p < 0.20$). Ten percent of start-up firms' projects make use of research findings, in comparison to 16% of non-start-up firms' projects (n.s.). Five percent of start-ups use prototypes versus seven percent for non-start-ups. And, only 5% of start-up projects use instruments/techniques, compared to 15% for non-start-up projects (n.s.).[29]

Thus, our comparison of start-ups and other firms in three high-tech industries shows that start-ups are most clearly tied to public research in the pharmaceutical industry. This should not be surprising, since the biotech start-up is the archetypical university, license-based spinoff. For medical equipment and computers, the difference between start-up firms and established firms is less manifest. Across all three industries, as in the full sample regressions, start-ups are, however, more likely than established firms to use public research to complete existing projects than as a source of new project ideas.

of their use of basic sciences. While large firms use basic research more, the start-up effect is close to zero.

[26] One might conjecture that, since we do not control for firm size in the analyses of the role of start-ups in the three industries, the association between start-ups and the influence of public research may appear to be weaker since start-ups tend to be small and larger firms tend to use public research more intensively (per our regressions). The qualitative results for the industry analyses are robust, however, even when we do control for firm size in logistic regressions run for each of the three industry samples.

[27] Regarding the relative importance of the different channels through which knowledge may flow from public research to firms' R&D (e.g., publications, conferences, licenses, etc. per Table 4), all score the same or higher for start-ups, reflecting the greater overall impact of public research on start-ups in the drug industry. There is, however, an especially notable difference with regard to the role of licensing, which scores on average 3.0 on a four-point scale versus a score of only 1.9 for non-start-up drug companies ($p < 0.05$). Of the other channels of knowledge flow, joint and cooperative R&D projects and contract research also scored notably higher for start-ups, but with differences that were only weakly significant (e.g., $p < 0.15$).

[28] Regarding the importance of the channels through which public research may affect industrial R&D information, there is no clear difference in scores between start-ups and established firms in the medical equipment industry.

[29] With regard to information channels, eight of the ten are more important to start-ups than to established firms, but only temporary personal exchanges is significant (1.8 v. 1.2, $p < 0.15$), and the importance of licensing is nearly equal (1.5 for start-ups; 1.4 for non-start-ups).

The overall picture, therefore, is that larger firms are more likely to make greater use of public research, though start-up firms also benefit from public research, especially in pharmaceuticals.

10. Conclusion

The responses to our questionnaire suggest, on balance, that public research importantly affects industrial R&D in a broad range of industries, though often in different ways. It is sometimes a difficult matter to judge importance of such an effect, and we do so in a number of ways. First, our data shows that a large share—almost a third of industrial R&D projects for our sample firms—made use of research findings from public research, and over a fifth made use of instruments and techniques. Other evidence from our survey further suggests that the knowledge originating from public research is often conveyed outside of these particular forms of output, such as through consulting or informal communications, implying that the share of R&D projects affected by public research is likely even greater than that which makes use of either the research findings or the techniques and instruments generated by public research. We also observed that the effect of the knowledge originating from public research on industrial R&D appears to be at least as great as the effect of that originating from rival R&D, and, as noted above, the empirical economics literature on R&D spillovers suggest intraindustry spillovers to be important to productivity growth. Most summaries of respondents' subjective evaluations of the importance of public research, broken down by field, also suggest that while public research is critical to a small number of industries, it is "moderately important" across a broad swath of the manufacturing sector. Although public research may be important in some absolute sense, we also find—as do Klevorick et al. (1995)—that its direct and immediate impact is much less, however, than sources of knowledge that lie more directly in the vertical chain of production and sale, including suppliers, buyers, and the firms' own manufacturing operations.

Our survey also lends considerable support to the more nuanced, interactive conception of the innovation process advanced by Gibbons and Johnston (1975), Kline and Rosenberg (1986), von Hippel (1988), and Nelson (1990), where public research will sometimes lead technological development, but, more typically, downstream research and development or buyer input provide the impetus and guidance for what industrial R&D labs do. Consistent with this picture is the dominant role, noted above, of buyers and the firm's own manufacturing operations in either suggesting new industrial R&D projects or contributing to the completion of existing projects. Also consistent with this characterization is our finding that public research provides ways of solving problems at least as often as it suggests new project ideas.

Our survey yields other findings similar to those of the Yale survey. For example, we observe that the impact of public research, at least in most industries, is exercised through engineering and applied science fields—especially materials and computer science—rather than through basic sciences. Reflecting on the earlier Yale survey findings, Nelson (1986) and Klevorick et al. (1995) have cautioned, however, that the greater importance of more applied fields does not mean that basic science has little impact, but that its impact may be mediated through the more applied sciences or through the application of industrial technologists' and scientists' basic scientific training to the routine challenges of conducting R&D.

We also find that there are some clear differences in the impacts of public research across industries. In this regard, the pharmaceutical industry stands out as an anomaly along many dimensions. There is no other industry where public research—and particularly a basic science (i.e., biology)—is thought to be so relevant. Also, knowledge from buyers and firms' own manufacturing operations are less important to R&D in pharmaceuticals than in other industries, suggesting that the linear model may characterize the innovation process better in this industry than in others. Moreover, patents and licenses are key means of conveying information from public research to industry in pharmaceuticals, partly reflecting the fact that patents are more effective in protecting inventions in drugs than in any other manufacturing industry (cf. Levin et al. 1987, Cohen et al. 2000). Finally, the drug industry is one where start-ups clearly are more tied to public research than are firms in general.

In contrast to prior work (e.g., Acs et al. 1994), we do not find that small firms disproportionately benefit from public research. Our regression results robustly show that large firms are much more likely to use public research. Although small firms in general exploit public research less, start-ups appear to use it more. While start-ups appear to differ from other small, established firms in the degree to which they benefit, our survey data do not tell us why. Perhaps some start-ups originate from universities and thus have strong pre-existing ties, or perhaps they are spinoffs from large firms and consequently benefit from the stronger ties that our data suggest exist between larger firms and the institutions of public research.

Two policy questions that come up repeatedly regarding university research in particular is how much government support should it receive, and, second, should particular kinds of bridging mechanisms between universities and industry—such as university-industry cooperative ventures or technology licensing—be encouraged.

Although we cannot address the first question directly, our research suggests that the contribution of public research to industrial R&D is considerable and pervasive. Regarding the second question, our results on the channels of information flow between public research and industry suggest that the decentralized and longstanding channels of publications, conferences, informal exchange, and consulting are the most important. We would suggest that, even when university-industry cooperative ventures or technology licensing between public research institutions and industry support technology transfer in an immediate way, encouragement of such bridging mechanisms should not come at the expense—as they occasionally do—of the other more important channels of open science noted above (cf. Cohen et al. 1998).

The limitations of our study are numerous. First, our survey was designed to convey a sense of the more immediate impacts of public research. It was not our intent to discern the long-run effects of science or the training effects that may be universities' greatest contribution to technical advance. Thus, to the degree that we have identified impacts, our estimates should be considered conservative. Second, we have relied only upon one method in this study—a survey-based description. To arrive at more accurate and robust estimates of the impacts of public research, one should deploy numerous data sources and methods of analysis, including surveys, field studies, and econometric analyses. Also, our analysis of start-ups is preliminary. Further research explicitly designed to compare start-ups to established firms is needed to examine their role. Also, based on a survey of firms at a point in time, our analysis cannot consider the role of public research in affecting the long-run evolution of industries, no less how industry itself shapes public research over time.

Overall, our results suggest that university research has a substantial impact on industrial research, that this impact is primarily through public and personal channels, and that university research contributes to project completion as well as suggesting new projects. We also find substantial cross-industry differences, though few systematic differences between high-tech and more mature industries. These results suggest that both public research and industry product and process development progress through complex, intertwined processes, with public research sometimes driving industry R&D, but also providing knowledge that abets the progress of projects initiated due to information, needs and opportunities that originate from buyers, the firm's own manufacturing operations, and other sources. We suspect that universities play this role not simply because they produce knowledge, but because, as pointed out by Goto (2000), they are also repositories thereof.

Acknowledgments

The authors thank Diana Hicks, David Hounshell, Rob Lowe, David Mowery, Scott Shane, Jerry Thursby, and other participants in the Georgia Institute of Technology Workshop on University Technology Transfer and Entrepreneurship for their comments. Principal research support was provided by the Alfred P. Sloan Foundation, and the Center for Global Partnership. Additional support was provided by Hitotsubashi University's Institute of Innovation Research.

References

Acs, Z. J., D. B. Audretsch, M. P. Feldman. 1994. R&D spillovers and recipient firm size. *Rev. Econom. Statist.* **76** 336–340.

Adams, J. D. 1990. Fundamental stocks of knowledge and productivity growth. *J. Political Econom.* **98** 673–702.

Association of University Technology Managers (AUTM). 1999. AUTM Licensing Survey, Fiscal Year 1999 Survey Summary. Available at ⟨http://www.autm.net/survey/99/survey99A.pdf⟩.

Bernstein, J. I., M. I. Nadiri. 1989. Research and development and intraindustry spillovers: An empirical application of dynamic duality. *Rev. Econom. Stud.* **56** 249–267.

Bowker Publishing. 1994. *Directory of American Research and Technology.*

Bush, V. 1945. *Science the Endless Frontier: A Report to the President on a Program for Postwar Scientific Research.* U.S. Government Printing Office, Washington, D.C.

Cohen, W. M., R. Florida, L. Randazzese, J. Walsh. 1998. Industry and the academy: Uneasy partners in the cause of technological advance. R. Noll, ed. *Challenges to Research Universities.* Brookings Institution Press, Washington, D.C.

——, S. Klepper. 1996. A reprise of size and R&D. *Econom. J.* **106** (July) 925–951.

——, R. R. Nelson, J. Walsh. 2000. Protecting their intellectual assets: appropriability conditions and why U.S. manufacturing firms patent (or not). Working paper no. 7552, National Bureau of Economic Research, Washington, D.C.

Faulkner, W., J. Senker. 1995. *Knowledge Frontiers: Public Sector Research and Industrial Innovation in Biotechnology, Engineering Ceramics, and Parallel Computing.* Oxford University Press, New York.

Feller, I. 1990. Universities as engines of R&D-based economic growth: They think they can. *Res. Policy* **19** 335–348.

Gibbons, M., R. Johnston. 1975. The roles of science in technological innovation. *Res. Policy* **3** 220–242.

Goto, A. 2000. Japan's national innovation system. *Oxford Rev. Econom. Policy* **16** 103–113.

Griliches, Z. 1992. The search for R&D spillovers. *Scand. J. Econom.* **94** (3, Supplement) 529–547.

Hounshell, D. 1996. The evolution of industrial research in the United States. R. S. Rosenbloom, W. J. Spencer, eds. *Engines of Innovation.* Harvard Business School Press, Boston, MA.

Jaffe, A. 1989. Real effects of academic research. *Amer. Econom. Rev.* **79** 957–970.

Jensen, R., M. Thursby. 2001. Proofs and prototypes for sale: the licensing of university inventions. *Amer. Econom. Rev.* **91** 240–259.

Klevorick, A. K., R. Levin, R. R. Nelson, S. Winter. 1995. On the sources and significance of interindustry differences in technological opportunities. *Res. Policy* **24**(2) 195–205.

Kline, S. J., N. Rosenberg. 1986. An overview of innovation. R. Landau, N. Rosenberg, eds. *The Positive Sum Strategy.* National Academy Press, Washington, D.C.

Levin, R., A. Klevorick, R. R. Nelson, S. G. Winter. 1987. Appropriating the returns from industrial R&D. *Brookings Papers Econom. Activity* (Special Issue) 783–820.

Link, A. L., J. Rees. 1990. Firm size, university based research, and the returns to R&D. *Small Bus. Econom.* **2** 25–31.

Mansfield, E. 1991. Academic research and industrial innovation. *Res. Policy* **20** 1–12.

Mazzoleni, R., R. R. Nelson. 1998. Economic theories about the benefits and costs of patents. *J. Econom. Issues* **32**(4) 1031–1052.

Mowery, D. C., R. R. Nelson, B. N. Sampat, A. A. Ziedonis. 2000. The effects of the Bayh-Dole Act on U.S. university research and technology transfer: An analysis of data from Columbia University, the University of California, and Stanford University, Mimeo, Haas School of Business, University of California, Berkeley, CA.

——, N. Rosenberg. 1979. The influence of market demand upon innovation: A critical review of some recent empirical studies. *Res. Policy* **8** 102–153.

Narin, F., K. S. Hamilton, D. Olivastro. 1997. The increasing link between U.S. technology and public science. *Res. Policy* **26**(3) 317–330.

National Science Board. 1998. Science and engineering indicators—1998. National Science Foundation, Arlington, VA.

——. 2000. Science and Engineering Indicators—2000. National Science Foundation, Arlington, VA.

Nelson, R. R. 1986. Institutions supporting technical advance in industry. *Amer. Econom. Rev. Proc.* **76** 186–189.

——. 1990. Capitalism as an engine of progress. *Res. Policy* **19** 193–214.

Price, D. de S. 1984. The science/technology relationship, the craft of experimental science, and policy for the improvement of high technology innovations. *Res. Policy* **13** 3–20.

Rosenberg, N. 1992. Scientific instrumentation and university research. *Res. Policy* **21** 381–390.

——, R. R. Nelson. 1994. American universities and technical advance in industry. *Res. Policy* **23**(3) 323–348.

Thursby, J. G., M. C. Thursby. 2001. Industry perspectives on licensing university technologies: Sources and problems. *J. Assoc. Univ. Tech. Managers* **13**

von Hippel, E. 1988. *The Sources of Innovation.* Oxford University Press, New York.

Walsh, J. P., T. Bayma. 1996. Computer networks and scientific work. *Soc. Stud. Sci.* **26** 661–703.

Accepted by David C. Mowery and Scott Shane; received December 2000. This paper was with the authors 6 months for 3 revisions.

[23]

UNIVERSITIES AS RESEARCH PARTNERS

Bronwyn H. Hall, Albert N. Link, and John T. Scott*

Abstract—Universities are a key institution in the U.S. innovation system, and an important aspect of their involvement is the role they play in public-private partnerships. This note offers insights into the performance of industry-university research partnerships, using a survey of precommercial research projects funded by the Advanced Technology Program. Although results must be interpreted cautiously because of the small size of the sample, the study finds that projects with university involvement tend to be in areas involving new science and therefore experience more difficulty and delay, yet are more likely not to be aborted prematurely. Our interpretation is that universities are contributing to basic research awareness and insight among the partners in ATP-funded projects.

I. Introduction

MANY OBSERVERS have emphasized the importance of research partnerships for U.S. innovative capacity (Council on Competitiveness, 1996, pp. 3–4). Indeed, industry-university research relationships appear to have strengthened over the past few decades. University participation in formal research joint ventures (RJVs) has increased steadily since the mid-1980s (Link, 1996), and the number of industry-university R&D centers increased by more than 60% during the 1980s. Cohen et al. (1997) and a recent survey of U.S. science faculty revealed

Received for publication April 21, 2000. Revision accepted for publication January 7, 2002.

* University of California at Berkeley and NBER, University of North Carolina at Greensboro, and Dartmouth College, respectively.

We are grateful for comments on earlier versions from William L. Baldwin, Adam Jaffe, Don Siegel, participants at the ASSA 2000 meetings in Boston and the Wake Forest University economics workshop, and an anonymous referee. Also, we appreciate the suggestions and guidance of Rosalie Ruegg and Jeanne Powell, both of the Advanced Technology Program, during the data collection stage.

The Review of Economics and Statistics, May 2003, 85(2): 485–491

TABLE 1.—DISTRIBUTION OF ATP-FUNDED PROJECTS BY TYPE OF UNIVERSITY INVOLVEMENT

Type of University Involvement (Dummy)	Number of Projects	Filtered Projects	Sample Projects	Sampling Probability	Number Responding
Joint venture:	118	81	36	44.4%	29
No university involvement (*jv*)	47	31	9	29.0%	8
Universities involved as subcontractors (*jvs*)	42	28	9	32.1%	8
Universities involved as research partners (*jvu*)	16	11	9	81.8%	8
Universities involved as both partner and sub. (*jvus*)	13	11	9	81.8%	5
Single applicant:	234	111	18	16.2%	18
No university involvement (*s*)	106	45	9	20.0%	9
Universities involved as a subcontractor (*ss*)	128	66	9	13.6%	9
Total	352	192	54	28.1%	47

Filtered projects are projects that have been active one year or more and are still active in the beginning of 1998.
Sampled projects were selected from the filtered project universe to ensure an equal number in each category.

that many universities desire even more partnerships with industry (Morgan, 1998). Yet, surprisingly, very little is known about the types of roles that universities play in such research partnerships or about the economic consequences associated with those roles. Our investigation is a first effort—an exploratory inquiry—to provide empirical information about universities as research partners.

Previous research falls broadly into examinations either of industry motivations or of university motivations for engaging in an industry-university research relationship. That research has not investigated the economic effects of university participation as thoroughly as the motivations for it. Hall, Link, and Scott (2000) review the literature and document its identification of two broad *industry motivations* for industry-university RJVs. The first is access to complementary research activity and research results; the second is access to key university personnel. The literature shows *university motivations* to be largely financially based. This note reports on the results of a small survey of such partnerships that focuses on their performance rather than the reasons underlying their formation.

II. An Analysis of the Data

A. *Identifying an Appropriate Database*

Given the potentially heterogeneous research role for universities, we developed and analyzed project-level data from the Advanced Technology Program (ATP), established within the National Institute of Standards and Technology (NIST) through the Omnibus Trade and Competitiveness Act of 1988. The ATP combines public funds with private investments to create and apply generic technology needed to commercialize new technology rapidly; it received its first appropriation from Congress in FY 1990.

ATP projects will not give a complete picture of industry-university R&D interaction. These projects are likely to have high social value and high risk, involve largely generic rather than largely proprietary technology, and be at such an early stage in development that the technology is not easily appropriable. Nonetheless, describing universities' roles in RJVs spawned by the ATP provides long-overdue insights into universities as research partners with industry.

Hall, Link, and Scott (2000) describe the population of ATP projects in terms of the numbers of awards since the program began, status of the projects (active, completed, or terminated), organizational types (single participants or RJVs), funding characteristics (mean total funding and the proportions from public and private sources), project durations, technology areas, and university involvement by technology area. We refer herein to some of the population characteristics, but the complete overview provides important additional detail.

In addition to classifying each project into a unique technology area, ATP classifies each project by lead participant. Each lead participant is placed in one of four ATP-defined type-size categories, including not-for-profit organizations.[1] Among the for-profit organizations, a small organization is defined as one with fewer than 500 employees; a large one is defined as a *Fortune* 500 or equivalent organization (a moving definition, requiring $2.6 billion in annual revenue at the time of our analysis). All others are medium. More than one-half of the lead participants are small.

B. *Selecting a Sample of ATP-Funded Projects*

We selected our sample of projects from the population of 352 projects funded by ATP from April 1991, when its first awards were made, through the start of our study in October 1997, using a series of filters, some under our control and others not. The first filter was that 21 projects terminated early (analyzed below) and were therefore unavailable for sampling. The second filter was that each project must be active and have been active for at least one year, to help ensure the respondent's use of a research project history when responding to the survey. These two filters reduced the population of 352 projects to 192. The 192 projects were then grouped by the six types of university involvement, listed in table 1 (column 3).[2] From each of the categories, a sample of nine projects was selected (column 4), with attention given to technology areas, sizes of lead participants, lengths of time the projects had been active, and total proposed research budgets of the projects. Also reported in table 1 are the sampling probabilities by type of university involvement.[3] This process of random stratified sampling yielded 54 projects.

Separate and distinct survey instruments were designed to obtain information about the nine projects selected in each of the six

[1] Nonprofit lead partners are not classified with respect to size, nor do we have revenue for these partners. This fact has two implications: (1) all of the size effects we measure are for for-profit lead partners, and (2) it is essential to include the nonprofit dummy when we include size in the regression in order to avoid misleading conclusions.

[2] In an industry-university relationship, either the university is a subcontractor, or it is a research partner, which means that it is a formal member of the joint venture.

[3] Variability in these probabilities reflects that the sample size is constant at 9 and that the size of the population of appropriate projects to sample, by category type, varies (column 3).

TABLE 2.—DETERMINANTS OF THE PROBABILITY OF EARLY TERMINATION

Variable	(1) Coefficient (s.e.)	(2) Coefficient (s.e.)	(3) Coefficient (s.e.)
D (university involvement)	−0.434 (0.258)*	−0.537 (0.269)**	−0.478 (0.249)*
ATP share of funding	−1.783 (0.943)*	−1.472 (0.957)	−1.374 (0.899)
Time trend	−0.112 (0.082)	−0.112 (0.084)	−0.079 (0.075)
Small lead participant	−0.716 (0.317)**	−0.818 (0.326)**	−0.914 (0.302)***
Large lead participant	−0.929 (0.348)***	−0.943 (0.351)***	−0.848 (0.335)***
Nonprofit lead participant	−0.401 (0.466)	−0.337 (0.467)	−0.516 (0.419)
Chi-squared for 3 size vars. (prob.)	8.47 (0.037)**	9.47 (0.024)**	10.50 (0.015)**
Information technology	0.025 (0.338)	−0.074 (0.347)	
Electronics	−0.488 (0.465)	−0.478 (0.389)	
Biotechnology	−0.533 (0.455)	−0.510 (0.569)	
Chemicals, energy, and environ.	−0.039 (0.387)	−0.022 (0.457)	
Chi-squared for 4 tech. vars. (prob.)	2.90 (0.575)	2.16 (0.675)	
Intercept	0.738 (0.655)	0.662 (0.664)	0.285 (0.569)
Number of observations	313	312	351
Log likelihood	−67.33	−64.42	−67.89
Scaled R^2	0.126	0.133	0.115
Chi-squared (d.f.)	19.38 (10)	19.75 (10)	17.67 (6)

Probit estimates: dependent variable = 1 if project terminated early.
Column (1) includes the full sample excluding projects in other manufacturing (none of which were terminated).
Columns (2) and (3) delete a single observation for a project that was terminated prior to starting.
The excluded category is a project in materials with no university participation and where the lead participant is of medium size.
Coefficient significance levels are denoted by * (10%), ** (5%), and *** (1%).
The scaled R^2 is a measure of goodness of fit relative to a model with only a constant term, computed as a nonlinear transformation of the LR test for zero slopes (see Estrella, 1998).

categories of university involvement.[4] Table 1 (column 6) shows the number of surveys received.[5] With 7 nonrespondents, our sample for analysis is 47.[6]

By ATP guidelines, universities cannot be lead partners. RJVs where the lead partner is a not-for-profit organization do have an affinity for inviting universities to be research partners. Yet the sample distinguishes such cases from RJVs with a university partner yet no not-for-profit lead partner. Thus, the effects of nonprofit lead partners and the effects of universities on RJV performance can be distinguished in our sample: there are 10 RJVs with a university and a nonprofit lead partner, 8 with a university but no nonprofit lead partner, 1 with a nonprofit lead partner but no university, 17 with no university and no nonprofit partner, 9 non-RJV projects with a university subcontractor, and 9 non-RJV projects with no university subcontractor. In sum, our sample has no nonprofits (as the main participant apart from any subcontractor) in the non-RJV observations, and nonuniversity nonprofits are mainly involved in RJVs with universities. However, the two—universities and nonprofit lead partners—are not the same thing (note the number of RJVs with universities without a nonprofit lead partner), and sometimes in our exploratory analyses of performance, coefficients for both universities and for nonprofit lead partners are identified.

[4] Copies of the survey instruments are available upon request from the authors.

[5] Because there are multiple dimensions of ATP-funded projects, we do not claim that our sample of 47 respondents is representative of the filtered population or of the whole population in all dimensions. We offer our sample as one sample to consider, and possibly to generalize about, given the stated filtering and selection process. More detailed information about the representativeness of the sample by other characteristics of ATP-funded projects is available from the authors.

[6] We are aware of the limitations of the self-reported data that will be analyzed below. Although our survey instruments were pretested, the possibility that our primary data reflect the personal attitudes of the respondents as well as objective characterizations of their projects is still present. This study is one of the first of its kind in attempting to quantify the role of universities in research partnerships. Efforts to generalize from our findings should be made with the utmost caution.

C. *Analysis of Terminated Projects in the Population*

We investigated the reasons for the early termination of 21 projects among the population of 352. These reasons ranged from the financial health of the participant(s) to lack of research success in the early part of the project. We estimated a probit model of termination probability conditional on ATP's share of funding, involvement of a university, type of project, size of the lead participant, and technology area.[7] A time variable denoting the year in which each project was initially funded was also included.

The probit estimates from alternative specifications are reported in table 2. Our particular interest is the nature of the relationship between university involvement and termination. The results imply that the projects with university involvement as either a research partner or subcontractor have a lower probability of early termination. Also, the probability of early termination decreases as ATP's share of funding increases, although the effect is barely significant, and only for the specification used to simulate the predicted probabilities shown in table 3. The termination rate does not vary across technology area,[8] but projects where the lead partner is of medium size are more likely to terminate early than the others.

The upper portion of table 3 presents the calculated probabilities for a project terminating early, by size of lead participant. For this example (information technology projects begun in 1991), the calculated probability of early termination is lower for each size category when a university is involved in the project. Similarly (lower portion of table 3), the calculated probability of early termination is lower for each discrete level of ATP's share of funding when a university is

[7] To be precise, we estimated the following model: Pr(project i terminates early) = $F(X_i, \beta)$, where F is the cumulative normal probability function, X_i is a vector of variables that characterizes project i, and β is the conformable vector of coefficients.

[8] This conclusion needs to be qualified slightly: because no projects in other manufacturing terminated early, these projects could not be included in the models estimated in the first two columns of table 2 (where we use technology dummies). Clearly, projects in this technology area have a lower early termination rate than projects in the other technology areas.

TABLE 3.—SIMULATION OF PROBABILITY OF TERMINATION: ATP INFORMATION TECHNOLOGY PROJECTS BEGUN IN 1991

	University Involved	No University Involved
Size of lead participant (50% ATP share):		
Small	0.036	0.094
Medium	0.189	0.344
Large	0.042	0.106
Not-for-profit	0.081	0.179
ATP share of funding (medium lead part.):		
0%	0.423	0.612
25%	0.296	0.477
50%	0.189	0.344
75%	0.111	0.228
100%	0.059	0.138

This simulation is based on specification (1) in table 2.

involved in the project.[9] In the population of ATP projects, university involvement is clearly associated with a lower probability of early termination.[10]

D. *Estimating the Probability of Response to the Sample Survey*

Only two of the six categories of university involvement listed in table 1 (column 6) had a 100% response rate. Contacts in joint ventures were less likely to respond, with the least responsive category being joint ventures with universities as both partners and subcontractors—only five of nine surveys were returned. We examined the probability of survey response using a probit model. When we include all of the right-side variables, nothing is very significant. The only variable that is even marginally informative is the dummy for joint ventures with universities as both partner and subcontractor (*jvus*), arguably the most complex arrangement contractually. Other factors held constant, joint ventures with universities as research partners and as subcontractors have a lower probability of response.

In the results presented below, we test and correct for response bias simply by including the *jvus* dummy in our estimations.[11] The implication of this strategy will be that we cannot identify the direct effects on performance of being a joint venture with a university participating as a partner and as a subcontractor separately from the effect of such a joint venture on the probability of survey response.

III. Role of Universities in ATP-Funded Projects

We ask three general questions about the roles of universities as research partners:

1. What roles do universities play in research partnerships in general?
2. Do universities enhance the research efficiency of research partnerships?
3. Do universities affect the development and commercialization of industry technology?

A. *Role Played by Universities in ATP-Funded Projects*

What research role do universities play in ATP-funded projects? At one level, the answer to this question comes from the organizational or administrative role that universities have in various projects. In a joint venture, the research role of a university is either as a research partner in the joint venture or as a subcontractor to the joint venture. In single-applicant projects, by definition, the research role of a university is only as a subcontractor.[12]

At a second level, we explored the role played by universities by asking each contact person to indicate, using a seven-point Likert scale, the extent to which the research project experienced difficulties acquiring and assimilating basic knowledge necessary for the project's progress. Strong agreement that such difficulties had been experienced was indicated with a 7, the responses ranging downward to 1 to indicate strong disagreement. Ordered probit models were estimated to explain interproject differences in responses about the difficulties. Held constant in these models are several characteristics of the project as determined from ATP information and from survey responses.

The estimates are in table 4. In column 1 we include the hazard rate for nontermination (the conditional probability density that the project will go forward to completion) and the proxy for the survey response hazard (*jvus*) in the model. Neither of these enters into the equation significantly, implying that selection bias is unlikely to be a problem for our estimates.[13]

We have five observations about the estimates in table 4:

[9] Similar relationships, available from the authors, exist across other research technology areas.

[10] The information in table 2 is used to calculate a hazard rate for the probability that a project does not terminate early for use in the subsequent statistical analyses of a sample of ATP-funded projects to control for possible sample selection bias. To anticipate the use of this variable in later survey-question equations, it is important to note that its inclusion in an ordered probit is not really econometrically correct if it actually enters. That is, if the probability distribution in the termination equation and the distribution in the survey question equation are dependent, then the appropriate method is to specify a full maximum likelihood model for the two random variables and estimate jointly [such a model is outlined in Hall, Link, and Scott (2000)]. In fact, we found that the termination hazard and the sample response hazard never entered significantly, and that joint maximum likelihood estimates did not differ significantly from our single-equation estimates, which implies that sample selection is unlikely to produce significant bias in our estimates. However, our sample size is small, so the power of all these tests is low.

[11] As with our analysis of the probability of early termination, the results from the response probit model could be used to calculate a survey hazard rate for the statistical analyses that follow. However, in practice, the only variable that predicted response or nonresponse in a simple probit model was *jvus*. We therefore used a simpler and more robust method to correct for response bias, by including the *jvus* dummy directly in our estimated model. Unlike the use of a hazard rate, this correction does not require normality of the response probability equation to be valid. For a single-dummy-variable predictor, of course, the two approaches for converting any response bias would be equivalent if normality held.

[12] Related to this organizational or administrative research role that universities have is another level at which to answer the first research question. Four of the six groups of contact persons for the survey were asked why the university subcontractors on this project were selected. For joint ventures where a university is only involved as a subcontractor and for single participants where the university is only involved as a subcontractor, the most frequent response was that the subcontractor was selected to gain access to eminent researchers. Joint ventures in which the university is only involved as a research partner reported that the university was most commonly invited to participate because of previous research interactions with other members of the joint venture. And, finally, the dominant response when universities are involved in a joint venture as research partners and as subcontractors (*jvus*) was that each was selected because of their overall research reputation.

[13] For completeness, we have estimated the full model for sample selection (an ordered probit equation plus an equation for the probability that the survey was returned), and the selection into the sample does not appear to be important for our results (Hall, Link, and Scott, 2000).

TABLE 4.—DETERMINANTS OF DIFFICULTY ACQUIRING BASIC KNOWLEDGE

Variable	(1) Ordered Probit Coefficient (s.e.)	(2) Ordered Probit Coefficient (s.e.)
Log of total project budget	−0.76 (0.36)**	−0.68 (0.30)**
ATP share (fraction)	−0.92 (3.00)	
D (university participant)	1.16 (0.61)*	1.12 (0.49)**
D (no prior experience)	1.16 (0.50)**	1.12 (0.48)**
log(revenue of lead part., $1,000)	0.12 (0.07)*	0.12 (0.06)**
Nonprofit lead part.	1.98 (1.22)	2.15 (1.09)**
Information technology	0.03 (0.64)	0.04 (0.57)
Manufacturing	−1.17 (0.94)	−1.24 (0.92)
Electronics	3.03 (1.07)***	3.03 (1.08)***
Biotechnology	0.07 (0.62)	0.06 (0.61)
Chemicals, energy, and environ.	−0.97 (0.84)	−0.92 (0.82)
Chi-squared for 5 tech. vars. (prob.)	12.7 (0.027)**	12.8 (0.025)**
Nontermination hazard	−0.34 (1.00)	
jvus	−0.29 (0.75)	
Number of observations	47	47
Pseudo R^2	0.216	0.209
Chi-squared (*p*-value)[a]	24.16 (0.030)	23.40 (0.009)

[a] This *chi*-squared is for the joint test that all coefficients except the intercept are zero.
The categories have been collapsed from seven to five, using the groupings (1&2), 3, 4, 5, (6&7).
The excluded category is a project in materials with no university participant.
Coefficient significance levels are denoted by * (10%), ** (5%), and *** (1%).

1. Respondents with a university participant (as a research partner or as a subcontractor) systematically agreed that the project had experienced difficulties acquiring and assimilating basic knowledge necessary for progress toward completion. Joint-venture projects are larger than others, which tends to lower difficulty in general but raise it if a partner is a university. That is probably also consistent with such projects being more difficult or closer to new science than are others, so that the university partner was chosen in anticipation of the difficulties. Or the university's presence may create a greater awareness that such difficulties exist.
2. Prior experience working with a university as a research partner or as a subcontractor is a significant factor in decreasing the difficulty of acquiring and assimilating basic knowledge.
3. Acquisition and assimilation difficulties with basic knowledge decrease slightly as overall project size increases.
4. Projects in the electronics area have substantially more difficulty in acquiring and assimilating basic knowledge than do projects in other technology areas.
5. Projects with larger for-profit lead partners or nonprofit lead partners have experienced difficulties acquiring and assimilating basic knowledge.

B. *Research Efficiencies from Universities in ATP-Funded Projects*

Are there systematic differences in the research efficiency of ATP-funded projects that have universities involved and those that do not? We addressed this question of research efficiency by asking each contact person to respond to a series of five statements. The first three of these statements investigate unexpected research problems encountered relative to expectations when the project began. Had the number of problems encountered been more than, less than, or about the same as what had been anticipated at the project's outset? The three types of research problems investigated are conceptual problems, equipment-related problems, and personnel problems.

TABLE 5.—DETERMINANTS OF THE PROBLEMS IN THE PROJECT: ORDERED PROBIT ESTIMATES

Variable	Personnel-Related Coefficient (s.e.)
Log of total project budget	−0.04 (0.34)
D (university participant)	0.89 (0.54)*
log(revenue of lead part., $1,000)	0.15 (0.08)*
Nonprofit lead partner	−0.57 (1.27)
Information technology	1.26 (0.77)
Manufacturing	1.30 (1.09)
Electronics	1.84 (1.26)
Biotechnology	1.53 (0.86)*
Chemicals, energy, and environ.	1.61 (0.93)*
Chi-squared for 5 tech. vars. (prob.)	4.61 (0.465)
Number of observations	44
Pseudo R^2	0.244
Chi-squared (*p*-value)[a]	14.51 (0.105)

[a] This *chi*-squared is for the joint test that all coefficients except the intercept are zero.
The excluded category is a project in materials with no university participant.
Coefficient significance levels are denoted by * (10%), ** (5%), and *** (1%).

To evaluate the responses to the first three statements, ordered probit models were estimated. Held constant in these models are several characteristics of the project as determined from ATP information and from survey responses.[14] In the specifications for conceptual problems and for equipment problems, none of the individual variables was significant in explaining the existence of unexpected conceptual or equipment-related research problems, apart from the perhaps obvious fact that problems with research about information technology were not equipment-related. Because only a few projects had fewer problems of any type than expected, we experimented with collapsing the responses from the original three categories (the number of problems encountered was more than, less than, or about the same as what had been anticipated at the project's outset) into two categories. Even when reestimated in this form in probit models, essentially no identifiable individual variable effects explained the existence of unexpected research problems. The presence of unexpected problems is perhaps random or a complex result of many factors that we cannot disentangle—truly unexpected given the information available to the firm (and to us).

Table 5 shows the specification for personnel problems, and the estimates suggest that the presence of unexpected personnel-related problems is associated somewhat with the technology field. Lead-partner size is a marginally significant explanatory variable in explaining the presence of unexpected personnel problems; projects with nonprofit lead partners are less likely to experience this kind of problem, although the effect is not significant. Joint ventures with university partners are both more likely to have personnel-related problems and, as we saw above, also less likely to respond to the survey, so we cannot disentangle these two effects.

The fourth and fifth statements addressed aspects of research efficiency related to the productive use of complementary research resources. The first of these asks for the approximate percentage of the project's research time that, in retrospect, had been unproductive. The

[14] Ordered probit models that allowed for sample selection were also estimated, but proved to be very difficult to identify because of the small sample. Therefore we rely mainly on the ad hoc correction terms discussed above; hence the effect for university participation cannot be disentangled from the selection effect.

TABLE 6.—PERCENTAGE OF UNPRODUCTIVE RESEARCH TIME AND COST: ORDERED PROBIT ESTIMATES

	Dependent Variable[a]:	
Variable	(1) Research Time Coefficient (s.e.)	(2) Research Cost Coefficient (s.e.)
D (university participant)	0.41 (0.39)	0.98 (0.41)**
Log (revenue of lead part., $1,000)	−0.16 (0.06)***	−0.14 (0.06)**
Nonprofit lead participant	−1.35 (0.83)	−1.08 (0.83)
Information Technology	−1.07 (0.50)**	−0.77 (0.51)
Manufacturing	−1.73 (0.73)**	−1.73 (0.74)**
Electronics	2.19 (0.98)**	2.87 (0.99)***
Biotechnology	−0.21 (0.53)	−1.26 (0.57)**
Chemicals, energy, and environ.	1.31 (0.62)**	0.93 (0.62)
Chi-squared for 5 tech. vars. (prob.)	20.8 (0.001)***	19.7 (0.001)***
No. of observations	42	42
Pseudo R^2	0.162	0.172
Chi-squared (*p*-value)[b]	24.86 (0.002)	23.95 (0.002)

[a] The dependent variable takes on the values 0%, 5%, 10%, 15%, 20%, 25%, 30%.
[b] This *chi*-squared is for the joint test that all coefficients except the intercept are zero.
The excluded category is a project in materials.
Coefficient significance levels are denoted by * (10%), ** (5%), and *** (1%).

second asks for the approximate percentage of the project's financial resources that, in retrospect, had been unproductive. These two statements are analyzed together because of the high correlation between responses. Of 42 contact persons, 22 responded to both questions with the same percentage.

Table 6 uses an ordered probit model to evaluate the responses across the categories of approximate percentage response.[15] Although we originally included all variables in the estimation, only the size of the lead partner and the technology variables were significant, and table 6 presents the model with just those effects and the effects of a university participant or a nonprofit lead participant. Unproductive time and cost is associated most with electronics projects and associated least with information technology and manufacturing projects. Projects with university participants are more likely to report unproductive costs.

Comparing further the estimates in the two columns of table 6, projects in electronics have the largest share of time and money that is unproductively used, whereas projects in manufacturing have the least. Unproductive research time and money in electronics may be related to projects in this field also having difficulty acquiring and assimilating the basic research they need. Biotechnology projects have relatively little unproductive research expenditure, although somewhat more unproductive research time. Larger (profit-making) lead partners are better at making productive use of research time and expenditure, or at least they perceive that to be the case.

C. *Accelerated Development and Commercialization of Technology from Universities in ATP-Funded Projects*

Are there systematic differences in the ability of ATP-funded projects to accelerate the development and commercialization of technology when universities are involved in the project and when they are not? We addressed this question by asking each contact person to provide a seven-point Likert-scale (7 denoting strong agreement and 1 denoting strong disagreement) response to two statements.

[15] Note that this survey statement addresses realized unproductive research time and not expected unproductive research time. The same is true for the unproductive use of financial resources.

TABLE 7.—PERFORMANCE DETERMINANTS: ORDERED PROBIT ESTIMATES

	Dependent Variable:	
Variable	(1) New applications of technology developed[a] Coefficient (s.e.)	(2) Commercialized sooner than expected Coefficient (s.e.)
Log of total project budget		−0.95 (0.26)***
ATP share (fraction)	2.97 (1.68)*	
D (university participant)	0.02 (0.34)	−0.78 (0.37)**
D (no prior experience)		−0.98 (0.44)**
Small lead participant		−1.52 (0.57)***
Large lead participant		−1.96 (0.65)***
Nonprofit lead participant		0.39 (0.65)
Chi-squared for size vars. (prob.)		15.04 (0.002)***
Information technology		1.08 (0.43)**
Manufacturing		
Electronics		
Biotechnology		
Chemicals, energy, and environ.		1.13 (0.65)*
Materials		1.69 (0.52)**
Chi-squared for tech. vars. (prob.)		12.49 (0.006)***
No. of observations	47	47
Pseudo R^2	0.024	0.167
Chi-squared (*p*-value)[b]	3.29 (0.193)	28.05 (0.001)

[a] The dependent variable takes on only six values, because one of the cells ($y = 3$) is empty.
[b] This *chi*-squared is for the joint test that all coefficients except the intercept are 0.
The excluded category in column 2 is a project where the lead participant is of medium size.
Coefficient significance levels are denoted by * (10%), ** (5%), and *** (1%).

The first statement posed to the lead participant assessed whether potential new applications of the technology being developed had been recognized over the course of the project. Ordered probit estimates for this question were for the most part insignificant; column 1 of table 7 shows a minimal specification of the model. It may be that the generation of new applications from a project in process cannot be attributed to any particular individual project characteristics and is essentially unpredictable regardless of the technology area. The results do however suggest that projects with a larger ATP share are more likely to develop unanticipated applications for the technology. Perhaps a larger ATP share brings greater resources for ATP monitoring or imparts to the research performers a greater leveraging effect to search for or to recognize new applications of the technology. University participation shows no effect on the generation of new technology applications.

The second statement assessed whether the lead participant—at the stage of the research reached at the time of the survey—believed that the technology would be developed and commercialized sooner than expected when the project began. The ordered probit estimates are shown in column 2 of table 7. A number of variables are significant, leading to five interesting conclusions:

1. Projects involving universities as partners are less likely to develop and commercialize technology sooner than expected, perhaps reflecting that universities are involved in more dif-

ficult projects to begin with, namely projects with a lower probability of early completion.

2. Large projects and/or projects with large lead participants are less likely to expect to develop and commercialize their technology sooner than expected than are projects with nonprofit or medium-size lead participants. Perhaps such larger projects reveal a whole new set of research insights. To the extent that larger research budgets are associated with research projects that can stretch the frontiers of knowledge then less time will be devoted toward looking for early-on commercialization opportunities of the technology.
3. Projects with a small lead participant are less likely to expect to develop and commercialize technology sooner than expected. Recall that this group is very small firms, and this may reflect resource constraints they face in development when the project budget does not cover the full cost of making the technology commercially viable.
4. Lack of experience with a university partner reduces the expectation of early commercialization, as does university involvement, perhaps because of lack of market pressure and focus on the particular project by the university participant, or perhaps simply because some adjustment costs are included as the participants learn to work with a university.
5. Projects in information technology, chemicals, energy, the environment, and materials are significantly more likely to commercialize earlier than expected than are projects in manufacturing, electronics, and biotechnology.

IV. Concluding Observations

The focus of this survey-based study of ATP-funded research projects is on universities as research partners.[16] Our analyses of the survey data allow us to set forth in this concluding section a consistent and illuminating story about their research role. Nonetheless, all of the results from the descriptive analyses and qualitative choice models presented should be interpreted cautiously, given the need for theoretical foundation, understanding of causality, and increased sample size. Given the caveats, we conclude the paper by emphasizing two themes that are consistent with our data.

A. *Universities Create Research Awareness in ATP-Funded Projects*

Our first conclusion is that universities create research awareness among the research partners in the ATP-funded projects studied. The qualitative models estimated suggest that projects with university involvement, either as a research partner or as a subcontractor, are (1) experiencing difficulties acquiring and assimilating basic knowledge for the project's progress (table 4), and (2) also not anticipating being able to develop and commercialize technology sooner than expected when the project began (table 7).

At one level, these two findings could be interpreted to mean that university involvement is creating research problems acquiring and assimilating basic knowledge and commercializing technology rapidly. We eschew that interpretation, because projects with university involvement are less likely to terminate early than are projects without university involvement (table 2). We conclude, albeit cautiously, that university involvement is creating a greater awareness of research problems.

We offer a possible interpretation of the research role of a university.[17] Universities are included (invited by industry) in those research projects that involve what we have called "new" science. Industrial research participants perceive that the university could provide research insight that is anticipatory of future research problems and that it could be an ombudsman anticipating and communicating to all parties the complexity of the research being undertaken. Thus, one finds universities purposively involved in projects that are characterized as problematic with regard to the use of basic knowledge. Because of the type of project into which a university is likely to be invited as a research partner, the research will not move faster than expected toward a commercial application of the resulting technology. Universities are more likely to partner in new technological fields where R&D is closer to science, and such fields can be more uncertain and difficult.

B. *Research Funding Influences the Scope of the Research*

We infer from the findings that projects with larger research budgets undertake research of broader scope, as opposed to researching narrower projects in greater detail. Projects with larger budgets are less likely to commercialize their technology ahead of schedule (table 7). That is not inconsistent with such projects attempting to foster newer frontiers of research. It is, however, also true that projects with larger budgets have fewer problems acquiring and assimilating basic knowledge (table 4). Thus, if the larger budgeted projects are broader, the scope and breadth may address new applications (new generic technology across many industries, for example) rather than fundamental research.

We do not speculate as to the extent to which our findings can be generalized either to other projects that are partially publicly funded, or to private sector joint ventures with and without university research interactions. As more research is conducted on this topic, the wider applicability of our observations will be tested.

REFERENCES

Cohen, Wesley M., Richard Florida, Lucien Randazzese, and John Walsh, "Industry and the Academy: Uneasy Partners in the Cause of Technological Advance," in R. Noll (Ed.), *Challenge to the University* (Washington: Brookings Institution Press, 1997).

Council on Competitiveness, *Endless Frontiers, Limited Resources: U.S. R&D Policy for Competitiveness* (Washington: Council on Competitiveness, 1996).

Estrella, Arturo, "A New Measure of Fit for Equations with Dichotomous Dependent Variables," *Journal of Business and Economic Statistics* (April 1998), 198–205.

Hall, Bronwyn H., Albert N. Link, and John T. Scott, "Universities as Research Partners," NBER working paper no. 7643 (April 2000).

Link, Albert N., "Research Joint Ventures: Patterns from Federal Register Filings," *Review of Industrial Organization* 11 (1996), 617–628.

Morgan, Robert P., "University Research Contributions to Industry: The Faculty View" (pp. 163–170), in Peter D. Blair and Robert A. Frosch (Eds.), *Trends in Industrial Innovation: Industry Perspectives & Policy Implications,* (Research Triangle Park: Sigma Xi, 1998).

[16] We set out to develop understanding of universities as research partners, designing our survey instruments to gather information pertinent to that understanding. Now that we have the results, as the referee for our note has observed, we have several findings about the importance of the characteristics of the RJV's lead partner. In that light, and at the suggestion of the referee, we gathered additional data about the lead partners beyond the characteristics used in our original statistical analyses and reported here. The additional information did not add to the explanatory power of our models. We believe that our small sample makes impractical the idea of pursuing in the present paper the importance of detailed characteristics—beyond size and profit versus nonprofit organization—of both the lead partner and all of the other partners to a RJV. We do believe the pursuit is an important one for future research.

[17] Absent baseline information about the technical difficulty of the projects or their closeness to new science other than the technology field, this interpretation is offered cautiously.

Name Index